Lecture Notes in Computer Science 1920

Edited by G. Goos, J. Hartmanis and J. van Leeuwen

Springer

Berlin
Heidelberg
New York
Barcelona
Hong Kong
London
Milan
Paris
Singapore
Tokyo

Aberto H.F. Laender   Stephen W. Liddle
Veda C. Storey (Eds.)

Conceptual Modeling – ER 2000

19th International Conference on Conceptual Modeling
Salt Lake City, Utah, USA, October 9-12, 2000
Proceedings

 Springer

Series Editors

Gerhard Goos, Karlsruhe University, Germany
Juris Hartmanis, Cornell University, NY, USA
Jan van Leeuwen, Utrecht University, The Netherlands

Volume Editors

Alberto H.F. Laender
Universidade Federal de Minas Gerais
Departamento de Ciência da Computação
31270-010 Belo Horizonte - MG, Brasil
E-mail: laender@dcc.ufmg.br

Stephen W. Liddle
Brigham Young University
Marriott School, School of Accountancy and Information Systems
585 TNRB, P.O. Box 23087, Provo, UT 84602-3087, USA
E-mail: liddle@byu.edu

Veda C. Storey
Georgia State University, College of Business Administration
Department of Computer Information Systems
Atlanta, Georgia 30302-4015, USA
E-mail: vstorey@gsu.edu

Cataloging-in-Publication Data applied for

Die Deutsche Bibliothek - CIP-Einheitsaufnahme

Conceptual modeling : proceedings / ER 2000, 19th International
Conference on Conceptual Modeling, Salt Lake City, Utah, USA, October
9 - 12, 2000. Alberto H. F. Laender ... (ed.). - Berlin ; Heidelberg ;
New York ; Barcelona ; Hong Kong ; London ; Milan ; Paris ; Singapore ;
Tokyo : Springer, 2000
  (Lecture notes in computer science ; Vol. 1920)
  ISBN 3-540-41072-4

CR Subject Classification (1998): H.2, H.4, F.4.1, I.2.4, H.1, J.1

ISSN 0302-9743
ISBN 3-540-41072-4 Springer-Verlag Berlin Heidelberg New York

Springer-Verlag Berlin Heidelberg New York
a member of BertelsmannSpringer Science+Business Media GmbH
© Springer-Verlag Berlin Heidelberg 2000

Printed on acid-free paper   SPIN: 10722816   06/3142   5 4 3 2 1 0

ER2000 Conference Organization

Conference Chair
David W. Embley, Brigham Young University, USA

Program Co-Chairs
Alberto H. F. Laender, Federal University of Minas Gerais, Brazil
Veda C. Storey, Georgia State University, USA

Workshops Chair
Bernhard Thalheim, Brandenburg Technical University at Cottbus, Germany

Tutorials Chair
Ling Liu, Georgia Institute of Technology, USA

Industrial Chair
Terry Halpin, Microsoft Corporation, USA

Publicity Chair
Il-Yeol Song, Drexel University, USA

Panels Chair
John F. Roddick, Flinders University of South Australia, Australia

DAMA International Liaison
Davida Berger, Roche Labs Inc., USA

Local Organization Committee
Local Arrangements: Scott N. Woodfield, Brigham Young University, USA
Registration: Yiu-Kai (Dennis) Ng, Brigham Young University, USA
Webmaster: Stephen W. Liddle, Brigham Young University, USA
Social: Douglas M. Campbell, Brigham Young University, USA
Treasurer: Dan Johnson, Brigham Young University, USA
Treasurer: Jennifer Shadel, Brigham Young University, USA

Steering Committee Representatives
Chair: Bernhard Thalheim, BTU Cottbus, Germany
Vice-Chair and ER2000 Liaison: Tok Wang Ling, NUS, Singapore
Emeritus: Peter P. Chen, Louisiana State University, USA

Area Liaisons
Asia: Tok Wang Ling, National University of Singapore, Singapore
Australia: John F. Roddick, Flinders University of South Australia, Australia
North America: Sudha Ram, University of Arizona, USA
South America: José Palazzo de Oliveira, Federal University of Rio Grande do Sul, Brazil
Europe: Stefano Spaccapietra, Swiss Federal Institute of Technology, Switzerland

Program Committee

Jacky Akoka, National Conservatory of Industrial Arts and Crafts, France
Hiroshi Arisawa, Yokohama National University, Japan
Paolo Atzeni, University of Rome, Italy
Terry Barron, University of Toledo, USA
Dinesh Batra, Florida International University, USA
Joachim Biskup, University of Dortmund, Germany
Mokrane Bouzeghoub, University of Versailles, France
Marco A. Casanova, Catholic University of Rio de Janeiro, Brazil
Tiziana Catarci, University of Rome, La Sapienza, Italy
Stefano Ceri, Milan Polytechnic, Italy
Roger H.L. Chiang, University of Cincinatti, USA
Wesley Chu, University of California at Los Angeles, USA
Yves Dennebouy, Swiss Federal Institute of Technology, Switzerland
Deb Dey, University of Washington, USA
Ramez Elmasri, University of Texas at Arlington, USA
Donal Flynn, University of Science and Technology in Manchester, UK
Antonio Furtado, Catholic University of Rio de Janeiro, Brazil
Paulo Goes, University of Connecticut, USA
Jean-Luc Hainaut, University of Namur, Belgium
Igor Hawryszkiewycz, University of Technology, Sydney, Australia
Matthias Jarke, Technical University of Aachen, Germany
Paul Johannesson, Stockholm University, Sweden
Yahiko Kambayashi, Kyoto University, Japan
Hideko S. Kunii, Ricoh Co. Ltd., Japan
Maurizio Lenzerini, University of Rome, La Sapienza, Italy
Stephen W. Liddle, Brigham Young University, USA
Ee-Peng Lim, Nanyang Technological University, Singapore
Tok Wang Ling, National University of Singapore, Singapore
Salvatore T. March, Vanderbilt University, USA
Claudia Bauzer Medeiros, University of Campinas, Brazil
Elisabeth Metais, University of Versailles, France
Takao Miura, Hosei University, Japan
Renate Motschnig-Pitrik, University of Vienna, Austria
John Mylopoulos, University of Toronto, Canada
Sham Navathe, Georgia Institute of Tecnnology, USA
Daniel O'Leary, University of Southern California, USA
Antoni Olivé, University of Catalunya, Spain
Maria Orlowska, University of Queensland, Australia
Jose Palazzo de Oliveira, Federal University of Rio Grande do Sul, Brazil
Guenther Pernul, University of Essen, Germany
Javier Pinto, Catholic University of Chile, Chile
Alain Pirotte, Catholic University of Louvain, Belgium
Niki Pissinou, University of Southwestern Lousiana, USA
Sandeep Purao, Georgia State University, USA

Sudha Ram, University of Arizona, USA
John Roddick, Flinders University of South Australia, Australia
Sumit Sarkar, University of Texas at Dallas, USA
Arne Solvberg, Norwegian Institute of Technology, Norway
Il-Yeol Song, Drexel University, USA
Vijayan Sugumaran, Oakland University, USA
Mohan Tanniru, Oakland University, USA
Toby Teorey, University of Michigan, USA
Bernhard Thalheim, Brandenburg Technical University at Cottbus, Germany
Olga De Troyer, Tilburg University, The Netherlands
Kyu-Young Whang, Korea Advanced Institute of Science and Technology, Korea
Michael Williams, Manchester Metropolitan University, UK

External Referees

Stephen Arnold
Maria Bergholtz
Stephane Bressan
Diego Calvanese
Cecil Eng Huang Chua
Angelo E. M. Ciarlini
Robert M. Colomb
Gillian Dobbie
Laurent Ferier
Thomas Feyer
Renato Fileto
Rakesh Gupta
Hyoil Han
Gaby Herrmann
Patrick Heymans
Ryosuke Hotaka
David Johnson
Masayuki Kameda
Gilbert Karuga
Hirofumi Katsuno
Hiroyuki Kitagawa
Birgitta Koenig-Ries
Takeo Kunishima
Dongwon Lee
Joseph Lee
Weifa Liang
Victor Liu
Wenlei Mao

Mihhail Matskin
Giansalvatore Mecca
Paolo Merialdo
Sophie Monties
Luciana Porcher Nedel
Sanghyun Park
Ilias Petrounias
Alysson Bolognesi Prado
Torsten Priebe
Ivan Radev
Srinivasan Raghunathan
Wasim Sadiq
Giuseppe Santucci
Klaus-Dieter Schewe
Torsten Schlichting
Joachim W. Schmidt
Junho Shim
Isamu Shioya
Roderick Son
Eng Koon Sze
Hideyuki Takada
Hiroki Takakura
Babis Theodoulidis
Hallvard Traetteberg
Meng-Feng Tsai
Vassilios S. Verykios
Bing Wu
Yusuke Yokota

Tutorials

The Unified Modeling Language
Scott N. Woodfield, Brigham Young University, USA

Conceptual Modeling of Internet Sites
Bernhard Thalheim, Brandenburg Univ. of Technology at Cottbus, Germany
Klaus-Dieter Schewe, Massey University, New Zealand

Utilizing Ontologies in eCommerce
Avigdor Gal, Rutgers University, USA

Conceptual Modeling for Multimedia Applications and the Role of MPEG-7 Standards
Uma Srinivasan, CSIRO Mathematical and Information Sciences, Australia
John R. Smith, IBM T.J. Watson Research Center, USA

Logical Design for Data Warehousing
Il-Yeol Song, Drexel University, USA

Application and Process Integration
Benkt Wangler, University of Skövde, Sweden
Paul Johannesson, Stockholm University, Sweden

Workshops

eCOMO2000
International Workshop on Conceptual Modeling Approaches for E-Business
Chair: Heinrich C. Mayr, University of Klagenfurt, Austria

WCM2000
2nd International Workshop on the World Wide Web and Conceptual Modeling
Chair: Stephen W. Liddle, Brigham Young University, USA

See *LNCS vol. 1921* for the workshop proceedings.

Table of Contents

Invited Papers

Database Integration

Temporal and Active Database Modeling

Database and Data Warehouse Design Techniques

Analysis Patterns and Ontologies

Web-Based Information Systems

Object-Oriented Modeling

Applying Object-Oriented Technology

Quality in Conceptual Modeling

Application Design Using UML

DAMA International Industrial Abstracts

Preface

This volume provides a comprehensive, state-of-the-art survey of conceptual modeling. It includes invited papers, research papers, and abstracts of industrial presentations given at ER2000, the 19th International Conference on Conceptual Modeling, held in Salt Lake City, Utah. Continuing in its long tradition of attracting the leading researchers and practitioners in advanced information systems design and implementation, the conference provided a forum for presenting and discussing current research and applications in which the major emphasis was on conceptual modeling. The conference topics reflected this strong conceptual-modeling theme while recognizing important, emerging developments resulting from recent technological advances.

The call for papers for the research track resulted in the submission of 140 papers from researchers around the world. Of these, 37 were selected for inclusion in the program. The authors of these papers are from 14 countries. These papers represent a variety of topics including:

- Database integration
- Temporal and active database modeling
- Database and data warehouse design techniques
- Analysis patterns and ontologies
- Web-based information systems
- Business process modeling
- Conceptual modeling and XML
- Engineering and multimedia application modeling
- Object-oriented modeling
- Applying object-oriented technology
- Quality in conceptual modeling
- Application design using UML

Three internationally recognized scholars in the area of conceptual modeling also submitted papers and delivered keynote speeches:

- John Mylopoulos: *From Entities and Relationships to Social Actors and Dependencies*
- Salvatore T. March: *Reflections on Computer Science and Information Systems Research*
- Philip A. Bernstein: *Generic Model Management—Why We Need It and How to Get There*

In addition to the research papers and invited papers, the conference included two workshops, two pre-conference full-day tutorials, four short tutorials, four industrial sessions, and two provocative panel discussions. The workshops addressed emerging topics in conceptual-modeling research. One workshop was entitled "Conceptual Modeling Approaches for E-Business (eCOMO2000)," and the other was entitled "The World Wide Web and Conceptual Modeling (WCM2000)." The pre-conference tutorials allowed participants to learn the latest information about using conceptual-modeling techniques for software development using UML and for Internet site development using co-design. Scott N. Woodfield taught the UML software development tutorial, while Bernhard Thalheim and Klaus-Dieter Schewe taught the Internet site development tutorial. Short tutorials were taught by Avigdor Gal on utilizing ontologies in e-commerce, by Uma Srinivasan and John R. Smith on conceptual modeling for multimedia applications, by Il-Yeol Song on logical design for data warehousing, and by Benkt Wangler and Paul Johanneson on application and process integration. The industrial sessions were largely sponsored in conjunction with DAMA International and addressed pertinent issues relating to the application of conceptual modeling to solving real-world problems.

Authors who submitted their work and the program committee (PC) members and additional reviewers who carefully reviewed the papers and provided their professional assessments deserve appreciation and recognition. A number of the PC members participated in a rather lengthy virtual meeting to finalize the paper selection and diligently reviewed additional papers on demand, all in an effort to create the best program possible. The PC chairs, Alberto H.F. Laender and Veda C. Storey, labored diligently to provide the best possible research program for ER2000. This volume is a tribute to their efforts. Stephen W. Liddle deserves special thanks for creating the Web-based conference management software that enabled the online submission of papers and reviews, and that made the virtual PC meeting possible. He also worked closely with the PC chairs to manage the review process and assemble the conference proceedings.

Many others deserve appreciation and recognition. The steering committee and conference chair provided advice and vision. Bernhard Thalheim directed the search for timely workshops. Selected workshops were chaired by Heinrich C. Mayr (eCOMO2000) and Stephen W. Liddle (WCM2000) who both worked tirelessly with their program committees to obtain good papers for their respective workshops. (A companion volume, LNCS 1921, contains these papers.) Ling Liu enticed 15 submissions for short tutorials and then had to make hard decisions to cut this list of submissions to the four accepted for the conference. Terry Halpin worked diligently as industrial chair, eventually joining forces with Davida Berger, Vice President of Conference Services for DAMA International, to jointly come up with an outstanding industrial track. Il-Yeol Song continuously provided publicity for the conference, John F. Roddick worked on panels, Scott N. Woodfield handled local arrangements, and Stephen W. Liddle kept the Web pages up to date and artistically presentable. Yiu-Kai (Dennis) Ng (registration), Douglas M. Campbell (socials), Dan Johnson and Jennifer Shadel (treasurers), and several Brigham Young University students also helped make the conference a success.

October 2000   David W. Embley

Data Warehouse Scenarios for Model Management

Philip A. Bernstein and Erhard Rahm[1]

Microsoft Corporation, One Microsoft Way, Redmond, WA 98052-6399 U.S.A.
philbe@microsoft.com, rahm@informatik.uni-leipzig.de

Abstract. Model management is a framework for supporting meta-data related applications where models and mappings are manipulated as first class objects using operations such as Match, Merge, ApplyFunction, and Compose. To demonstrate the approach, we show how to use model management in two scenarios related to loading data warehouses. The case study illustrates the value of model management as a methodology for approaching meta-data related problems. It also helps clarify the required semantics of key operations. These detailed scenarios provide evidence that generic model management is useful and, very likely, implementable.

1 Introduction

Most meta-data-related applications involve the manipulation of models and mappings between models. Such applications include data translation, data migration, database design, schema evolution, schema integration, XML wrapper generation, message mapping for e-business, schema-driven web site design, and data scrubbing and transformation for data warehouses. By "model," we mean a complex discrete structure that represents a design artifact, such as an XML DTD, web-site schema, interface definition, relational schema, database transformation script, semantic network, or workflow definition. One way to make it easier to develop meta-data related applications is to make *model* and *mapping* first-class objects with generic high-level operations that simplify their use. We call this capability *model management* [1,2].

There are many examples of high-level algebraic operations being used for specific meta-data applications [4, 7, 10, 11, 14]. However, these operations are not defined to be generic across application domains. Our vision is to provide a truly generic and powerful model management environment to enable rapid development of meta-data related applications in different domains. To this end we need to define operations that are generic, powerful, implementable, and useful.

In this paper, we take a step toward this goal by investigating the detailed semantics of some of the operations proposed in [2]. We do this by walking through the design of two specific data warehouse scenarios. In addition to providing evidence that our model management approach can solve realistic problems, these scenarios also demonstrate a methodology benefit: Reasoning about a problem using high-level model management operations helps a designer focus on the overall strategy for manipulating models and mappings — the choice of operations and their order. We

---

[1] On leave from University of Leipzig (Germany), Institute of Computer Science.

A.H.F. Laender, S.W. Liddle, V.C. Storey (Eds.): ER2000 Conference, LNCS 1920, pp. 1-15, 2000.

believe that solution strategies similar to the ones developed in this paper can be applied in other application domains as well.

We begin in Section 2 with definitions of the model management operations. Sections 3 and 4 describe applications of these operations to two data warehouse scenarios. Section 5 summarizes what we learned from this case study.

2 Model Representation and Operations

This section summarizes the model management approach introduced in [2]. We represent models by objects in an object-oriented database. Some of the relationships in the database are distinguished as *containment relationships* (e.g., by a "containment flag" on the relationship). A *model* is identified by a root object r and consists of r plus the objects that are reachable from r by following containment relationships.

A mapping, *map*, is a model that relates the objects of two other models, M_1 and M_2. Each object in *map*, called a *mapping object*, has two properties, domain and range, which point to objects in M_1 and M_2 respectively. It may also have a property expr, which is an expression whose variables include objects of M_1 and M_2 referenced by its domain and range; the expression defines the semantics of that mapping object.

For example, Fig. 1 shows two Customer relations represented as models M_1 and M_2. Mapping *map₁* associates the objects of the two models. Mapping object m_1 has domain {C#}, range {CustID}, and expr "Cust.C# = Customer.CustID" (not shown). Similarly for m_2. For m_3, the domain is {FirstName, LastName}, range is {Contact}, and expr is "Customer.Contact =Concatenate(Cust.FirstName, Cust.LastName)".

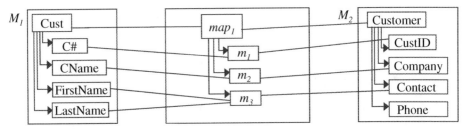

Fig. 1. A simple mapping *map₁* between models M_1 and M_2

Models are manipulated by a repertoire of high-level operations including
- Match – create a mapping between two models
- ApplyFunction – apply a given function to all objects in a model
- Union, Intersection, Difference –applied to a set of objects
- Delete – delete all objects in a model
- Insert, Update – applied to individual objects in models

Unless a very controlled vocabulary is used in the models, the implementation of a generic Match operation will rely on auxiliary information such as dictionaries of synonyms, name transformations, analysis of instances, and ultimately a human arbiter. Approaches to perform automatic schema matching have been investigated in [3, 5, 7, 8, 9, 10, 12, 13].

By analogy to outer join in relational databases, we use OuterMatch to ensure that all objects of an input model are represented in the match result. For instance, Right-

OuterMatch(M_1, M_2) creates and returns a mapping *map* that "covers" M_2. That is, every object o in M_2 is in the range of at least one object m in *map*, e.g., by matching o to the empty set, if o doesn't match anything else (i.e., range(m) = {o}, domain(m) = ∅). For example, in Fig. 1, to make map_1 a valid result of RightOuterMatch(M_1, M_2), we need to add a node m_4 in map_1 with range(m_4) = PhoneNo and domain(m_4) = ∅.

Since mappings are models, they can be manipulated by model operations, plus two operations that are specific to mappings:

- Compose – return the composition of two mappings
- Merge – merge one model into another based on a mapping.

Compose, represented by •, creates a mapping from two other mappings. If map_1 relates model M_1 to M_2, and map_2 relates M_2 to M_3, then the composition map_3 = map_1 • map_2 is a mapping that relates M_1 to M_3. That is, given an instance x of M_1, (map_1 • map_2)(x) = map_2(map_1(x)), which is an instance of M_3. There are right and left variations, depending on which mapping drives the composition and is completely represented in the result; we define RightCompose here.

The definition of composition must support mapping objects whose domains and ranges are sets. For example, the domain of a mapping object m_2 in map_2 may be covered by a proper subset of the range of a mapping object m_1 in map_1 (e.g., Fig. 2a). Or, a mapping object in map_2 whose domain has more than one member may use more than one mapping object in map_1 to cover it. For example, in Fig. 2b the domain of m_2 is covered by the union of the ranges of m_{1a} and m_{1b}.

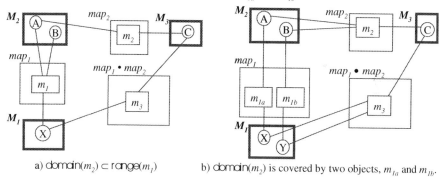

a) domain(m_2) ⊂ range(m_1)   b) domain(m_2) is covered by two objects, m_{1a} and m_{1b}.

Fig. 2. Composition with set-oriented domains and ranges (Examples)

In general, multiple mapping objects in map_1 may be able to provide a particular input to a mapping object in map_2. For example, in Fig. 2a, a second object m_1' in map_1 may have A in its range, so either m_1 or m_1' could provide input A to m_2. In the examples in this paper, each input of each M_2 object, m, is in the range of at most one map_1 object, so there is never a choice of which map_1 object should provide input to m. However, for completeness, we give a general definition of composition that handles cases where a choice of inputs is possible.

In the general case, the composition operation must identify, for each object o in the domain of each mapping object m in map_2, the mapping objects in map_1 that provide input to o. We define a function f for this purpose,

$$f: \{m \in map_2\} \times \cup_{m \in map2} \text{domain}(m) \to \{m' \in map_1\}$$

such that if $f(m, o) = m'$ then $o \in$ range(m') (i.e., m' is able to provide input o to m). Given f, we create a copy of each mapping object m in map_2 and replace its domain by its "new-domain," which is the domain of the map_1 objects that provide m's input. More precisely, for $m \in map_2$, we define the set of objects that provide input to m:

input$(m) = \{ f(m, o) \mid o \in$ domain$(m) \}$

based on which, we define the new-domain(m) as follows:

if domain$(m) \subseteq \cup_{m' \in input(m)}$ range(m') and domain$(m) \neq \varnothing$

   then new-domain$(m) = \cup_{m' \in input(m)}$ domain(m') else new-domain$(m) = \varnothing$.

So, the *right composition* of map_1 and map_2 with respect to f, represented by $map_1 \bullet_f map_2$, is defined constructively as follows:

1. Create a shallow copy map_3 of map_2 (i.e., copy the mapping objects and their relationships, but not the objects they connect to)

2. For each mapping object m'' in map_3, replace domain(m'') by newdomain(m), where m is the map_2 object of which m'' is a copy.

This definition would need to be extended to allow $f(m, o)$ to return a set of objects, that is, to allow an object in domain(m) to take its input from more than one source. We do not define f explicitly in later examples, since there is only one possible choice of f and the choice is obvious from the context.

The above definitions leave open how to construct the expression for each mapping object in the result of a composition, based on the expressions in the mapping objects being composed. Roughly speaking, in step (2) of the above definition, each reference to an object o in m''.domain should be replaced in m''.expr by the expression in the map_1 object that produces o. For example, in Fig. 2b, replace references to A and B in m_2.expr by m_{1a}.expr and m_{1b}.expr, respectively. However, this explanation is merely intuition, since the details of how to do the replacement depend very much on the expression language being used. In this paper, we use SQL.

The Merge operation copies some of the objects of one model M_2 into another M_1, guided by a mapping, map. We finesse the details here, as they are not critical to the examples at hand. As discussed in [2], a variety of useful semantics is possible.

3 Data Warehouse Scenario 1: Integrating a New Data Source

A data warehouse is a decision support database that is extracted from a set of data sources. A data mart is a decision support database extracted from a data warehouse. To illustrate model management operations, we consider two scenarios for extending an existing data warehouse: adding a new data source (Section 3) and a new data mart (Section 4). These are challenging scenarios that commonly occur in practice.

We assume a simple data warehouse configuration covering general order processing. It has a relational data source described by schema rdb1 (shown in Fig. 3), a relational warehouse represented by star schema dw1 (Fig. 4), and mapping map_1 between rdb1 and dw1 (Fig. 5). We note the following observations about the configuration:

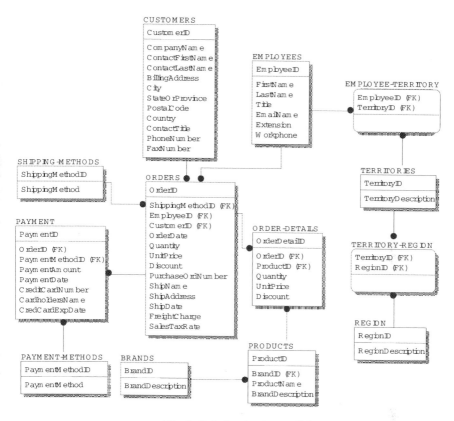

Fig. 3. Relational schema rdb1

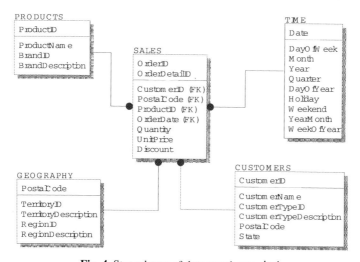

Fig. 4. Star schema of data warehouse dw1

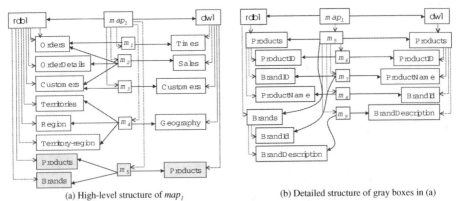

(a) High-level structure of map_1 (b) Detailed structure of gray boxes in (a)

Fig. 5. The structure of map_1. Dotted lines are containment relationships. Solid lines are relationships to the domain and range of mapping objects.

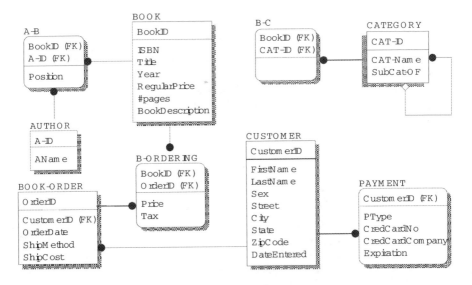

Fig. 6. Relational schema rdb2 (Book Orders)

Table 1. SQL statements defining the semantics of two mappings into **dw1**

map_1 (rdb1 → dw1)	map_2 (rdb2→dw1)
create view dw1.Sales (OrderID,OrderDetail-ID, CustomerID, PostalCode, ProductID, OrderDate, Quantity, UnitPrice, Discount) as select O.OrderID, D.OrderDetailID, O.CustomerID, C.PostalCode, D.ProductID, O.OrderDate, D.Quantity,D.UnitPrice, D.Discount from rdb1.Orders O, rdb1.Order-details D, rdb1.Customers C where O.OrderID=D.OrderID and O.CustomerID=C.CustomerID order by O.OrderId, D.OrderDetailID	create view dw1.Sales (OrderID, OrderDetail-ID, CustomerID, PostalCode, ProductID, OrderDate, Quantity, UnitPrice, Discount) as select O.OrderID, D.BookID, O.CustomerID, C.ZipCode, D.BookID, O.OrderDate, 1, D.Price, 0 // Default-Settings quantity=1,discount=0 from rdb2.Book-orders O, rdb2.B-ordering D, rdb2.Customer C where O.OrderID=D.OrderID and O.CustomerID=C.CustomerID order by O.OrderId, D.OrderDetailID
create view dw1.Customers (CustomerID, CustomerName, CustomerTypeID, Customer-TypeDescription, PostalCode, State) as select C.CustomerID, C.CompanyName, C.CustomerID%4, case (C.CustomerID%4) when 0 then 'Excellent' when 1 then 'Good' when 2 then 'Average' when 3 then 'Poor' else 'Average' end, C.PostalCode, C. StateOrProvince from rdb1.Customers C	create view dw1.Customers (CustomerID, ... State) as select C.CustomerID, Concatenate (C.FirstName,C.LastName), C.CustomerID % 4, case (C.CustomerID % 4) when 0 then 'Excellent' when 1 then 'Good' when 2 then 'Average' when 3 then 'Poor' else 'Average' end, C. ZipCode, C. State from rdb2.Customer C
create view dw1.Times (Date, DayOfWeek, Month, Year, Quarter, DayOfYear, Holiday, Weekend , YearMonth, WeekOfYear) as select distinct O.OrderDate, DateName (dw, D.OrderDate), DatePart(mm ,O.OrderDate), DatePart(yy ,O.OrderDate), DatePart(qq, O.OrderDate), DatePart(dy,O.OrderDate),'N', case DatePart(dw,O.OrderDate) when (1) then'Y' when (7) then 'Y' else 'N' end, DateName(month, O.OrderDate) + '_' + DateName(year,O.OrderDate), DatePart(wk,O.OrderDate) from rdb1.Orders O	create view dw1.Times(Date,...,WeekOfYear) as select distinct O.OrderDate, DateName (dw, D.OrderDate), DatePart(mm,O.OrderDate), DatePart(yy ,O.OrderDate), DatePart(qq, O.OrderDate), DatePart(dy,O.OrderDate), 'N', case DatePart(dw,O.OrderDate) when (1) then 'Y' when (7) then 'Y' else 'N' end, DateName(month, O.OrderDate) + '_' + DateName(year,O.OrderDate), DatePart(wk,O.OrderDate) from rdb2.Book-orders O
create view dw1.Geography (PostalCode, TerritoryID, TerritoryDescription, RegionID, RegionDescription) as select T.TerritoryID, T.TerritoryID, T.TerritoryDescription, R.RegionID, R.RegionDescription from rdb1.Territories T, rdb1.Region R, rdb1.Territory-region TR where T.TerritoryID=TR.TerritoryID and TR.RegionID=R.RegionID	create view dw1.Geography (PostalCode, ... RegionDescription) as select distinct C.ZipCode, C.ZipCode, NULL, NULL, NULL from rdb2.Customer C // Where clause dropped because required attributes not existing
create view dw1.Products (ProductID, ProductName, BrandID, BrandDescription) as select P.ProductID, P.ProductName, B.BrandID, B.BrandDescription from rdb1.Brands B, rdb1.Products P where B.BrandID=P.BrandID	create view dw1.Products(ProductID, ...) as select B.BookID, B.Title, NULL, NULL from rdb2.Book B // Where clause dropped because required attributes not existing

- We chose to write the expressions for map_1 as SQL view definitions, shown in column 1 of Table 1. There is one statement for each of the 5 mapping objects in Fig. 5a (one per table in dw1). To create dw1, simply materialize the views.
- Only 8 out of 13 tables in rdb1 take part in domain(map_1). In addition, only a subset of these tables' attributes are mapped to dw1, as is typical for data warehousing. This is different from other areas, such as schema integration in federated databases, where one strives for complete mappings to avoid information loss.
- Range(map_1) fully covers dw1, since map_1 is the only source of data for dw1.
- The SQL statements in map_1 perform 1:1 attribute mappings (e.g., name substitution and type conversion) and complex transformations involving joins and user-defined functions (in map_1 for date transformations and customer classification). Although all mappings in this example are invertible, this is not true in general, e.g., if aggregate values are derived and mapped to the warehouse.

Suppose we want to integrate a second source into the warehouse. The new source covers book orders and is described by a relational schema rdb2 (see Fig. 6). The integration requires defining a mapping from rdb2 to the existing warehouse schema dw1 and possibly changing dw1 to include new information introduced by rdb2. To simplify the integration task, we want to re-use the existing mappings as much as possible. The extent to which this can be achieved depends on the degree of similarity between rdb2 and rdb1. Some of rdb2's tables and attributes are similar to rdb1 and dw1, but there are also new elements, e.g., on authors and categories. We present two solutions for the integration task.

3.1 First Solution

Figure 7 illustrates the model management steps of our first solution. The elements shown in boldface (rdb1, map_1, dw1, rdb2) are given. A Venn-diagram-like notation is used to show subsets. E.g., rdb1' ⊆ rdb1 means every row of table rdb1' is in rdb1.

The first solution exploits the similarities between rdb1 and rdb2 by attempting to re-use map_1 as much as possible. This requires a match between rdb2 and rdb1, to identify which elements of map_1 can be reused for rdb2. The match result is then composed with map_1, thereby reusing map_1 to create a mapping between rdb2 and dw1.

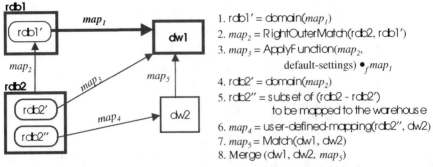

1. rdb1' = domain(map_1)
2. map_2 = RightOuterMatch(rdb2, rdb1')
3. map_3 = ApplyFunction(map_2, default-settings) $\bullet_f map_1$
4. rdb2' = domain(map_2)
5. rdb2'' = subset of (rdb2 - rdb2') to be mapped to the warehouse
6. map_4 = user-defined-mapping(rdb2'', dw2)
7. map_5 = Match(dw1, dw2)
8. Merge (dw1, dw2, map_5)

Fig. 7. Sequence of model management operations to integrate a new data source.

Match and RightOuterMatch
For the match between rdb2 and rdb1, it is unnecessary to consider all of schema rdb1, but only the part that is actually mapped to dw1, namely rdb1' = domain(map_1). (The

latter assignment is just a macro for notational convenience. I.e., the program need not construct a physical representation of rdb1'.) This avoids identifying irrelevant rdb1-rdb2 overlaps (e.g., w.r.t. payment attributes) that are not used in the warehouse and thus need not be mapped. In our example, rdb1' is easy to identify: it simply consists of the rdb1 tables and attributes being used in the SQL statements of map_1.

Object matching is driven by the correspondence table in Table 2, which specifies equivalence of attribute names or attribute expressions. The table consists mostly of 1:1 attribute correspondences (e.g., rdb2.Book.BookID matches rdb1.Products.ProductID, etc.). In one case, two rdb2 attributes are combined: concatenate (rdb2.Customer.FirstName, rdb2.Customer.LastName) matches rdb1.Customers.CompanyName.

Table 2. Correspondence table specifying equivalence of attributes in rdb2 and rdb1'

rdb2	rdb1'
Customer.CustomerID	Customers.CustomerID
Catenate(Customer.FirstName, Customer.LastName)	Customers.CompanyName
Customer.ZipCode	Customers.PostalCode
Customer.State	Customers.StateOrProvince
Book-Orders.OrderID	Orders.OrderID
Book-Orders. CustomerID	Orders.CustomerID
Book-Orders.OrderDate	Orders.OrderDate
B-Ordering.OrderID	Order-Details.OrderID
B-Ordering.BookID	Order-Details.ProductID
B-Ordering.Price	Order-Details.UnitPrice
B-Ordering.BookID	Order-Details.OrderDetailID
Book.BookID	Products.ProductID
Book.Title	Products.ProductName

We want to compose the result of the match operation between rdb2 and rdb1' with map_1. However, not all rdb1' elements have matching counterparts in rdb2, i.e., rdb1' is a proper superset of range(Match(rdb2, rdb1')). For instance, rdb2 has no equivalent of the Quantity and Discount attributes in the Orders table or of the Brands and Region tables, which are in rdb1'. Without this information, three of the five SQL statements in map_1 cannot be used, although only a few of the required attributes are missing.

To ensure that the match captures all of rdb1', we use a RightOuterMatch of rdb2 and rdb1', i.e., map_2 = RightOuterMatch(rdb2, rdb1') in step (2) of Fig. 7. We explain in the next section what to do with objects in map_2 that have an empty domain.

An alternative strategy is to perform a RightOuterMatch of rdb2 and rdb1. This would allow the match to exploit surrounding structure not present in rdb1', but produces a larger match result that would need to be manipulated later, an extra expense.

Composition

The next step is to compose map_2 with map_1 to achieve the desired mapping, map_3, from rdb2 to dw1. There are several issues regarding this composition. First, it needs a to work for mapping objects that have set-valued domains. For example, the last Create View statement in map_1 represents a mapping object m_5 in Fig. 5 with multiple attributes for each of the tables in its domain. When composing map_2 with map_1, we need "enough" mapping objects in map_2 to cover domain(m), for each mapping object m in map_1. This is analogous to m_{1a} and m_{1b} covering A and B in Fig. 2b.

Second, the composition must create an expression in each mapping object that combines the expressions in the mapping objects it is composing. This requires substituting objects in the mapping expressions (i.e., SQL statements) of map_1. That is, it replaces each rdb1′ attribute and its associated table by its rdb2 counterpart defined by map_2. The right column of Table 1 shows the resulting SQL statements that make up map_3, which can automatically be generated in this way. For example, in the Sales query, since map_2 maps B-Ordering.BookID in rdb2 to Order-Details.OrderDetailID in rdb1, it substituted D.BookID for D.OrderDetailID in the Select clause.

Third, since map_2 is the result of a RightOuterMatch, we need to deal with each object m_2 in map_2 where domain(m_2) is empty. The desired outcome is to modify the SQL expression from map_1 to substitute either NULL or a user-defined value for the item in range(m_2). One way to accomplish this is to extend map_2 by adding dummy objects with all the desired default values (e.g., "NULL") to rdb2, and adding a dummy object to domain(m_2) for each m_2 in map_2 where domain(m_2) is empty. The latter can be done by using the model management ApplyFunction operation to apply the function "set domain(m_2) = {dummy-object} where domain(m_2) = ∅" to map_2. This makes the substitution of default values for range(m_2) automatic (step (3) of Fig. 7).

As shown in the first Create View of Table 1, we use default values 1 and 0 for attributes Quantity and Discount (resp.), which were not represented in rdb2. All other unmatched attributes from rdb1′ are replaced by NULL. Note that this allows two queries to be simplified (eliminating joins in the Geography and Products queries).

While the query substitutions implementing the composition are straightforward in our example, problems arise if more complex match transformations have to be incorporated, such as aggregates. This is because directly replacing attributes with the equivalent aggregate expression can lead to invalid SQL statements, e.g., by using an aggregate expression within a Where clause. Substitution is still possible, but requires more complex rules than simple variable substitution.

Re-using existing transformations may not always be desirable, as these transformations may only be meaningful for a specific source. For instance, the customer mapping entails specific expressions for customer classification (second SQL statement in Table 1), which may not be useful for a different set of customers. Such situations could be handled by allowing the user to define new transformations.

Final Steps

The final integration steps check whether any parts of rdb2 not covered by the previous steps should be incorporated into the warehouse. In our example one might want to add authors as a new dimension to the data warehouse. Determining the parts to be integrated obviously cannot be done automatically. Hence, we require a user-defined specification of the additional mapping (step (6) of Fig. 7). Merging the resulting warehouse elements with the existing schema dw1 may require combining tables in a preliminary Match (step (7)) followed by the actual Merge (step (8)).

Observations

Obviously, Match and Compose are the key operations in the proposed solution to achieve a re-use of an existing mapping. The use of SQL as the expression language requires that these operations support mapping objects with set-valued domains and ranges. The use of RightOuterMatch in combination with ApplyFunction to provide default values allowed us to completely re-use the existing mapping

The power and abstraction level of the model management operations resulted in a short solution program, a huge productivity gain over the specification and programming work involved with current warehouse tools. This is especially remarkable given the use of generic operations, not tailored to data warehousing. The main remaining manual work is in supporting the Match operations (although its implementation can at least partially be automated) and in specifying new mapping requirements that cannot be derived from the existing schemas and mappings. Of course, more effort may be needed at the data instance level for data cleaning, etc.

3.2 Alternative Solution

An alternative solution to integrate rdb2 is illustrated in Fig. 8. In contrast to the previous solution, it first identifies which parts of rdb2 can be directly matched with the warehouse schema. It tries to re-use the existing mapping map_1 only for the remaining parts of rdb2.

In step (1), we thus start by matching rdb2 with the warehouse schema dw1, resulting in a mapping map_2 that identifies common tables and attributes of rdb2 and dw1. This gives a direct way to populate range(map_2), called dw1', by copying data from domain(map_2). Note that we do not have a RightOuterMatch since we can expect that only some parts of dw1 can be derived from rdb2.

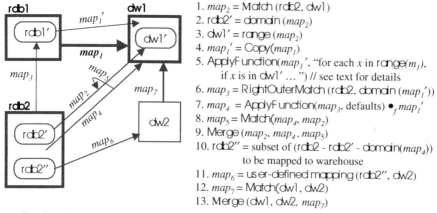

1. map_2 = Match (rdb2, dw1)
2. rdb2' = domain (map_2)
3. dw1' = range (map_2)
4. map_1' = Copy(map_1)
5. ApplyFunction(map_1', "for each x in range(m_1),
    if x is in dw1' ... ") // see text for details
6. map_3 = RightOuterMatch (rdb2, domain (map_1'))
7. map_4 = ApplyFunction(map_3, defaults) $\bullet_f map_1'$
8. map_5 = Match(map_4, map_2)
9. Merge (map_2, map_4, map_5)
10. rdb2" = subset of (rdb2 - rdb2' - domain(map_4))
    to be mapped to warehouse
11. map_6 = user-defined mapping (rdb2", dw2)
12. map_7 = Match(dw1, dw2)
13. Merge (dw1, dw2, map_7)

Fig. 8. Alternative sequence of model management operations to integrate a new source

For those parts of the warehouse schema that cannot be matched directly with rdb2 (i.e., dw1 – dw1'), we try to re-use the existing mapping map_1. We therefore create a copy map_1' of map_1 and, in step(5), use ApplyFunction to remove objects from the range of map_1' that are in dw1'. That is, for each object m_1 in map_1', "for each x in range(m_1), if x is in dw1' and not part of a primary key, then remove x from range(m_1) and from the SQL statement associated with m_1." We avoid deleting primary key attributes so that the mapping produced in steps (6)-(7) can be merged with existing tables in steps (8)-(9). Deleting x from the SQL statement involves deleting x from the Create View and deleting the corresponding terms of rdb1 from the Select clause, but not, if present, from the Where clause, since its use there indicates that x is needed to define a relevant restriction or join condition. After all such x are deleted from the statement, additional equivalence-preserving simplifications of the statement may be possible. In particular, if a dw1 table T is completely in dw1', then the map_1 SQL statement for T will be eliminated from the result mapping map_1'. The model management algebra needs to be structured in a way that allows the SQL inferencing plug-in to make such modifications to the SQL statement.

Next, we match rdb2 with the domain of map_1', called rdb1' (step (6) of Fig. 8). It is not sufficient to perform the match for rdb2 – rdb2', even though rdb2' has already been mapped to dw1. This is because for some objects m_1 in map_1', there may be an object x in domain(m_1) that maps to an object in rdb2' but not to one in rdb2 – rdb2'. There is no problem using x as input to m_1 as well as mapping x directly to dw1 using map_2. As in the first solution, we use RightOuterMatch to ensure the resulting map includes all elements of domain(map_1').

As in the previous solution, we use ApplyFunction to add default mappings for elements of domain(map_1') that do not correspond to an element of rdb2 via map_3. And then we compose map_3 and map_1', resulting in map_4 (step (7)).

The mapping between rdb2 and dw1 computed so far consists of map_2 and map_4, which we match and merge in steps (8) and (9). If map_2 and map_4 populate different tables of dw1 then Merge is a simple union. However, if there is a table that they both populate, more work is needed; hence the need for the preliminary Match forming map_5. For tables common to both maps, the two Create View statements need to be combined. This may involve non-trivial manipulation of SQL w.r.t. key columns.

As in steps (5)-(8) of the first solution, there may be a user-defined mapping for other rdb2 elements to add to the warehouse (steps (10)-(13) in Fig. 8). If there is any overlap with previous maps, then these mappings too must be merged with other Create View statements.

In our example, in step (1) we can directly match the dw1 tables Products, Customers and Geography with rdb2 tables Book and Customer as only 1:1 attribute relationships are involved. Among other things, this avoids the unwanted re-use of CustomerTypeDescription, applied for rdb1. For the two other warehouse tables, Time and Sales, we match rdb2 with rdb1 in step (6) to re-use the corresponding mapping expressions in map_1, particularly the time transformations and join query. We thus have two mappings referring to different tables; their union in step (9) provides the complete mapping from rdb2 to dw1.

Alternatively, instead of deriving the Sales table in steps (6)-(7), we could match three of its attributes, OrderID, CustomerID, and OrderDate, with table Book-Orders when creating map_2 in step (1), and using map_1 for the remaining attributes in steps (6)-(7). We thus would use ApplyFunction in step (5) to eliminate the three attributes from the Create View and Select clauses of the Sales statement in map_1 and keep the reduced query in map_1' (together with the Time query). We would leave the OrderID and CustomerID attributes in the Where clause of the modified Sales query in step (5) to perform the required joins. We thus obtain these two mapping statements for Sales:

Map₂:
create view dw1.Sales (OrderID, CustomerID, OrderDate) **as**
select   B.OrderID, B.CustomerID, B.OrderDate
from   rdb2.Book-Orders B

Map₄:
create view dw1.Sales1 (OrderID, OrderDetailID, PostalCode, ProductID, Quantity,
                          UnitPrice, Discount) **as**
select   D.OrderID, D.BookID, C.ZipCode, D.BookID, 1, D.Price, 0
from   rdb2.Book-Orders O, rdb2.B-ordering D, rdb2.Customer C
where   O.OrderID = D.OrderID **and** O.CustomerID = C.CustomerID
order by O.OrderId, D.BookID

Notice that we retain OrderID in Sales1, so we can match *map₂* and *map₄* in step (8) to drive a Merge in step (9). The result corresponds to the SQL statement in the right column of row 1 in Table1.

Observations

This approach applied similar steps to the first solution, in particular for RightOuter-Match, RightCompose and ApplyFunction. Its distinguishing feature is the partial re-use of an existing mapping, which is likely to be more often applicable than a complete re-use. The new source was matched against both the warehouse and the first source, leading to the need to merge mappings. The solution can be generalized for more than one preexisting data source. In this case, multiple mappings to the warehouse schema may be partially re-used for the integration of a new data source.

4 Data Warehouse Scenario 2: Adding a New Data Mart

The usage of model management operations described in Section 3 seems to be typical, at least for data warehouse scenarios. To illustrate these recurring patterns, we briefly consider a second scenario. We assume a given star schema dw, an existing data mart dm1 and a mapping *map₁* from dw to dm1, where range(*map₁*) = dm1. We want to add a second data mart dm2. The task is to determine the mapping from dw to dm2. Obviously this mapping must be complete with respect to dm2.

To solve the problem we can use solution patterns similar to Section 3, allowing us to give a compact description. Three possibilities are illustrated in Fig. 9. Solution 1 is the simplest approach; just apply RightOuterMatch to dw and dm2. This is possible if the two schemas differ little in structure, e.g., if dm2 is just a subset of dw.

Solution 2 is useful if some but not all of dm2 can be matched with dw. We first match dw with dm2 and then match the unmatched parts of dm2 with dm1 to re-use the associated parts of *map₁*. Remaining parts of dm2 are derived by a user-specified mapping *map₆* and then merged in.

Solution 3 tries to maximally re-use the existing mapping *map₁* as in Section 3.1. This is appropriate if the data marts are similarly structured and *map₁* contains complex transformations that are worth re-using. We first compose *map₁* with the match of dm1 and dm2. The rest of dm2 not covered by this mapping is matched with dw. Any remaining dm2 elements are derived by a user-specified mapping *map₆*.

Solution 1: map_2 = RightOuterMatch (dw, dm2)

Solution 2: (no OuterMatch, but partial re-use of map_1)

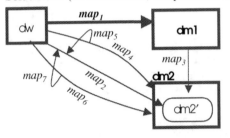

1. map_2 = Match(dw, dm2)
2. dm2′ = range(map_2)
3. map_3 = Match(dm1, dm2 - dm2′)
4. map_4 = map_1 •$_f$ map_3
5. map_5 = Match(map_4, map_2)
6. Merge(map_2, map_4, map_5)
7. map_6 = user-defined-mapping(dw,
   dm2 - dm2′ - range(map_3))
8. map_7 = Match(map_2, map_6)
9. Merge (map_2, map_6, map_7)

Solution 3: (maximal re-use of map_1)

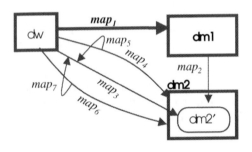

1. map_2 = Match(dm1, dm2)
2. map_3 = map_1 •$_f$ map_2
3. dm2′ = range(map_2)
4. map_4 = Match(dw, dm2 - dm2′)
5. map_5 = Match(map_4, map_3)
6. Merge(map_3, map_4, map_5)
7. map_6 = user-defined-mapping(dw,
   dm2 - dm2′- range(map_3))
8. map_7 = Match(map_6, map_3)
9. Merge (map_3, map_6, map_7)

Fig. 9. Three alternatives to add a new data mart

5 Conclusions

We evaluated the application of a generic model management approach for two data warehouse scenarios which used relational sources, star schemas, and SQL as an expression language for mappings. We devised several alternatives for solving typical mapping problems in a generic way: integrating a new data source and adding a new data mart. The solutions re-use existing mappings to a large extent and combine model operators in different ways. User interaction may be required to provide semantic equivalence information for match operations and to specify new mapping requirements that cannot be derived from existing models (mappings, schemata).

The study has deepened our understanding of two key operators: Match and Compose. In particular, we introduced the notion of OuterMatch. We showed the need for composition semantics to cover mapping objects with set-valued domains and ranges. We also proposed a general way to provide default values by employing the ApplyFunction operation. We expect this idiom will be commonly used when composing mappings.

We would like the expression manipulation associated with Compose to be managed by a module that can plug into the algebraic framework. One such module would handle SQL. The examples in this paper show what such a module must be able to do.

We found the model management notation to be a useful level of abstraction at which to consider design alternatives. By focusing on mappings as abstract objects, the designer is encouraged to think about whether a mapping is total, is onto, has a set-valued domain, can be composed with another mapping, and has a range entirely contained within the set of interest. In this paper's examples, at least, these were the main technical factors in deriving the solution. Moreover, we introduced a Venn-diagram-like notation, which enables quick comparisons between design choices, such as Figures 7 and 8 and Solutions 2 and 3 of Fig. 9. These examples show that the notation is a compact representation of each solution's approach, highlighting how the approaches differ.

Altogether, the study has provided evidence of the usefulness of a general model management approach to manage models and mappings in a generic way. Furthermore, the considered level of detail suggests that model management is more than a vision but likely to be implementable in an effective way.

Acknowledgments

We thank Alon Levy, Jayant Madhavan, Sergey Melnik, and Rachel Pottinger for many suggested improvements to the paper. We are also grateful to the Microsoft SQL Server group for providing the schemas used in the examples.

References

1. Bernstein, P.A.: Panel: Is Generic Metadata Management Feasible? VLDB 2000
2. Bernstein, P.A., Levy, A., Pottinger, R.: A Vision for Management of Complex Models. MSR-TR-2000-53, http://www.research.microsoft.com/pubs/, June 2000
3. Doan, AH., Domingos, P., Levy, A.: Learning Source Descriptions for Data Integration. Proc. WebDB 2000, pp. 81-92
4. Jannink, J., Mitra, P., Neuhold, E., Pichai, S., Studer, R., Wiederhold, G.: An Algebra for Semantic Interoperation of Semistructured Data. Proc. 1999 IEEE Knowledge and Data Engineering Exchange Workshop (KDEX'99), Nov. 1999.
5. Li, W., Clifton, C.: Semantic Integration in Heterogeneous Databases using Neural Networks. Proc. VLDB94
6. Li, W., Clifton, C.: SEMINT: A Tool for Identifying Attribute Correspondences in Heterogeneous Databases Using Neural Network. Data and Knowledge Engineering, 33 (1), 2000
7. Miller, R., Ioannidis, Y.E., Ramakrishnan, R.: Schema Equivalence in Hetereogeneous Systems: Bridging Theory and Practice. Information Systems 19(1), 3-31, 1994
8. Milo, T., Zohar, S.: Using Schema Matching to Simplify Heterogeneous Data Translation. Proc. VLDB98
9. Mitra, P., Wiederhold , G., Jannink, J.: Semi-automatic Integration of Knowledge Sources. Proc. of Fusion '99, Sunnyvale, USA, July 1999
10. Mitra, P., Wiederhold, G., Kersten, M.: A Graph-Oriented Model for Articulation of Ontology Interdependencies; Proc. Extending DataBase Technologies, EDBT 2000, LNCS Springer Verlag.
11. Mylopoulos, J., Motschnig-Pitrik, R.: Partitioning Information Bases with Contexts. Proc. 3rd CoopIS, Vienna, pp. 44-54, May 1995.
12. Palopoli, L., Sacca, D., Ursino, D.: Semi-automatic, semantic discovery of properties from database schemas. Proc. IDEAS, 1998.
13. Palopoli, L., Sacca, D., Ursino, D.: An automatic technique for detecting type conflicts in database schemas. Proc. CIKM, 1998
14. Shu, N.C., Housel, B.C., Taylor, R.W., Ghosh, S.P., Lum, V.Y.: EXPRESS: A Data EXtraction, Processing and REStructuring System. ACM TODS 2,2: 134-174, 1977.

Reflections on Computer Science and Information Systems Research

Salvatore T. March

David K. Wilson Professor of Management
Owen Graduate School of Management
Vanderbilt University
Nashville, TN 37203
Sal.March@owen.vanderbilt.edu

Abstract. Computer and related information and communication technologies have profoundly affected the shape of modern society. Shepherding the creation and utilization of effective and efficient computational technologies are the joint tasks of Computer Science and Information Systems researchers. Their realm is to understand and explicate the nature of those technologies and how and why they come into existence. This knowledge forms the foundation of theories to explain and, hopefully, to predict their impacts on individuals, groups, organizations, and society as a whole. The very creation of an innovative technology focused on a specific problem in a specific context can have far reaching effects, completely unpredicted and unintended by the innovator. We argue that researchers in Computer Science and Information Systems must be cognizant of the broader implications of their work and encourage their interaction with practitioners and researchers in a variety of disciplines to identify fruitful areas of scientific inquiry.

1. Introduction

The Computer Science discipline deals with "the phenomena surrounding computers ... not just the hardware, but the programmed, living machine" [Newell and Simon, 1976]. Research in Computer Science is concerned with the theory, experimentation, and engineering that form the basis for the design and use of computing technologies within a context. It is concerned with expanding, "the frontiers of computer and information systems by pioneering the designs of more complex, reliable, and powerful computers; enabling networks of computers to efficiently exchange vast amounts of information; and seeking ways to make computers behave intelligently" [Weems, 1998].

The phenomena of interest include the structure and operation of computer systems, principles underlying their design and programming, effective methods for utilizing their capabilities, and theoretical characterizations of their properties and limitations. In short, the discipline deals with understanding the nature of computing technologies, how they come into existence, and how they are utilized to address problems and tasks in specific environments [Denning, 1998].

A.H.F. Laender, S.W. Liddle, V.C. Storey (Eds.): ER2000 Conference, LNCS 1920, pp. 16-26, 2000.
© Springer-Verlag Berlin Heidelberg 2000

The Information Systems discipline, termed Management Information Systems (MIS) by Davis [1974], deals with computer-based systems designed to support organizational processes. An MIS is, "an integrated, user-machine system for providing information to support operations, management, analysis and decision-making functions in an organization. The system utilizes computer hardware and software; manual procedures; models for analysis, planning, control and decision-making; and a database" [Davis and Olson, 1985]. Research in Information Systems is concerned with identifying, measuring, explaining, and predicting the impacts of computer technologies on individuals, groups, and organizations, including society as a whole, with an aim toward improving organizational processes. The MIS discipline deals with the application and utilization of technical capabilities that Computer Science has made possible and with the interactions among those technologies and the social settings in which they are implemented [Lee, 1999].

If the goal of Computer Science research can be characterized as *creating* innovations [Tsichritzis, 1997], the goal of Information Systems research can be characterized as *evaluating* those innovations, explaining and predicting their utilization, diffusion, and impact within a given social context. If the main question of Computer Science research is, "What *can* be (efficiently) automated?" Information Systems research addresses the question, "What *should* be (effectively) automated (within a given social context)?"

Clearly there are areas of overlap between Computer Science and Information Systems research activities and each may be considered to be involved in adding to fundamental knowledge that will ultimately "improve society." A number of Computer Science researchers have advocated taking a broad view of Computer Science that includes the behavioral and organizational issues traditionally addressed by Information Systems researchers [Tsichritzis, 1997; Denning, 1997; Brooks, 1996]. Historically, however, research in Computer Science (apart from that in Human-Computer Interaction) has focused on the *design* rather than the *use* or *impacts* of computer technologies, leaving the study of the behavioral, social, and organizational phenomena surrounding computers to Information Systems researchers. Computer Science researchers have not considered "MIS" to be a part of their domain of interest [Wulf, 1995]. This is not surprising or problematic. A discipline must define its boundaries.

Computer Science researchers, being concerned with the design and construction of technologies, must have a strong background in computation, mathematics, problem solving, and engineering. They must understand the capabilities and limitations of the technologies with which they work. The artifacts upon which they focus are mechanistic and algorithmic, constructed from hardware and software components. Their theories, experiments, and evaluations have a computational, mathematical, or "design science" basis [March and Smith, 1995].

Conversely, Information Systems researchers, being concerned with the impacts of technologies on individuals, groups, and organizations, must have a strong background in cognitive science, social and organizational theory, economics, and empirical research methodologies. The artifacts upon which they focus contain both artificial and human or organizational components. Their theories, experiments, and evaluations have a behavioral, economic, or "social science" basis.

Both computer science researchers [Tsichritzis, 1997; Denning, 1997; Brooks, 1996] and Information Systems researchers [Madnick, 1992; Benbasat and Zmud, 1999] argue that, research must be relevant. Tsichritziz [1997] and Denning [1997] argue that the goal of research is innovation defined as "shifting the standard practices of a community of people so that they are more effective at what they do." Brooks [1996] argues that computer scientists are "are concerned with *making things*;" in particular, things that service a community of users and are evaluated by "their usefulness and their costs." Benbasat and Zmud [1999] argue that, "Information Systems researchers should look to practice to identify research topics."

Both Computer Science and Information Systems struggle with the question of theoretical underpinnings and the construction of a cumulative tradition. Computer Science clearly has theoretical roots in mathematics and engineering [Hartmanis, 1994]. Information Systems has its theoretical roots in organizational and social science and in economics [Davis and Olson, 1985]. Yet neither Computer Science [Stewart, 1995] nor Information Systems [Weber, 1987] has yet developed a theory-base unique to the discipline.

Furthermore, the major contributions of Computer Science have addressed the efficiency and internal workings of computer systems. These include, Algorithms, Complexity Theory, Operating Systems, Languages, Software Development Environments, Database Management Systems, Compilers, Artificial Intelligence, Computer Networks, and Security. Complexity Theory, for example, is a mechanism for comparing the efficiency of alternate algorithms. While mathematically sound, Tarjan [Tarjan, 1987], for one, argues that it is insufficient for practice and that empirical data must be carefully gathered to compare the performance of alternate algorithms within appropriate contexts. The emphasis on contextual analysis of empirical data is a fundamental characteristic of Information Systems research.

2. The Nature of Science and Scientific Research

In its broadest sense, the term **Science** (Latin *scientia*, from *scire*, "to know") is used to denote systematized knowledge in any field. However, it is usually applied to "the organization of objectively verifiable sense experience" [Encarta, 1998]. The pursuit of knowledge without direct regard for how it may be used is known as *pure* science. It is distinguished from *applied* science, which is the search for practical uses of scientific knowledge, and from *technology*, which is the mechanism through which applications of knowledge are realized.

Webster's New World Dictionary [Webster, 1998] defines **Science** as, "The state or fact of knowledge" and as, "systematized knowledge derived from observation, study, and experimentation carried on in order to determine the nature or principles of what is being studied." It defines **Research** as, "Careful, systematic, patient study and investigation in some field of knowledge, undertaken to discover or establish facts or principles." In contrast, it defines **Engineering** as, "The science concerned with putting scientific knowledge to practical uses." The distinction between science and engineering posed by the noted scientist and engineer von Karman [Petroski, 2000] is particularly enlightening, "a scientist studies what is, whereas an engineer creates what never was." In the sense that an Information Systems researcher, "studies what is" and a Computer Scientist "creates what never was," this distinction ironically designates the Computer *Scientist* as an "engineer" and the Information Systems researcher as a "scientist." This designation is consistent with that made by Brooks [1996].

The unifying task of science, as articulated by the Greek philosopher Thales, in the 6th century BC, is to seek the fundamental causes of natural phenomenon. The basic approaches of scientific inquiry were developed by the 4th century BC. They are comprised of *deductive reasoning* and *mathematical representation* attributed to the Academy of Plato and *inductive reasoning* and *qualitative description* attributed to the Lyceum of Aristotle. Fundamentally, the earliest researchers sought to explain why phenomenon occurred, either through observation, deduction, and mathematical modeling or through action, induction, and symbolic description.

Modern scientific methods are attributed to Galileo, who, in the 17th century added systematic verification through planned experiments to the ancient methods of induction and deduction. Galileo used newly created scientific instruments such as the telescope, the microscope, and the thermometer to enable him to observe phenomenon and measure those observations. Such direct observations of nature lead, for example, to the development and testing of theories such as Newton's Universal Law of Gravitation, later superceded by Einstein's Theory of Relativity.

It can be argued that, although technically inaccurate, Newton's Universal Law of Gravitation have had more impact on society than Einstein's Theory of Relativity (aside, perhaps from its effects on second year Physics students). That is, within the realm of everyday life, Einstein's work in this area, although a better explanation than Newton's, is not as relevant. Throughout the history of science, researchers have struggled with the dual objectives of truth and relevancy, sometimes by categorizing research work as being "pure" or "applied." Pure research seeks truth and general laws or principles of nature, while applied research seeks utility and useful innovations to a certain supporting community.

Truth and utility are complementary in nature. The pragmatists [Aboulafia, 1991] argue that truth implies utility and that scientific research should be evaluated in light of its practical implications. It must also be recognized that the results of scientific inquiry can, and do, change the realm of every day life, although the transformation can take many years and typically require engineering innovations. One can visualize, for example, a world in which space travel at speeds approaching the speed of light would bring Einstein's aforementioned work into everyday life.

Several recent articles have addressed the distinction between science and engineering within the Computer Science discipline [Denning, 1995; Stewart, 1995; Wulf, 1995]. These make the critical observation that very often the construction of an artifact precedes the scientific knowledge required to understand why it works. In fact, the construction of the artifact is, itself, the motivating or enabling factor for studying phenomena or for its very existence. In like manner, scientific discoveries can result in knowing *that* rather than knowing *why*. Research in medicine, for example, utilizing the hypothetico-deductive methods may first establish *that* a procedure or drug or treatment works and later address *why* it works. A danger with such a pragmatic approach is, of course, that failure to understand *why* it works may lead to undesirable long-term side effects or inappropriate application of that artifact or discovery.

Among the dangers with the design science approach are the tendency to solve solvable problems rather than real problems and the inability to discover general principles or laws. Creating an artifact is not sufficient; the utility of that artifact must be demonstrated within its intended context. The created artifact may, in fact work, but what have we learned? We have learned that it is feasible to construct such an artifact. Its existence demonstrates this. However, the question remains if there are fundamental principles or laws that can be learned through design science. Hartmanus [1994] argues that in Computer Science there are such laws – mathematical principles of computability. Brooks [1996] argues that there are not – artificial phenomena are not amenable to natural laws.   Hartmanus points to Information Theory and Complexity Theory as fundamental principles; however, these were not discovered through the construction of artifacts. They were derived mathematically.

Among the dangers with the social science approach are the tendency to misunderstand or misrepresent the technological environment and the failure to adequately identify the components of the technology that differentiate one environment from another.   Hence, research efforts can be expended fruitlessly studying existing theory in an environment in which the technological component does not differentiate it from previous environments.   What, for example, differentiates human communication mediated through an electronic channel from face-to-face communication? What existent theories in face-to-face communication can be applied directly to such studies and which come into question?

For the Computer Scientist (engineer) the question remains, is design science [Simon, 1996], really "science?" Alternately one might ask, is design science research really "research?"   It depends, of course, on your definition of science and of research.   Brooks argues that Computer Science is not science but creation (engineering), an even nobler activity. Science, according to Newton, seeks to "think God's thoughts after Him."   It seeks to discover and understand the underlying principles of nature. It asks the question "what are the laws governing the behavior of some 'natural' system?" Creation (engineering) seeks to "imitate the activity of God."   It seeks to construct artifacts capable of doing things that have never been done before. It asks the question "can an artifact be constructed to achieve a desired result?"

For the Information Systems researcher (scientist) the question remains, are the results relevant? While one might argue that rigorous scientific inquiry always results in the discovery of truth and that all truth is relevant, it is clear that many disagree, the Accounting discipline being a clear case in point [Johnson and Kaplan, 1987]. Hence, not only must the Information Systems researcher apply appropriate research methodology, the questions asked, theories posed, and hypotheses tested must, themselves, be important and relevant to an identified constituency.

Since the Renaissance, scientific knowledge has been transmitted mainly through journals published by scientific societies. The oldest such society, to which Galileo belonged and which still survives today, is the Accademia del Lincei. This society was established in 1603 to promote the study of mathematical, physical, and natural sciences. The Royal Society of London founded in 1662 published *Philosophical Transactions* and the Académie des Sciences de Paris founded in 1666 published *Mémoires.*

Since that time literally thousands of scientific societies have been form, typically producing journals within specialized areas of interest.

In the U.S., a club organized in 1727 by Benjamin Franklin became, in 1769, the American Philosophical Society for "promoting useful knowledge." In 1780, the American Academy of Arts and Sciences was organized by John Adams, who became the second U.S. president in 1797. In 1831 the British Association for the Advancement of Science met for the first time, followed in 1848 by the American Association for the Advancement of Science, and in 1872 by the Association

Française pour l'Avancement des Sciences. These national organizations issue the journals *Nature, Science,* and *Compte-Rendus,* respectively. [Encarta, 1998].

Today, publication has become the currency of research activity at universities, international societies, industrial research centers, and government agencies. Scientific research journals typically rely on peer review to select articles for publication. Similarly, funding agencies rely on publication records and peer reviews to select research projects to support. This leads to public policy issues with respect to the role of governments and other funding agencies in the support of scientific research and the notion of a "social contract for research" [Denning, 1997].

Not unreasonably, funding agencies expect a "return" on their investment in research activities. Government agencies expect those returns in the form of improvements to society. Industrial research centers expect them in the form of products. However, it is difficult to assess the "scientific value per dollar spent" [Zak, 1999] when the full impacts of scientific discoveries may take years or even decades to be realized. Furthermore, as discussed above, the impacts of scientific discoveries may be both short and long term. Research activities that are profitable in the short term may be quite unprofitable in the long term and conversely. Science fiction writers have long used the storyline of scientific knowledge that the researcher wished could be "un-discovered," such as those underlying chemical and biological weapons and the capability to travel in n-dimensional space.

Computer science and information systems researchers are faced with their own set of controversial issues, particularly with respect to privacy and social concerns. "Big Brother" cannot watch without significant information technology capabilities. The Internet and the World Wide Web have created an environment in which information and mis-information can be spread virtually at the speed of light. Credit verification agencies have created an environment in which an individual is judged based on the (questionable) accuracy of essentially uncontrolled data. Online and programmed securities trading have created an environment in which billions of dollars of equity can be added or removed from an economy in a matter of minutes (seconds? milliseconds? nanoseconds?). They have collapsed worldwide markets and created an information elite with the potential to manipulate markets to their own advantage.

Advances in the Computer Science and Information Systems disciplines have enabled the automation of more and more business, engineering, and scientific activities. While such automation can make organizations more productive, it also has the potential to create a bi-modal society, eliminating middle level jobs whose main functions are the collection, interpretation, analysis, integration, and dissemination of information. Hence, Computer Science and Information Systems researchers must also be concerned with the impacts and utilization of the technology.

3. Research in Entity-Relationship Modeling

The Entity-Relationship (ER) Model [Chen, 1976] is a conceptual artifact – the result of a design science effort. It is a set of constructs used to represent and communicate the data requirements of a system, typically one that will utilize a database. As the first data modeling formalism published in the archival literature it significantly changed the environment for Information System development. It was an "innovation" and, in that sense, a significant research effort worthy of publication.

If there is no question about the feasibility of building an artifact, then doing so is not considered to be viable design science research. It is simply development. Hence, building "the first" artifact capable of solving a problem is considered to be research but building "the second" is not, unless "the second" one "substantially improves" upon the first. More likely the research involved in creating "the first" artifact increases our understanding of "the problem" and possible solution approaches. As with all good research, it answers some questions and raises others. Hence, "the second" has significant opportunity to improve upon the first. Furthermore if there is no utility from building an artifact, then its contribution to design science is questionable.

Herein lie the problems faced by design science researchers and reviewers. How can you decide if an artifact will "significantly change the environment" when it is first proposed? How can you assess its utility if it has not yet been used? Given an existing artifact, what does it mean to "substantively improve upon" an artifact? How can a researcher demonstrate that his improvement is "substantial?" How can its performance be measured?

With respect to the ER model, Chen's original work [Chen, 1976] demonstrated the feasibility and utility of using a graphical notation with a mathematical basis to represent the conceptual data requirements of a system, separate from its physical implementation. He also demonstrated a method that could be used to transform that conceptual representation into a physical implementation in a variety of database management systems. Subsequent design science work (e.g., [Teorey, Yang, and Fry, 1986] among many others) enhanced and extended the basic model adding constructs to meet specific application domains and requirements. Issues such as constraints, abstraction mechanisms, temporal-spatio requirements, and evaluation metrics were addressed [Bubenko, 1977; Ferg, 1985; Ling, 1985; Peckam and Maryanski, 1988; McCarthy, 1982; Wand and Weber, 1995; Lee and Elmasri, 1998]. Each project was put to the test of utility and significance by peer review.

An artifact cannot violate natural laws; hence it is important to understand the natural laws governing the environment in which the artifact is to be implemented. The ER model and its extensions cannot violate laws governing the mechanisms by which humans communicate and interact. On the other hand, such artifacts often define environments or make it possible to observe situations in which previously accepted natural "laws" are no longer sufficient. Prior to the introduction of the ER model data representations were based on physical or at best logical structures rather than conceptual semantics. The ER model enabled an environment in which managers and end users can interact with developers on a semantic level. This constitutes a new environment for system development that has been studied [Batra, Hoffer, and Bostrom, 1990; Shoval and Frumermann, 1994; Kim and March 1995] and, indeed, needs to be studied further.

Conceptual modeling, as the Computer Science and Information Systems disciplines, has matured over its short lifetime. Tsichritzis [1999] argues that the exciting areas for research in Computer Science lie at its boundaries with other disciplines such as Biology, Medicine, Physics, and Chemistry. Brooks [1996] argues that the boundaries are the only place where Computer Science will survive as a viable research discipline. Similar arguments are made for Information Systems research [Madnick, 1992]. Exciting areas lie at its boundaries with disciplines such as Psychology, Sociology, Economics, Cognitive Science, and Organizational Theory.

It can be argued that research in conceptual modeling is similarly positioned. As Madnick [1992] challenges Information Systems researchers to identify and address key problems found in practice, researchers in conceptual modeling should look to practice for important new areas of research. Solving these problems will lead not only to innovations in practice, but will facilitate the identification of fundamental research questions. When (if) a solution works, the researcher must ask, "why does it work?" The answers to questions of that type lie at the boundaries of conceptual modeling and a host of other disciplines.

4. What, then Should We Study?

Specifically, then, what areas of practice are ripe for design science and social science research in conceptual modeling? We propose using a simple a research cycle, much like the problem solving process discussed in [March and Smith, 1995]. Given a fundamental problem within a domain, problem-solving activities begin by creating a representation of the problem using the best available conceptual framework and notation. That representation or model must be tested for fidelity with the problem environment. Such testing may result in the identification of weaknesses within the conceptual framework or notation utilized and their subsequent revision. Given an accurate representation of the problem domain, potential solutions can be created and evaluated, resulting in a solution that can then be constructed. Using the constructed artifact in the real world is likely to identify inaccuracies or required extensions in the representation, resulting it its revision, and so forth. This process is not unlike the scientific process of observing, theorizing, generating and testing hypotheses, and subsequent theory revision.

Critical to this research cycle are (1) identifying important problems within significant application domains, (2) utilizing appropriate frameworks and notations, (3) applying rigorous testing and validation process on posed representations and models, and (4) development and astute observation of the resulting solutions. The fourth stage is, in fact, equivalent to the first. That is, if they are significant, the artifacts we create change the world in some way. Astute observation of the phenomenon surrounding these artifacts can yield significant insight into the important problems in significant domains. Much like the phenomena in natural science research where analysis of data collected to test a theory results in the recognition of significant opportunities in other areas. Numerous examples of this phenomenon exist in medical research and the hard sciences, including the proposition of germ theory by Louis Pasteur and the discovery of the Van Allen Radiation belt.

From a design science perspective, researchers should look to emerging types of "data" that must be conceptually represented and create solutions to the problems these types of data present. Several areas are apparent including, multi-media, real-time sensors and control systems, continuous process management, temporality, medical applications, design data, knowledge, and wisdom. Design science research must create constructs, models, methods, and instantiations to enable the development of computer-based systems in these areas. These must have appropriate capabilities and must be understandable, usable, and effective. Understandability, usability, and effectiveness, of course, are with respect to a context in which the artifacts are applied. This is the realm of social science research. Simply constructing an artifact is no longer sufficient. That artifact must be evaluated, tested, and exercised in real world conditions. A population of users must be educated, and practices standardized for it use. These activities correspond to the processes focusing on generating new practices and new products [Tsichritzis, 1997], enabling an artifact to become an "innovation." We must understand why the artifact works (or does not work) and under what conditions.

The body of knowledge in Information Systems can serve as a basis for these investigations. To effectively determine if and how an artifact can be effectively utilized, an organization must assess its costs and benefits, many of which are "intangible," yet crucial to the success or failure of that technology within that organization. An organization must identify the factors affecting success and failure, determine if and how those factors interact, define measures for those factors and appropriately gather data to assess them. This is the fundamental Information Systems research process. Depending on the technology there may be different factors at the individual, group, organization, and societal levels. Results of such research provide guidance to organizations in planning for, introducing, evaluating, and replacing various information technologies.

Evaluating a "new" conceptual modeling formalism in terms of "completeness," for example, may simply not be appropriate. Arguing that a new conceptual modeling formalism is "a significant improvement" over existing formalisms based on an assertion that the new formalism is "more complete" than any previous formalism fails to recognize the purpose for which a conceptual modeling formalism is created. It is a means of communication and a plethora of constructs may serve to confuse rather than clarify [Rossi and Brinkkemper, 1996].

While conceptual modeling is routinely taught in systems analysis and design courses and described in basic database textbooks, very little research has investigated how they should be taught or the impact of conceptual modeling in the real world. Given the large number of conceptual modeling formalisms and the amazing research effort being poured into the creation of additional formalisms as well as methods and tools based on them, it is crucial that we understand which existing conceptual models are being used, which constructs within those formalisms are being used, and what problems are being encountered by students and practitioners in learning and using them.

As practices become accepted, standards are developed for teaching and evaluating competence in that area. Effectively a community of competent practitioners emerges. In economic terms, this phenomenon creates "externality" effects [Shapiro and Varian, 1999] such that an organization can engage in the innovation confident that sufficient expertise exists to address problems in its implementation. Research is needed to understand how people learn and use conceptual modeling formalisms. This corresponds to what Tsichritzis [1997] calls people innovations.

Finally, business innovations change the way in which business is done. These may obviate some information system capabilities and demand others. The premise of business system reengineering [Hammer and Champy, 1994] is that business organizations must constantly assess their design in light of industry, market, and technological changes. The popular press has even picked up on the concept citing companies that have "re-invented themselves" in response to the opportunities presented by the World Wide Web and other information technologies. Researchers in conceptual modeling formalisms must similarly examine, "the way business is done" with respect to the utilization of conceptual modeling formalisms – their constructs, models, methods, and instantiations.

References

Aboulafia, M., *Philosophy, Social Theory, and the Thought of George Herbert Mead* (SUNY Series in Philosophy of the Social Sciences), State University of New York Press, 1991.

Batra, D., Hoffer, J. A. and Bostrom, R. P., "A Comparison of User Performance Between the Relational and the Extended Entity-Relationship Models in the Discovery Phase of Database Design," *Communications of the ACM*, (33, 2) February 1990.

Benbasat, I. And Zmud, R. W., "Empirical Research in Information Systems: The Practice of Relevance," *MIS Quarterly*, Vol. 25, No. 1, March 1999, pp. 3-16.

Brooks, F. P., "The Computer Scientist as Toolsmith II," *Communications of the ACM*, Vol. 39, No. 3, March 1996, pp. 61-68.

Bubenko, J., "The Temporal Dimension in Information Modeling," in Nijssen, G (ed) *Architecture and Models in Data base Management Systems*, North-Holland, 1977.

Chen, P. P-S. "The Entity-Relationship Model—Toward a Unified View of Data." *ACM Transactions on Database Systems*, (1, 1) 1976.

Davis, G. B., *Management Information Systems*, McGraw-Hill Book Company, New York, 1974.

Davis, G. B. and Olson, M. H., *Management Information Systems*, (second edition), McGraw-Hill Book Company, New York, 1985.

Denning, Peter J., "Can There Be a Science of Information?" ACM Computing Surveys, Vol. 27, No. 1, March 1995.

Denning, Peter J., "The New Social Contract for Research," *Communications of the ACM*, Vol. 40, No. 2, February 1997, pp. 132-134.

Denning, Peter J., "Computing the Profession," *Educom Review*, Vol. 33, No. 6, November/December, 1998, pp. 26-30, 46-59.

Encarta, "Science," *Microsoft Encarta 98 Encyclopedia*, Microsoft Corporation 1998.

Ferg, S., "Modeling the Time Dimension in an Entity-Relationship Diagram,"*Proceedings of the 4th International Conference on Entity-Relationship Approach*, Chicago, IEEE Computer Society Press, Silver Spring, MD, 1985.

Hammer, M. and Champy, J., *Reengineering the Corporation: A Manifesto for Business Revolution,* Harperbusiness, reprint 1994.

Hartmanis, J., "On the Complexity and the Nature of Computer Science," *Communications of the ACM*, Vol. 37, No. 10, October 1994, pp. 37-43.

Johnson, H. T. and Kaplan, R. S., *Relevance Lost The Rise and Fall of Management Accounting*, Harvard Business School Press, Boston, MA 1987.

Kim, Y-G. and March, S. T., "Comparing EER and NIAM Data Modeling Formalisms for Representing and Validating Information Requirements," *Communications of the ACM*, December 1995.

Lee, A., "Inaugural Editor's Comments," *MIS Quarterly*, Vol. 23, No. 1, March, 1999, pp. v-xi.

Lee, J. Y. and Elmasri, R. A., "An EER-Based Conceptual Model and Query Language for Time-Series Data, *Proceedings of the 17th International Conference on Entity-Relationship Approach*, Singapore, Springer, 1998.

Ling, T., "A Normal Form for Entity-Relationship Diagrams," *Proceedings of the 4th International Conference on Entity-Relationship Approach*, Chicago, IEEE Computer Society Press, Silver Spring, MD, 1985.

Madnick, S. E., "The Challenge: To Be Part of the Solution Instead of Being the Problem," *Proceedings of the Workshop on Information Technologies and Systems*, Dallas, TX, Dec. 12-13, 1992.

March, S. T. and Smith, G. F., "Design and Natural Science Research on Information Technology," *Decision Support Systems*, Vol. 15, No. 4, 1995, pp 251-266.

McCarthy, W. E. "The REA Accounting Model: A Generalized Framework For Accounting Systems In A Shared Data Environment," *The Accounting Review*. (58, 3) 1982.

Newell, A and Simon, H. A., "Computer Science as Empirical Inquiry: Symbols and Search," *Communications of the ACM*, Vol 19, No 3, March 1976, pp. 113-126.

Peckam, J. and Maryanski, F., "Semantic Data Models," *ACM Computing Surveys*, (20, 3) September 1988.

Petroski, H., "Making Headlines," *American Scientist*, Volume 88, May-June 2000.

Rossi, M. and Brinkkemper, S. "Complexity Measures for Systems Development Methods and Techniques," *Information Systems*, Vol 21, No 2, pp. 209-227, 1996.

Shapiro, C. and Varian, H. R., *Information Rules: A Strategic Guide to the Network Economy*, Harvard Business School Press, 1999.

Shoval, P. and Frumermann, I., "OO and ER Conceptual Schemas: A Comparison of User Comprehension," *Journal of Database Management*, (5, 4) Fall 1994.

Simon, H. A., *The Sciences of the Artificial*, Third Edition, Cambridge, MA: The MIT Press, 1996.

Stewart, N. F., "Science and Computer Science," ACM Computing Surveys, Vol. 27, No. 1, March 1995.

Tarjan, R. E., "Algorithm Design," *Communications of the ACM*, Vol 30, No 3, March 1987, pp. 205-212.

Teorey, T., Yang, D., and Fry, J. P., "A Logical Design methodology for Relational Databases Using the Extended Entity-Relationship Model," *ACM Computing Surveys*, (18, 2) June 1986.

Tsichritzis, D., "The Dynamics of Innovation," *Beyond Calculation: The Next Fifty Years of Computing*, Copernicus, 1997, pp. 259-265.

Tsichritzis, D., "Reengineering the University," *Communications of the ACM*, Vol 42, No 6, June 1999.

Wand, Y. and Weber, R., "On the Ontological Expressiveness of Information Systems Analysis and Design Grammars," *Journal of Information Systems*, (5, 3) July 1995.

Weber, R., "Toward a Theory of Artifacts: A Paradigmatic Base for Information Systems Research," Journal of Information Systems, Spring 1987, pp 3-17.

Webster's New World Dictionary, IDG Books Worldwide, 1998.

Weems, Jr., C. C., "Computer Science," *Microsoft Encarta 98 Encyclopedia*, Microsoft Corporation, 1998.

Wulf, W. A., "Are We Scientists or Engineers?" ACM Computing Surveys, Vol. 27, No. 1, March 1995.

Zak, A., " Musings on Space Mission Development and Information Systems Support," in Zupancic, J., Wojtkowsji, W., Wojtkowsji, W. G., and Wrycza, S. (eds) Evolution in System Development, Kluwer Academic / Plenum Publishers, New York, 1999.

Zelkowitz, M. and Wallace, D., "Experimental Models for Validating Technology," *IEEE Computer*, Vol. 31, No. 5, May 1998.

From Entities and Relationships to Social Actors and Dependencies

John Mylopoulos[1], Ariel Fuxman[1], and Paolo Giorgini[2]

[1] Department of Computer Science, University of Toronto,
6 King's College Road, Toronto, Canada M5S 3H5
{jm, afuxman}@cs.toronto.edu
[2] Department of Computer Science, University of Trento,
Via Sommarive 14, 38050, Povo, Italy
pgiorgini@science.unitn.it

Abstract. Modeling social settings is becoming an increasingly important activity in software development and other conceptual modeling applications. In this paper, we review i* [Yu95], a conceptual model specifically intended for representing social settings. Then, we introduce Tropos, a formal language founded on the primitive concepts of i*, and demonstrate its expressiveness through examples. Finally, we give an overview of a project which uses Tropos to support software development from early requirements analysis to detailed design.

Keywords: conceptual models, semantic data models, entity-relationship model, requirements models, enterprise models, software development methodologies.

1 Introduction

The Entity-Relationship (hereafter E-R) model was proposed by Peter Chen at the first VLDB conference [Che75] as a modeling framework for capturing the meaning of data in a database. Since then, the model has been widely taught and used during structured information system analysis and design. It was also extended to support abstraction mechanisms such as generalization and aggregation [BCN92]. Elements from this extended version can be found in various object-oriented analysis techniques, e.g., OMT [RBP+91], as well as the ever-popular UML [BRJ99]. Unlike many other concepts in Computer Science that enjoyed a short lifespan of practical use, the E-R model has had an enduring impact on software engineering research and practice. Its use spans 25 years, as well as the two dominant software development methodologies of this period (structured and object-oriented software development). There are good reasons for this longevity. The world of software applications has revolved around the notions of static entities and dynamic processes, and the E-R model offers a simple, yet powerful means for modeling the former.

We argue in this paper that the notions of social actor and dependency (among actors) will become increasingly important in software development and

A.H.F. Laender, S.W. Liddle, V.C. Storey (Eds.): ER2000 Conference, LNCS 1920, pp. 27–36, 2000.

conceptual modeling. In software development, agent-oriented programming is gaining popularity because agent-oriented software offers (or, at least, promises) features such as software autonomy, evolvability and flexibility – all of them much-needed in the days of the Internet and e-commerce. In Requirements Engineering, agents and goals are explicitly modeled and analyzed during early requirements phases in order to generate functional and non-functional requirements for a software system (e.g., [DvLF93]). In Conceptual Modeling, more and more organizations invest in models of their business processes, organizational structure and organizational function. Enterprise resource Planning (ERP) technology, as offered by SAP, Oracle, PeopleSoft, Baan et al, includes enterprise models which serve as blueprints for an organization, as well as a starting points for customizing ERP systems. For enterprise models and agent-oriented software alike, the notions of social actor and dependency constitute modeling cornerstones.

This paper reviews a particular social actor model, i*, originally introduced in Eric Yu's PhD thesis [Yu95]. In addition, we propose a formal language, called Tropos, which extends i* and makes it more expressive and amenable to analysis. Section 2 reviews the i* model and discusses similarities and differences with E-R and UML class diagrams. In Section 3, we introduce Tropos and demonstrate its expresiveness through examples. Finally, Section 4 summarizes the contributions of this paper and suggests directions for further research.

2   A Model of Distributed Intentionality

i* offers a conceptual framework for modeling social settings. The framework is founded on the notions of *actor* and *goal*. i* (which stands for "distributed intentionality") assumes that social settings involve social actors who depend on each other for goals to be achieved, tasks to be performed, and resources to be furnished. The i* framework includes the *strategic dependency model* for describing the network of relationships among actors, as well as the *strategic rationale model* for describing and supporting the reasoning that each actor goes through concerning its relationships with other actors. These models have been formalized using intentional concepts from AI, such as goal, belief, ability, and commitment (e.g., [CL90]). The framework has been presented in detail in [Yu95] and has been related to different application areas, including requirements engineering [Yu93], business process reengineering [YML96], and software processes [YM94].

A strategic dependency diagram consists of actors, represented by circles, and social dependencies, represented as directional links. The actors are the stakeholders relevant to the social setting being modelled. Every dependency link has a *depender* actor, who needs something to be done, and a *dependee* actor, who is willing and able to deliver on that something. Thus, dependencies represent commitments of one actor to deliver on what another actor needs. There are four types of dependencies: *goal*, *softgoal*, *task* and *resource*. A goal dependency implies that one actor wants a goal to be achieved, e.g., RepairCar, and another actor is willing and able to fulfill this goal. Softgoals are goals that are not formally definable, such as SecureEmployment, or FairEstimate. Unlike goals, softgoals do not have a well-defined criterion as to whether they are fulfilled. A task dependency implies that one actor wants a task to be performed, e.g., DoAppraisal, and another actor is willing and able to carry it out. Finally, a resource dependency means that one actor needs a resource, such as Premium, and another actor can deliver on it.

Figure 1 presents a strategic dependency diagram for insurance claims. The diagram includes four actors, named respectively BodyShop, Customer, InsuranceCo and Appraiser. These are the relevant stakeholders to the task of handling insurance claims. The diagram also shows the dependencies among these actors. For instance, Customer depends on BodyShop to repair her car (RepairCar, a goal dependency), while BodyShop depends on Customer for the repairs to be paid (RepairCosts, a resource dependency). In addition, Customer depends on BodyShop to maximize the repairs done to her car (MaxRepairs), while BodyShop depends on Customer for keeping her clients (KeepClient). These are both softgoal dependencies. Turning to the dependencies between the customer and the insurance company, Customer depends on InsuranceCo to cover the repairs (CoverRepairs, goal dependency) and pay damage costs (DamageCosts, resource dependency); InsuranceCo depends on Customer to pay the insurance premium (Premium) and for continued business (ContinuedBusiness, softgoal dependency). Customer also depends on Appraiser for a fair estimate of the damages on her car (FairEstimate). Finally, InsuranceCo depends on Appraiser to carry out an appraisal (DoAppraisal, task dependency), while Appraiser depends on InsuranceCo for secure employment (SecureEmployment).

Superficially, one can view strategic dependency diagrams as variations on entity and relationship diagrams. After all, they are graph-based and nodes represent particular kinds of things while edges represent particular kinds of relationships. However, strategic dependency diagrams come about by asking entirely different questions about the application being modelled. For E-R diagrams the basic question is "what are the relevant entities and relationships?". For strategic dependency diagrams, on the other hand, the basic questions are "who are the relevant stakeholders" and "what are their obligations to other actors?". During analysis, we might also want to answer questions such as "is it possible that an insurance claim will never be served?" or "what could possibly happen during the lifetime of a claim?".

Similar comments apply in comparing strategic dependency diagrams with UML class diagrams. However, we do propose to adopt some of the notational conveniences of class diagrams, such as min/max cardinalities on dependencies and specialization relationships among actors. For instance, we would like to be able to declare different kinds of customers: CorporateCustomer, IndividualCustomer, SpecialCustomer, ValuedCustomer, etc. For each of these customer classes, the insurance company may have different insurance claim handling procedures in place, and also different expectations.

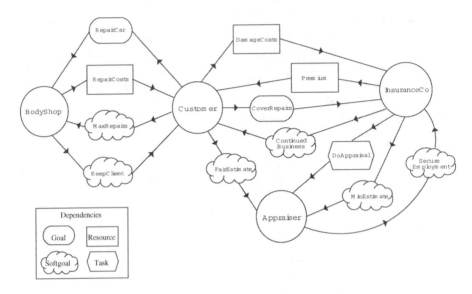

Fig. 1. Handling insurance claims

3   Formal Specification of i* Diagrams

The diagram of Figure 1 only provides a sketch of the social setting being modeled. In order to produce analyzable models, we propose a (new) specification language, called Tropos, founded on the primitives of i*. The language provides constructs for the specification of actors, social dependencies, entities, relationships and actions. Some of its features are inspired by the formal requirements specification language KAOS [DvLF93]. However, KAOS is based on different primitive concepts; in particular, it does not support the notion of social dependency.

The language is structured in two layers. The outer layer declares concepts and has an entity-relationship flavour; the inner layer expresses constraints on those concepts in a typed first-order temporal logic with real time constructs [Koy92]. The notation for temporal operators is summarized in Figure 2.

To begin the formal specification of our example, we note that many of the dependencies of Figure 1 relate to a particular claim. Claim, therefore, is an important entity of our model, and can be formulated as follows:

Entity Claim
> **Has** claimId: Number, date: Date, claimant: Customer, quote[0..n]: Quote,
> insP: InsPolicy
> **Invariant**
> $date \le insP.expirDate$

P	P holds in the *current* state
$\bullet P$	P holds in the *previous* state
$\circ P$	P holds in the *next* state
$\blacklozenge P$	P holds in *current or some past* state
$\lozenge P$	P holds in *current or some future* state
$\blacksquare P$	P holds in *current and all past* states
$\square P$	P holds in *current and all future* states

Fig. 2. Temporal operators

This entity has a number of attributes, such as identification number (claim-Id), date, claimant and a list of quotes from the bodyshops. It is also associated to a valid insurance policy (... "no valid policy, no claim" your friendly insurance agent would say...):

Entity InsPolicy
 Has insNo: Number, expirDate: Date, car: Car, customer: Customer,
  policyType: PolicyType, premiumAmount: Currency

Actors are defined in terms of their attributes, their initial goals and the actions they are capable of. For instance, the specification for Customer is:

Actor Customer
 Has name: Name, addr: Address, phone: Number
 Capable of MakeClaim, Pay
 Goal
  $\forall cl : Claim(cl.claimant = self$
    $\Rightarrow (\exists cover : CoverRepairs[cl](\lozenge Fulfil(cover))))$

This actor can perform the actions MakeClaim and Pay. Its goal is that all its claims be eventually covered by the insurance company. There are several details to note about the specification. First, the style is "object-oriented", in the sense that Customer is considered a class and self refers to one of its instances. Second, the variable cover is quantified over the set CoverRepairs[cl], which defines the set of all instances of the goal dependency CoverRepairs that have the entity instance cl among their attributes. Finally, Fulfil is a special predicate that states that a goal has been achieved; we will explain it shortly when we introduce the notion of goal modalities.
Similarly, the actor InsuranceCo is defined as follows:

Actor InsuranceCo
    Has name: Name, addr: Address, phone: Number
    Capable of AcceptClaim, Pay
    Goal $\forall customer : Customer, \exists business : ContinuedBusiness$
                     $(business.Dependee = customer \wedge Fulfil(business))$

The goal of this actor is that all its clients continue their business with the company. Therefore, the softgoal dependency `ContinuedBusiness` should be fulfilled for all the customers. Note that it is possible to refer to a depender or a dependee just as if they were any other attribute of a dependency.

Goal dependencies are defined in terms of their modality, attributes, involved agents and constraints. The most important goal dependency in our example is `CoverRepairs`: the customer depends on the insurance company to cover the cost of repairing her car.

GoalDependency CoverRepairs
    Mode Fulfil
    Depender Customer
    Dependee InsuranceCo
    Has cl: Claim
    Defined
       $Claim(cl) \wedge \bullet \neg Claim(cl) \wedge \Diamond_{\leq 6mo} RunsOk(cl.insP.car)$
    Necessary
       $\exists repair : RepairCar[cl](Fulfil(repair))$
       $\exists damage : DamageCost[cl] \Rightarrow (Fulfil(damage)$
                $\wedge(\exists cost : RepairCost[cl](Fulfil(cost)$
                         $\Rightarrow damage.amount \geq cost.amount)))$

The modality of this dependency is `Fulfil`, which means that it should be achieved at least once. There are other modalities in our language, such as `Maintain` (continuing fulfilment) and `Avoid` (continuing prevention of a goal from being fulfilled). The **Defined** clause gives a formal definition of the goal; **Necessary** specifies necessary conditions for it to be achieved. According to its modality, we interpret the definition of `CoverRepairs` as

$$Fulfil(self) \Leftrightarrow \blacklozenge(Claim(cl) \wedge \bullet \neg Claim(cl) \wedge \Diamond_{\leq 6mo} RunsOk(cl.insP.car))$$

This definition states that the customer expects that the car should start running OK in no more than 6 months after the claim is made. We capture the moment in which the claim is made (`Claim(cl)` $\wedge \bullet \neg$ `Claim(cl)`) by using a fluent `Claim(cl)` that is true if and only if `cl` is an instance of the class `Claim` at a particular point in time. The first necessary condition can be interpreted as

$$Fulfil(self) \Rightarrow \blacklozenge(\exists repair : RepairCar[cl](Fulfil(repair)))$$

It states that the car must be repaired by a bodyshop before the customer considers that the claim has been covered. The second necessary condition states that the customer expects to receive the damage costs from the insurance company, and they should be enough to pay the repair costs to the bodyshop.

Resource dependencies are specified in a similar way. The following are the definitions for the resources `DamageCosts` and `Premium`:

ResourceDependency DamageCosts
    Mode Fulfil
    Depender Customer
    Dependee InsuranceCo
    Has cl: Claim, amount: Currency
    Necessary

$$\exists app : DoAppraisal[cl], \exists min : MinEstimate[app](Fulfil(app)$$
$$\wedge Fulfil(min) \wedge \exists quote : Quote(quote \in cl.quote$$
$$\wedge quote.amount \leq app.amount))$$

ResourceDependency Premium
    Mode Fulfil
    Depender InsuranceCo
    Dependee Customer
    Has insP: insurancePolicy, dueDate: Date
    Necessary

$$\forall cl : Claim(\exists cover : CoverRepairs[cl]$$
$$(cl.insP = insP \Rightarrow Fulfil(cover)))$$

The necessary condition of `DamageCosts` states that `InsuranceCo` will deliver the resource only if there is an appraisal that minimizes the estimate on the cost of the damages. Furthermore, there must be a quote from a bodyshop that is lower or equal than the value appraised. In the second dependency, the `Customer` will only pay the `Premium` if all its claims have been covered by the insurance company.

As softgoal dependencies cannot be formally defined, we only attempt to characterize them with necessary conditions. For instance, the following is the specification for `MinEstimate`:

SoftGoalDependency MinEstimate
    Mode Fulfil
    Depender InsuranceCo
    Dependee Appraiser
    Has appraisal : DoAppraisal
    Necessary

$$\neg \exists otherApp : DoAppraisal[appraisal.cl]$$
$$(otherApp.amount < appraisal.amount)$$

We state that the insurance company expects that the estimate be minimized, in the sense that there should be no other appraisal for the same claim estimating a lower amount. Note that this condition does not fully characterize the softgoal since we are actually not quantifying over all possible appraisals, but rather on the appraisals that exist in our system. Another important softgoal dependency in our example is ContinuedBusiness:

SoftGoalDependency ContinuedBusiness
    Mode Maintain
    Depender InsuranceCo
    Dependee Customer
    Necessary
        $\exists pr : Premium(pr.Dependee = self.Dependee$
$$\wedge pr.insP.expirDate > now)$$

This softgoal has a different modality from the dependencies presented so far. The necessary condition, which states that the insurance company expects the customer to always be up to date with her payments, is interpreted as follows:

$Maintain(self)$
$$\Rightarrow \exists t : Time(\blacksquare_{\geq t}(\exists pr : Premium(pr.Dependee = self.Dependee$$
$$\wedge pr.insP.expirDate > now))$$

Finally, actions are input-output relations over entities. They are characterized by pre- and post-conditions; action applications define state transitions. For instance, the action MakeRepair performed by the bodyshop defines a transition from a state in which the car is not working well to another state in which it starts running:

Action MakeRepair
    Performed By BodyShop
    Refines RepairCar
    Input cl : Claim
    Output cl : Claim
    Pre $Claim(cl) \wedge \neg RunsOK(cl.insP.car)$
    Post $RunsOK(cl.insP.car)$

4   Conclusions and Directions for Further Research

We have presented a formal model of actors and social dependencies, intended for modeling social settings. We have also argued that models of social settings will become increasingly important as agent-oriented software becomes more prevalent, and organizational models become more widely used in corporate practice.

    We are currently working on a software development methodology which is founded on the Tropos model. The methodology supports the following phases:

- *Early requirements*, concerned with the understanding of a problem by studying an existing organizational setting; the output of this phase is an organizational model which includes relevant actors and their respective dependencies;
- *Late requirements*, where the software system-to-be is described within its operational environment, along with relevant functions and qualities; this description models the system as a (small) number of actors which have a number of dependencies with actors in their environment; these dependencies define the system's functional and non-functional requirements;
- *Architectural design*, where the system's global architecture is defined in terms of subsystems, interconnected through data and control flows; within our framework, subsystems are represented as actors and data/control interconnections are represented as (system) actor dependencies;
- *Detailed design*, where each architectural component is defined in further detail in terms of inputs, outputs, control, and other relevant information; our framework adopts elements of AUML [OPB99] to complement the features of i*;
- *Implementation*, where the actual implementation of the system is carried out, consistently with the detailed design; we use a commercial agent programming platform, based on the BDI (Beliefs-Desires-Intentions) agent architecture for this phase.

[MC00] and [CKM00] present the motivations behind the Tropos project and offer an early glimpse of how the methodology would work for particular examples. We are also exploring the application of model checking techniques [CGP99] in analyzing formal specifications such as those presented in this paper. For example, we are studying the possibility of "simulating" formal specifications in order to establish that certain properties will hold for all possible futures, or some future. We are also looking at analysis techniques which can facilitate the discovery of deviations between a formal specification and the modeller's understanding of what is being modelled.

Acknowledgements. We are grateful to our colleagues Eric Yu, Jaelson Castro, Manuel Kolp, Raoul Jarvis (University of Toronto), Yves Lesperance (York University), Fausto Giunchiglia (University of Trento), and Marco Pistore, Paolo Traverso, Paolo Bresciani and Anna Perini (IRST) for contributing ideas and helpful feedback to this research.

This work is funded partly by the Communications and Information Technology Ontario (CITO) and the Natural Sciences and Engineering Research Council (NSERC) of Canada. Also, by the University of Trento, and the Institute of Research in Science and Technology (IRST) of the province of Trentino, Italy.

References

[BCN92]   C. Batini, S. Ceri, and S. Navathe.   *Database Design: An Entity-Relationship Approach.* Benjamin/Cummings, 1992.

[BRJ99]   G. Booch, J. Rumbaugh, and I. Jacobson. *The Unified Modeling Language User Guide*. The Addison-Wesley Object Technology Series. Addison-Wesley, 1999.

[CGP99]   E. Clarke, O. Grumberg, and D. Peled. *Model Checking*. MIT Press, 1999.

[Che75]   P. Chen. The Entity-Relationship model: Towards a unified view of data. In D. Kerr, editor, *Proceedings of the International Conference on Very Large Data Bases*, September 1975.

[CKM00]   J. Castro, M. Kolp, and J. Mylopoulos. Developing agent-oriented information systems for the enterprise. In *Second International Conference on Enterprise Information Systems*, July 2000.

[CL90]   P. Cohen and H. Levesque. Intention is choice with commitment. *Artificial Intelligence*, 32(3), 1990.

[DvLF93]   A. Dardenne, A. van Lamsweerde, and S. Fickas. Goal directed requirements acquisition. *Science of Computer Programming*, 20:3–50, 1993.

[Koy92]   R. Koymans. Specifying message passing and time-critical systems with temporal logic. In *Springer-Verlag LNCS 651*. Springer-Verlag, 1992.

[MC00]   J. Mylopoulos and J. Castro. Tropos: A framework for requirements-driven software development. In J. Brinkkemper and A. Solvberg, editors, *Information Systems Engineering: State of the Art and Research Themes*. Springer-Verlag, 2000.

[OPB99]   J. Odell, H. Van Dyke Parunak, and B. Bauer. Representing agent interaction protocols in UML. To be published, 1999.

[RBP+91]   J. Rumbaugh, M. Blaha, W. Premerlani, F. Eddy, and W. Lorensen. *Object-Oriented Modelling and Design*. Prentice Hall, 1991.

[YM94]   E. Yu and J. Mylopoulos. From E-R to A-R – modeling strategic actor relationships for business process reengineering. In P. Loucopoulos, editor, *Proceedings Thirteenth International Conference on the Entity-Relationship Approach*. Springer-Verlag, December 1994.

[YML96]   E. Yu, J. Mylopoulos, and Y. Lesperance. AI models for business process reengineering. *IEEE Expert*, 1996.

[Yu93]   E. Yu. Modelling organizations for information systems requirements engineering. *First IEEE Int. Symposium on Requirements Engineering*, January 1993.

[Yu95]   E. Yu. *Modelling Strategic Relationships for Process Reengineering*. PhD thesis, University of Toronto, Toronto, Canada, 1995.

A Pragmatic Method for the Integration of Higher-Order Entity-Relationship Schemata

Thomas Lehmann[1] and Klaus-Dieter Schewe[2]

[1] Volkswagen AG, K-DOE 2, IS Produktmanagement
P.O.Box 1837, 38436 Wolfsburg, Germany
thomas2.lehmann@volkswagen.de

[2] Massey University, Department of Information Systems
Private Bag 11 222, Palmerston North, New Zealand
K.D.Schewe@massey.ac.nz

Abstract. One of the challenges in practical information system development is to find suitable methods for schema integration. Schema integration aims at replacing a set of existing schemata by a single new one. In this case there is a need to guarantee that with respect to information capacity the new schema dominates or is equivalent to the old ones. We develop formal transformation rules for schema integration. These rules rely on the Higher-order Entity-Relationship model and its theory of schema equivalence and dominance. The rules are embedded in a pragmatic method telling how they should be applied for integration. The method has been applied to various schemata of realistic size.

1 Introduction

The starting point for schema integration is a set of schemata over some data models. Usually the focus is on two schemata over the same data model. If the underlying data models differ, then we may assume some preprocessing transforming both schemata—or one of them, if this is sufficient—into an almost equivalent schema over a data model with higher expressiveness.

The work in [3,4,8] is based on the flat Entity-Relationship model. Larson et al. [4] consider containment, equivalence and overlap relations between attributes and types that are defined by looking at "real world objects". Equivalence between types gives rise to their integration, containment defines a hierarchy with one supertype and one subtype, and overlapping gives rise to new common supertype. The work by Spaccapietra and Parent [8] considers also relationships, paths and disjointness relations betwen types. The work by Koh et al. [3] provides additional restructuring rules for the addition or removal of attributes, generalization and specialization, and the introduction of surrogate attributes for types.

The work by Biskup and Convent in [1] is based on the relational data model with functional, inclusion and exclusion dependencies. The method is based on the definition of integration conditions, which can be equality, containment, disjointness or selection conditions. Transformations are applied aiming at the

A.H.F. Laender, S.W. Liddle, V.C. Storey (Eds.): ER2000 Conference, LNCS 1920, pp. 37–51, 2000.

elimination of disturbing integration conditions. In the same way as our work it is based on a solid theory. On the other hand, it has never been applied to large systems in practice. The approach by Sciore et al. in [7] investigates conversion functions on the basis of contexts added to values. These contexts provide properties to enable the semantic comparability of values.

In our approach we shall assume that both schemata are defined on the basis of the higher-order Entity-Relationship model (HERM) [9] which is known to provide enough expressiveness such that schemata existing in practice can be easily represented in HERM. The integration problem then consists in finding a single new schema such that the semantics of both old schemata is preserved within the new one.

Ideally the information capacity of the new schema should be the same as the combined information capacity of the old schemata. From a theoretical point of view it has to be clarified how the term "information capacity" is understood and what it means to have equivalent or augmented capacity. We shall briefly outline that there exist different and to some extent not even comparable notions of schema equivalence and schema dominance [2,6,9]. We argue that for our purposes the best choice is to rely on the notions developed in connection with HERM [9].

Our approach is based on transformation rules to refine HERM schemata. Roughly speaking we consider the disjoint union of the original schemata and outline the dependencies between its parts. Each application of a transformation rule guarantees that the old schema will appear as a view on the new one. This allows at least to preserve existing functionality.

One group of rules addresses the restructuring of the complex attributes, entity types, relationship types and clusters. Another group of rules considers the shifting of attributes over hierarchies and paths. A third group of rules deals with selected integrity constraints such as keys, functional, inclusion and join dependencies, cardinality constraints and path constraints. A fourth group of rules is devoted to aggregation, decomposition, specialization and generalization.

The rules are chosen in such a way that the most important problems in practical systems could be addressed [5]. To that end our approach is pragmatic with respect to the technical information systems that exist in that work environment. We outline how the framework of rules should be used. This sets up a pragmatic method for schema integration. We describe the specific problems addressed, the method and its core, the transformation rules. Finally, we shall take a look at the application of our method to selected schemata arising from practice.

The novelty of our work is not only due to the fact that it is based on the more sophisticated HERM which provides nested structures and clusters. Up to a certain degree we also deal with various classes of integrity constraints. Furthermore, the approach is theoretically founded on dominance and equivalence for HERM schemata and we could clarify the relation of these notions to the work by Hull [2] and Qian [6]. Finally, our work is not just a paper exercise. It has been applied to schemata of reasonable size as they occur in practice.

2   Higher-Order Entity-Relationship Structures

Let us start with a brief outline of the Higher-order Entity-Relationship model (HERM) [9]. The major extensions of the HERM compared with the flat ER model concern nested attribute structures, higher-order relationship types and clusters, a sophisticated theory of integrity constraints, operations, dialogues and their embedding in development methods. Here we only sketch some of the structural aspects. For a detailed description of HERM we refer to [9].

Let $\mathcal{A}$ be some set of *simple attributes*. We assume that each simple attribute $A \in \mathcal{A}$ is associated with a base data type $dom(A)$. In the HERM it is allowed to define nested attributes.

A *nested attribute* is either a simple attribute, the null attribute $\perp$, a tuple attribute $X(A_1, \ldots, A_n)$ with pairwise different nested attributes A_i and a label X or a set attribute $X\{A\}$ with a nested attribute A and a label X. Let $\mathcal{NA}$ be the set of all nested attributes.

We may extend dom to nested attributes in the standard way, i.e. a tuple attribute will be associated with a tuple type, a set attribute with a set type, and the null attribute with $dom(\perp) = \mathbb{1}$, where $\mathbb{1}$ is the trivial type with only one value.

On nested attributes we have a partial order $\leq$ with $A \leq \perp$ for all $A \in \mathcal{NA}$, $X\{A\} \leq X\{A'\} \Leftrightarrow A \leq A'$ and $X(A_1, \ldots, A_n) \leq X(A'_1, \ldots, A'_m) \Leftrightarrow \bigwedge_{1 \leq i \leq m} A_i \leq A'_i$. It is easy to see that $A \leq A'$ gives rise to a canonical mapping $\pi^{(A)}_{A'} : dom(A) \to dom(A')$.

A *generalized subset* of a set F of nested attributes is a set G of nested attributes such that for each $A' \in G$ there is some $A \in F$ with $A \leq A'$.

A *level-k-type* R consists of a set $comp(R) = \{r_1 : R_1, \ldots, r_n : R_n\}$ of labelled components, a set $attr(R) = \{A_1, \ldots, A_m\}$ of nested attributes and a key $key(R)$. Each component R_i is a type or cluster of a level at most $k-1$. At least one of the R_i must be level-$(k-1)$-type or -cluster. The labels r_i are called *roles*. Roles can be omitted in case the components are pairwise different. For the key we have $key(R) = comp'(R) \cup attr'(R)$ with $comp'(R) \subseteq comp(R)$ and a generalized subset $attr'(R)$ of the set of attributes.

A *level-k-cluster* is $C = R_1 \oplus \cdots \oplus R_n$ with components R_i, each of which is a type or cluster of a level at most k. At least one of the R_i must be level-k-type or -cluster. The definition of semantics leads to disjoint unions.

A level-0-type E—here the definition implies $comp(E) = \emptyset$—is usually called an *entity type*, a level-k-type R with $k > 0$ is called a *relationship type*. A *HERM schema* is a closed finite set $\mathcal{S}$ of entity types, relationship types and clusters together with a set Σ of integrity constraints defined on $\mathcal{S}$. The semantics is defined as usual [9].

For querying a HERM schema we concentrate on the algebraic approach. The *HERM algebra* $\mathcal{H}$ provides the operations σ_φ (selection) with a selection formula φ, $\pi_{A_1, \ldots, A_m}$ (projection) with a generalized subset $\{A_1, \ldots, A_m\}$, δ_f (renaming) with a renaming function f, $\bowtie_G$ (join) with a common generalized subset G, $\cup$

(union), − (difference), $\nu_{X:A_1,\ldots,A_n}$ with attributes $A_1,\ldots,A_n$ (nest), and μ_A (unnest) with a set attribute A. We omit the details and refer to [9].

The algebra may be extended with assignment, non-deterministic value creation and WHILE, in which case we talk of the *extended HERM algebra* $\mathcal{H}_{ext}$.

3 Schema Dominance and Equivalence

Let us compare the theory of schema equivalence and dominance in HERM [9] with the work by Hull [2] and Qian [6].

We say that a HERM schema $(\mathcal{S}', \Sigma')$ *dominates* another HERM schema $(\mathcal{S}, \Sigma)$ (notation: $(\mathcal{S}, \Sigma) \sqsubseteq (\mathcal{S}', \Sigma')$) iff there are mappings f and g taking instances of $(\mathcal{S}, \Sigma)$ to instances of $(\mathcal{S}', \Sigma')$ and the other way round, respectively, such that $g \circ f$ is the identity and both mappings are expressed by $\mathcal{H}$.

If we have $(\mathcal{S}, \Sigma) \sqsubseteq (\mathcal{S}', \Sigma')$ as well as $(\mathcal{S}', \Sigma') \sqsubseteq (\mathcal{S}, \Sigma)$, we say that the two schemata are *equivalent* (notation: $(\mathcal{S}, \Sigma) \cong (\mathcal{S}', \Sigma')$).

If in both cases the HERM algebra is replaced by the extended HERM algebra, we talk of dominance and equivalence in the extended sense. Note that the notions of dominance and equivalence formalize "information capacity". A dominating schema is able to represent more information; it has an augmented information capacity. Schema equivalence means that we have equal information capacity.

In the literature there are different notions of schema dominance and equivalence. Hull introduces four different notions on the basis of the relational data model [2]. In all these cases the definition is analogous to the one given for HERM schemata. The differences concern the conditions on the mappings f and g.

In the case of *calculus dominance* $\sqsubseteq_{calc}$ the mappings f and g must be defined by safe relational calculus or equivalently the relational algebra. In the case of *generic dominance* $\sqsubseteq_{gen}$ the mappings f and g must be generic in the sense that they commute with permutations of domain values fixing only a finite set Z of values. In the case of *internal dominance* $\sqsubseteq_{int}$ the mappings f and g may only introduce a finite set of new values. Finally, for *absolute dominance* $\sqsubseteq_{abs}$ there are no restrictions on f and g. Thus, it is sufficient to have an injective mapping f.

In [2] it has been shown that calculus dominance implies generic dominance, which itself implies internal dominance, which implies absolute dominance. All these implications are strict. In [5] it has been shown that calculus dominance implies HERM dominance, which implies generic dominance. HERM dominance in the extended sense implies absolute dominance, but not internal dominance. Of course, this only holds for relational HERM schemata.

ADT dominance as defined by Qian [6] is based on order-sorted signatures and algebras. A schema transformation, i.e. our f, must be defined as a signature interpretation. It has been shown in [6] that calculus dominance implies ADT dominance, which implies absolute dominance. These implications are strict. Furthermore, ADT dominance is incomparable with the other notions of dominance as defined by Hull. In [5] it has been shown that ADT dominance and HERM dominance are also incomparable.

4 The Method for Integration

In this section we describe the transformation rules and their embedding in a pragmatic method for their application. We adopt the following strategy. The details are filled by transformation rules described in the following subsection.

1. The first step is the homogenization of the schemata. This includes the renaming of synonymous attributes and the introduction of equal keys. Further we restructure the schemata turning attributes into entity types, entity types into relationship types and vice versa. Furthermore, we add attributes and shift attributes along hierarchies and paths. All these individual paces correspond to the application of transformation rules.
   For homogenization it turned out that the approach in [1]—i.e., the removal of disturbing integration conditions, if these are implied by the other constraints—is not sufficient, because there are transformations for type homogenization that do not rely on such integration conditions. In particular, this applies to the common case of artificial keys.
2. The second step consists in the analysis of the schema overlap which is formally described by integrity constraints on the disjoint union.
3. The third step is only a preparation of the following steps. Due to the expected large size of the schemata, these are divided into modules, each of which describes a subschema. Corresponding modules are identified in order to approach the integration of modules first.
4. Next we consider the integration of types on level 0, 1, etc., i.e. we start with entity types and level-0-clusters, then proceed with relationship types and clusters on level 1, then relationship types and clusters on level 2, etc. For each level we integrate corresponding types or clusters with respect to equality, containment, overlap and disjointness conditions. Note that this step is similar to the work done in [3,4,8].
5. The fifth step deals with the integration of paths.
6. Finally, we consider remaining integrity constraints such as (path) functional dependencies, path inclusion dependencies and join dependencies.

Let us proceed with the details of our method and describe the transformation rules. In order to avoid blown-up formalism, all rules are presented in the same way. We assume we are given a HERM schema $(\mathcal{S}, \Sigma)$, but we only indicate some parts of it. The resulting schema will be $(\mathcal{S}_{new}, \Sigma_{new})$. The new types in the new schema will be marked with a subscript $_{new}$. With these conventions the rules will be self-explaining.

Whenever we talk of schema equivalence or dominance we refer to HERM equivalence $\cong$ and HERM dominance $\sqsubseteq$.

The first group of rules addresses the aspect of schema restructuring which will be used in the homogenization step 1 of our method.

Rule 1. Replace a tuple attribute $X(A_1,\ldots,A_m)$ in an entity or relationship type R by the attributes $A_1,\ldots,A_m$. The resulting type R_{new} will replace R. For $X(A'_1,\ldots,A'_n) \in key(R)$ with $A_i \le A'_i$ we obtain $A'_1,\ldots,A'_n \in key(R')$. $\square$

In [5] the simple case, where R is an entity type was handled by a separate rule.

Rule 2. Replace a component $r : R'$ in a relationship type R by lower level components and attributes. Let the new type be R_{new}. For $comp(R') = \{r_1 : R_1,\ldots,r_n : R_n\}$ we get $comp(R_{new}) = comp(R) - \{r : R'\} \cup \{r_1^{(r)} : R_1,\ldots,r_n^{(r)} : R_n\}$ with new role names $r_i^{(r)}$ composed from r_i and r and $attr(R_{new}) = attr(R) \cup attr(R')$. In the case $r : R' \in key(R)$ and $key(R') = \{r_{i_1} : R_{i_1},\ldots,r_{i_k} : R_{i_k}, A_1,\ldots,A_m\}$ we obtain $key(R_{new}) = key(R) - \{r : R'\} \cup \{r_{i_1}^{(r)} : R_{i_1},\ldots,r_{i_k}^{(r)} : R_{i_k}, A_1,\ldots,A_m\}$, otherwise we have $key(R_{new}) = key(R)$. $\square$

It is easy to see how to simplify this rule in the case, where R' is an entity type. These two cases were treated by two separate rules in [5].

Rule 3. Replace a cluster $C = C_1 \oplus \cdots \oplus C_n$ with a cluster component $C_i = C_{i_1} \oplus \cdots \oplus C_{i_m}$ by a new cluster $C = C_1 \oplus \cdots \oplus C_{i-1} \oplus C_{i_1} \oplus \cdots \oplus C_{i_m} \oplus C_{i+1} \oplus \cdots \oplus C_n$. $\square$

Rule 4. Replace a relationship type R with a cluster component $r : C$ ($C = C_1 \oplus \cdots \oplus C_n$) by a new cluster $C_{new} = R_{1,new} \oplus \cdots \oplus R_{n,new}$ and new relationship types $R_{i,new}$ with $comp(R_{i,new}) = comp(R) - \{r : C\} \cup \{r_i : C_i\}$ and $attr(R_{i,new}) = attr(R)$. For $r : C \in key(R)$ we obtain $key = key(R) - \{r : C\} \cup \{r : C_i\}$, otherwise take $key = key(R)$. $\square$

In the case of the restructuring rules 1 – 4 we can always show that the original schema and the resulting schema are equivalent. The next rule only guarantees that the resulting new schema dominates the old one.

Rule 5. Replace a key-based inclusion dependency $R'[key(R_i)] \subseteq R[key(R)]$ by new relationship types R'_{new} with $comp(R'_{new}) = \{r' : R', r : R\} = key(R'_{new})$ and $attr(R'_{new}) = \emptyset$ together with participation cardinality constraints
$card(R'_{new}, R) = (0,1)$ and $card(R'_{new}, R') = (1,1)$. $\square$

The last two restructuring rules allow to switch between attributes and entity types and between entity and relationship types. These rules 6 and 7 guarantee schema equivalence.

Rule 6.   Replace an entity type E with $A \in attr(E)$ by E_{new} such that $attr(E_{new}) = attr(E) - \{A\}$ holds. Furthermore, introduce an entity type E'_{new} with $attr(E'_{new}) = \{A\} = key(E'_{new})$ and a new relationship type R_{new} with $comp(R_{new}) = \{r_{new} : E_{new}, r'_{new} : E'_{new}\} = key(R_{new})$ and $attr(R_{new}) = \emptyset$. Add the cardinality constraints $card(R_{new}, E_{new}) = (1,1)$ and $card(R_{new}, E'_{new}) = (1, \infty)$.   □

Rule 7.   Replace an relationship type R with $comp(R) = \{r_1 : R_1, \ldots, r_n : R_n\}$ and the cardinality constraints $card(R, R_i) = (x_i, y_i)$ by a new entity type E_{new} with $attr(E_{new}) = attr(R) = key(E_{new})$ and n new relationship types $R_{i,new}$ with $comp(R_{i,new}) = \{r_i : R_i, r : E_{new}\} = key(R_{i,new})$ and $attr(R_{i,new}) = \emptyset$. Replace the cardinality constraints by
$card(R_{i,new}, R_i) = (1, y_i)$ and $card(R_{i,new}, E_{new}) = (1, \infty)$.   □

In the case of rule 7 explicit knowledge of the key of R allows to sharpen the cardinality constraints.

The second group of rules deals with the shifting of attributes. This will also be used in the homogenization step 1 of our method. Rule 8 allows to shift a synonymous attribute occurring in two subtypes—i.e., whenever tuples agree on the key, they also agree on that attribute—to be shifted to a supertype. This rule leads to a dominating schema. Conversely, rule 9 allows to shift an attribute from a supertype to subtypes, in which case schema equivalence can be verified.

Rule 8.   For $comp(R_i) = \{r_i : R\}$ and $A_i \in attr(R_i) - key(R_i)$ $(i = 1, 2)$ together with the constraint $\forall t, t'.t[key(R_1)] = t'[key(R_2)] \Rightarrow t[A_1] = t'[A_2]$ replace the types R, R_1 and R_2 such that $attr(R_{new}) = attr(R) \cup \{A_i\}$, $comp(R_{i,new}) = \{r_i : R_{new}\}$ and $attr(R_{i,new}) = attr(R_i) - \{A_i\}$ hold.   □

Rule 9.   For $comp(R_i) = \{r_i : R\}$ $(i = 1, \ldots, n)$ and $A \in attr(R) - key(R)$ together with the constraint $\forall t \in R.\exists t' \in R_i.t'[r_i] = t$ replace the types such that $attr(R_{new}) = attr(R) - \{A\}$, $comp(R_{i,new}) = \{r_i : R_{new}\}$ and $attr(R_{i,new}) = attr(R_i) \cup \{A\}$ hold.   □

The next two rules 10 and 11 concern the reorganization of paths and the the shifting of attributes along paths. In both cases we obtain a dominating schema. In [5] rule 10 was split into two rules dealing separately with binary and unary relationship types R_n.

Rule 10.   For a path $P \equiv R_1 - \cdots - R_n$ and a relationship type R with $r_n : R_n \in comp(R)$ together with path cardinality constraints $card(P, R_1) \leq (1, 1) \leq card(P, R_n)$ replace R such that $comp(R_{new}) = comp(R) - \{r_n : R_n\} \cup \{r_{1,new} : R_1\}$ with a new role $r_{1,new}$ holds.   □

Rule 11.   For a path $P \equiv R_1 - \cdots - R_n$ with $A \in attr(R_n)$ and path cardinality constraints $card(P, R_1) \leq (1, 1) \leq card(P, R_n)$ replace R_1, R_n such that $attr(R_{1,new}) = attr(R_1) \cup \{A\}$ and $attr(R_{n,new}) = attr(R_n) - \{A\}$ hold.   □

The next group of rules deal with the extension of the schema. This either concerns new attributes, new subtypes or the simplification of hierarchies. These rules are needed in step 1 of our method.

Rule 12. Add a new attribute A to the type R, i.e. $attr(R_{new}) = attr(R) \cup \{A\}$. In addition, the new attribute may be used to extend the key, i.e. we may have $key(R_{new}) = key(R) \cup \{A\}$.   □

If the new attribute A introduced by rule 12 does not become a key attribute, we obtain a dominating schema. If it becomes a key, HERM dominance turns out to be too weak, but we can verify HERM dominance in the extended sense. To be precise, we need the facility of $\mathcal{H}_{ext}$ to create new values. In [5] these cases have been treated by separate rules and only absolute dominance was shown to hold.

The next two rules allow to introduce a new subtype via selection or projection on non-key-attributes. In both cases we have schema equivalence.

Rule 13. For a type R introduce a new relationship type R'_{new} with $comp(R'_{new}) = \{r : R\} = key(R'_{new})$ and add a constraint $R'_{new} = \sigma_\varphi(R)$ for some selection formula φ.   □

Rule 14. For a type R and attributes $A_1, \ldots, A_n \in attr(R)$ such that there are no $B_i \in key(R)$ with $A_i \leq B_i$ introduce a new relationship type R'_{new} with $comp(R'_{new}) = \{r : R\} = key(R'_{new})$ and $attr(R'_{new}) = \{A_1, \ldots, A_n\}$, and add a constraint $R'_{new} = \pi_{A_1, \ldots, A_n}(R)$.   □

The last rule 15 in this group allows to simplify hierarchies. Here again we must exploit $\mathcal{H}_{ext}$ to obtain dominance.

Rule 15. Replace types R, $R_1, \ldots, R_n$ with $comp(R_i) = \{r_i : R\} = key(R_i)$ and $card(R, R_i) = (0, 1)$ $(i = 1, \ldots, n)$ by a new type R_{new} with $comp(R_{new}) = comp(R)$, $attr(R_{new}) = attr(R) \cup \bigcup_{i=1}^{n} attr(R_i)$ and $key(R_{new}) = key(R)$.   □

The next group of rules deals with the integration of types in step 4 of our method. Rule 16 considers the equality case, rule 17 considers the containment case, and rule 18 covers the overlap case. Note that these transformation rules cover the core of the approaches in [3,8,4].

Rule 16. If R_1 and R_2 are types with $key(R_1) = key(R_2)$ and we have the constraint $R_1[key(R_1) \cup X] = f(R_2[key(R_2) \cup Y])$ for some $X \subseteq comp(R_1) \cup attr(R_1)$, $Y \subseteq comp(R_2) \cup attr(R_2)$ and a bijective mapping f, then replace these types by R_{new} with $comp(R_{new}) = comp(R_1) \cup (comp(R_2) - Y - key(R_2))$, $attr(R_{new}) = attr(R_1) \cup (attr(R_2) - Y - key(R_2)) \cup \{D\}$ and $key(R_{new}) = key(R_1) \cup \{D\}$ and an optional new distinguishing attribute D.   □

Rule 17. If R_1 and R_2 are types with $key(R_1) = key(R_2)$ and the constraint $R_2[key(R_2) \cup Y] \subseteq f(R_1[key(R_1) \cup X]$ holds for some $X \subseteq comp(R_1) \cup attr(R_1)$, $Y \subseteq comp(R_2) \cup attr(R_2)$ and a bijective mapping f, then replace R_1 by $R_{1,new}$ with $comp(R_{1,new}) = comp(R_1)$, $attr(R_{new}) = attr(R_1) \cup \{D\}$ and $key(R_{new}) = key(R_1) \cup \{D\}$ and an optional new distinguishing attribute D. Furthermore, replace R_2 by $R_{2,new}$ with $comp(R_{2,new})\{r_{new} : R_{1,new}\} \cup comp(R_2) - Y - key(R_2)$, $attr(R_{2,new}) = attr(R_2) - Y - key(R_2)$ and $key(R_{2,new}) = \{r_{new} : R_{1,new}\}$. □

Rule 18. Let R_1 and R_2 are types with $key(R_1) = key(R_2)$ such that for $X \subseteq comp(R_1) \cup attr(R_1)$, $Y \subseteq comp(R_2) \cup attr(R_2)$ and a bijective mapping f the constraints

$$R_2[key(R_2) \cup Y] \subseteq f(R_1[key(R_1) \cup X] \quad ,$$
$$R_2[key(R_2) \cup Y] \supseteq f(R_1[key(R_1) \cup X] \quad \text{and}$$
$$R_2[key(R_2) \cup Y] \cap f(R_1[key(R_1) \cup X] = \emptyset$$

are not satisfied. Then replace R_1 by $R_{1,new}$ with $comp(R_{1,new})\{r_{1,new} : R_{new}\} \cup comp(R_1) - X - key(R_1)$, $attr(R_{1,new}) = attr(R_1) - X - key(R_1)$ and $key(R_{1,new}) = \{r_{1,new} : R_{new}\}$, replace R_2 by $R_{2,new}$ with $comp(R_{2,new})\{r_{new} : R_{1,new}\} \cup comp(R_2) - Y - key(R_2)$, $attr(R_{2,new}) = attr(R_2) - Y - key(R_2)$ and $key(R_{2,new}) = \{r_{new} : R_{1,new}\}$ and introduce a new type R_{new} with $comp(R_{new}) = comp(R_1) \cup comp(R_2)$, $attr(R_{new}) = attr(R_1) \cup attr(R_2) \cup \{D\}$ and $key(R_{new}) = key(R_1) \cup \{D\}$ and an optional new distinguishing attribute D. □

In [5] the rules 16 – 18 were split into several rules depending on f being the identity or not and the necessity to introduce D or not. Without the new attribute D we obtain dominance. However, if D is introduced, we only obtain dominance by exploiting $\mathcal{H}_{ext}$.

Rule 19 considers the case of a selection condition, in which case schema equivalence holds.

Rule 19. If R and R' are types with $comp(R') \cup attr(R') = Z \subseteq comp(R) \cup attr(R)$ such that the constraint $R' = \sigma_\varphi(\pi_Z(R))$ holds for some selection condition φ, then omit R'. □

The next group of rules to be applied in step 5 of our method concerns transformations originating from path inclusion constraints. Rule 20 allows to change a relationship type. This rule leads to equivalent schemata. Rule 21 allows to introduce a relationship type and a join dependency. Finally, rule 22 handles a condition under which a relationship type may be omitted. Both rules 21 and 22 guarantee dominance.

Rule 20. If there are paths $P \equiv R_1 - R - R_2$ and $P' \equiv R_2 - R' - R_3$ with $comp(R) = \{r_1 : R_1, r_2 : R_2\}$ and $comp(R') = \{r_3 : R_3, r'_2 : R_2\}$ such that the constraint $P[R_2] \subseteq P'[R_2]$ holds, then replace R in such a way that $comp(R_{new}) = \{r_1 : R_1, r_{new} : R'\}$, $attr(R_{new}) = attr(R)$ and $key(R_{new}) = key(R) - \{r_2 : R_2\} \cup \{r_{new} : R'\}$ hold. □

Rule 21. If there are paths $P \equiv R_1 - R - R_2$ and $P' \equiv R_2 - R' - R_3$ with $comp(R) = \{r_1 : R_1, r_2 : R_2\}$ and $comp(R') = \{r_3 : R_3, r'_2 : R_2\}$ such that the constraint $P[R_2] = P'[R_2]$ holds, then replace R and R' by R_{new} such that $comp(R_{new}) = \{r_1 : R_1, r_{2,new} : R_2, r_3 : R_3\}$, $attr(R_{new}) = attr(R) \cup attr(R')$ and $key(R_{new}) = (key(R) - \{r_2 : R_2\}) \cup (key(R') - \{r'_2 : R_2\}) \cup \{r_{2,new} : R_2\}$ hold. Add the join dependency
$$R_{new}[r_1, r_{2,new}] \bowtie R_{new}[r_{2,new}, r_3] \subseteq R_{new}[r_1, r_{2,new}, r_3].$$
□

Rule 22. If there are paths $P \equiv R_1 - R_2 - \cdots - R_n$ and $P' \equiv R_1 - R - R_n$ with $comp(R) = \{r_1 : R_1, r_n : R_n\}$ such that the constraint $P[R_1, R_n] = P'[R_1, R_n]$ holds, then omit R.
□

The final group of transformation rules 23 – 26 allows to handle remaining constraints such as functional dependencies, path functional dependencies, and join dependencies. All these constraints are described in detail in [9]. The rules refer to step 6 of our method.

Rule 23 handles vertical decomposition in the presence of a functional dependency. Rule 24 allows to simplify a key in the presence of a path functional dependency. Rule 25 introduces a new entity type in the presence of a path functional dependency. Finally, rule 26 replaces a multi-ary relationship type by binary relationship types in the presence of a join dependency. The four rules lead to dominating schemata.

Rule 23. If a functional dependency $X \rightarrow A$ with a generalized subset X of $attr(E)$ and an Attribute $A \in attr(E) - X$ holds on an entity type E, but $X \rightarrow key(E)$ does not hold, then remove A from $attr(E)$ and add a new entity type E'_{new} with $attr(E'_{new}) = X \cup \{A\}$ and $key(E'_{new}) = X$.
□

Rule 24. For a path $P \equiv R_1 - R - R_2$ with $comp(R) = \{r_1 : R_1, r_2 : R_2\}$ such that the path functional dependency $X \rightarrow key(R_2)$ holds for a generalized subset X of $attr(R_1)$ replace $key(R)$ by $\{r_1 : R_1\}$.
□

Rule 25. For a path $P \equiv R_1 - \cdots - R_n$ such that the path functional dependency $X \rightarrow A$ holds for a generalized subset X of $attr(R_1)$ and $A \in attr(R_n)$ add a new entity type E_{new} with $attr(E_{new}) = X \cup \{A\}$ and $key(E_{new}) = X$. □

Rule 26. If R is an n-ary relationship type with $comp(R) = \{r_1 : R_1, \ldots, r_n : R_n\}$ and $attr(R) = \emptyset$ such that the join dependency $R[r_1, r_2] \bowtie \cdots \bowtie R[r_1, r_n] \subseteq R[r_1, \ldots, r_n]$ holds, then replace R by n new relationship types $R_{1,new}, \ldots, R_{n,new}$ with $comp(R_{i,new}) = \{r_1 : R_1, r_i : R_i\} = key(R_{i,new})$ and $attr(R_{i,new}) = \emptyset$.
□

5   Case Study

In this section we want to apply our method and integrate two schemata S_1 and S_2 represented in Figures 1 and 2. We have omitted many details. Key attributes have been emphasized in the diagrams. Furthermore, we always have $comp(R) \subseteq key(R)$. Both schemata have been adapted from [5]. They provide typical development system schemata.

First concentrate on the types **vehicle**, **load_param**, **air_param**, **weight** in S_1 and **vehicle** in S_2. For each of the non-key attributes in one schema we either find a synonymous attribute in the other or we may add those attributes (rule 12) to homogenize the types.

The same applies to types centered around **engine**. In particular, we must add types **rule** and **status** to S_1. This implies to turn the type **full_throttle** into a relationship type (rule 7). The inverse of rule 2 leads to the "duplication" of the types **full_thr** and **char_curve** in S_2. The corresponding new types are **c_full_thr** and **e_full_thr** with components **comb_engine** and **el_engine**, respectively. Combined with rule 7 we obtain binary relationship types **el_full_thr** and **rcm_full_thr**.

The same applies to types centered around **transmission**, **converter** and **coupling**, or **wheel**, respectively. This also gives rise to the separation of modules, which we denote as VEHICLE, ENGINE, TRANSMISSION, FORCE_TRANSFER and WHEEL. Next we look at each of these modules.

Starting with VEHICLE we find types **vehicle** in both schemata. According to the equality-rule 16 we obtain a single type—by abuse of notation we still call it **vehicle**—in the integrated schema. This type will have a distinguishing attribute 'dif'. We keep the key attributes 'id' and 'date'. Other attributes originating from schema S_2 can be shifted to the subtypes **air_param**, **weight** and **load_param** according to rule 9. The result is the subschema of S_3 shown in Figure 3 consisting of the types **vehicle**, **air_param**, **weight** and **load_param**.

For the module ENGINE we proceed analogously, i.e. we identify **engine** in both schemata and combine the complete paths originating from **comb_engine**. The result is the subschema of schema S_3 in Figure 3 consisting of the types **engine**, **comb_engine**, **red_cyl_eng**, **rcm_full_thr**, **c_full_thr**, **c_char_curve**, **el_engine**, **el_full_thr**, **e_full_thr**, **el_char_curve**, **status** and **rule**.

For the module TRANSMISSION we may again proceed analogously. The only interesting point to be mentioned concerns the shifting of the type **transformation**. This is due to an application of rule 20 for paths. The result gives the subschema of S_3 consisting of the types **transmission**, **mech_transm**, **cvt_transm**, **dist_transm** and **transformation**.

The handling of the modules FORCE_TRANSFORMER and WHEEL is simple. It directly leads to the three entity types **converter**, **coupling** and and **wheel** in Figure 3.

Finally, we have to consider the connections between these modules. In schema S_2 we have an n-ary relationship type **type_data** for this. In schema S_2 we have five binary relationship types instead. These are **inst_engine**, **inst_conv**, **inst_transm**, **inst_wheel** and **inst_coupling**. According to rule 26 we have to

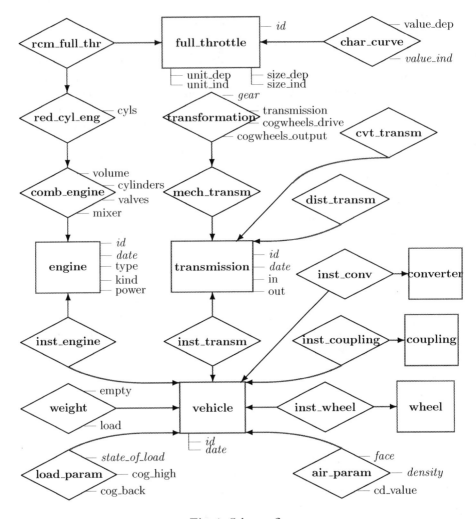

Fig. 1. Schema $\mathcal{S}_1$

choose the binary relationship types for the integration. The final result of the integration is the schema $\mathcal{S}_3$ shown in Figure 3.

6   Conclusion

We presented a pragmatic method for schema integration. The method relies on formal transformation rules defined on HERM schemata. The rules deal with restructuring, shifting of attributes and selected integrity constraints. For each of the rules it can be shown that it leads to schema dominance, in most cases even equivalence in the formal sense of the HERM.

The pragmatics of the method is due to the fact that it has been developed in close connection with the needs of schema integration in technical environ-

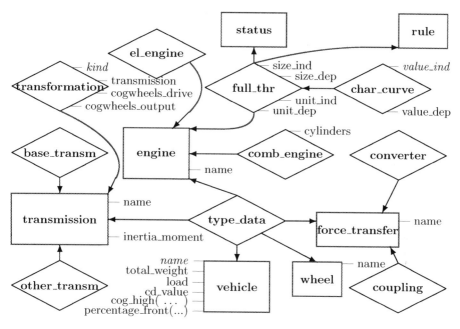

Fig. 2. Schema S_2

ments. In addition, it has been applied to large schemata arising from practical information systems in this area.

Most of the existing work on schema integration has influenced the piece of work presented in this paper. The novelty of our approach is due to it being based on the highly expressive HERM, dealing not only with structures, but also with integrity constraints, being theoretically founded on the theory of schema dominance and equivalence for HERM, and being embedded in a development method, which has been applied to schemata of realistic size. Furthermore, it has been shown in [5] how to adapt the approach to the related problem of schema cooperation.

Acknowledgement. Thomas Lehmann likes to thank his colleagues at Volkswagen for the motivating working atmosphere and the fruitful discussions on the topic of this work.

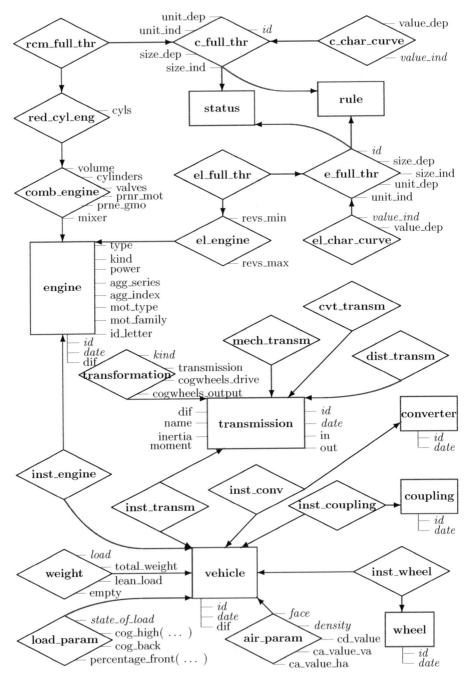

Fig. 3. Schema $\mathcal{S}_{new}$ resulting from the integration of $\mathcal{S}_1$ and $\mathcal{S}_2$

References

1. J. Biskup, B. Convent. A Formal View Integration Method. *Proc. ACM SIGMOD* 1986, 398-407.
2. R. Hull. Relative Information Capacity of Simple Relational Database Schemata. *SIAM Journal of Computing* vol. 15 (3), 1986, 856-886.
3. J.L. Koh, A.L.P. Chen. Integration of Heterogeneous Object Schemas. *Proc. ER'93*, 1993, 297-314.
4. J. Larson, S.B. Navathe, R. Elmasri. A Theory of Attribute Equivalence in Databases with Application to Schema Integration. *IEEE Transactions on Software Engineering* vol. 15 (4), 1989, 449-463.
5. T. Lehmann. *Ein pragmatisches Vorgehenskonzept zur Integration und Kooperation von Informationssystemen.* Ph.D. Thesis, TU Clausthal, 1999.
6. X. Qian. Correct Schema Transformations. *Proc. EDBT'96*, Springer LNCS 1057, 1996, 114-126.
7. E. Sciore, M. Siegel, A. Rosenthal. Using Semantic Values to Facilitate Interoperability Among Heterogeneous Information Systems. *ACM TODS* vol. 19 (2), 1994, 254-290.
8. S. Spaccapietra, C. Parent. View Integration – A Step Forward in Solving Structural Conflicts. *IEEE Transactions on Knowledge and Data Engineering* vol. 6 (2), 1994, 258-274.
9. B. Thalheim. *Entity-Relationship Modeling – Foundations of Database Technology.* Springer 2000.

Explicit Modeling of the Semantics of Large Multi-layered Object-Oriented Databases

Christoph Koch[1], Zsolt Kovacs[1], Jean-Marie Le Goff[1], Richard McClatchey[2], Paolo Petta[3], and Tony Solomonides[2]

[1] EP Division, CERN, CH-1211 Geneva 23, Switzerland
{Christoph.Koch, Zsolt.Kovacs, Jean-Marie.Le.Goff}@cern.ch
[2] Centre for Complex Cooperative Systems, Univ. West of England,
Bristol BS16 1QY, UK
Richard.McClatchey@cern.ch, Tony.Solomonides@uwe.ac.uk
[3] Austrian Research Institute for Artificial Intelligence, Schottengasse 3,
A-1010 Vienna, Austria
paolo@ai.univie.ac.at

Abstract. Description-driven systems based on meta-objects are an increasingly popular way to handle complexity in large-scale object-oriented database applications. Such systems facilitate the management of large amounts of data and provide a means to avoid database schema evolution in many settings. Unfortunately, the description-driven approach leads to a loss of simplicity of the schema, and additional software behaviour is required for the management of dependencies, description relationships, and other Design Patterns that recur across the schema. This leads to redundant implementations of software that cannot be handled by using a framework-based approach. This paper presents an approach to address this problem which is based on the concept of an ontology of Design Patterns. Such an ontology allows the convenient separation of the structure and the semantics of database schemata. Through that, reusable software can be produced which separates application behaviour from the database schema.

1 Introduction

Object-oriented database systems (OODBMS) are applicable for storing and efficiently accessing highly structured data, which renders them well-suited for large data-intensive applications in technical and scientific domains. Nearly transparent object persistency is achieved through programming language bindings, which are offered with virtually all object-oriented database systems. Nevertheless, developers of truly large-scale object-oriented database applications experience the problem that due to the size of the database schema (i.e., the number of classes), such databases are hard to maintain, in particular when it comes to schema evolution. Also, because of the programming language binding, there is a strong coupling of the database schema and the application code. This leads to a lack of reusability of software components, and the coupling imposes many dependencies in the code that have to be coped with when the schema evolves.

A.H.F. Laender, S.W. Liddle, V.C. Storey (Eds.): ER2000 Conference, LNCS 1920, pp. 52–65, 2000.

Multi-layered description-driven systems (with two model layers) provide a partial solution to the database maintenance and schema evolution problems [11]. Through the introduction of meta-classes, the complexity of the database schema is reduced and schema evolution can be evaded in many cases. In those cases, though, where schema evolution cannot be prevented, the action to be taken becomes more complicated through the loss of simplicity of the schema that was incurred by adopting the description-driven approach. The programming language binding, a feature of object-oriented database systems which is quite useful for small applications or for the implementation of highly specific behaviours, becomes counterproductive for very large applications that have to deal with complex relationships and large amounts of data. There, application code which is strongly coupled with the database schema has to be amended every time the database schema changes, often requiring a substantial amount of re-engineering. This is unacceptable if software components have to be reusable or schema-independent. This is the case for many software components in large database applications, such as those entailed by the description-driven approach, which take care of versioning and updating while keeping layers consistent, as well as for query processing and data mining.

Framework-based approaches (e.g [1]), in which the application behaviour is implemented as a reusable set of classes to be inherited from in the database schema, entail the binding of the application components with the database schema at compile time. Because of that, they are unsuitable in certain cases:

- Frameworks impose dependencies between the framework code and the application code which uses it. If both are under development, some very close cooperation of application developers and developers of the database schema is needed. Whenever the source code of the framework is changed, all applications using it have to be recompiled. This is often inacceptable in large and mission-critical distributed applications.
- Frameworks often require the database schema to be amended in unnatural ways that carry it away from the initial conceptualization. Framework-based approaches require the application behaviour to be decomposed into interoperable methods of the framework classes. This in some cases leads to additional complexity or to a counterintuitive design of the framework. In other words, the granularity of the building blocks of the behaviours – classes and their methods – is too small, the behaviour really belongs to an entity – a pattern – which may consist of several classes.
- Finally, a binding of components and a schema at runtime is often desirable if the interface of the components to others itself is dynamic and operates at a high level.

For all these reasons, a decoupling of the database schema from the application code is desirable. This paper outlines an approach which uses both a meta-model and meta-data for the description of the domain and an ontology [8] for making the semantics of the schema explicit. It will be argued that this approach allows to deal efficiently with very large amounts of data as well as to avoid the above-mentioned problems.

The remainder of this paper is structured as follows. Section 2 introduces description-driven systems and discusses their merits and restrictions. Section 3 elaborates on the restrictions of object-oriented data models and justifies the need to combine the two approaches, i.e. description-driven systems and ontologies, rather than choosing between them. Section 4 describes how to introduce the missing semantics through an ontology that joins the model layer and the meta-model layer of a description-driven system and how this logically leads to an ontology of design patterns. Section 5 discusses the CRISTAL project at CERN in which this work is applied, and finally, Section 6 gives some conclusions.

2   Multi-layered Description-Driven Systems

In software projects in which very large amounts of persistent data have to be managed, developers often experience several problems, including a complexity and maintenance problem due to the size of database schemata, the need for schema evolution when requirements change, and (as a consequence) restricted flexibility and scalability. A popular approach in systems that follow the object-oriented paradigm, which aims at alleviating these problems, is the use of an additional meta-layer (Figure 1) [10]. Meta-classes are introduced and used to abstract from the commonalities of related classes, and individual classes are then represented by meta-objects. While in principle meta-objects correspond to and should replace classes, in practical systems some model-layer classes firstly have to exist and secondly have to be adjusted to be sufficiently general to be able to represent objects that comply with all possible descriptions. This approach attempts to circumvent the restriction of statically typed programming languages which do not allow classes to be treated as objects and to change them at runtime or, in other words, it resolves the problem of defining instances of instances. In the description-driven approach, meta-objects will store all the data of a number of "similar" objects.

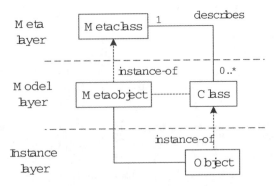

Fig. 1. The three layers of a description-driven system.

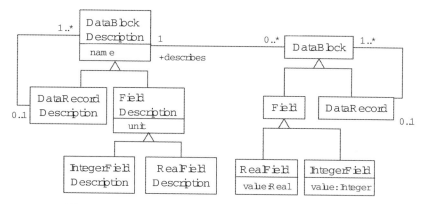

Fig. 2. The DataBlockDescription and DataBlock trees.

For an example, consider Figure 2 which shows a simplified UML class diagram of how characteristics of objects can be stored in an object-oriented database. `DataBlockDescription` trees represent data structures as dynamically changeable trees which specify the way data of a certain type are stored. If a schema now has to store information about certain objects, the database designer can choose to associate `Datablock` with his class rather than adding a number of fixed attributes to it. Through this general method, it is clear that schema evolution can often be avoided.

Schema evolution in description-driven systems (DDS) only becomes necessary if ontological commitments of the (already very) flexible data models become invalid at some point. To really avoid this, the data model would have to be as flexible as the object-oriented design methodology itself. An example of such a data model is the UML Meta-model [13]. In a practical large-scale system, however, a trade-off is necessary between flexibility on one hand and efficiency and run-time complexity on the other.

To summarize, the description-driven approach minimizes the amount of data that has to be stored in the databases because data common to many objects are stored only once. It also reduces the complexity and error-proneness of the system since, through the added layer of abstraction, the order in the system is improved, and it promotes the reuse of information, since every description can be instantiated many times and from many applications. Finally, it often makes schema evolution unnecessary, since many changes can be performed at runtime. In these cases, no new classes have to be added to the system and no application code has to be changed as the existing system is flexible enough to abstractly describe the new class of objects so that software need not be re-engineered.

The description-driven approach, however, does have limitations. Firstly, DDS are initially more difficult to implement than conventional software systems. DDS always have to consider the relationship between the two layers of abstraction, the model layer and the meta-model layer, such that they never become inconsistent. Secondly, when changes become necessary which are beyond the scope of the DDS, or when the system is extended into different application domains, the system is harder to maintain and to evolve due to the additional dependencies incurred by the description-driven approach. Therefore, in the case of schema evolution, the software that manages the dependencies between the two layers usually also has to be changed. Thirdly, DDS are often implemented using object-oriented database systems, which, lacking specific support for meta-classes, cannot express the semantics of the relationship between the model and the meta-model layer. While these relationships are implemented as associations in the database, their special meaning is nowhere recorded. Because of that, the database cannot distinguish them from other, simple associations. The relationship is only really taken care of in the application code. This leads to software development which can be rather ad-hoc, and, as a consequence, implies maintenance problems.

The remainder of this paper will discuss the applicability of ontologies for integrating the layers of a description-driven system to avoid the need to evolve some of the common software components operating in such an environment. It can be argued that by separating the evolution of the schema from that of the software, the complexity of the problem of schema evolution is also reduced.

3   Object-Oriented Models and Semantic Expressiveness

The drawbacks of DDS presented in the previous section can be summarized as the loss of semantic expressiveness that is incurred when the conceptual model of the domain is mapped to an object-oriented model at design time. This mapping induces some dilution of the semantics of the relationship between the layers as well as inside the layers, in some contexts. For example, the item description pattern [9] (i.e., the relationship between a class and its meta-class) is usually mapped to an association between two classes in an object-oriented model, which does not conserve its special meaning.

The strong coupling of the database schema and application code in DDS is due to the need to build the missing semantics into the source code of components. This can only be avoided by making some description of these semantics available, which allows software components to obtain this knowledge at run-time. Such an explicit account of the semantics of a schema (an ontology [8]) is the topic of this paper.

3.1   Design Patterns

In Software Engineering, the contexts of a certain semantics (as discussed above) are called Design Patterns [7]. Such design patterns can be complex, consisting of many classes, attributes and methods, inheritance relationships, and associations. Design patterns are static design elements and can be specified on the class level rather than on the object level. The notation that is used for design patterns in the following examples is UML.

Complete specifications of design patterns are usually only achieved through class diagrams, static and run-time constraints, pseudocode, state or sequence diagrams, and a verbal description. A certain class of patterns, the so-called structural patterns, particularly those that represent data structures, rely more heavily on the static aspects than others. This class of patterns includes tree and graph patterns and is of particular importance in the context of this paper, since data structures are the main building blocks of a database schema. As will be demonstrated later in this section, even patterns for data structures are only at first sight fully specified by class diagrams. There is plenty of need to add semantic knowledge to the schema in this area too. Behavioural patterns [7] on the other hand rely particularly strongly on the semantics of certain class attributes and methods.

Design patterns can be seen as building blocks of an object-oriented model or database schema, like classes and associations are building blocks on a lower level. By allowing them to overlap, complete models and database schemata can be built out of them. Unfortunately, there are often several ways to represent the same things. See for example Figure 3, which shows three ways to implement a tree. While the pattern in the middle is the "standard", most intuitive tree pattern, all three could be used in a database schema, and although they have slightly different characteristics, there are many commonalities. It is useful to generalize the design patterns so far that the ontology to be created really represents all the design knowledge about the database schemata, and to allow applications themselves to find out which patterns they are dealing with.

Among the most heavily used design patterns in multi-layered database schemata are the item description pattern, tree and graph patterns, the versioning pattern, and the pattern for handling update dependencies between meta-objects [9]. Modelling design patterns with UML allows patterns to be specified more thoroughly, but this information is not accessible in the application and is often specified on a case-by-case basis, preventing the formal deduction of semantics. UML extends beyond the "static" aspects, i.e. the class diagrams, and supports state diagrams, sequence diagrams, and the specification of use cases of software. While the usefulness of these tools in their semiformal settings for the design of software is not in question, they are not necessarily applicable for semantic modelling.

3.2   Justification for a Combined Meta-model/Ontology Approach

In this paper, an architecture that combines a multi-layered object-oriented database schema with an ontology for modeling the schema semantics is proposed.

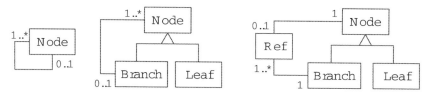

Fig. 3. Three ways to specify trees in UML.

This allows full semantic description of the schema to be achieved while supporting very large amounts of data.

This section shortly discusses why an alternative approach in which one subsumes the meta-layer of a description-driven system by a domain ontology is not preferable (i.e., why the DDS approach is still valuable). This would be possible since object-oriented models can be seen as restricted versions of more expressive models, and an ontology can be seen as a generalization of the simple description-driven systems approach that was discussed in the previous section. A purely ontology-based approach has the advantage that it does not require a meta-model as a hard-wired database schema which may be subject to evolution. However, an approach which totally replaces the meta-layer by an ontology has several drawbacks:

- Such an ontology can grow large, as a DDS may work with thousands or even millions of meta-objects. Since the schema semantics are exclusively defined on classes rather than instances (meta-objects), it would be overkill to apply the powerful reasoning capabilities of common knowledge representation formalisms to the full meta-level database.
- While the meta-data (which are edited by expert users) evolve quite often, the ontology only changes through the hand of the software developers whenever a new domain is integrated into the system. A small and self-contained ontology of design patterns is also eligible for reuse.
- Current trends in ontological engineering assume that ontologies are maintained by humans, who are assumed to put a high density of knowledge into ontologies with a comparably small number of concepts. An ontology that is built from a large repository of meta-data may lead to a serious maintenance problem, as ordinary users of a system are in practice required to change meta-data, while more general and high-level ontological terms should be left untouched.
- Legacy system constraints. The approach which adds an ontology while preserving the existing layers is also preferable to a pure ontology-based approach without a meta-layer because it facilitates the integration of existing DDS.

4   An Ontology of Design Patterns

The first step in modifying software to achieve schema independence is to make the binding between application software and the database dynamic. Apart from the requirement that the database system supplies a convenient way to access its data dynamically, one has to be able to read the schema description as data. In relational databases, where dynamic access is virtually omnipresent, such a schema description that can be queried is provided through the Data Dictionary. In object-oriented database applications, dynamic binding is much less widely used, although the object database standard [12] specifies such an interface, which is itself meta-class-based.

4.1   The Need to Have an Explicit Account of Schema Semantics

Having access to a database schema description which includes classes, class attributes, associations, and subclass relationships plus an interface for dynamic access to the data allows the sharing of simple behaviours. For example, suppose it is known from querying the database schema that several classes each have an attribute called `Timestamp` in a data format which allows a date and a time to be stored. By making the assumption that from this automatically follows that this attribute represents the time of the last change of the object, one could write some code so that every time an object is changed which has an attribute called `Timestamp`, this attribute is set to the current time. This code could, as a consequence, be shared among all those classes that have such an attribute.

In simple cases like the one above this may work, but in general, such an approach in which the semantics are followed from syntactic elements of a database structure is impractical for many reasons. The most important reason why the semantic description should be separated from the database structure is probably that object-oriented modelling mechanisms like UML class diagrams have very restricted modelling power with regard to semantic content.

What is required is the addition of a semantic description which contains all the concepts of the schema description, like classes and associations, as well as specializations of them (which are introduced by the knowledge engineer) as additional semantic content. Consider a class `MyClass` that has an attribute called `MyTimestamp`. Figure 4 shows what its semantic description could look like. The primitive concepts `Class` and `Attribute` of our ontology describe standard classes and attributes of the database schema. Therefore, the connections between the elements of the schema and these concepts are comprehensible without introducing further knowledge; they can be established automatically. The fact that `MyTimestamp` is a `ChangeTimestamp`, though, is semantic knowledge that cannot be automatically derived from the schema. By explicitly introducing a new concept, which is called `ChangeTimestamp`, into the ontology and by specializing that `MyTimestamp` adheres to it, one can now assign behaviour to the concept that can then be shared by many occurrences of timestamps having these semantics in the schema. The dotted lines in this example denote the semantic information that was introduced by the knowledge engineer. Apart from the

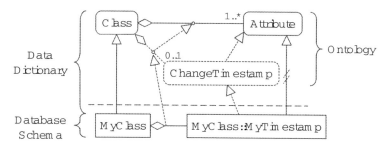

Fig. 4. An example in which the semantic description is separated from the schema.

specialized concept `ChangeTimestamp`, Figure 4 also shows a specialized aggregation relationship between `Class` and `ChangeTimestamp`. If this were missing, a
class could contain several of these attributes through the normal (one-to-many)
aggregation of attributes.

The notation of Figure 4, which will be used and extended throughout this
section, is basically due to [4]. It was altered to resemble UML more closely
where possible. To distinguish concepts from UML classes more easily concepts
are represented by rectangles with rounded corners. The little circles that are
centered on some of the relationships in the ontology are used to denote that
these are first-class-objects with identity that can be referred to and inherited
from. The links between the concepts of the ontology and the elements of the
database schema – which were called "Data Dictionary" in Figure 4 – can be
represented as instances of the concept classes in the knowledge base that maintains the ontology.

The remainder of this section discusses some important concepts that have
to be modelled in our ontology.

4.2   The Ontology of the Item Description Pattern

The item description pattern, the main semantic building block of description-
driven systems, was shown in Figure 1. It deserves special treatment since it is
one of the main reasons for using an ontology. It embodies the correspondence
relationships between classes and their meta-classes, i.e., it links the model and
the meta-model layer. While associations might exist that link classes with their
meta-classes in a database schema, the special semantics of this relationship that
distinguished it from other associations cannot be expressed. Therefore, this has
to be done externally, in the ontology. Figure 5 shows how this pattern could be
described in an ontology using the example of a `Part` class and its meta-class.

The item description pattern has also some implications that stem from the
homomorphism that it induces [3] (and which e.g. can be used for semantic
query optimization). Figure 6 shows the conceptualization of homomorphisms

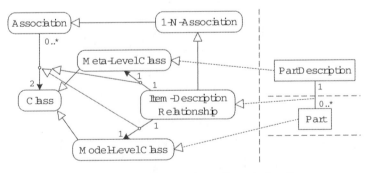

Fig. 5. The Ontology of the Item Description Pattern.

in our simple graphical notation. Two associations, one on the meta-level and one on the model-level which semantically correspond to each other (that is, they link classes resp. meta-classes that correspond to each other and have the same meaning) entail a homomorphism from the model-level to the meta-level.

Consider the example of Figure 6. Assume that C1 is a specific class of your schema and C2 models dynamically assignable attributes (together with their values) to the objects of C1, as discussed in Section 2. Then, MC1 is the description of C1 and the associated objects of MC2 represent all those attributes that may be assigned to C1. The homomorphism says that for every link between objects of C1 and C2 there must be a link between their meta-objects that semantically corresponds to it, which "describes" it.

The homomorphism semantics is transitive; if there is a homomorpism between associations A_1 and MA_1 as well as between A_2 and MA_2, and A_1 and A_2 link to a common class, then there is also a homomorphism between $A_1 \circ A_2$ and $MA_1 \circ MA_2$. Therefore one has to define "generalized relationships" in the ontology which can be simple associations or paths of several associations via intermediate classes.

4.3 Patterns for Dynamic Data Structures

Earlier in this paper, structural patterns have been introduced and it was emphasized that even in the apparently simple cases of data structures, class diagrams insufficiently specify the semantics of the patterns.

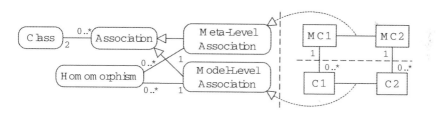

Fig. 6. Making Homomorphisms explicit in the ontology.

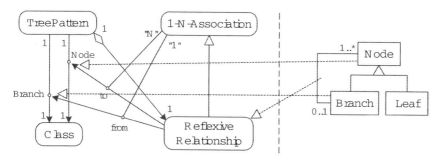

Fig. 7. The Ontology of the tree pattern.

Now consider again Figure 3, which shows three class diagrams that represent trees in UML. It can be easily seen that a tree pattern can be defined as a structure that consists of classes and a one-to-many association between a class and itself or a superclass of it. Such an association is called a `ReflexiveRelationship` in Figure 7. To cover also the right tree with this generalization, one has to generalize associations, so that they can be more that just simple associations but may consist of several classes and associations. The two associations and the class `Ref` together constitute a one-to-many relationship between `Branch` and `Node` (see Figure 7). Note that although all this information about the tree pattern is available in the schema, it is not possible to derive the existence of this pattern automatically from a database schema. It is possible that each node has one incoming link, therefore, one cannot enforce that there is a root node. Instead, there could be a cycle from a node somewhere down in the tree up to the root node, which would make a data structure instantiated from this pattern something other than a tree.

There are two solutions to this problem. Firstly, one could model the object level in the ontology as well. This would allow the description of the concept of cycles and root nodes and would put the whole problem on a sound foundation from a graph-theoretic point of view. The two drawbacks of this approach are that (i) this may require reasoning with a very large amount of data, i.e., the whole database, and (ii) it might only reveal at the current point in time if a certain pattern is a tree. This is obvious, because the real semantic information expressing whether a data structure is used as a tree is in the application code, which one assumes to be inaccessible for ontological reasoning tasks. The second and preferred solution is to put in semantic knowledge manually, such that the above conditions for a tree pattern are necessary but not sufficient conditions for something to constitute a tree pattern.

Similar issues apply to graph patterns, which again can be modelled in many different ways in a database schema. While a certain data structure for graphs in a schema will usually be very generic, it will normally be populated only with a certain more restricted class of graphs (e.g. it is again not possible to model a class structure in UML which can store all directed acyclic graphs but only these). Again, additional semantic knowledge is needed to optimize application behaviour.

4.4   Discussion

Earlier in this paper, the solution of the `Timestamp` example of Figure 4 indicated how the problem of under-specification at the semantic level of a domain in behavioural patterns can be solved in general by modelling it in ontologies. Furthermore, in structural patterns some additional information for the applications, which are implemented as specialized concepts, are often additionally needed. This is e.g. the case for the tree patterns aforementioned, where it is impossible to enforce in an object-oriented schema that really only trees are stored.

For reusability reasons, the semantic specializations that are application-specific should be conceptually separated from the core ontology of design patterns and be moved into an application-specific ontology. The reusable concepts can be put into three categories. Firstly, there are the concepts from the UML Meta-model, the primitives of object-oriented schemata such as `Class` or `Association`. Secondly, and based on these, there are some basic reusable concepts like `GeneralizedRelationship` and `ReflexiveRelationship` that facilitate the modelling of complex semantics. Finally, there are the conceptualizations of general-purpose Design Patterns (a catalog of which can be found e.g. in [7]).

5   The CRISTAL Project

The research reported in this paper is currently being carried out in the CRISTAL (Concurrent Repositories/Information System for Tracking Assembly Life cycles) project [2] at the European Organization for Nuclear Research (CERN). The CRISTAL project aims at providing the Compact Muon Solenoid (CMS) experiment [6], a large particle detector currently being built at CERN, with a system that satisfies its product data management and workflow management needs. CRISTAL is now in production, managing the construction of an increasing number of subsystems of CMS. Following a multi-layered description-driven approach, CRISTAL is collecting and persistently storing data up to the order of one TeraByte per subdetector in a federation of globally distributed object-oriented databases. Due to the experimental nature of the CMS project and its once-off production, the flexibility of the description-driven approach is tested nearly daily.

The integration of the layers of description-driven systems as presented in this paper is evaluated in two areas within the CRISTAL project:

- A query facility which uses meta-data to optimize the access to very large databases is currently being extended by capabilities for semantic query optimization (SQO) [5]. By using item-description patterns and homomorphisms that can be derived from the database schema and the ontology of design patterns, the available meta-data can be used to optimize the queries, since the amount of meta-data is much smaller than the amount of instance-level data that are described by them. That is, restricting selection constraints can be refined or propagated between classes by using these relationships. Also, recursive queries that are in many cases necessary due to the flexible data structures common in description-driven systems (see Figure 2) can often be grounded into much restricted nonrecursive queries by consulting only the meta-data.
- Domain-independent database management components. This includes the software components for populating the meta-data and instance-level data, respectively, those for managing the object lifecycles, performing tasks like versioning, and the management of dependencies within and between layers. These components implement a number of behaviours (as code) that are each known to be valid for a specific concept in the ontology; i.e., a publish-subcribe pattern [7] in the ontology would be based on concepts denoting

e.g. subscription that directly map to application code. That is, the classes in the database schema do not provide these methods of the patterns in their own interfaces; instead, there is software code mapped to these classes via the ontology.

By making the data-management components schema-independent, one can better cope with open environments, as the high-level design moves away from distributed object computing and into the area of semantic capability descriptions, where the basic granularity of software components is outlined by the Design Patterns.

6   Conclusions

This paper has advocated an ontology which describes the relationship between the layers of abstraction of description-driven systems, as well as some important design patterns that are popular in software engineering. Using such an approach, one is able to separate certain software components sufficiently so that they do not have to evolve when the database schemata are changed or extended. Following this approach, the software maintenance problem is solved and software components can be implemented in a cleaner, more lightweight fashion. This also allows a system to grow more easily. The resulting ontology tends to be a powerful abstraction of important concepts of the system that reveals new and interesting facts about the description-driven approach.

As always, the main concern about the usefulness of a formalized and computerized body of knowledge is its reusability. One of the strong points of an ontology of design patterns is that it seems to be true that because this domain is highly formalized and standardized, confusion among ontology developers and users – who are themselves application developers – does not occur to the same extent as it might occur in other domains where ontologies are applied. Also, such an ontology seems to be quickly understood by newcomers and because of its standardization and restricted scope it appears to be a good candidate for reuse. This reuse seems to be possible although in principle the ontology is quite specialized. In this paper, pointers have been given on how to structure such an ontology into smaller packages to promote reuse and to supply a range of ontology packets with different degrees of specialization.

An ontology of design patterns seems to be a viable approach for sharing knowledge about database schemata which can be made operational easily and can be used for a variety of tasks, including semantic query optimization, operation of description-driven systems (including the enforcement of consistency constraints and version management), and the reverse-engineering of database applications (as this approach facilitates the addition of semantic context to existing legacy databases).

Moreover, this research created a motivation to extend UML by a semantic modelling perspective. This paper also gave explanation of a way the notation of such a facility could be specified such that it extends UML gracefully. Instead of being part of an actual architecture, the ontology could be a starting point for the design of software. Existing patterns could be reused more easily, and pre-assembled code libraries could plug in seamlessly into object-oriented applications. Existing concepts could be used and specialized, making software engineering more an instance of knowledge engineering.

This also has interesting implications on automatic code generation and template libraries when semantic terms become standardized and agreed. A conceptualization of design patterns as the building blocks of object-oriented database schemata does not only facilitate the decoupling of software components from the database structure, it also seems to provide the appropriate building blocks for the design and the decomposition of data-intensive components in their systems, as well as the concepts on which a description of the capabilities and interfaces of components can be reasonably based. Through this, the present work enables software components to provide better support for open environments.

References

1. Baumer, D. et al., "Framework Development for Large Systems". *Communications of the ACM 40(10)*, October 1997 pp. 52-59.
2. Bazan, A. et al., "The Use of Production Management Techniques in the Construction of Large Scale Physics Detectors". In *Proceedings of the IEEE Nuclear Science Symposium and Medical Imaging Conference*, Toronto, Canada. November 1998.
3. Blaha, M., and Premerlani, W., *Object-Oriented Modeling and Design for Database Applications*. Prentice Hall Publishers, 1998.
4. Brachman, R.J., and Schmolze, J.G. "An Overview of the KL-ONE Knowledge Representation System". *Cognitive Science* 9(2), 1985.
5. Chakravarthy, U.S., Grant, J., and Minker, J. "Logic-based Approach to Semantic Query Optimization". *ACM TODS* 15(2):162-207, June 1990.
6. CMS Collaboration, "CMS Technical Proposal". January 1995. Available from `ftp://cmsdoc.cern.ch/TPref/TP.html`
7. Gamma, E., Helm, R., Johnson, R., and Vlissides, J., *Design Patterns – Elements of Reusable Object-Oriented Software*. Addison-Wesley Longman Publishers, 1995.
8. Huhns, M., and Singh, M. 1997. "Ontologies for Agents". In *IEEE Internet Computing* November-December 1997, pp. 81-83.
9. Kovacs, Z., "The Integration of Product Data with Workflow Management Systems through a Common Data Model". PhD thesis, Univ. of the West of England, 1999.
10. Laddaga, R., and Veitch, J. (ed.), "Dynamic Object Technology", *Communications of the ACM* 40(5):37–69, May 1997.
11. McClatchey, R. et al., "The Role of Meta-Objects and Self-Description in an Engineering Data Warehouse". *Proc. 3rd IEEE Int'l Database Engineering & Applications Symposium (IDEAS'99)*, pp. 342-350. Montreal, Canada. August 1999.
12. ODMG. *The Object Database Standard: ODMG 2.0*. Morgan Kaufmann, 1997.
13. UML Semantics. Version 1.1, 1 September 1997. Rational Software Corporation, 18880 Homestead Rd, Cupertino, CA 95014 `http://www.rational.com/uml`

Declarative Mediation in Distributed Systems*

Sergey Melnik

Stanford University, Stanford CA 94305 USA**
melnik@db.stanford.edu

Abstract. The mediation architecture is widely used for bridging heterogeneous data sources. We investigate how such architecture can be
extended to embrace information processing services and suggest a framework that supports declarative specification of mediation logic. In
this paper we show how our framework can be applied to enrich interface descriptions of distributed objects and to integrate them with other
client/server environments.

1 Introduction

More and more information processing services are becoming available online.
Such services accept data, process it, and return results. A variety of services like
summarizers, indexers, report generators, calendar managers, visualizers, databases, and personalized agents are used in today's client/server systems. As more
such components are deployed for use, the diversity of program-level interfaces
is emerging as an important stumbling block. Interoperation of heterogeneous
information processing services is hard to achieve even within a given domain
like digital libraries [13].

The mediation architecture [17] has often been used for leveraging solutions
for the interoperability problem. It introduces two key elements, wrappers and
mediators. The wrappers hide a significant portion of the heterogeneity of services, whereas the mediators perform a dynamic brokering function in a relatively
homogeneous environment created by the wrappers.

Frequently, mediation is implemented on top of distributed object architectures like CORBA or DCOM. Typically, a wrapper acts as a server object and
provides a standard interface through which mediators can access heterogeneous
components. This solution works well in environments targeted at querying of
data sources. The reason for this is that it is relatively easy to develop a common querying interface that has to be supported by all wrapped sources. Serious
complications arise, however, when the underlying components support a rich
set of interfaces and protocols. In this case, even if the individual components
are wrapped by distributed objects, their interfaces remain very diverse. Thus,
mediators become more detailed and complex, expensive to create and maintain.

---

* This work was supported by a German Research Council fellowship and by the NSF
  grant 9811992
** Permanent address: Leipzig University, Augustusplatz 10, D-04109 Germany

A.H.F. Laender, S.W. Liddle, V.C. Storey (Eds.): ER2000 Conference, LNCS 1920, pp. 66–79, 2000.
© Springer-Verlag Berlin Heidelberg 2000

In this paper we describe a framework tailored for declarative specification of mediation logic that is required to integrate heterogeneous information processing services. We examine an environment in which services expose rich interface descriptions and mediators are specified using declarative languages. Such environment promises significant advantages over hard-coded mediators [14,18]. In fact, developing mediators for disparate systems becomes an engineering task leveraging established formal methods as opposed to error-prone programming.

Although declarative mediation promises substantial benefits, it may introduce penalties in efficiency and additional complexity. Nevertheless, our initial experience suggests that the framework offers substantial flexibility that can be exploited in different application scenarios. In [10] we describe one such scenario based on heterogeneous retrieval services. In this paper we investigate how our approach can be applied to enhancing the distributed object technology and bridging it with other client/server environments.

The next section introduces a sample scenario that we use throughout the paper to illustrate the major tasks needed to implement mediator systems based on declarative specifications. Sect. 3 introduces canonical wrappers that provide mediators with logical abstractions of the components. Sect. 4 gives an overview of our approach to declarative mediation. In Sect. 5 we elaborate on the techniques that can be used to manipulate the content of the messages exchanged by heterogeneous components. Sect. 6 sketches our approach to representing the dynamic aspect of mediation, i.e. how message sequences originating at one component can be translated into message sequences expected by another component. Interface descriptions of services are examined in Sect. 7. Sect. 8 summarizes the challenges of building declarative mediators. Sect. 9 describes the execution environment used for our running example. Related work is discussed in Sect. 10.

2 Running Example

A typical operation needed for a digital library is the document conversion between different formats like PostScript, PDF, plain text etc. A rudimentary conversion model can be described by an operation which accepts source and destination format specifications and a sequence of bytes as the content of the document. The result of the conversion is a byte sequence in the destination format. In following, we examine two rather obvious implementations of the conversion, one as a CORBA object and another as a Web form. A CORBA client intending to use an HTTP-based service faces a number of obstacles that we sketch below.

For the CORBA implementation of the service it may make sense to provide a BLOB interface to large binary objects in order to enable the server to determine the size of the object to be converted in advance and fetch its pieces incrementally. A likely CORBA specification of the conversion service comprises two interfaces:

```
interface Converter {
 BLOB convert(in BLOB doc, in int sourceFormatID, in int destFormatID);}
```

```
interface BLOB {
  long getSize();
  sequence<octet> getBytes(in long start, in long end); }
```

A conventional Web form for an HTTP-based conversion service includes two fields, say `from` and `to` identifying the source and destination format and a `file` field which allows the user to upload a file to be converted from the local disk.

To enable the CORBA-based client to utilize the HTTP-based server, an intermediate component (mediator) is required. In our example, such mediator translates requests between two services. The translation could be achieved using the following algorithm:

1. Receive the request parameters `doc`, `sourceFormatID` and `destFormatID` via the `Converter` interface
2. Translate the format identifiers `sourceFormatID` and `destFormatID` into corresponding format strings `from` and `to` for the Web-based service
3. Retrieve the size of the source object using `doc.getSize()`
4. Retrieve the binary content of the source object via `doc.getBytes()` to fill the `file` field of the HTML form
5. Emit an HTTP POST request after completing the appropriate form fields
6. Create a BLOB instance for the binary data contained in the HTTP reply.

Our goal is to capture the translation between the CORBA client and the HTTP service in a declarative fashion. Such declarative specification would describe how the messages originating from the client are transformed into the messages understood by the server, and the other way around. For that, the mediator needs to be able to manipulate the content of the messages and the order in which they are exchanged.

3   Canonical Wrappers

Mediators need a convenient way to manipulate the content of the messages passed back and forth between different components. For our purposes, convenient means that the message manipulation operations can be described in a declarative fashion, ideally without using a programming language like C++ or Java. To do that, we use logical descriptions of the messages encoded as directed labeled graphs. Represented as a labeled graph, the message content can be manipulated using algebraic operations, transformation rules etc.

Logical descriptions of messages exchanged by the components can often be derived from their informal descriptions in a straightforward way. Consider how one could formulate a conversion request message as a CORBA invocation:

'This is a conversion request specifying which BLOB object to convert (`obj`), and what the source format (`sourceFormatID`) and the destination format (`destFormatID`) of the conversion are.'

This sentence can be represented using the following five logical statements of the kind 'subject predicate object':

```
CR  is-a                 ConvertRequest
CR  object-to-convert     obj
CR  source-format         sourceFormatID
CR  destination-format    destFormatID
obj is-a                  BLOB
```

The entity CR designates an instance of a CORBA conversion request. The logical description of the request can be represented graphically as a directed labeled graph depicted in Fig. 1. The object reference of obj is IOR:XYZ, and the source and destination format IDs are 10 and 11 respectively. Ovals represent any entities that might have relationships with other entities. In the figure, such entities are, for example, the concrete BLOB object identified by its object reference, or the type of the object (BLOB), or the concept ConvertRequest. Arrows represent relationships among entities. They might be conceptual relationships, such as the is-a, or 'has-property' relationships, such as destination-format. Literals (string values) are depicted in rectangles. The representation used in the figure is similar to an entity-relationship diagram that includes instances of entity types.

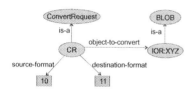

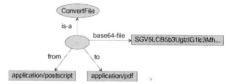

Fig. 1. Logical representation of the CORBA conversion request

Fig. 2. Logical representation of the HTTP-based conversion request

A logical description of the request makes it possible to abstract out factors irrelevant to the purpose of the service e.g. whether it is implemented as a distributed object or a CGI script. The corresponding HTTP request is represented as shown in Fig. 2. The name of the entity in the middle is not relevant and is omitted for clarity. Note that using logical descriptions allows us to talk about things as different as CORBA calls and HTTP requests in a uniform language.

The mediator and the wrappers used in our running example are depicted on the left-hand side of Fig. 3. Wrapper A acts as a CORBA conversion server for the CORBA client. Logical descriptions of CORBA invocations are passed by wrapper A as object graphs to the mediator. The mediator irons out semantic incompatibilities in the structure and ordering of the messages received from wrapper A and forwards the translated logical representations (graphs) to wrapper B which invokes the corresponding CGI script. In this way, the wrappers A and B allow the mediator to treat the interfaces of the components uniformly. Recall that our example is very simple; a realistic mediation environment contains many more components.

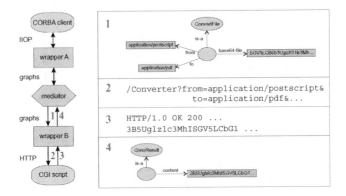

Fig. 3. Canonically wrapped HTTP-based converter

In our approach, wrappers are as simple as possible, and all data conversion and protocol translation logic is embodied in mediators. We call a wrapper *canonical* if it faithfully preserves the semantics of the component and captures information contained in the messages of the component by means of logical descriptions. A canonical wrapper is not required to do any processing beyond trivial syntactic transformation of the messages of the wrapped component.

The right-hand side of the Fig 3. illustrates a request-reply interaction with the canonically wrapped HTTP-based conversion service. The wrapper accepts a logical description of the message, translates it into a native HTTP request and delivers the response of the component as a set of facts. Note that the wrapper performs no semantic translation of data and protocols used by the component and does not commit to any high-level query language. In this sense, the wrapper is 'thin' and requires a minimal implementation effort.

4   Declarative Mediation

Canonical wrappers like the one described above allow the mediator to manipulate the messages exchanged between the components as directed labeled graphs. To illustrate, consider the following example. In our scenario, one complication the mediator has to face in order to reconcile the representations of the CORBA and HTTP requests are incompatible format specifications. On the Web, Post-Script is identified using the MIME type 'application/postscript' whereas for our CORBA service it is the integer 10.

The mapping of the format IDs can be formalized using one rule and two facts illustrated in Fig. 4. Every expression in parentheses is again of the form (subject predicate object) where variables are in boldface. Two top statements are facts representing the mapping between the integers and MIME types. The body of the rule (right-hand side) is a conjunction of expressions that produces a set of variable substitutions. The head of the rule (left-hand side) describes how new statements are obtained from the variable substitutions.

Applying this rule to the CORBA conversion request (Fig. 1) delivers almost the desired HTTP request (Fig. 2). The missing element are the bits and bytes needed to fill the `base64-file` property in the HTTP request from the description (`IOR:XYZ is-a BLOB`). To retrieve the binary content the mediator has to issue at least two additional CORBA invocations, `getSize()` and `getBytes()`.

Hence, to complete the translation we need to specify the dynamic behavior of the mediator, i.e. how a series of messages of one component are mapped into series of messages of another component. We call such mappings *dynamic transformations* as opposed to the manipulations of the content of the messages, which we call *static transformations*.

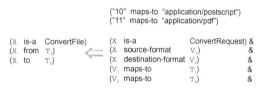

Fig. 4. Specification of the format translation from CORBA to HTTP

The need for dynamic and static transformations suggests that, in general, a number of different formalisms is required to describe the mediator logic. For example, manipulations of the content of the messages can be expressed using Datalog rules like the one depicted in Fig. 4. Transformations of message sequences can be described using finite-state machines, Petri nets etc. Thus, in order to build a fully functional declarative mediator, a variety of questions need to be answered:

- How do we express dynamic and static transformations performed by the mediator? Which formalisms are appropriate for that?
- How can different formalisms be used in the same mediator specification? How can we represent expressions that use different languages?
- How can declarative mediators be executed? Which middleware and APIs are necessary to build wrappers and mediators?

To facilitate mixing and reuse of different formalisms, the declarative languages used by our mediators are represented using a *meta-model* that is capable of capturing and linking expressions in different languages. To illustrate the role of the meta-model, consider that we choose to map the message sequences of the CORBA component to that of the HTTP component using a finite-state automaton. Let the automaton specification be encoded in some XML-based syntax. On transitions from one state to another, the automaton executes actions that manipulate the content of the messages. These actions could be formalized using rules that are stored in an ASCII file. Further imagine that the rule language does not support regular expressions or arithmetic operations, but we need both of them to transform one message represented as a graph into another. Regular expressions happen to use yet another syntax, and the arithmetic operations are implemented by a Java program.

A common meta-model allows us to represent expressions in all these disparate languages in a uniform way and to link them with each other. Our meta-model is based on directed labeled graphs, similarly to how the messages themselves are encoded. To represent expressions in different languages we deploy the Resource Description Framework [8] that has been designed for the use on the Web. Using a Web-ready meta-model has the advantage that the mediator specifications can be disseminated on the Web and refer to each other. All graph representations used in the figures in this paper are based on the RDF model.

To execute mediators that deploy different formalisms and languages, a comprehensive runtime environment is required. Such environment has, for instance, to make sure that the appropriate interpreters are invoked for the declarative languages used in the specifications, or that the whole specification can be compiled into executable code. We describe the details of language mixing and the runtime environment elsewhere [10]. The following two sections demonstrate different formalisms for implementing static and dynamic transformations that can be used in our sample scenario. We also illustrate how expressions in different languages can be represented using a meta-model. After that, we describe the architecture of the prototype system that supports execution of declaratively specified mediators like the one used in our sample scenario.

5   Static Transformations

To illustrate how different static transformations can be described declaratively, let us return to our running example. After receiving the initial CORBA conversion request, the mediator first needs to retrieve the size of the binary object to be converted. The only parameter required for the `getSize()` invocation is the object reference received in the conversion request. Thus, the corresponding CORBA `getSize()` message can be generated by applying a static transformation T to the received request.

This static transformation T can be expressed using the following rule (let GSR be a constant that represents an instance of a `GetSizeRequest`):

```
(GSR is-a GetSizeRequest)
(GSR blob B)                 <=  (B is-a BLOB)
(B   is-a BLOB)
```

This rule finds an instance B of type `BLOB` and constructs a new message that contains a reference to B. A possible encoding of the above rule in the meta-model is shown in the top part of Fig. 5. We omitted some portions of the graph like (B is-a t:Variable) for better readability. Prefix t: denotes specific language elements (vocabulary) used for representing static transformations of this kind. The bottom part of the figure shows the result of applying transformation T to the received request.

The language we use to specify transformations like the one above is a variant of Datalog with negation and allows us to encode arbitrary rules and facts. In this and further examples we are using this and other formalisms like finite-state machines merely for illustrating our approach to declarative mediation. They might not even be the best choice for our running example.

Using the Datalog-based language, more complex functions can be defined which can still be analyzed automatically. For instance, a request that may have an optional parameter of type `PType` can be described using the expression shown below:

```
(X is-a Request) & ( not exists Y (X opt-param Y)  ||
                   ((X opt-param Z) & (Z is-a PType)) )
```

For a graph representation of a message, this condition is satisfied only if every node of type Request either has no arc labeled with opt-param, or the node reachable via opt-param is marked to be of type PType.

Although rule-based descriptions provide a powerful instrument for specifying static transformations, many useful transformations cannot not be easily expressed. For example, consider removing certain facts from a message. In this case, however, we can obtain the result as a set difference $m - T(m)$, where T selects the records to be removed from the message m. Therefore, we generalize static transformations as functions $F(m_1, \ldots, m_n)$ applied to sets of facts m_i. Many frequently used n-ary functions can be expressed as a composition of Datalog-based unary functions like the ones presented above and the conventional binary set operators ∩, ∪, and −. Together, the rule-based unary functions and the set operators form an expressive algebra for static transformations of logical descriptions.

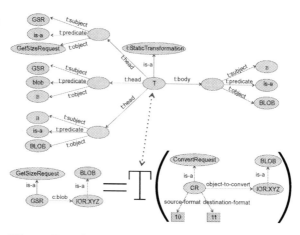

Fig. 5. Transformation T of the CORBA conversion request into a getSize() invocation

In the next section we describe a language that can be used in our running example for describing the dynamic behavior of the mediator. We show how we represent the expressions in this language using the same RDF-based meta-model, and how we combine it with the static transformations to produce complete mediator descriptions.

6   Dynamic Transformations

In our example, after receiving a CORBA conversion request, the mediator has to perform a callback doc.getSize() asking for the size of the binary object and then retrieve the object content via one or more invocations of doc.getBytes() before forwarding it to the HTTP-based service. These interactions constitute a part of the dynamic transformation realized by the mediator.

A mediator compensating the discrepancies in behavior between these two interfaces can be modeled in different ways. For our scenario, it is convenient to use a simple finite-state machine with additional memory. Thus, we can use a mediator that has a number of internal states and moves from one state to another until a final state is reached. The state transitions are triggered by external events like sending and receiving a message and may cause actions, e.g. storing the received message in a memory cell. Fig. 6 depicts a state machine that carries out the dynamic transformation between the CORBA client interface and the HTTP server interface. Assume that the mediator communicates with these components via channel c and channel h, respectively. In the notation we use, ?c ConvertRequest means that a message of type ConvertRequest is expected via channel c whereas !h ConvFile denotes sending a message of type ConvFile over channel h.

Upon receiving a conversion request from the CORBA client (1-2), the mediator retrieves the size of the binary object to be converted (2-3-4). Knowing the size, the mediator fetches the binary content of the object[1] (4-5-6). In state 6, the mediator has obtained all information that is necessary to send a conversion request to the HTTP server. After receiving the content of the converted object (7-8), the mediator informs the CORBA client that the conversion has been completed (8-9). In state 9, the mediator gives the client an opportunity to inquire about the size (9-10-9) and the content (9-11-9) of the converted object. The mediation protocol terminates in state 12.

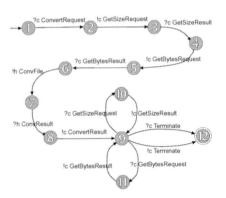

Fig. 6. Finite-state automaton of the mediator

Note that in our model a message is not atomic but consists of a set of facts. Every such message can be characterized by a 'classificator' function delivering true or false depending on whether the message is of a certain type. The state automaton needs such functions to recognize the types of the received messages and to trigger the appropriate transitions. Both classificator functions and manipulations on the content of the messages can be realized using static transformations discussed in the previous section. We implement classificator functions as rule-based transformations that produce some substitutions for true and deliver the empty set for false. For example, a simple classificator for the CORBA conversion request can be specified as follows:

```
(X is-a              ConvertRequest) &
(X source-format     V1)             &
(X destination-format V2)            &
(X object-to-convert B)              &
(B is-a              BLOB)
```

---

[1] This operation could be split in multiple requests to retrieve very large objects if needed.

The format parameters contained in the initial CORBA request (transition 1-2) have to be stored until the binary content of the object is retrieved (2-3-4-5-6), and merged with the content in a combined HTTP request sent in (6-7). The actions used in our example are storing a message in a memory cell, reading from a cell, and arbitrary static transformations of the messages.

7   Interface Descriptions

The interface definition languages used in today's distributed object systems (e.g. CORBA IDL) are remarkably limited. For instance, given a file interface definition in CORBA IDL that has the methods open and write, it is not possible to specify that write can be called only after open. Or take our converter service as an example. The service comprises two interfaces, and it is not possible to derive valid interaction patterns directly from those interfaces. Rich interface descriptions and the ability to discover and compare interfaces of the components are, however, essential in a mediation environment with a variety of complex information processing services.

Similarly to the mediator specification, interface descriptions of the canonical wrappers for the CORBA client and HTTP service can be captured using finite-state automata (see Fig. 7). Having interface descriptions of the wrapped components has a number of notable advantages. First, they can be used to support the wrapper designer by generating wrapper skeletons automatically. The wrapper implementor solely has to provide the native code to perform component-specific actions for every transition. The wrapper skeleton can also enforce the correctness of the protocol. Secondly, for any pair of interface descriptions is it possible to determine algorithmically whether they are compatible [19]. Analogously, we can find all available mediator specifications that are capable of translating between the given server interface and other client interfaces. Finally, mediator specifications can be simplified by including by reference portions of the interface descriptions of the components. Evident candidates for that are classificator functions that determine the types of the messages.

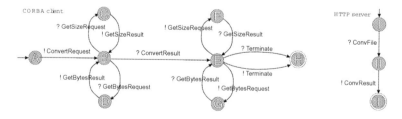

Fig. 7. Interface descriptions of the CORBA conversion client and the HTTP-based conversion service

In distributed object environment, there can be further immediate benefits of having rich interface descriptions of services. For example, the CORBA conversion service returns a reference to a BLOB object that contains the converted file. For the server it is not possible to determine when it is safe to release the converted object without cooperation from the client (e.g. using special protocols for asynchronous garbage collection). However, if the server has an interface description for the client like the one presented in the top part of Fig. 7, it can dispose the converted object whenever the client reaches the state H.

8   Challenges of Integrating Distributed Services

The idea of using declarative specifications for mediators that integrate information processing services seems promising since it makes developing mediators an engineering task. However, it also presents a number of challenges.

One of the biggest challenges is harnessing the complexity of declarative specifications. Even for the simple scenario presented in this paper the complete specification of the mediator comprises about than 350 statements (i.e. contains as many arcs in the meta-model representation). Additionally, is uses interfaces descriptions of the wrapped components that contain about 270 and 190 statements for the CORBA and HTTP wrapper, respectively (they are so verbose because the structure of every message needs to be completely specified). Fortunately, this is merely the internally used encoding of the mediators that is not intended to be manipulated by human engineers directly. Ideally, every formalism used for mediation will have specialized graphical editor that allows to 'click' mediators together using graphical representations similar to that in Fig. 6. The rules used for static transformations can be input in their textual notation and translated into the meta-model representation automatically. To handle the complexity of descriptions, support for mediator composition is required. Supporting composition is non-trivial, since multiple formalisms may be used throughout the mediator. For example, a complex mediator may be specified as a finite-state machine on the high-level and may invoke Petri nets for executing subtasks that require concurrency control.

Another concern is the efficiency of mediation. For simplicity, we implemented the sample scenario described in the paper using a set of interpreters for the finite-state machines, Datalog-rules etc. (more implementation details are described in the next section). In turned out that runtime interpretation of state machine transitions and execution of the rules has a noticeable performance impact as compared to a hard-coded Java implementation. To deploy declarative mediators in realistic environments, more work on optimization is needed. On-demand compilation of mediators into bytecode may be a viable strategy.

Furthermore, we experienced a number of difficulties specific to the distributed object environments. Examples include adequate handling of streams or large binary objects, dealing with exceptions and mapping of complex nested parameters used in IDL specifications to our declarative descriptions. For instance, the complexity of mediator descriptions rapidly increases if all exceptions need to be represented. We found that it would be beneficial to use a representation where the exception handling does not belong to the core interface.

Moreover, to enable a closer integration with the existing distributed object systems the rich interface descriptions should reuse and extend the existing IDL specifications. For this, a meta-model representation for IDL is needed. Finally, a mediation environment of a realistic scale needs to address the problem of management of declarative specifications in database systems. The specifications need to be stored, queried, and analyzed. Powerful manipulations of complex declarative specifications will require use of advanced database techniques.

9   Architecture of the Prototype

In this section we briefly describe a prototype implementation of the elements of the mediation infrastructure that have been introduced in our running example. In our implementation declarative specifications of mediators are interpreted at runtime. The complete runtime environment for our sample scenario is presented in Fig. 8. The native components, i.e. the CORBA client and the CGI script, are presented on the bottom of the figure. The canonical wrappers are depicted above the native components. Both wrappers consist of two generic modules and one component-specific module. The component-specific module accesses the native API of the component. The generic modules are a finite-state machine (FSM) language interpreter and an inference engine that supports extended Datalog descriptions. These interpreters process interface descriptions of the components that are encoded in RDF and stored on the Web. Both interface descriptions contain exactly the FSM specifications presented in Fig. 7 plus the classificator functions of the messages.

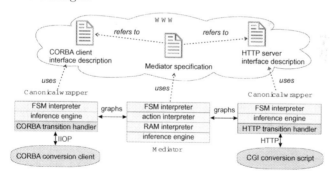

Fig. 8. Runtime environment for the sample scenario

Similarly to the wrappers, the mediator consists of four generic interpreter modules. In addition to the FSM and the inference engine, the mediator contains an 'action interpreter' for built-in and custom actions that can be incorporated into state machines. A random memory access (RAM) interpreter processes the descriptions of store and read memory operations. We use a generic executable module to launch the mediator and the wrappers. The parameters of this module are a reference to a declarative specification and a list of interpreters and native components. This module loads the declarative specification and tries to find suitable interpreters that can process it.

10   Related Work

Integration of heterogeneous systems raised a variety of questions both in data-intensive and interaction-intensive application domains. Diversity of data-intensive
systems inspired a number of projects focused on the integration of heterogeneous data sources. Examples include TSIMMIS [4], Information Manifold [9],
Garlic [15], InfoSleuth [3], Infomaster [5], and OBSERVER [11], just to name a
few. In essence, integration of data sources deals with translation and routing of
queries and reconciling heterogeneous data structures returned by the sources.
Some attempts has been made to cover database operations like updates [6].

Recently, incorporating declarative capability descriptions into query evaluation [9] has gained interest reaching beyond the scope of generation of executable
query plans. For example, work presented in [20] examines computing capabilities of mediators based on the capabilities of sources they integrate. On the
other hand, interaction-intensive applications urged research targeted at bridging the differences between applications that have functionally compatible but
protocol-incompatible interfaces [19,1]. The difficulties that arise upon tackling
both data and protocol integration became tangible during building of the Stanford InfoBus [12].

In approaching this combined integration problem we tried to select the tools
developed for data-intensive and interaction-intensive applications and to adjust
them to our needs. In particular, data integration was addressed in-depth in
MedMaker/MSL [14] and YATL [2]. Both approaches use nested data structures, Skolem functions, path expressions and powerful restructuring primitives
that are desirable in the general case of data integration but may be an overkill
for manipulation of logical descriptions of requests and replies of information
processing services. Our decision to use logical languages for both capturing the
mediation and representing the information flow between heterogeneous components was motivated by the work like [14,7,16].

Conclusion

We take a step towards integration of heterogeneous information processing services by developing a framework in which mediators are specified in a declarative fashion. To make declarative mediation feasible, we introduced canonical
wrappers that provide logical abstractions of the underlying components. Furthermore, we examined how different formalisms can be deployed for mediator
specifications and how the static and dynamic transformations performed by
the mediators can be formalized. We demonstrated how our framework can be
applied to addressing some of the mediation problems in distributed object systems, and discussed the benefits and challenges of our approach.

References

1. R. Allen and D. Garlan. A formal basis for architectural connection. *ACM Transactions on Software Engineering and Methodology*, 6(3):213–249, July 1997.
2. S. Cluet, C. Delobel, J. Simeon, and K. Smaga. Your Mediators Need Data Conversion! In *ACM SIGMOD Int. Conf.*, pages 177–188, 1998.
3. R. Bayardo et al. InfoSleuth: Semantic Integration of Information in Open and Dynamic Environments. In *Proc. ACM SIGMOD Conf.*, pages 195–206, Tucson, Arizona, 1997.
4. H. García-Molina et at. The TSIMMIS Approach to Mediation: Data Models and Languages. *Journal of Intelligent Inf. Systems 8:2*, pages 117–132, 1997.
5. M. R. Genesereth, A. M. Keller, and O. M. Dushka. Infomaster: An Information Integration System. In *In Proc. ACM SIGMOD Conference*, Tucson, 1997.
6. T. Härder, G. Sauter, and J. Thomas. The Intrinsic Problems of Structural Heterogeneity and an Approach to their Solution. *The VLDB Journal 8:1*, 1999.
7. V. Kashyap and A. Sheth. Semantic Heterogeneity in Global Information Systems: The Role of Metadata, Context and Ontologies. *In M. Papazoglou and G. Schlageter (Eds.), Boston: Kluwer Acad. Press*, 1997.
8. O. Lassila and R. Swick. Resource Description Framework (RDF) Model and Syntax Specification. http://www.w3.org/TR/REC-rdf-syntax/, 1998.
9. A. Levy, A. Rajaraman, and J. Ordille. Querying heterogeneous information sources using source descriptions. In *Proc. of the 22nd VLDB Conference*, pages 251–262, Bombay, India, September 1996.
10. S. Melnik, H. Garcia-Molina, and A. Paepcke. A Mediation Infrastructure for Digital Library Services. In *Proc. ACM Digital Libraries 2000*, June 2000.
11. E. Mena, A. Illarramendi, V. Kashyap, and A. Sheth. OBSERVER: An Approach for Query Processing in Global Information Systems based on Interoperation across Pre-existing Ontologies. *Distributed and Parallel Databases Journal*, 1999.
12. A. Paepcke, M. Baldonado, C. Chang, S. Cousins, and H. Garcia-Molina. Using Distributed Objects to Build the Stanford Digital Library Infobus. *IEEE Computer*, February 1999.
13. A. Paepcke, K. Chang, H. García-Molina, and T. Winograd. Interoperability for Digital Libraries Worldwide. *Communications of the ACM*, 41(4):33–43, April 1998.
14. Y. Papakonstantinou, S. Abiteboul, and H. Garcia-Molina. Object fusion in mediator systems. In *Proc. of the 22nd VLDB Conference*, pages 413–424, Bombay, India, September 1996.
15. M. T. Roth and P. M. Schwarz. Don't Scrap It, Wrap It! A Wrapper Architecture for Legacy Data Sources. In *Proc. 23rd VLDB Conf.*, Athens, Greece, 1997.
16. K. Shah and A. Sheth. Logical Information Modeling of Web-accessible Heterogeneous Digital Assets. In *Proc. of the Forum on Research and Technology Advances in Digital Libraries (ADL'98)*, Santa Barbara, CA, 1998.
17. G. Wiederhold. Mediators in the Architecture of Future Information Systems. *IEEE Computer 25:38-49*, 1992.
18. G. Wiederhold and M. Genesereth. The Conceptual Basis for Mediation Services. *IEEE Expert*, 12(5):38–47, 1997.
19. D. M. Yellin and R. E. Strom. Protocol Specifications and Component Adaptors. *ACM Transactions on Programming Languages and Systems*, 19(2):292–333, Mar 1997.
20. R. Yerneni, Ch. Li, H. Garcia-Molina, and J. D. Ullman. Computing Capabilities of Mediators. In *Proc. of ACM SIGMOD*, pages 443–454, Philadelphia, PA, 1999.

Temporal Constraints for Object Migration and Behavior Modeling Using Colored Petri Nets

Hideki Sato[1] and Akifumi Makinouchi[2]

[1] School of Community Policy, Aichi Gakusen University,
1 Shiotori, Oike-cho, Toyota 471-8532, Japan
hsato@gakusen.ac.jp
[2] Graduate School of Information Science and Electrical Engineering,
Kyushu University, 6-10-1 Hakozaki, Higashi-ku, Fukuoka 812, Japan
akifumi@is.kyushu-u.ac.jp

Abstract. In databases based on a multi-aspects object-oriented data model which enables multiple aspects of a real-world entity to be represented and to be acquired/lost dynamically, Object Migration (OM) updating membership relationships between an object and classes occurs, as the properties of the object evolve in its lifetime. To keep an object consistent in OM, this paper introduces temporal consistency constraints such as temporal transitional constraints and temporal multiplicity constraints by extending OM consistency constraints for snapshot databases. To this end, a temporal interval is attached to each aspect of an object for representing its duration in the real world. Then, temporal transitional constraints are represented by transitional rules with conditions referring to temporal intervals. Additionally, temporal multiplicity constraints are represented by object-schemas. Furthermore, the paper proposes OM behavior modeling using Colored Petri Nets (CPN) based on temporal consistency constraints.

1   Introduction

In an object-oriented database [1], it is assumed that an object belongs to exactly one specific class (if we don't count its superclasses), namely the one in which the object was created, and the class membership is unchangeable in its lifetime. Such an assumption, however, imposes some serious restrictions in modeling a real-world entity with dynamic nature. For example, one person might become a university student at one time. After graduating from the university, the person may not be a student any longer but become a graduate. At the same time, the person may also be a worker during this period. Thus, a real-world entity has multiple aspects simultaneously and acquires/losses such aspects in its lifetime. Conventional object-oriented data models are insufficient, since they cannot model such phenomena even with the multiple inheritance capability.

To resolve the problems mentioned above, researches have been carried out on multi-aspects data models [2], [3], [4], [5], [6], [7], [8] which enable multiple aspects of a real-world entity to be represented and to be acquired/lost dynamically. Additionally, related researches have been done on Object Migration (OM) [9], [10], [11], [12] updating membership relationships between an object and classes. In the research fields, we have proposed MAORI (Multi-Aspects Object-Oriented Data

A.H.F. Laender, S.W. Liddle, V.C. Storey (Eds.): ER2000 Conference, LNCS 1920, pp. 80-95, 2000.

Model) [13], OM framework [14], and OM behavior modeling [15], [16] using Petri Nets [17]. To make an object satisfy the requirements which are imposed on by applications and vary with time, MAORI enables an object to dynamically change its class memberships. As for OM framework, it prevents meaningless and/or erroneous OM from occurring based on OM consistency constraints. Also, OM behavior modeling is organized based on the constraints. However, the descriptive power of the constraints is restricted, since they cannot handle temporal conditions among the aspects of an object.

Allen introduced the interval-based temporal logic [18] to represent temporal knowledge and temporal reasoning. Let R be the real line. In the logic, a temporal interval I is defined as an ordered pair of time points stp and etp; $[stp,etp] (\in R \times R)$ with the first point (=stp) less than the second (=etp). The start (end) time point of an interval I is denoted by I.stp (I.etp). Allen showed that there are thirteen temporal relations at most that could hold between two temporal intervals (see Table 1). Note that they are mutually exclusive by definition in Table 1. Also, note that six out of them except relation *equal* are the inverses of other six relations. For example, relation *before* is the inverse of *after*.

Table 1. Temporal relation definition

Time interval relation between I_1 and I_2	Equivalence relations on endpoints
equals(I_1, I_2)	$I_1.stp = I_2.stp$ and $I_1.etp = I_2.etp$
before(I_1, I_2)	$I_1.etp < I_2.stp$
after(I_1, I_2)	$I_1.stp > I_2.etp$
during(I_1, I_2)	$I_1.stp > I_2.stp$ and $I_1.etp < I_2.etp$
contains(I_1, I_2)	$I_1.stp < I_2.stp$ and $I_1.etp > I_2.etp$
overlaps(I_1, I_2)	$I_1.stp < I_2.stp$ and $I_1.etp > I_2.stp$ and $I_1.etp < I_2.etp$
overlapped_by(I_1, I_2)	$I_1.stp > I_2.stp$ and $I_1.etp < I_2.stp$ and $I_1.etp > I_2.etp$
meets(I_1, I_2)	$I_1.etp = I_2.stp$
met_by(I_1, I_2)	$I_1.stp = I_2.etp$
starts(I_1, I_2)	$I_1.stp = I_2.stp$ and $I_1.etp = I_2.etp$
started_by(I_1, I_2)	$I_1.stp = I_2.stp$ and $I_1.etp > I_2.etp$
finishes(I_1, I_2)	$I_1.stp > I_2.stp$ and $I_1.etp = I_2.etp$
finished_by(I_1, I_2)	$I_1.stp < I_2.stp$ and $I_1.etp = I_2.etp$

This paper introduces temporal consistency constraints such as temporal transitional constraints and temporal multiplicity constraints by extending OM consistency constraints for snapshot databases. To this end, a temporal interval is attached to each aspect of an object for representing its duration in the real world. Then, temporal transitional constraints are represented by transitional rules with conditions referring to temporal intervals. Additionally, temporal multiplicity constraints are represented by object-schemas. Furthermore, the paper proposes OM behavior modeling using Colored Petri Nets (CPN) [19] based on temporal consistency constraints.

The remainder of the paper is organized as follows. Sect. 2 discusses OM and OM consistency constraints. Sect. 3 proposes representation for OM temporal consistency constraints. Sect. 4 explains CPN and presents OM behavior modeling using CPN. Sect. 5 compares related works with ours. Sect. 6 concludes the paper.

2   Consistency Constraints for Object Migration

2.1 Object Migration

A real-world entity is modeled as an object with multiple aspects in MAORI [13]. Each aspect of an object is an instance of the class defining a list of attributes/methods as the type of instances. There exist superclass (subclass) relationships among classes. A subclass inherits the definition of its superclass. An object is a member of all the classes defining its aspects. An object is uniquely identified using its object identifier and an aspect is identified using its aspect identifier within an object. Thus, an object can have more than one aspect defined by a class.

Membership relationships between an object and classes vary with time in MAORI. Change of the relationships is called OM. OM is caused by a sequence of OM events. Fig. 1 shows an exemplary OM event that object "Smith" is promoted to a full professor from an associate professor. Object "smith" has an associate professor aspect before the event occurs. After the event occurs, object "Smith" acquires a full professor aspect and loses the associate professor one.

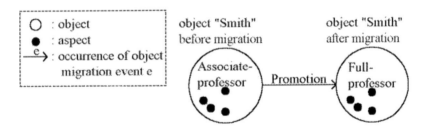

Fig. 1. An example of object migration event

In this paper, it is assumed that a temporal interval is attached to each aspect of an object for representing its duration in the real world. The start (end) time point of a temporal interval is *valid time* [20], [21]. However, it is not supposed that a history of an attribute value is handled, since a temporal interval is not attached to an attribute value of an aspect. The temporal interval of aspect A is denoted by interval(A). It is assumed that the OM event of Fig. 1 occurs at interval I_e that is instantaneous in the real world. Furthermore, assume that the temporal interval of the associate professor aspect is $[t_0,\infty]$ before the event occurs. Note that ∞ is a special time point meaning forever. Under the assumptions, the temporal interval of the associate professor aspect to be lost is changed into $[t_0,I_e.stp]$ and the temporal interval of the full professor to be added is initially $[I_e.etp,\infty]$ just after the event occurs. Hereafter, let temporal interval NOW denote an instant at which an OM event occurs.

2.2 Consistency Constraints

OM consistency constraints are used to prevent meaningless and/or erroneous OM from occurring. They originate from characteristics such as multiplicity (i.e., representation for multiple aspects of a real-world entity) and dynamism (i.e., dynamic acquisition/loss of aspects) which a multi-aspects object-oriented data model possesses. The former characteristic introduces multiplicity constraints for restricting possible aspect sets that an object can possess. The latter characteristic introduces transitional constraints for restricting possible changes for a set of aspects. In the following, requirements of transitional constraints and multiplicity constraints are examined for snapshot databases where temporal concepts are not considered. Then, the requirements are extended to ones for temporal consistency constraints.

The requirements of transitional constraints are as follows.

[TC1] The transitional direction among aspects of an object may be restricted. A high school student may become a university student, while a university student cannot become a high school student.

[TC2] Transition among aspects of an object may be context-sensitive. For example, an associate professor may be promoted to a full professor, if he/she is a Ph.D. The Ph.D. aspect plays the role of the context for the transition.

The requirements of multiplicity constraints are as follows.

[MC1] The type of an aspect that an object can possess is restricted. A dog may have a pet aspect but a person cannot.

[MC2] Each aspect type has the maximum number of instances that an object can possess. For example, a worker may concurrently hold up to two distinct manager positions in a company.

[MC3] There may exist a group of aspect types whose instances an object cannot possess at the same time. A person cannot be a student and a graduate of a university at the same time. It is supposed that a person cannot become a student again after having graduated from a university.

The following [TTC1], [TTC2], and [TTC3] are the requirements of temporal transitional constraints. [TTC1] and [TTC2] extend [TC1] and [TC2] respectively to the temporal dimension. [TTC3] is also a requirement for temporal transitional constraints.

[TTC1] A temporal condition referring to duration of aspects may be imposed on transition among aspects of an object. For example, in the case that a person becomes a university student, the duration of a high school student must precede that of a university student. This does not necessarily mean that both duration occur consecutively.

[TTC2] Transition among aspects may be temporally context-sensitive. A vice-chairperson may be raised to a chairperson in a committee, if he/she was a secretary in the past. In the case, the secretary aspect plays the role of the temporal context for the transition.

[TTC3] Condition referring to the temporal interval of an aspect may be imposed on transition. For example, an assistant professor having been engaged for at least 5 years may be promoted to an associate professor in a university.

In [TC1] and [TC2], the aspects relating to an OM event must exist just before and/or after it occurs. The condition is, however, abandoned in [TTC1] and [TTC2], for [TC1] and [TC2] are implied by [TTC1] and [TTC2] respectively. As for multiplicity constraints, they must be satisfied throughout the lifetime of an object.

Thus, the requirements [MC1], [MC2], and [MC3] are ones for temporal multiplicity constraints as they are. As a result, [TTC1], [TTC2], [TTC3], [MC1], [MC2], and [MC3] are chosen to investigate OM temporal consistency constraints.

3   Consistency Constraints Representation for Object Migration

3.1 Temporal Transitional Constraints Representation

From [TTC1], [TTC2], and [TTC3], temporal transitional constraints require to express (1) temporal condition among aspects, (2) temporal context, and (3) condition referring to the temporal interval of an aspect, in relation to OM events. A transitional rule is used for the purpose.

[Definition1] A transitional rule is defined as follows.

$$p: \quad c(\alpha)$$

$$\alpha \longrightarrow \beta$$

α ($\in N^*$; N denoting a set of class names) is a string of class names representing input-aspects of the rule. β ($\in N^*$) is also a string of class names representing output-aspects of the rule. $c(\alpha)$ is transitional condition and specifies temporal condition among input-aspects, condition referring to the temporal interval of an input-aspect, or composite condition given as a result of boolean operations on the former conditions. Multiple occurrences of a class name appearing in α and β indicate the same aspect. On the other hand, each of distinct aspects defined by a class is distinguished by a suffix attached to the class name.

[Definition2]Let the input-aspect set of transitional rule p be IA_p and the output-aspect set be OA_p.

1. An element of $(IA_p - OA_p)$ is a deletion-aspect, the end time point of whose temporal interval is to be changed to NOW.stp just after p is invoked.
2. An element of $(OA_p - IA_p)$ is an addition-aspect, whose temporal interval is to be set [NOW.etp,∞] just after p is invoked.
3. An element of $(IA_p \cap OA_p)$ is a context-aspect, whose temporal interval is not changed just before and after p is invoked.

Examples are presented to show features of a transitional rule as follows. Note that a class name starts with "@".

[Exemplary rule1] In the case that a person becomes a university student, the duration of a high school student must precede that of a university student.

p_{11}: $cond_{11}$
@High-School-Student———>@University-Student
$cond_{11}$: *meets*(interval(@High-School-Student), NOW)

p_{12}: $cond_{12}$
@High-School-Student———>@High-School-Student @University-Student

$cond_{12}$: *before*(interval(@High-School-Student), NOW)

The example is divided into two cases depending on whether or not the duration of a university student is consecutive to that of a high school. Rule p_{11} corresponds to the first case and rule p_{12} corresponds to the second case.

[Exemplary rule2] A vice-chairperson may be raised to a chairperson in a committee, if he/she was a secretary in the past.

p_2: $cond_2$

@Secretary @Vice-Chairperson ———>@Secretary @Chairperson

$cond_2$: *before*(interval(@Secretary), NOW) and

$$meets(\text{interval}(@\text{Vice-Chairperson}), \text{NOW})$$

Aspect @Secretary plays the role of the temporal context for transition from aspect @Vice-Chairperson to aspect @Chairperson. According to $cond_2$ using temporal interval NOW, the duration of aspect @Secretary is before that of aspect @Chairperson to be added.

[Exemplary rule3] An assistant professor having been engaged for at least five years may be promoted to an associate professor in a university.

p_3: $cond_3$

@Assistant-Professor ———>@Associate-Professor

$cond_3$: (interval(@Assistant-Professor).etp–

$$\text{interval}(@\text{Assistant-Professor}).stp) \geq period'05:00:00'$$

The $cond_3$ specifies condition referring to the temporal interval of aspect @Assistant-Professor. It is required that an assistant professor has been engaged for at least five years. Note that a single quoted character string following keyword period stands for a literal whose format is *years:months:days*.

3.2 Temporal Multiplicity Constraints Representation

From [MC1], [MC2], and [MC3], temporal multiplicity constraints require to express (1) possible type of an aspect that an object can possess, (2) the maximum number of instances that an object can possess for each type at any time, and (3) a group of aspect types whose instances an object cannot possess at the same time.

[Definition3] Object-schema OS is defined to be $\{T_1, \ldots, T_n\}$. T_i $(1 \leq i \leq n)$ is $(C_{i(1)}^{r(i(1))} | \ldots | C_{i(k(i))}^{r(i(k(i)))})$. $C_{i(j)}$ $(1 \leq j \leq k(i))$ stands for a class name. $r(i(j))$ specifies the maximum number of instances which class $C_{i(j)}$ or its subclasses define. In T_i, each term $C_{i(j)}$ is mutually exclusive. Also, note that $C_{i(p)}$ is neither a superclass nor a subclass of $C_{i(q)}$ $(1 \leq p, q \leq k(i), p \neq q)$.

T_i $(1 \leq i \leq n)$ defined above specifies aspects that an object can possess. Each term $C_{i(j)}^{r(i(j))}$ of T_i $(1 \leq j \leq k(i))$ indicates that an object can possess at most '$r(i(j))$' number of aspects that class $C_{i(j)}$ or its subclasses define. Because $C_{i(p)}$ and $C_{i(q)}$ are mutually exclusive, a constraint is imposed on an object, that it cannot possess their instances at the same time. As class $C_{i(p)}$ is neither a superclass nor a subclass of class $C_{i(q)}$, aspects defined by $C_{i(p)}$, $C_{i(q)}$, or their subclasses don't violate a constraint on a group of aspect types whose instances an object cannot possess at the same time. As class $C_{i(j)}$ includes designation of its subclasses, a concise object-schema can be defined by designating an aspect with a class of a highly abstract level. Furthermore, an object-schema using classes of a highly abstract level prevents the total number of object-schemas from increasing, since it is potentially able to define constraints applicable to a broad range of objects. The following is an example of an object-schema. Note that an object-schema name starts with "#".

[Exemplary object-schema]

$$\text{\#PERSON}=\{(@STUDENT^1|@GRADUATE^1),(@WORKER^1)|(@MANAGER^2)\}$$

An object may possess as its aspects an instance defined by class @STUDENT, @GRADUATE, @WORKER, @MANAGER, or their subclasses, if it is obeyed to object-schema #PERSON. However, aspect @STUDENT and aspect @GRADUATE don't exist at the same time. As for the number of aspects, the object can possess at most two aspects @MANAGER and at most one aspect for each of @STUDENT, @GRADUATE, and @WORKER.

4   Object Migration Behavior Modeling

4.1 Colored Petri Nets

OM behavior is defined by both states of an aspect set of an object and events changing states. We make use of CPN [19] to construct OM behavior models, since it can inherently represent a type of an aspect, relationships between states and events, and a condition on which an event occurs. It is also noted that CPN is an executable modeling formalism. Fig. 2 shows CPN for the so-called philosophers' dinner problem. CPN is a bipartite directed graph, with mutually disjoint sets of nodes (i.e., a set of places and a set of transitions) and a set of arcs among nodes. A place is represented as a circle and a transition as a rectangle or a bar. An arc is (i) a directed edge from a place to a transition, or (ii) a directed edge from a transition to a place. An arc is represented as an arrow. In case (i), the arc is called input-arc and the place is called input-place. In case (ii), the arc is called output-arc and the place is called output-place.

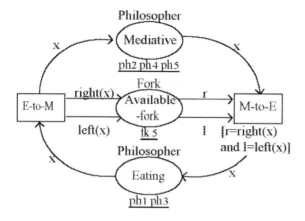

Fig. 2. CPN for philosphers' dinner problem

A character string inscribed in a place or a transition is a label for distinguishing it from others. A character string in the outside of a place is called color. It specifies the type of tokens that a place can possess. An underlined character string in the outside of a place is called initial marking. It specifies tokens to be initially assigned to the place. Tokens to be assigned to a place is also represented as circles blacked out. An expression attached to an arc is called arc-expression. It specifies tokens which can pass through the arc. An expression in the outside of a transition is called guard. A transition can fire, (i) if each input-place possesses a token specified by the arc-expression of the input-arc, and (ii) if its guard is satisfied. If a guard does not exist, the condition of (ii) is always satisfied. A token passing through an input-arc (output-arc) is called input-token (output-token). On transition firing, an input-token is removed from the input-place and an output-token specified by the arc-expression of the output-arc is put into the output-place.

Transitions are divided into immediate firing ones and unimmediate firing ones, depending on how they fire. An immediate firing transition fires at the instant when it becomes enabled. It is represented as a rectangle blacked out. On the contrary, an unimmediate firing transition does not necessarily fire at the instant when it becomes enabled. It is represented as a rectangle not blacked out or a bar, as explained before. Enabling to fire is only a necessary condition for an unimmediate transition. Firing sequence is non-deterministic, if there exist multiple immediate firing transitions being enabled to fire.

Labels of places/transitions, colors, arc-expressions, and guards may be omitted to simplify CPN diagrams. A label of a place may be also specified in its outside. Furthermore, if an input-arc and an output-arc between a place and a transition have the same arc-expression, a single arc with double heads may be substituted for them.

4.2 Modeling Overview

Table 2 shows relationships between object-oriented concepts and CPN concepts. An object is a basic unit of an OM behavior model using CPN. The specifications of temporal consistency constraints relating to an object are gathered into a single CPN. A class defining an aspect corresponds to a color. An aspect, being an instance of a class, is represented by a color token and put in a place with the corresponding color. For a class, there exists at most one place whose color corresponds to it. Accordingly, all aspects defined by a class are put in the same place. Also, for a class whose instance an object cannot possess as its aspect, there does not exist a place whose color corresponds to it. An OM event is represented by a transition, and relationship between the event and an aspect is represented by arc. An invalid state of aspects accompanying an OM event is also represented by a transition, and relationship between the state and an aspect is represented by arc. A set of aspects that an object possesses is represented by a so-called marking.

Table 2. Relationship between Object-Oriented concepts and CPN concepts

Object-Oriented concept	CPN concept
Object	CPN
Class	Color
Aspect	Token
Object migration event	Transition
Invalid state of aspects accompanying an object migration event	Transition
Repository of aspects	Place
Relationship of an OM event or an accompanying invalid state to aspects	Arc
State of an aspect set	Marking

An arc-expression of an input-arc is materialized by a variable. On transition firing, a token in an input-place is bound to the variable attached to an input-arc. A transition and arc-expressions attached to inputs-arcs or output-arcs form a scope of variables. A variable can be used inside a guard and an arc-expression of an output-arc, to refer to an aspect bound to it and/or its temporal interval. In addition, a temporal interval can be changed by an arc-expression of an output-arc. Furthermore, a free variable, not attached to an input-arc, can be used as an arc-expression of an output-arc. This is the case of adding an aspect to an object.

4.3 Temporal Transitional Constraints Modeling

Temporal transitional constraints restrict possible changes for a set of aspects possessed by an object. They provide each OM event with its behavioral structure. To specify the behavioral structure of an OM event, CPN is constructed based on transitional rules as follows.

- For a rule, an unimmediate firing transition is provided, whose guard specifies the transitional condition. It fires in response to each occurrence of an OM event.

- For a deletion-aspect of the rule, let a place to put a token denoting the aspect be both input-place and output-place of the transition. The same variable is attached to the input-arc and the output-arc as each arc-expression. Furthermore, the arc-expression of the output-arc specifies that the end time point of the temporal interval of an aspect bound to the variable is changed into NOW.stp.
- For an addition-aspect of the rule, let a place to put a token denoting the aspect be output-place of the transition. A free variable is attached to the output-arc as its arc-expression. The temporal interval of an aspect to be assigned to the variable is set to be [NOW.etp,∞].
- For a context-aspect of the rule, let a place to put a token denoting the aspect be both input-place and output-place of the transition. The same variable is attched to the input-arc and the output-arc as each arc-expression.

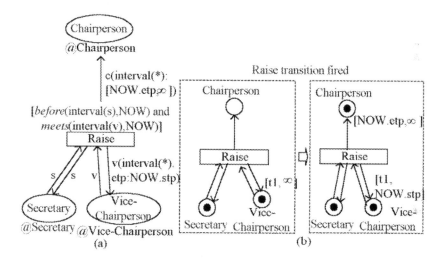

Fig. 3. CPN for temporal transitional constraint

Fig. 3(a) shows CPN for a conditional and context-sensitive temporal transitional constraint on the raise from a vice-chairperson to a chairperson (see [TTC2]). Fig. 3(b) is an exemplary state transition of the CPN. If transition Raise fires in the state that the guard is satisfied, a token is removed from place Vice-Chairperson and a token with temporal interval [NOW.stp,∞] is put into place Chairperson. And a token is put into place Vice-Chairperson. The temporal interval of the token is gotten by changing the end time point of the temporal interval of the removed token into NOW.etp. While a token is removed from place Secretary, the same token is put into the place again. Namely, aspect @Secretary, denoted by the token, plays the role of the temporal context for the transition. In addition, the guard specifies the condition that a vice-chairperson must be a secretary in the past.

4.4 Temporal Multiplicity Constraints Modeling

Temporal multiplicity constraints restrict possible aspect sets to be possessed by an object. They specify conditions to be always satisfied throughout the lifetime of an object, differently from temporal transitional constraints that aim at each occurrence of an OM event.   To prevent an object from violating temporal multiplicity constraints, CPN is constructed for detecting constraint violation immediately after an OM event occurs. To this end, temporal predicate *exists* is defined to see if an aspect exists immediately after an OM event occurs.

[Definition4] Temporal predicate *exists* is defined as follows. Note that A denotes an aspect.

$$exists(A)=during(NOW,interval(A)) \text{ or } meets(NOW,interval(A))$$

The first disjunct *during*(NOW,A) of predicate *exists*(A) indicates that aspect A exists before and after an OM event occurs.   Also, the second disjunct *meets*(NOW,A) indicates that aspect A is an addition-aspect accompanying an OM event.   Namely, these two kinds of aspects do exist immediately after an OM event occurs (see Fig. 4).

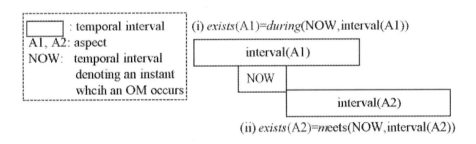

Fig. 4. Diagrammatic representation of temporal predicate *exists*

CPN for detecting constraint violation is based on the negated conditions of constraints and constructed as follows.

- For a constraint, an immediate firing transition is provided, which fires immediately after the constraint is violated.
- Let a place, into which a token denoting an aspect relating to the constraint is put, be both input-place and output-place of the transition. In addition, let the guard of the transition be condition that each relating aspect exists immediately after an OM event occurs.
- A place is provided to put a control token for detecting constraint violation. Also, a transition is provided to put the token into the place.
- A place is provided to put a token denoting constraint violation. Also, a transition is provided to remove the token from the place.

Fig. 5(a) shows CPN for temporal multiplicity constraint on the maximum number of aspects (see [MC2]). Transition SC-check2 is an immediate firing one. It can immediately fire, if more than two tokens exist in place Manager at the same time and if a token exists in control place C2. After it fires, a token denoting constraint violation is put into control place V2. Place Manager is both input-place and output-place of SC-check2. There exist between them three pairs of arcs with variable m1, m2, and m3 as each arc-expression. Thus, @MANAGER token is removed from place Manager, but is returned to the place as it is. A token in C2 controls to fire SC-check2. As it can prevent SC-chec2 from firing infinitely, the number of tokens to be put into C2 does not increase infinitely. Fig. 5(b) is an exemplary state transition of the CPN. Suppose that the third token is put into place Manager in the state that two tokens already exist in the place. Then, SC-check2 fires, since the guard becomes satisfied. As a result, a token denoting constraint violation is put into V2.

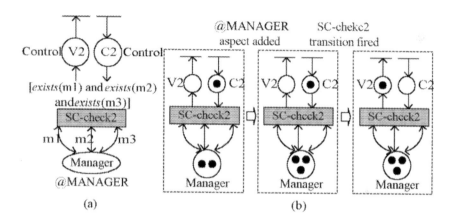

Fig. 5. CPN for temporal multiplicity constraint type1

Fig. 6(a) shows CPN for temporal multiplicity constraint on exclusive aspects (see [MC3]). Transition SC-check3 is an immediate firing one. It can immediately fire, if a token exists in place Student and a token exist in place @GRADUATE at the same time and if a token exists in control place C3. After it fires, a token denoting constraint violation is put into V3. Place Student and Graduate are both input-place and output-place of SC-check3. A pair of arcs with variable 's' as each arc-expression exists between place Student and the transition. Also, another pair of arcs with variable 'g' as each arc-expression exists between place Graduate and the transition. Fig. 6(b) is an exemplary state transition of the CPN. Suppose that a token is put into place Graduate in the state that a token already exists in place Student. Then, SC-check3 fires, since the guard becomes satisfied. As a result, a token denoting constraint violation is put into V3.

As mentiond in Sect. 4.2, there does not exist a place with a color corresponding to a class whose instance an object cannot possess as its aspect. Thus, requirement [MC1] is necessarily satisfied.

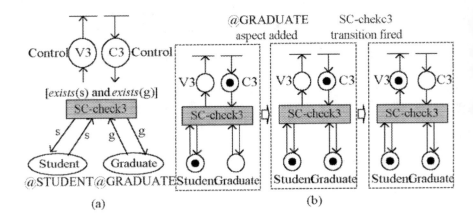

Fig. 6. CPN for temporal multiplicity constraint type2

5   Related Works

This section refers to related works of OM consistency constraints and OM behavior modeling. To the best of our knowledge, there has not been any works relating to OM temporal consistency constraints. Therefore, related works are confined to OM consistency constraints for snapshot databases.

In the reference [9], an object may belong to multiple classes (called role set) at any time. A set of possible sequences of role sets presents dynamic constraints on OM. A theoretical study has been conducted on analysis of OM sequences. Namely, dynamic constraints for three kinds of transactional languages to update role sets have been characterized. This is a study to make a theoretical foundation on OM. In the contrary, our study is concerned with constraint description at a system level and with construction of OM behavior modeling based on the description.

In the reference [12], OM is called object evolution. Control rules and the control mechanism for OM are presented based on classification of OM patterns. Rule description from the side of an object referring to another object is provided differently from ours. Though relationships between an object and classes are changeable, an object is required to belong to only one class at any time.

In the reference [10], rules are used to describe OM consistency constraints. However, their rules are neither conditional nor context-sensitive differently from transitional rules of our study. Furthermore, multiplicity constraints are not considered in their study. Totally speaking, the framework is not enough to describe OM consistency constraints.

In the reference [11], the conflict problem among type constraints such as "Property P of an object typed with class C takes a value typed with class D" is dealt with. To resolve the kind of conflicts, a mechanism is proposed to adapt a database by updating either a class or a property value of an object, based on the distance among database states. While our study is concerned with a problem to prevent meaningless and/or erroneous OM from occurring, their mechanism is concerned with a problem to repair database states produced by invalid OM. The investigation of repairing problem with accompanying adaptation mechanism is our future work.

A place-transition net [17], a kind of Petri net, has been also applied to behavior modeling in object-oriented databases [22]. However, the net is low level and cannot model a complex structured object. On the contrary, a color of CPN allows for representing it. Thus, we adopt CPN to construct OM behavior models in multi-aspects object-oriented databases.

OMT diagram [23] is a diagrammatic notation for developing object-oriented software. Comparison must be carefully done between OMT diagram and OM behavior modeling proposed in this paper, since the former is applicable to a wide area of applications. Roughly speaking, however, OM behavior model corresponds to both dynamic model and functional model in OMT diagram. In other words, states, state transitions, and transitional conditions correspond to markings of tokens, transitions, guards in OM behavior model based on CPN respectively. Also, functional model in OMT diagram is specified as arc-expressions of CPN. UML [24] offers many kinds of diagrams. Among them, the state diagram corresponds to OM behavior model. As CPN usage proposed in this paper is specialized to OM behavior modeling as shown in Table 2, it cannot be applied to all the phases for developing object-oriented software. An OM behavior model deals with the description which dynamic model and functional model of OMT diagram express, or state diagram of UML expresses. And specification is gathered into a single CPN. Accordingly, it is concluded that an OM behavior model can express related specification in a concise and uniform way, in comparison with OMT diagram and UML.

6 Conclusions

This paper has introduced temporal consistency constraints such as temporal transitional constraints and temporal multiplicity constraints by extending OM consistency constraints for snapshot databases. This paper has also proposed OM behavior modeling using CPN, based on the constraints. We summarize our proposal as follows.

- OM temporal consistency constraints originate from characteristics such as multiplicity and dynamism which a multi-aspects object-oriented data model possesses. To the best of our knowledge, this is the only proposal for OM temporal consistency constraints.

- Temporal transitional constraints are represented by transitional rules with conditions referring to temporal intervals. Temporal multiplicity constraints are represented by object-schemas. Both of them are powerful for representing the constraints.

- OM behavior modeling provides us with models to understand object behavior at the time when an OM event occurs. As the specifications of temporal consistency constraints relating to an object are gathered into a single CPN, its diagrammatic notation makes mutual relationships among the constraints concise, clear and easy to understand. OM behavior models are executable according to the firing rule of CPN. Simulation of OM behavior models using software such as Design/CPN [25] requires further investigation.

A set of consistency constraints introduced in Sect. 2.2 is not complete. Therefore, our future work includes extension of transitional condition so that it can refer to (i) aggregate values on temporal intervals of aspects, (ii) attribute/method values of aspects, and so on. The above (ii) involves investigation of a temporal multi-aspects object-oriented data model. Another work should be done for re-formalizing OM framework which includes temporal consistency constraints, by extending our previous formalization [14].

References

1. Cattel, R. G. G., Barry, D. G. (eds.): The Object Database Standard: ODMG2.0. Morgan Kaufmann (1997)
2. Sciore, E.: Object Specialization. ACM Trans. Office Information Systems, Vol.7, 2 (1989) 103-127
3. Steing, L. A., Zdonik, S. B.: Clovers: The Dynamic Behavior of Type and Instances. Brown University Technical Report, No.CS-89-42 (1989)
4. Richardson, J., Schwarz, P.: Aspects: Extending Object to Support Multiple, Independent Roles. Proc. ACM International Conference on Management of Data (1991) 298-307
5. Tsukada, H., Sugimura, T.: MAC-model: An Extended Object-Oriented Data Model for Multiple Classification. Computer Software, Vol.11, 5 (1994) 44-57
6. Ishimaru, T., Uemura, S.: An Object Oriented Data Model for Multiple Representation of Object Semantics. IEICE Transaction D-I, Vol.J78-D-I, 3 (1995) 349-357
7. Aritsugi, M., Makinouchi, A.: Design and Implementation of Multiple Type Objects in a Persistent Programming Language. Proc. COMPSAC 95 (1995) 70-76
8. Gottlob, G., Schrefl, M., Rock, B.: Extending Object-Oriented System with Roles. ACM Trans. Information Systems, Vol.14, 3 (1996) 268-296
9. Su, J.: Dynamic Constraints and Object Migration. Proc. International Conference on Very Large Data Bases (1991) 233-242
10. Qing, L., Guozhu, D.: A Framework for Object Migration in Object-Oriented Databases. Data and Knowledge Engineering, Vol.12, (1994) 221-242
11. Mendelzon, A. O., Milo, T., Walker, E.: Object Migration. Proc. PODS94 (1994) 232-242
12. Onizuka, M., Yamamuro, M., .Ishigaki, S: A Class-Based Object-Oriented Database Design to Support Object Evolution. IEICE Transaction D-I, Vol.J79-D-I, 10 (1996) 803-810
13. Sato, H., Ikeda, M., Funahashi, S., Hayashi, T.: MAORI: A Multi-Aspects Object-Oriented Data Model. IEICE Transaction D-I, Vol.J79-D-I, 10 (1996) 781-790
14. Sato, H., Funahashi, S., Hayashi, T.: A Framework of Object Migration in Multi-Aspects Object-Oriented Databases. IEICE Transaction D-I, Vol.J81-D-I, 3 (1998) 271-282
15. Sato, H., Hayashi, T.: Object Migration Behavior Modeling with Petri-Nets. Proc. IASTED International Conference, Artificial Intelligence and Soft Computing (1998) 250-253
16. Sato, H., Hayashi, T.: Object Migration Behavior Modeling Using Colored Petri Nets. IPSJ Transaction on Databases, Vol.40, No.SIG8 (1999) 13-28

17. Peterson, J. L.: Petri Net Theory and the Modeling of Systems. North-Holland (1981)
18. Allen, J. F.: Maintaining Knowledge about Temporal Intervals. Communications of the ACM, Vol.26, 11 (1983) 832-843
19. Jensen, K.: Coloured Petri Nets: A High Level Language for System Design and Analysis. In: Rozenberg, G. (ed.): Petri Nets 1990. Lecture Notes in Computer Science, Vol.483. Springer-Verlag (1990) 342-416
20. Snodgrass, R. T.: Temporal Object-Oriented Databases: A Critical Comparison. In: Kim, W. (ed): Modern Database Systems. Addison-Wesley (1995) 386-408
21. Snodgrass, R. □T.: Temporal Databases. In: Zaniolo, C. (ed.): Advanced Database Systems. Morgan Kaufmann (1997) 97-126
22. Sakai, H.: Object Oriented Database Design. Journal of IPSJ, Vol.32, 5 (1991) 568-576
23. Rumbaugh, J., Blaha, M., Premerlani, W., Eddy, F., Lorensen, W.: Object-Oriented Modeling and Design., Prentice-Hall (1991)
24. Fowler, M., Scott, K.: UML DISTILLED: Applying the Standard Object Modeling Language. Addison-Wesley (1997)
25. Design/CPN Reference Manual for X-Windows Version2.0. Meta Software Corporation (1993)

SQLST: A Spatio-Temporal Data Model and Query Language

Cindy Xinmin Chen and Carlo Zaniolo

Computer Science Department
University of California at Los Angeles
Los Angeles, CA 90095, U.S.A.
{cchen,zaniolo}@cs.ucla.edu

Abstract. In this paper, we propose a query language and data model for spatio-temporal information, including objects of time-changing geometry. Our objective is to minimize the extensions required in SQL, or other relational languages, to support spatio-temporal queries. We build on the model proposed by Worboys where each state of a spatial object is captured as a snapshot of time; then, we use a directed-triangulation model to represent spatial data, and a point-based model to represent time at the conceptual level. Spatio-temporal reasoning and queries can be fully expressed with no new constructs, but user-defined aggregates, such as AREA and INSIDE for spatial relationships, DURATION and CONTAIN for temporal ones, and MOVING_DISTANCE for spatio-temporal ones. We also consider the implementation problem under the assumption that, for performance reasons, the representation at the physical level can be totally different from the conceptual one. Thus, alternative physical representations and mappings between conceptual and physical representations are discussed.

1 Introduction

Spatio-temporal data models and query languages have received much attention in the database research community because of their practical importance and the interesting technical challenges they pose. Because of space limitation, we will only discuss previous research that most influenced our approach.

Much previous work focuses on either temporal information or spatial information, rather than both. For instance in the temporal domain, interval-based time models [17] were followed by TSQL2's implicit-time model [21], and point-based time models [22,23]. SQL extensions to express spatial queries were proposed by several authors, including [8] and [9].

In a seminal paper, Worboys [25] defines a spatio-temporal object as a unified object which has both spatial and temporal extents, and is represented by attaching a temporal element to the components of a collection of non-overlapping spatial objects that include points, straight line segments and triangular areas. This model is extended in [5] to allow the vertices of triangles to be linear functions of time. The constraint database framework in [16] is used to characterize the expressiveness of the data model.

A.H.F. Laender, S.W. Liddle, V.C. Storey (Eds.): ER2000 Conference, LNCS 1920, pp. 96–111, 2000.

In [12] and [13], a spatio-temporal data model also based on linear constraints is proposed. The model restricts the orthographic dimension of an object, then it processes queries independently on each dimension of the components. A d-dimensional object is stored as constraints on d variables, with an upper bound on the number of variables that can occur in a single constraint. A spatio-temporal query language based on this model will have to add many new constructs to existing query languages such as SQL.

A topic of growing interest is that of modeling and storing moving objects. A framework for specifying spatio-temporal objects is presented in [6]. The paper defines a number of classes of spatio-temporal objects and studies their closure properties. However, implementation requires the database to store functions as tuple components (function objects). In [4], a model based on parametric rectangles is proposed for representing spatio-temporal objects, where the vertices of triangles are linear functions of time. In [10], a design of moving points and moving regions is discussed, which focuses on generality, closure and consistency, but only discusses the abstract level of modeling. The approach discussed in [10] introduces new spatio-temporal data types, such as *mregion* and *mpoint*, for moving region and moving points, respectively.

In this paper, we propose a minimalist's solution to the problem of supporting spatio-temporal data models and queries in databases, insofar as we want to achieve the desired functionality and expressive power with minimal extensions to current SQL3 standards. Indeed, we want to minimize the effort needed to implement spatio-temporal extensions on the new generation of object relational databases, and thus facilitate the incorporation of such extensions into commercial systems and SQL standards. Toward this goal, we apply several lessons learned with SQL^T, where we were able to support valid-time queries by minimal extensions of SQL [7]—specifically, by extensions that could be supported in current Object-Relational systems with the help of user-defined functions and aggregates [15].

Thus our paper is organized as follows. In the next section we give an overview of our data model SQL^{ST}. Then, in Sections 3 and 4, we introduce the temporal, spatial and spatio-temporal operators supported in SQL^{ST} in a way that is conducive to their implementation through user-defined aggregates or user-defined functions. Then, in Section 5, we show how these operators are used to provide a simple and expressive formulation for complex spatio-temporal queries. The final two sections discuss implementation issues and the opportunities for further research.

2   The Data Model of SQL^{ST}

Many applications only use temporal information, other only use spatial information, finally many use both. Therefore, following Worboys' suggestion [25], we define an SQL^T component that is effective at supporting temporal information, and a SQL^S component that is effective at supporting spatial information; then, we combine the two representations into SQL^{ST} and show that this is effective in supporting spatio-temporal information, including two-dimensional spatial objects whose geometry and shape change with time.

To model time at the conceptual level, we use a point-based time model [22], where information is repeated for each time granule where it is valid, thus eliminating the need for temporal coalescing that besets interval-based approaches [17]. Moreover, temporal aggregates can easily express interval operators in a point-based model [7].

A point-based representation of two-dimensional spatial objects was initially considered for SQL^S, but a polygon-oriented representation was finally selected for two reasons. One is that coalescing is needed much less frequently than in temporal queries. The second is that two dimensional shapes offer a more natural representation for many application domains. For instance a region in a GIS system can be drawn, enlarged, moved, or split; it is hard to associate these behaviors and operations with the points in the region. On the other hand, temporal intervals do not represent any concrete application object; in fact, an interval-based representation is often less appealing than a point-based representation (that models snapshots of reality), or an event-based representation (that models transitions between two successive states of reality). Therefore, SQL^{ST} views reality as a sequence of snapshots of objects that are moving and/or changing in shape.

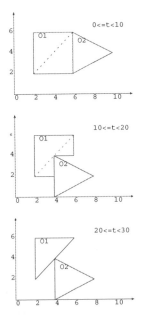

Figure 1 is an example of spatial objects changing with time. At time $t = 0$, there are two spatial objects in the graph, a square $O1$ and a triangle $O2$. At time $t = 10$, $O1$ changes its shape and $O2$ is moved to a new position. At time $t = 20$, $O1$ has some more changes in shape while $O2$ stays unchanged.

An internal representation of the spatio-temporal objects shown in Figure 1 could be as follows:

$(O1, [(2,6), (2,2), (6,2), (6,6)], [0,10))$
$(O2, [(6,6), (6,2), (10,4)], [0,10))$
$(O1, [(2,6), (2,2), (4,2), (4,4), (6,4), (6,6)],$
$\qquad [10,20))$
$(O2, [(4,4), (4,0), (8,2)], [10,20))$
$(O1, [(2,6), (2,2), (6,6)], [20,30))$
$(O2, [(4,4), (4,0), (8,2)], [20,30))$

Here, the regions are represented by a circular list of vertexes, and the time elements are stored as intervals. Mapping method between internal representation and the conceptual model will be discussed in Section 6.

Fig. 1. Graphs representing spatio-temporal data

Table 1 shows how the changes are recorded in the database at the conceptual level. From time $t = 0$ on, the square $O1$ is represented by two triangles and the triangle $O2$ is represented by one triangle. At time $t = 10$, changes of the shape of $O1$ and position of $O2$ occur and their representation also change accordingly. Now $O1$ is represented by three triangles while $O2$ is still represented by one triangle with changed coordinates of the vertexes. This representation is valid from time $t = 10$ till further change occurs. At time $t = 20$, the shape of $O1$ is changed further while $O2$ remains unchanged. Now $O1$ is represented by one triangle and $O2$ stays unchanged.

In Table 1, we also notice that the valid time of each fact is recorded as a time instant VTime.

Table 1. Conceptual model of the spatio-temporal data shown in Figure 1

ID	x_1	y_1	x_2	y_2	x_3	y_3	VTime
$O1$	6	2	2	2	6	6	0
...							...
$O1$	6	2	2	2	6	6	9
$O1$	6	6	2	2	2	6	0
...							...
$O1$	6	6	2	2	2	6	9
$O2$	6	6	6	2	10	4	0
...							...
$O2$	6	6	6	2	10	4	9
$O1$	6	2	2	2	6	6	10
...							...
$O1$	6	2	2	2	6	6	19
$O1$	6	6	4	4	6	4	10
...							...
$O1$	6	6	4	4	6	4	19
$O1$	4	4	2	2	4	2	10
...							...
$O1$	4	4	2	2	4	2	19
$O2$	4	4	4	0	8	2	10
...							...
$O2$	4	4	4	0	8	2	19
$O1$	6	2	2	2	6	6	20
...							...
$O1$	6	2	2	2	6	6	29
$O2$	4	4	4	0	8	2	20
...							...
$O2$	4	4	4	0	8	2	29

In our spatial model we use triangles to represent polygons; a similar approach was proposed in [11,18]. A polygon having n vertexes can be decomposed into $n - 2$ triangles in $O(n \log n)$ TIME. Decomposing a polygon into a set of triangles makes determine spatial relationships between two polygons easy to do. We extended the triangulation method to use sets of directed triangles to represent polygons at the conceptual level. The three edges of a triangle are directed lines and form a counterclockwise circle. The directed triangulation method not only makes testing whether a point is inside a triangle need fewer calculations than the method proposed in [18] but also can handle holes in polygons.

We use user-defined aggregates [24] to support spatial operators such as area, inside, etc., temporal operators such as duration, overlap, etc., and spatio-temporal operators such as moving_distance, etc.

3   Temporal Operators

As has been discussed in [7,15], an important requirement of all temporal languages is to support Allen's interval operators such as overlap, precede, contain, equal, meet, and intersect [1]. Figure 2 shows the meaning of these operators.

Temporal languages that are based on temporal intervals [17] use the overlap operator to express temporal joins. In a point-based temporal model, no explicit use of overlap is needed since two intervals overlap if and only if they share some

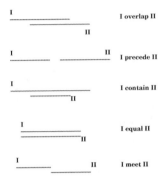

Fig. 2. Allen's interval operators

common points [2,22]. Moreover, since intervals are sets of contiguous points, set aggregates should be used to support interval-oriented reasoning [7]. For instance the duration operator that compute the total time span of a set of intervals is in fact equivalent to a count of time points (but a new name is used in SQL^T to improve its intuitive appeal and to allow its direct and more efficient implementation).

For example, consider the following SQL^T relation:

Example 1. Define the employ relation

```
CREATE TABLE employ
        (name CHAR(30), title CHAR(30), salary DECIMAL(2),
        mgrname CHAR(256), VTime DATE)
```

Then, a query such as: "find employees who for a time had the same manager as Melanie, and show their history during such time" can be expressed as follows:

Example 2. Employees who had the same manager as Melanie and their history during such time.

```
SELECT E2.name, E2.title, E2.VTime
FROM   employ AS E1 E2
WHERE E1.name = "Melanie"
        AND E1.mgrname = E2.mgrname AND E2.name <> "Melanie"
        AND E1.VTime = E2.VTime
```

This example illustrates that project-select-join queries can be expressed in a very natural fashion in SQL^T. The "same time" notion is simply expressed as the equality of the *points in time* in which E1 and E2 worked for the same manager. The same notion in an interval-oriented language would be expressed by a condition stating that the time intervals in which E1 and E2 worked for a same manager overlap. Supporters of interval-based approaches would hereby point that the notion of overlapping periods is quite intuitive; however, detractors would point out that the coalescing issue limits the appeal of this approach.

For instance if we modify Example 2 by dropping E.title from the SELECT clause, then histories of employees who had a change of title while working under Melanie's manager must be consolidated by coalescing their intervals into larger ones. Example 3 below, also discusses this important issue.

The desire of avoid coalescing was a motivation in *implicit valid time* approach followed by TSQL2. In TSQL2, there would be no "**VTime DATE**" column in Example 1, and an annotation will be used instead to denote that this is a valid-time relation, rather than an ordinary non-temporal relation. Likewise, there is no valid time attribute in TSQL2 queries; in fact if we take Example 2 and drop E2.VTime from the SELECT clause, and E1.VTime = E2.VTime from the WHERE clause, we obtain a correct TSQL2 formulation for our query. Therefore, "at the same time" becomes the default interpretation for all queries in TSQL2, and the coalescing operations needed to support this interpretation are implicitly derived by the system.

Consider now a query such as: "Find all the positions held by Melanie for more than 90 days." This can be expressed in SQL^T as follows:

Example 3. Melanie's positions of more than 90 consecutive days.

```
SELECT title
FROM employ
WHERE name = "Melanie"
GROUP BY title
HAVING DURATION(VTime) > 90
```

The fact that the time span of each position must be computed irrespective of the change of manager is expressed naturally and explicitly by the group-by qualifications attached to the new SQL^T aggregate duration. In fact, all period oriented operators, such as contain and precede from Allen's interval algebra, can be expressed naturally by new temporal aggregates. Rather than relying on constructs such as group-bys and aggregates already in SQL, TSQL2 introduces two new constructs to express the same query: one is the "snapshot" annotation in the "select" clause, and the other is the restructuring construct in the "from" clause. Furthermore, TSQL2 constructs cannot be extended to languages such as QBE or Datalog that do not have select clauses and from clauses— a lack of robustness called "lack of universality" in [7]. More sample queries of temporal queries using user-defined aggregates can be found in [7,15]

An efficient implementation for SQL^T called TEnORs (for Time Enhanced Object Relational system) is available for DB2 [15]. In TEnORs, the point-based representation for valid time is mapped into a modified internal model based on intervals that are segmented, indexed and allocated to temporal blocks to optimize temporal clustering and storage utilization. User defined aggregates [24] are then used to map the queries expressed on the point-based model into equivalent queries expressed on the segmented interval-based model. The design and implementation of SQL^{ST} discussed in this paper apply and extend to the spatio-temporal domain the lessons learned with TEnORs.

4   Spatial Operators

Our representation method of a spatial object is based on directed triangulation. The three basic spatial data types, i.e., points, lines (straight line segments), regions (polygons), are represented by triangles at the conceptual level as follows:

- point (x, y): is represented as $((x, y), (x, y), (x, y))$.
- line $((x_1, y_1), (x_2, y_2))$: is represented as $((x_1, y_1), (x_2, y_2), (\frac{x_1+x_2}{2}, \frac{y_1+y_2}{2}))$, where (x_1, y_1) is the start point and (x_2, y_2) is the end point of the line, and $(\frac{x_1+x_2}{2}, \frac{y_1+y_2}{2})$ is the center of mass of the line.
- region $[(x_1, y_1), (x_2, y_2), (x_3, y_3) \ldots, (x_{n-1}, y_{n-1}), (x_n, y_n)]$: is represented as a set of directed triangles, i.e., $\{((x_1, y_1), (x_2, y_2), (x_3, y_3)), \ldots, ((x_{n-2}, y_{n-2}), (x_{n-1}, y_{n-1}), (x_n, y_n))\}$. The region is represented as a circular list at the physical level and the line with two ends as (x_i, y_i) and (x_{i+1}, y_{i+1}) is an edge of the region. The algorithm to decompose a region into a set of triangles will be discussed in Section 6. For each triangle the region is decomposed into, the three vertexes are ordered according to a counterclockwise orientation.

If a region has a hole, the vertexes of the hole also form a circular list but prefixed with a negative sign. And the vertexes of each triangle in the set that the hole is decomposed into are clockwisely orientated.

The commonly used spatial predicates [19,5] are equal, disjoint, overlap, meet, contain, adjacent and common_border, etc. Spatial operations include intersect, area, perimeter, distance, etc.

4.1   Properties of Directed Triangles

First, we define the direction of border lines of a triangle as *counterclockwise*.

Definition 1: a triangle is a **counterclockwisely directed triangle** if its three vertexes, point1 (x_1, y_1), point2 (x_2, y_2), and point3 (x_3, y_3) are *counterclockwisely* orientated, i.e.,

$$\begin{vmatrix} x_1 & y_1 & 1 \\ x_2 & y_2 & 1 \\ x_3 & y_3 & 1 \end{vmatrix} > 0$$

Next, we define the following basic spatial predicates. According to [5], these are first-order queries with linear arithmetic constraints.

- **left(point, line):** a point (x_0, y_0) is on the *left* side of a line $((x_1, y_1), (x_2, y_2))$, which is an edge of a directed triangle, iff

$$\begin{vmatrix} x_1 & y_1 & 1 \\ x_2 & y_2 & 1 \\ x_0 & y_0 & 1 \end{vmatrix} \geq 0$$

A point is considered to be on the *left* side of a line if it is on the (extended) line, i.e., if the above determinant equals zero.

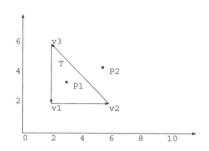

Fig. 3. An example of counterclock-wisely directed triangle

Figure 3 is an example of a directed triangle and the meanings of left and inside operators are showed. The vertexes of the triangle T — $v1$, $v2$ and $v3$ are *counterclockwisely* orientated. The edges $v1 \rightarrow v2$, $v2 \rightarrow v3$ and $v3 \rightarrow v1$ are directed lines and form the counterclockwisely directed triangle T. For example, point $P1$ is on the *left* side of all three edges, so $P1$ is *inside* T. On the other hand, point $P2$ is on the *left* side of edges $v1 \rightarrow v2$ and $v3 \rightarrow v1$ but not on the *left* side of edge $v2 \rightarrow v3$, so $P2$ is not *inside* T.

- **inside(point, triangle):** a point is *inside* a triangle iff the point is on the *left* side of all three edges of the triangle.
- **vertex(point, triangle)** — iff the point is one of the vertexes of the triangle
- **on(point, line)** — iff $\begin{vmatrix} x_0 & y_0 & 1 \\ x_1 & y_1 & 1 \\ x_2 & y_2 & 1 \end{vmatrix} = 0$ and $(min(x_1, x_2) \le x_0 \le max(x_1, x_2)$
  or $min(y_1, y_2) \le y_0 \le max(y_1, y_2))$
- **boundary(point, triangle)** — iff the point is on an edge of the triangle
- **equal(line1, line2)** — iff the set of the end points of line1 is equal to the set of the end points of line2
- **overlaps(line1, line2)** — iff for two lines, line1 $((x_1, y_1), (x_2, y_2))$ and line2 $((x_1', y_1'), (x_2', y_2'))$, $\frac{y_2 - y_1}{x_2 - x_1} = \frac{y_2' - y_1'}{x_2' - x_1'}$ and (x_1, y_1) is on line2.
- **boundary(line, triangle)** — iff the line overlaps with an edge of the triangle
- **edge(line, triangle)** — iff the two end points of the line are two neighboring vertexes of the triangle.
- **crosspoint(line1, line2)** — the cross point of two lines, line1 $((x_1, y_1), (x_2, y_2))$ and line2 $((x_1', y_1'), (x_2', y_2'))$ is point0 (x, y) with the coordinates as:

$$x = \frac{(x_1 - x_2)\begin{vmatrix} x_1' & y_1' \\ x_2' & y_2' \end{vmatrix} - (x_1' - x_2')\begin{vmatrix} x_1 & y_1 \\ x_2 & y_2 \end{vmatrix}}{(y_1 - y_2)(x_1' - x_2') - (y_1' - y_2')(x_1 - x_2)}$$

$$y = \frac{(y_1 - y_2)\begin{vmatrix} x_1' & y_1' \\ x_2' & y_2' \end{vmatrix} - (y_1' - y_2')\begin{vmatrix} x_1 & y_1 \\ x_2 & y_2 \end{vmatrix}}{(y_1 - y_2)(x_1' - x_2') - (y_1' - y_2')(x_1 - x_2)}$$

and $(min(x_1, x_2) \le x \le max(x_1, x_2)$ or $min(y_1, y_2) \le y \le max(y_1, y_2))$. When there is no solution to the above equations, we say point0 = *null*.
- **cross(line1, line2)** — iff the cross point of the two lines is not *null*

4.2   Spatial Relationships

Based on the basic operators discussed in the above section, we can compute relationships between triangles as follows:

- **equal(triangle1, triangle2)** — iff the set of the vertexes of triangle1 are equal to the set of the vertexes of triangle2
- **overlap(triangle1, triangle2)** — iff at least one vertex of triangle2 is *inside* triangle1
- **contain(triangle1, triangle2)** — iff the three vertexes of triangle2 are all *inside* triangle1
- **disjoint(triangle1, triangle2)** — iff non of the vertexes of triangle1 is *inside* triangle2 and vice versa; and non of the edges of triangle1 *crosses* with any edge of triangle2
- **adjacent(triangle1, triangle2)** — iff one edge of triangle1 *overlaps* with an edge of triangle2 and at least one vertex of triangle1 is not *inside* triangle2
- **commonborder(triangle1, triangle2)** — iff an edge of triangle1 is *equal* to an edge of triangle2 and one vertex of triangle1 is not *inside* triangle2
- **meet(triangle1, triangle2)** — iff one vertex of triangle1 is *on* an edge of triangle2 and two vertexes of triangle1 are not *inside* triangle2

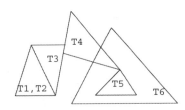

Fig. 4. Example of Relationships Between Triangles

Figure 4 illustrates the definition of these relationships between two triangles. For example, (i) $T1$ and $T2$ which have the same set of vertexes are *equal* to each other. (ii) $T1$ and $T3$ share a *commonborder*, so do $T2$ and $T3$; (iii) $T1$, $T2$, and $T3$ are *disjoint* with $T5$ and $T6$, also $T1$ and $T2$ are *disjoint* with $T4$; (iv) $T3$ and $T4$ are *adjacent* because an edge of $T3$ overlaps with an edge of $T4$ and two vertexes of $T3$ are not inside $T4$; (v) $T4$ *meets* $T5$ because one vertex of $T4$ is on an edge of $T5$ and the other two vertexes of $T4$ are not inside $T5$; (vi) $T6$ *overlaps* $T4$ because one vertex of $T4$ is inside $T6$; and (vii) $T6$ *contains* $T5$ because all three vertexes of $T5$ are inside $T6$.

Furthermore, the relationships between these spatial operators are:

$$
\begin{array}{rcl}
\text{equal} & \Longrightarrow & \text{contain} \\
\text{equal} & \Longrightarrow & \text{overlap} \\
\text{contain} & \Longrightarrow & \text{overlap} \\
\text{adjacent} & \Longrightarrow & \text{overlap} \\
\text{commonborder} & \Longrightarrow & \text{overlap} \\
\text{meet} & \Longrightarrow & \text{overlap} \\
\text{commonborder} & \Longrightarrow & \text{adjacent}
\end{array}
$$

For example, if equal(triangle1, triangles2) evaluates to TRUE, then it implies that contain(triangle1, triangle2) also evaluates to TRUE.

4.3 Spatial Operations

The main operations associated with spatial objects are as follows:

- **intersect(triangle1, triangle2, region)** — calculates the intersection form- ed by the cross points of the edges of the two triangles. The steps are showed in Algorithm 1 where $t1$ and $t2$ are the two input triangles and L is the circular list of the vertexes of the intersected region.

Algorithm 1 intersect two triangles into a region
Require: initially $t1$, $t2$, $L = \emptyset$.
1: **for** each edge $e1$ of $t1$ **do**
2:    **for** each edge $e2$ of $t2$ **do**
3:       $point0 = \text{crosspoint}(e1, e2)$
4:       **if** $point0 \neq null$ **then**
5:          append $point0$ to L
6:       **end if**
7:    **end for**
8: **end for**
9: **return** L

A triangle $((x1, y1), (x2, y2), (x3, y3))$ has the following properties [3]:

- **area** $a = \frac{1}{2} \begin{vmatrix} x_1 & y_1 & 1 \\ x_2 & y_2 & 1 \\ x_3 & y_3 & 1 \end{vmatrix}$

- **perimeter** $p = \sqrt{(x_2 - x_1)^2 + (y_2 - y_1)^2} + \sqrt{(x_3 - x_2)^2 + (y_3 - y_2)^2} + \sqrt{(x_1 - x_3)^2 + (y_1 - y_3)^2}$

- **centerofmass** $x_c = \frac{x_1 + x_2 + x_3}{3}$, $y_c = \frac{y_1 + y_2 + y_3}{3}$

The distance between two triangles are defined as the distance between their centers of mass:

- **distance** $d = \sqrt{(x_{2c} - x_{1c})^2 + (y_{2c} - y_{1c})^2}$

5 Spatio-Temporal Queries

In this section, we express various queries in SQL^{ST}. We use examples taken from [10].

The database contains three relations: (i) forest relation in which the location and the development of forests changing over time are recorded; (ii) forest_fire relation in which the evolution of forest fires are recorded; and (iii) fire_fighter relation which records the motion of fire fighters.

First, we define the schema in SQL^{ST} as follows:

Example 4. Define the forest relation

    CREATE TABLE forest (forestname CHAR(30), territory REGION, VTime DAY)

Example 5. Define the forest_fire relation

    CREATE TABLE forest_fire (firename CHAR(30), extent REGION, VTime DAY)

Example 6. Define the fire_fighter relation

    CREATE TABLE fire_fighter (fightername CHAR(30), location POINT, VTime DAY)

The columns *territory* and *extent* have a spatial data type as REGION and *location* has a type as POINT; temporal data column *VTime* has a granularity of DAY.

Example 7. When and where did the fire called "The Big Fire" reach what largest extent?

    SELECT F1.VTime, F2.extent, AREA(F1.extent)
    FROM forest_fire as F1 F2
    WHERE F1.firename = "The Big Fire" AND F2.firename = "The Big Fire"
        AND F1.VTime = F2.VTime
    GROUP BY F1.VTime
    HAVING AREA(F1.extent) = (SELECT MAX(AREA(extent))
                        FROM forest_fire WHERE firename = "The Big Fire")

We only use an user-defined spatial aggregate **area** and a built-in aggregate **max** to express the query. We group-by F1.VTime to calculate the area of the fire extent at each time instant. On the contrary, [10] introduces several special constructs plus a new clause LET to accomplish the same query.

Example 8. When and where was the spread of fires larger than 500 km^2?

    SELECT F1.VTime, F2.extent
    FROM forest_fire as F1 F2
    WHERE F1.VTime = F2.VTime AND F1.firename = F2.firename
    GROUP BY F1.VTime, F2.extent, F1.firename
    HAVING AREA(F1.extent) > 500

Example 9. How long was fire fighter Th. Miller enclosed by the fire called "The Big Fire" and which distance did he cover there?

```
SELECT DURATION(fire_fighter.VTime),
       MOVING_DISTANCE(fire_fighter.location, fire_fighter.VTime)
FROM forest_fire, fire_fighter
WHERE forest_fire.VTime = fire_fighter.VTime
   AND firename = "The Big Fire" AND fightername = "Th. Miller"
GROUP BY forest_fire.VTime
HAVING INSIDE(location, extent)
```

Example 10. Determine the times and locations when "The Big Fire" started.

```
SELECT VTime, extent
FROM forest_fire
WHERE firename = "The Big Fire"
   AND VTime = (SELECT MIN(VTime)
                FROM forest_fire WHERE firename = "The Big Fire")
```

These examples suggest that SQL^{ST} (i) provides a simple and expressive formulation for complex spatio-temporal queries, and (ii) represents a minimalist's extensions for SQL that preserves the syntax and the flavor of SQL3, since only new functions and aggregates are required.

6   Implementation of SQL^{ST}

As we have discussed before, at the conceptual level and physical level, the spatio-temporal objects should have different data models. At conceptual level, we have used point-based time model and directed-triangulation-based spatial data model. At physical level, an interval-based time model is used, and the same approach should be taken for spatial data, i.e., a region should be stored by its vertexes. Mapping methods between the conceptual level and the physical level representation is thus needed.

6.1   Mapping between Different Representations

Mapping to an interval-based time model at the physical level solves the space efficiency problem associated with the point-based time model used at the conceptual level. Then, tuples in our internal relations are timestamped with two time instants: one indicates the start-point and the other indicates the end-point of the interval.

Each temporal join is mapped into an intersect operation. A coales aggregation is required on the temporal argument when various columns have been projected. However, coalescing is not needed for those rules where (i) there is no temporal argument in the SELECT clause of a query, or (ii) all variables appearing in the WHERE clause of a query also appear in its SELECT clause. [15]

To map a set of triangles which have the same timestamp into one region, we use Algorithm 2. In Algorithm 2, the input T is a set of triangles and the output L is a circular list of the vertexes of the region that the triangles are merged into.

Algorithm 2 map a set of triangles into a region

Require: initially T, $L = \emptyset$.
1: **for** each triangle $t1 = (v1, v2, v3) \in T$ **do**
2:    **for** each triangle $t2 = (v4, v5, v6) \in T$ **do**
3:       **if** commonborder$(t1, t2)$ **then**
4:          $v = \{v1, v2, v3\} - \{v4, v5, v6\}$
5:          $v' = \{v4, v5, v6\} - \{v1, v2, v3\}$
6:          $v_s \in \{v1, v2, v3\} - \{v\}$
7:          $v_e \in \{v1, v2, v3\} - \{v\} - \{v_s\}$
8:          **if** left$(v, (v_s, v_e))$ and $\neg$ left$(v', (v_s, v_e))$ **then**
9:             append $[v, v_s, v', v_e]$ to L
10:         **end if**
11:      **end if**
12:   **end for**
13: **end for**
14: return L

To decompose a region R into a set of directed-triangles T, we use Algorithm 3 where the input and output are just opposite to that of Algorithm 2.

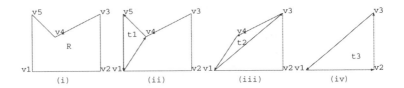

Fig. 5. Decomposing a region into triangles

Figure 5 is an example of decomposing a region R into set of triangles T. Step (i) showed the region with $v1, v2, v3, v4$ and $v5$ as its vertexes. So, initially, $L = [v1, v2, v3, v4, v5]$ and $T = \emptyset$.

In step (ii), we start with $v1$ which has the smallest x and y values, and get $v2 = next(v1)$ and $v5 = previous(v1)$. Since $v4$ is inside the triangle $(v1, v2, v5)$ so we move on to start with $v5$ and get $v1 = next(v5)$ and $v4 = previous(v5)$. Since no other vertex of R is inside the triangle $(v5, v1, v4)$ (namely $t1$) so we set $T = \{t1\}$, $L = [v1, v2, v3, v4]$.

Algorithm 3 decompose a region into a set of triangles

Require: initially L, $T = \emptyset$.

 1: **for** each element $v \in L$ **do**
 2:   **if** $x_v = \min(\text{all } x_{vi})$ and $y_v = \min(\text{all } y_{vi})$ **then**
 3:     $v0 = v$
 4:   **end if**
 5: **end for**
 6: $v1 = \text{next}(v0)$ and $v2 = \text{previous}(v0)$
 7: **if** length$(L) > 3$ **then**
 8:   **for** each element $v' \in L$ **do**
 9:     **if** inside$(v', (v0, v1, v2))$ and $v' \notin \{v0, v1, v2\}$ **then**
10:       $v0 = v2$
11:       goto line 6
12:     **end if**
13:   **end for**
14:   insert triangle $(v0, v1, v2)$ into T
15:   remove $v0$ from L
16: **end if**
17: insert triangle $(v0, v1, v2)$ into T
18: return T

In step (iii), we start with $v4$, and get a triangle $(v4, v1, v3)$. Since no other vertex of R is inside this triangle (namely $t2$), so we set $T = \{t1, t2\}$, $L = [v1, v2, v3]$.

Lastly, we get the triangle $(v3, v1, v2)$ (namely $t3$), and set $T = \{t1, t2, t3\}$. The procedure stops.

6.2   Spatial Operators for Regions

The spatial operators defined for triangles in Section 4 can be used towards regions with little change.

Let S and S' denote sets of triangles that two regions $R1$ and $R2$ are decomposed into, and t and t' denote an element in S and S', respectively. The operators proposed in Section 4 can be used on two regions in the following ways:

1. $R1$ is equal to $R2$ if $S = S'$.
2. $R1$ overlaps $R2$ if $\exists t \in S$, $t' \in S'$, such that t *overlaps* t'.
3. $R1$ contains $R2$ if $\forall t' \in S'$, $\forall$vertexes v of t', $\exists t \in S$, such that v is *inside* t.
4. $R1$ is disjoint with $R2$ if $\forall t \in S$, $\neg \exists t' \in S'$, such that t *overlaps* t'.
5. $R1$ is adjacent with $R2$ if $\exists t \in S$, $\exists t' \in S'$ where t is *adjacent* with t' and the two ends of the adjacent edge are *neighbouring vertexes* of both $R1$ and $R2$.
6. $R1$ and $R2$ have a commonborder if $\exists t \in S$, $\exists t' \in S'$ where t and t' have a *common border* and the two ends of the common border are *neighboring vertexes* of both $R1$ and $R2$.
7. $R1$ meets $R2$ if $\exists t \in S$, $\exists t' \in S'$, such that t *meets* t' and $\forall t1 \in S$, $\forall t1' \in S'$, $t1 \neq t$, $t1$ does not *overlap* $t1'$.

Since a region with n vertexes can be decomposed into $n - 2$ triangles while the relationships between two triangles can be determined in constant time, so the comparison of two regions can be done in $O(n^2)$ TIME.

To find the intersected part of two regions, we only need to find the intersected parts of the triangles decomposed from them. Similarly, The area of a region is simply the sum of the areas of all the triangles it decomposed into. The perimeter of a region is the sum of the length of all its edges. The center of mass of a region can be calculated in a similar way to the calculation of the center of mass of a triangle. The distance between two regions is also defined as the distance between their centers of mass.

7   Conclusion

In this paper, we propose a spatio-temporal data model and query language — SQL^{ST} that satisfies Worboys' prescription for spatio-temporal model. A cornerstone to the simplicity and generality of this approach, is the use of a point-based representation of time and directed-triangulation-based representation of spatial objects at the conceptual level. Whereas query languages proposed in the past rely on the introduction of new spatio-temporal constructs, we have taken a minimalist's approach, and showed that the basic syntactic constructs and semantic notions provided by current query languages are sufficient if user-defined functions and user-defined aggregates are supported [24]. With these minimal extensions, SQL^{ST} can express queries as powerful as those expressible in other works [14,10] in a simple and intuitive fashion.

Efficient support for spatio-temporal queries can be obtained by using internal representations that are different from the conceptual one, and then mapping conceptual queries into equivalent queries on the internal representations. This approach has already produced an efficient implementation for SQL^T. In this paper we have laid the foundations for efficient implementations of spatial structures by using directed triangular representations and defining equivalent operators on general polygons in terms of these. The subject of efficient implementation for objects whose shape or position continuously changes with time has been left for later research.

References

1. J.F. Allen. Maintaining knowledge about temporal intervals. In *Communications of the ACM*, Vol.26, No.11, pp.832-843, 1983
2. M.H. Bohlen, R. Busatto and C.S. Jensen. Point- Versus Interval-based Temporal Data Models. In *Proceedings of the 14th International Conference on Data Engineering*, pp.192-200, 1998
3. Y.S. Bugrov. *Fundamentals of linear algebra and analytical geometry*, Moscow : Mir Publishers, 1982
4. M. Cai, D. Keshwani, and P.Z. Revesz. Parametric Rectangles: A Model for Querying and Animation of Spatiotemporal Databases. In *Proceedings of the 7th International Conference on Extending Database Technology*, pp.430-444, 2000
5. J. Chomicki and P.Z. Revesz. Constraint-Based Interoperability of Spatiotemporal Databases. In *Advances in Spatial Databases*, LNCS 1262, pp.142-161, Springer, 1997

6. J. Chomicki and P.Z. Revesz. A Geometric Framework for Specifying Spatiotemporal Objects. In *Proceedings of the 6th International Workshop on Time Representation and Reasoning*, pp.1-46, 1999
7. C.X. Chen and C. Zaniolo. Universal Temporal Extensions for Data Languages. In *Proceedings of the 15th International Conference on Data Engineering*, pp.428-437, 1999
8. M.J. Egenhofer. Spatial SQL: A Query and Presentation Language. In *IEEE Transactions on Knowledge and Data Engineering* Vol.6. No.1. pp.86-95, 1994
9. R.H. Guting and M. Schneider. Realm-Based Spatial Data Types: The ROSE Algebra. In *VLDB Journal*, Vol.4, No.2, pp.243-286, 1995
10. R.H. Guting, M.H. Bohlen, M. Erwig, C.S. Jensen, N.A. Lorentzos, M. Schneider, and M. Vazirgiannis. A Foundation for Representing and Querying Moving Objects. To appear in *ACM Transactions on Database Systems*
11. M.R. Garey, D.S. Johnson, F.P. Preparata, and R.E. Tarjan. Triangulating a Simple Polygon. In *Information Processing Letters*, Vol.7, No.4, pp.175-179, 1978
12. S. Grumbach, P. Rigaux and L. Segoufin. Spatio-Temporal Data Handling with Constraints. In *Proceedings of ACM International Symposium on Geographic Information Systems*, pp.106-111, 1998
13. S. Grumbach, P. Rigaux and L. Segoufin. The DEDALE System for Complex Spatial Queries. In *Proceedings of ACM-SIGMOD International Conference on Management of Data*, pp.213-224, 1998
14. C. S. Jensen and R. T Snodgrass. Temporal Data Management. In *IEEE Transactions on Knowledge and Data Engineering*, Vol.11, No.1, pp.36-44, 1999
15. J. Kong, C.X. Chen, and C. Zaniolo. A Temporal Extension of SQL for Object Relational Databases. *submitted for publication*
16. P.C. Kanellakis, G. Kuper and P.Z. Revesz. Constraint Query Languages. In *Journal of Computer and System Sciences*, special issue edited by Y.Sagiv, Vol.51, No.1, pp.26-52, 1995
17. N.A. Lorentzos and Y.G. Mitsopoulos. SQL Extension for Interval Data. In *IEEE Transactions on Knowledge and Data Engineering, Vol.9, No.3*, pp.480-499, 1997
18. R. Laurini and D. Thompson. *Fundamentals of Spatial Information Systems*. Academic Press, 1992
19. M. Schneider. *Spatial Data Types for Database Systems*, LNCS 1288. Springer, 1997
20. R.T. Snodgrass. The Temporal Query Language TQuel. In *Proceedings of the 3rd ACM SIGACT-SIGMOD-SIGART Symposium on Principles of Database Systems* pp.204-213, 1984
21. R.T. Snodgrass, et al. *The TSQL2 Temporal Query Language*, Kluwer, 1995
22. D. Toman. Point vs. Interval-based Query Languages for Temporal Databases. In *Proceedings of the 15th ACM SIGACT-SIGMOD-SIGART Symposium on Principles of Database Systems*, pp.58-67, 1996
23. D. Toman. A Point-Based Temporal Extension of SQL. In *Proceedings of the 6th International Conference on Deductive and Object-Oriented Databases*, pp.103-121, 1997
24. H. Wang and C. Zaniolo. User Defined Aggregates in Object-Relational Systems. In *Proceedings of the 16th International Conference on Data Engineering*, pp.135-144, 2000
25. M.F. Worboys. A Unified Model for Spatial and Temporal Information. *Computer Journal*, Vol.37, No.1, pp.26-34, 1994
26. C. Zaniolo, S. Ceri, C. Faloutsos, R. Snodgrass, and R. Zicari. *Advanced Database Systems*, Morgan Kaufmann, 1997

TBE: Trigger-By-Example*

Dongwon Lee, Wenlei Mao, and Wesley W. Chu

Department of Computer Science
University of California, Los Angeles
Los Angeles, CA 90095, USA
{dongwon,wenlei,wwc}@cs.ucla.edu

Abstract. TBE (Trigger-By-Example) is proposed to assist users in writing trigger rules. TBE is a graphical trigger rule specification language and system to help users understand and specify active database triggers. Since TBE borrowed its basic idea from QBE, it retained many benefits of QBE while extending the features to support triggers. Hence, TBE is a useful tool for novice users to create simple trigger rules easily. Further, since TBE is designed to insulate the details of underlying trigger systems from users, it can be used as a universal trigger interface for rule formation.

1 Introduction

Triggers provide a facility to autonomously react to events occurring on the data, by evaluating a data-dependent condition, and by executing a reaction whenever the condition evaluation yields a truth value. Such triggers have been adopted as an important database feature and implemented by most major database vendors. Despite their diverse potential usages, one of the obstacles that hinder the triggers from its wide deployment is the lack of tools that aid users to create complex trigger rules in a simple manner. In many environments, the correctness of written trigger rules is very crucial since the semantics encoded in the trigger rules are shared by many applications [18]. Although the majority of the users of triggers are DBAs or savvy end-users, writing *correct* and *complex* trigger rules is still a daunting task.

On the other hand, QBE (Query-By-Example) has been very popular since its introduction decades ago and its variants are currently being used in most modern database products. As it is based on domain relational calculus, its expressive power has proved to be equivalent to that of SQL which is based on tuple relational calculus [2]. As opposed to SQL where users have to conform to the phrase structure strictly, QBE users may enter any expression as an entry insofar as it is syntactically correct. That is, since the entries are bound to the table skeleton, the user can only specify admissible queries [17].

In this paper, we propose to use the established QBE as a user interface for writing trigger rules. Since most trigger rules are complex combinations of SQL

---

* This research is supported in part by DARPA contract No. N66001-97-C-8601 and SBIR F30602-99-C-0106.

A.H.F. Laender, S.W. Liddle, V.C. Storey (Eds.): ER2000 Conference, LNCS 1920, pp. 112–125, 2000.

statements, by using QBE as a user interface for triggers, the user may create only admissible trigger rules. The main idea is to use QBE in a *declarative* fashion for writing the *procedural* trigger rules [6].

2 Background and Related Work

SQL3 Triggers: In SQL3, *triggers*, sometimes called *event-condition-action rules* or *ECA rules*, mainly consist of three parts to describe the event, condition, and action, respectively. Since SQL3 is still evolving at the time of writing this paper, albeit close to its finalization, we base our discussion on the latest ANSI X3H2 SQL3 working draft [13].

QBE (Query-By-Example): QBE is a query language as well as a visual user interface. In QBE, *programming* is done within two-dimensional skeleton tables. This is accomplished by filling in an example of the answer in the appropriate table spaces (thus the name "by-example"). Another kind of two-dimensional object is the *condition box*, which is used to express one or more desired conditions difficult to express in the skeleton tables. By QBE convention, variable names are lowercase alphabets prefixed with "_", system commands are uppercase alphabets suffixed with ".", and constants are unquoted. Let us see a QBE example. We use the following schema throughout the paper.

Example 1. Define the emp and dept relations with keys underlined. emp.DeptNo and dept.MgrNo are foreign keys referencing to dept.Dno and emp.Eno attributes, respectively.

emp(**Eno**, **Ename, DeptNo, Sal**),   dept(**Dno**, **Dname, MgrNo**)

Then, Example 2 shows two equivalent representations of the query in SQL3 and QBE, respectively.

Example 2. Who is being managed by the manager 'Tom'?

SELECT **E2.Ename**
FROM emp **E1**, emp **E2**, dept **D**
WHERE **E1.Ename** = 'Tom' AND **E1.Eno** = **D.MgrNo**
        AND **E2.DeptNo** = **D.Dno**

emp	Eno	Ename	DeptNo	Sal
	_e	Tom		
		P.	_d	

dept	Dno	Dname	MgrNo
	_d		_e

Related Work. Past active database research has focused on active database rule language (e.g., [1]), rule execution semantics (e.g., [6]), or rule management and system architecture issues (e.g., [15]). In addition, research on visual querying has been done in traditional database research (e.g., [7,17]). To a greater or lesser extent, all these research focused on devising novel visual querying schemes to replace data retrieval aspects of SQL language. Although some have considered data definition aspects [3] or manipulation aspects, none have extensively considered the *trigger* aspects for SQL, especially from the user interface point of view.

Other works (e.g., IFO_2 [16], IDEA [5]) have attempted to build graphical triggers description tools. Using IFO_2, one can describe how different objects interact through events, thus giving priority to an overview of the system. Argonaut from the IDEA project [5] focused on the automatic generation of active rules that correct integrity violations based on declarative integrity constraint specifications, and active rules that incrementally maintain materialized views based on view definitions. TBE, on the other hand, helps users to *directly* design active rules with minimal learning.

Other than QBE skeleton tables, *forms* have been popular building blocks for visual querying mechanisms as well. For instance, [7] proposes the NFQL as a communication language between humans and database systems. It uses forms in a strictly nonprocedural manner to represent queries. Other works using forms are mostly for querying aspect of the visual interface [3].

To the best of our knowledge, the only work that is directly comparable to ours is RBE [4]. Although RBE also uses the idea of QBE as an interface for creating trigger rules, there are the following differences:

- Since TBE is carefully designed with SQL3 triggers in mind, it is capable of creating all the complex SQL3 trigger rules. Since RBE's capability is, however, limited to OPS5-style production rules, it cannot express, for instance, the subtle difference of the trigger activation time nor granularity.
- No evident suggestion of RBE as a user interface for writing triggers is given. On the other hand, TBE is specifically aimed for that purpose.
- The implementation of RBE is tightly coupled with the underlying rule system and database so that it cannot easily support multiple heterogeneous database triggers. Since TBE implementation is a thin layer utilizing a translation from a visual representation to the underlying triggers, it is loosely coupled with the database.

The organization of this paper is as follows. Section 3 proposes our TBE for SQL3 triggers and discusses several related issues. A few complex SQL3 trigger examples are illustrated in Section 4. The preliminary implementation and potential applications of TBE are presented in Sections 5 and 6, respectively. Concluding remarks are given in Section 7.

3   TBE: Trigger-By-Example

We propose to use QBE as a user interface for writing trigger rules. Our tool is called Trigger-By-Example (TBE) which has the same spirit as that of QBE. The philosophy of QBE is to require the user to know very little in order to get started and to minimize the number of concepts that he or she subsequently has to learn to understand and use the whole language [17]. By using QBE as an interface, we attain the same benefits for creating trigger rules.

3.1   Difficulty of Expressing Procedural Triggers in Declarative QBE

Triggers in SQL3 are procedural in nature. Trigger actions can be arbitrary SQL procedural statements, allowing not only SQL data statements (i.e., select, project, join) but also transaction, connection, session statements[1]. Also, the order among action statements needs to be obeyed faithfully to preserve the correct semantics. On the contrary, QBE is a declarative query language. While writing a query, the user does not have to know if the first row in skeleton tables needs to be executed before the second row or not. That is, the order is immaterial. Also QBE is specifically designed as a tool for only 1) data retrieval queries (i.e., SELECT), 2) data modification queries (i.e., INSERT, DELETE, UPDATE), and 3) schema definition and manipulation queries. Therefore, QBE cannot really handle other procedural SQL statements such as transaction or user-defined functions in a simple manner. Thus, our goal is to develop a tool that can represent the *procedural* SQL3 triggers in its entirety while retaining the *declarative* nature of QBE as much as possible.

In what follows, we shall describe how QBE was extended to be TBE, what design options were available, and which option was chosen by what rationale, etc.

3.2   Trigger Name

A unique name for each trigger rule needs to be set in a special input box, called the *name box*, where the user can fill in an arbitrary identifier as shown below:

$$\boxed{<\text{TriggerRuleName}>}$$

Typically, the user first decides the trigger name and then proceeds to the subsequent tasks. There are often cases when multiple trigger rules are written together in a single TBE query. For such cases, the user needs to provide a unique trigger name for each rule in TBE query separately. In what follows, when there is only a single trigger rule in the example, we take the liberty of not showing the trigger name for briefness.

3.3   Event-Condition-Action Triggers

SQL3 triggers use the ECA model. Therefore, triggers are represented by mainly three isolated E, C, A parts. In TBE, each E, C, A part maps to the corresponding skeleton tables separately. To differentiate among three parts, three prefix flags, E., C., A., are introduced. That is, in skeleton tables, table name is prefixed with one of these flags. The condition box in QBE is also similarly extended. For instance, a condition statement is specified in the C. prefixed skeleton table and condition box below.

C.emp	Eno	Ename	DeptNo	Sal

C.conditions

---

[1] SQL3 triggers definition in [13] leaves it implementation-defined whether the transaction, connection, or session statements should be contained in the action part or not.

3.4 Triggers Event Types

SQL3 triggers allow only the INSERT, DELETE, and UPDATE as legal event types. Coincidentally, QBE has constructs I., D., and U. for each event type to describe the data manipulation query. The TBE uses these constructs to describe the trigger event types. Since the INSERT and DELETE always affect the whole tuple rather than individual columns, I. and D. must be filled in the leftmost column of skeleton table. When the UPDATE trigger is described as to particular column, then U. is filled in the corresponding column. Otherwise, U. is filled in the leftmost column. Consider the following example.

Example 3. Skeleton tables (1) and (2) depict INSERT and DELETE events on the dept table, respectively. (3) depicts UPDATE event of columns Dname and MgrNo. Thus, changes occurring on other columns do not fire the trigger. (4) depicts UPDATE event of any columns on the dept table.

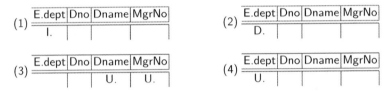

Note that since SQL3 triggers definition limits that only a *single* event be monitored per *single* rule, there can not be more than one row having I., D., or U. flag unless multiple trigger rules are written together. Therefore, same trigger actions for different events (e.g., "abort when either INSERT or DELETE occurs") need to be expressed as separate trigger rules in SQL3 triggers.

3.5 Triggers Activation Time and Granularity

The SQL3 triggers have a notion of the *event activation time* that specifies if the trigger is executed before or after its event and the *granularity* that defines how many times the trigger is executed for the particular event.

1. The activation time can have two modes, *before* and *after*. The *before* mode triggers execute before their event and are useful for conditioning the input data. The *after* mode triggers execute after their event and are typically used to embed application logic [6]. In TBE, two corresponding constructs, BFR. and AFT., are introduced to denote these modes. The "." is appended to denote that these are built-in system commands.
2. The granularity of a trigger can be specified as either *for each row* or *for each statement*, referred to as *row-level* and *statement-level* triggers, respectively. The row-level triggers are executed after each modification to tuple whereas the statement-level triggers are executed once for an event regardless of the number of the tuples affected. In TBE notation, R. and S. are used to denote the row-level and statement-level triggers, respectively.

Consider the following illustrating example.

Example 4. SQL3 and TBE representation for a trigger with *after* activation time and *row-level* granularity.

CREATE TRIGGER **AfterRowLevelRule**
AFTER UPDATE OF **Ename, Sal** ON **emp** FOR EACH ROW

E.emp	Eno	Ename	DeptNo	Sal
AFT.R.		U.		U.

3.6   Transition Values

When an event occurs and values change, trigger rules often need to refer to the *before* and *after* values of certain attributes. These values are referred to as the *transition values*. In SQL3, these transition values can be accessed by either transition variables (i.e., OLD, NEW) or tables (i.e., OLD_TABLE, NEW_TABLE) depending on the type of triggers, whether row-level or statement-level. Furthermore, in SQL3, the INSERT event trigger can only use NEW or NEW_TABLE while the DELETE event trigger can only use OLD or OLD_TABLE to access transition values. However, the UPDATE event trigger can use both transition variables and tables. We have considered the following two approaches to introduce the transition values in TBE.

1. *Using new built-in functions*: Special built-in functions (i.e., OLD_TABLE() and NEW_TABLE() for statement-level, OLD() and NEW() for row-level) are introduced. The OLD_TABLE() and NEW_TABLE() functions return a set of tuples with values before and after the changes, respectively. Similarly the OLD() and NEW() return a single tuple with value before and after the change, respectively. Therefore, applying aggregate functions such as CNT. or SUM. to the OLD() or NEW() is meaningless (i.e., CNT.NEW(_s) is always 1 or SUM.OLD(_s) is always same as _s). Using new built-in functions, for instance, the event "every time more than 10 new employees are inserted" can be represented as follows:

E.emp	Eno	Ename	DeptNo	Sal
AFT.I.S.		_n		

E.conditions
CNT.ALL.NEW_TABLE(_n) > 10

Also the event "when salary is doubled for each row" can be represented as follows:

E.emp	Eno	Ename	DeptNo	Sal
AFT.U.R.				_s

E.conditions
NEW(_s) > OLD(_s) * 2

It is illegal to apply the NEW() or NEW_TABLE() to the variable defined on the DELETE event. Likewise for the application of OLD() or OLD_TABLE() to the variable defined on the INSERT event. Asymmetrically, it is redundant to apply the NEW() or NEW_TABLE() the variable defined on the INSERT event. Likewise for the application of OLD() or OLD_TABLE() to the variable defined on the DELETE event. For instance, in the above event "every time more than 10 new employees are inserted", _n and NEW_TABLE(_n) are equivalent. Therefore, the condition expression at the condition box can be rewritten as "CNT.ALL._n > 10". It is ambiguous, however, to simply refer to the variable defined in the UPDATE event without the built-in functions. That is, in the event "when salary is doubled for each row", _s can refer to values both before and after the UPDATE. That is, "_s > _s * 2" at the condition box would cause an error due to its ambiguity. Therefore, for the UPDATE event case, one needs to explicitly use the built-in functions to access transition values.

2. *Using modified skeleton tables*: Depending on the event type, skeleton tables are modified accordingly; additional columns may appear in the skeleton tables[2]. For the INSERT event, a keyword NEW_ is prepended to the existing column names in the skeleton table to denote that these are newly inserted ones. For the DELETE event, a keyword OLD_ is prepended similarly. For the UPDATE event, a keyword OLD_ is prepended to the existing column names whose values are updated in the skeleton table to denote values before the UPDATE. At the same time, additional columns with a keyword NEW_ appear to denote values after the UPDATE. If the UPDATE event is for all columns, then OLD_column-name and NEW_column-name appear for all columns.

Consider an event "when John's salary is doubled within the same department". Here, we need to monitor two attributes – Sal and DeptNo. First, the user may type the event activation time and granularity information at the leftmost column as shown in the first table. Then, the skeleton table changes its format to accommodate the UPDATE event effect as shown in the second table. That is, two more columns appear and the U. construct is relocated to the leftmost column.

E.emp	Eno	Ename	DeptNo	Sal
AFT.R.			U.	U.

E.emp	Eno	Ename	OLD_DeptNo	NEW_DeptNo	OLD_Sal	NEW_Sal
AFT.U.R.						

Then, the user fills in variables into the proper columns to represent the conditions. For instance, "same department" is expressed by using same variable _d in both OLD_DeptNo and NEW_DeptNo columns.

---

[2] We have also considered modifying tables, instead of columns. For instance, for the INSERT event, a keyword NEW_ is prepended to the *table* name. For the UPDATE event, a keyword OLD_ is prepended to the *table* name while new table with a NEW_ prefix is created. This approach, however, was not taken because we wanted to express column-level UPDATE event more explicitly. That is, for an event "update occurs at column Sal", we can add only OLD_Sal and NEW_Sal attributes to the existing table if we use the "modifying columns" approach. If we take the "modifying tables" approach, however, we end up with two tables with all redundant attributes whether they are updated or not (e.g., two attributes OLD_emp.Ename and NEW_emp.Ename are unnecessarily created; one attribute emp.Ename is sufficient since no update occurs for this attribute).

E.emp	Eno	Ename	OLD_DeptNo	NEW_DeptNo	OLD_Sal	NEW_Sal
AFT.U.R.		John	_d	_d	_o	_n

E.conditions
_n > _o * 2

We chose the approach using new built-in functions to introduce transition values into TBE. Although there is no difference with respect to the expressive power between two approaches, the first one does not incur any modifications to the skeleton tables, thus minimizing cluttering of the user interface.

3.7   The REFERENCING Construct

SQL3 allows the renaming of transition variables or tables using the REFEREN-CING construct for the user's convenience. In TBE, this construct is not needed since the transition values are directly referred to by the variables filled in the skeleton tables.

3.8   Procedural Statements

When arbitrary SQL procedural statements (i.e., IF, CASE, assignment statements, etc.) are written in the action part of the trigger rules, it is not straightforward to represent them in TBE due to their procedural nature. Because their expressive power is beyond what the declarative QBE, and thus TBE described so far, can achieve, we instead provide a special kind of box, called *statement box*, similar to the condition box. The user can write arbitrary SQL procedural statements delimited by ";" in the statement box. Since the statement box is only allowed for the action part of the triggers, the prefix A. is always prepended. For example,

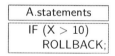

A.statements
IF (X > 10)
ROLLBACK;

3.9   The Order among Action Trigger Statements

SQL3 allows multiple action statements in triggers, each of which is executed according to the order they are written. To represent triggers whose semantics depend on the assumed sequential execution, TBE uses an implicit agreement; like prolog, the execution order follows from top to bottom. Special care needs to be taken in translation time for such action statements as follows:

- The *action* skeleton tables appearing before are translated prior to that appearing after.
- In the same *action* skeleton tables, action statements written at the top row are translated prior to that written at the bottom one.

3.10   Expressing Conditions in TBE

In most active database triggers languages, the event part of the triggers language is exclusively concerned with what has happened and cannot perform tests on values associated with the event. Some triggers languages (e.g., Ode [1], SAMOS [9], Chimera [5]), however, provide filtering mechanisms that perform tests on event parameters (see [14], chapter 4). Event filtering mechanisms can be very useful in optimizing trigger rules; only events that passed the parameter filtering tests are sent to the condition module to avoid unnecessary expensive condition evaluations.

In general, we categorize condition definitions of the triggers into 1) *parameter filter (PF)* type and 2) *general constraint (GC)* type. SQL3 triggers definition does not have PF type; event language specifies only the event type, activation time and granularity information, and all conditions (both PF and GC types) need to be expressed in the WHEN clause. In TBE, however, we decided to allow users to be able to differentiate PF and GC types by providing separate condition boxes (i.e., E. and C. prefixed ones) although it is not required for SQL3. This is because we wanted to support other trigger languages who have both PF and GC types in future.

1. *Parameter Filter Type*: Since this type tests the event parameters, the condition must use the transition variables or tables. Event examples such as "every time more than 10 new employees are inserted" or "when salary is doubled" in Section 3.6 are these types. In TBE, this type is typically represented in the E. prefixed condition box.

2. *General Constraint Type*: This type expresses general conditions regardless of the event type. In TBE, this type is typically represented in the C. prefixed condition boxes. One such example is illustrated in Example 5.

Example 5. When an employee's salary is increased more than twice within the same year (a variable CURRENT_YEAR contains the current year value), record changes into the log(Eno, Sal) table. Assume that there is another table sal-change(Eno, Cnt, Year) to keep track of the employee's salary changes.

```
CREATE TRIGGER TwiceSalaryRule AFTER UPDATE OF Sal ON emp
FOR EACH ROW
WHEN EXISTS (SELECT * FROM sal-change WHERE Eno = NEW.Eno
      AND Year = CURRENT_YEAR AND Cnt >= 2)
BEGIN ATOMIC
      UPDATE sal-change SET Cnt = Cnt + 1
          WHERE Eno = NEW.Eno AND Year = CURRENT_YEAR;
      INSERT INTO log VALUES(NEW.Eno, NEW.Sal);
END
```

E.emp	Eno	Ename	DeptNo	Sal
AFT.R.	_n			U._s

C.sal-change	Eno	Cnt	Year
	NEW(_n)	_c	CURRENT_YEAR

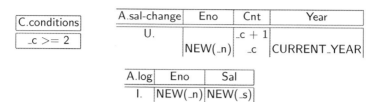

Here, the condition part of the trigger rule (i.e., WHEN clause) checks the Cnt value of the sal-change table to check how many times salary was increased in the same year, and thus, does not involve testing any transition values. Therefore, it makes more sense to represent such condition as GC type, not PF type. Note that the headers of the sal-change and condition box have the C. prefixes.

4 Complex SQL3 Triggers Examples

In this section, we show a few complex SQL3 triggers and their TBE representations. These trigger examples are modified from the ones in [8,18].

4.1 Integrity Constraint Triggers

A trigger rule to maintain the foreign key constraint is shown below.

Example 6. When a manager is deleted, all employees in his or her department are deleted too.

```
CREATE TRIGGER ManagerDelRule AFTER DELETE ON emp
FOR EACH ROW
    DELETE FROM emp E1 WHERE E1.DeptNo =
        (SELECT D.Dno FROM dept D WHERE D.MgrNo = OLD.Eno)
```

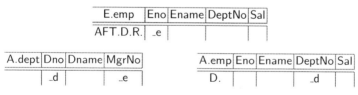

In this example, the WHEN clause is missing on purpose; that is, the trigger rule does not check if the deleted employee is in fact a manager or not because the rule deletes only the employee whose manager is just deleted. Note that how _e variable is used to join the emp and dept tables to find the department whose manager is just deleted. Same query could have been written with a condition test in a more explicit manner as follows:

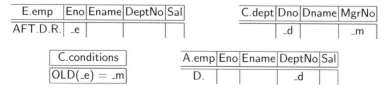

Another example is shown below.

Example 7. When employees are inserted to the emp table, abort the transaction if there is one violating the foreign key constraint.

```
CREATE TRIGGER AbortEmp AFTER INSERT ON emp
FOR EACH STATEMENT
WHEN EXISTS (SELECT * FROM NEW_TABLE E WHERE NOT EXISTS
                    (SELECT * FROM dept D WHERE D.Dno = E.DeptNo))
    ROLLBACK
```

E.emp	Eno	Ename	DeptNo	Sal		C.dept	Dno	Dname	MgrNo		A.statements
AFT.I.S.			_d			¬	_d				ROLLBACK

In this example, if the granularity were R. instead of S., then same TBE query would represent different SQL3 triggers. That is, row-level triggers generated from the same TBE representation would have been:

```
CREATE TRIGGER AbortEmp AFTER INSERT ON emp
FOR EACH ROW
WHEN NOT EXISTS
       (SELECT * FROM dept D WHERE D.Dno = NEW.DeptNo)
    ROLLBACK
```

We believe that this is a good example illustrating why TBE is useful in writing trigger rules. That is, when the only difference between two rules is the trigger granularity, simple change between R. and S. is sufficient in TBE. However, in SQL3, users should devise quite different rule syntaxes as demonstrated above.

4.2   View Maintenance Triggers

Suppose a company maintains the following view derived from the emp and dept schema.

Example 8. Create a view HighPaidDept that has at least one "rich" employee earning more than 100K.

```
CREATE VIEW HighPaidDept AS
        SELECT DISTINCT D.Dname
        FROM emp E, dept D
        WHERE E.DeptNo = D.Dno AND E.Sal > 100K
```

The straightforward way to maintain the views upon changes to the base tables is to re-compute all views from scratch. Although incrementally maintaining the view is more efficient than this method, for the sake of trigger example, let us implement the naive scheme below. The following is only for UPDATE event case.

Example 9. Refresh the HighPaidDept when UPDATE occurs on emp table.

```
CREATE TRIGGER RefreshView AFTER UPDATE OF DeptNo, Sal ON emp
FOR EACH STATEMENT
```

```
BEGIN ATOMIC
    DELETE FROM HighPaidDept;
    INSERT INTO HighPaidDept
        (SELECT DISTINCT D.Dname FROM emp E, dept D
            WHERE E.DeptNo = D.Dno AND E.Sal > 100K);
END
```

E.emp	Eno	Ename	DeptNo	Sal
AFT.S.			U.	U.

A.emp	Eno	Ename	DeptNo	Sal
			_d	> 100K

A.dept	Dno	Dname	MgrNo
	_d	_n	

A.HighPaidDept	Dname
D.	
I.	_n

By the implicit ordering of TBE, the DELETE statement executes prior to the INSERT statement.

5 Implementation

A preliminary version of the TBE prototype has been implemented using jdk 1.2. Although the underlying concept is the same as what we have presented so far, we added several bells and whistles (e.g., context sensitive pop-up menu) for better human-computer interaction. The algorithm to generate trigger rules from TBE is omitted due to space limitation. For details, please refer to [10].

The main screen consists of two sections – one for input and another for output. The input section is where the user creates trigger rules by QBE mechanism and the output section is where the interface generates trigger rules in the target trigger syntax. Further, the input section consists of three panes for event, condition, action, respectively. The main screen of the prototype is shown in Figure 1, where the query in Example 5 is shown.

6 Applications

Not only is TBE useful for writing trigger rules, but it can also be used for other applications with a few modifications. Two such applications are illustrated in this section.

Declarative Constraints in SQL3: SQL3 has the ASSERTION to enforce any condition expression that can follow WHERE clause to embed some application logic. The syntax of the ASSERTION is:

CREATE ASSERTION <assertion-name> CHECK <condition-statement>

Note the similarity between the assertion and triggers syntax in SQL3. Therefore, a straightforward extension of TBE can be used as a tool to enforce assertion constraints declaratively. In fact, since the ASSERTION in SQL3 only permits declarative constraints, TBE suits the purpose perfectly.

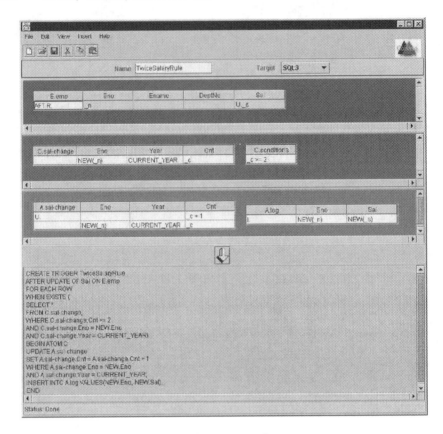

Fig. 1. Main screen dump.

Universal Triggers Construction Tool: Although SQL3 is close to its final form, many database vendors are already shipping their products with their own proprietary trigger syntaxes and implementations. When multiple databases are used together or one database needs to be migrated to another, these diversities can introduce significant problems. To remedy this problem, one can use TBE as a universal triggers construction tool. That is, the user creates triggers using TBE interface.When changing database from one to another (e.g., from Oracle to DB2), the user can simply reset one of the preference information of TBE to re-generate the new trigger rules. Extending TBE to support all unique features of diverse database products is not a trivial task. Nevertheless, we believe that retaining the visual nature of the triggers construction with TBE can be useful in coping with heterogeneous database systems.

7   Conclusion

A novel user interface called TBE for creating triggers is proposed. TBE borrows the visual querying mechanism from QBE and applies it to triggers construction application in a seamless fashion. An array of new constructs are introduced to extend QBE to support triggers semantics and syntaxes properly. To prove the concept, a prototype is implemented and demonstrated the feasibility and benefits of applying QBE in writing trigger rules.

References

[1] Agrawal, R., Gehani, N. "Ode (Object Database and Environment): The Language and the Data Model", *Proc. SIGMOD*, Portland, Oregon, 1989.

[2] Codd, E. F. "Relational Completeness of Data Base Languages", *Data Base Systems, Courant Computer Symposia Series*, Prentice-Hall, 6:65-98, 1972.

[3] Collet, C., Brunel, E. "Definition and Manipulation of Forms with FO2", *Proc. IFIP Visual Database Systems*, 1992.

[4] Chang, Y.-I., Chen, F.-L. "RBE: A Rule-by-example Action Database System", *Software – Practice and Experience*, 27(4):365-394, 1997.

[5] Ceri, S., Fraternali, P., Paraboschi, S., Tanca, L. "Active Rule Management in Chimera", In J. Widom and S. Ceri (ed.), *Active Database Systems: Triggers and Rules for Active Database Processing*, Morgan Kaufmann, 1996.

[6] Cochrane, R., Pirahesh, H., Mattos, N. "Integrating Triggers and Declarative Constraints in SQL Database Systems", *Proc. VLDB*, 1996.

[7] Embley, D. W. "NFQL: The Natural Forms Query Language", *ACM TODS*, 14(2):168-211, 1989.

[8] Embury, S. M., Gray, P. M. D. "Database Internal Applications", In N. W. Paton (ed.), *Active Rules In Database Systems*, Springer-Verlag, 1998.

[9] Gatziu, S., Dittrich, K. R. "SAMOS", In N. W. Paton (ed.), *Active Rules In Database Systems*, Springer-Verlag, 1998.

[10] Lee, D., Mao, W., Chiu, H., Chu, W. W. "TBE: A Graphical Interface for Writing Trigger Rules in Active Databases", *5th IFIP 2.6 Working Conf. on Visual Database Systems (VDB)*, 2000.

[11] Lee, D., Mao, W., Chu, W. W. "TBE: Trigger-By-Example (Extended Version)", *UCLA-CS-TR-990029*, 1999.
http://www.cs.ucla.edu/~dongwon/paper/

[12] McLeod, D. "The Translation and Compatibility of SEQUEL and Query by Example", *Proc. Int'l Conf. Software Engineering*, San Francisco, CA, 1976.

[13] Melton, J. (ed.), "(ANSI/ISO Working Draft) Foundation (SQL/Foundation)", *ANSI X3H2-99-079/WG3:YGJ-011*, March, 1999.
ftp://jerry.ece.umassd.edu/isowg3/dbl/BASEdocs/public/sql-foundation-wd-1999-03.pdf

[14] Paton, N. W. (ed.), "Active Rules in Database Systems", *Springer-Verlag*, 1998.

[15] Simon, E., Kotz-Dittrich, A. "Promises and Realities of Active Database Systems", *Proc. VLDB* 1995.

[16] Teisseire, M., Poncelet, P., Cichetti, R. "Towards Event-Driven Modelling for Database Design", *Proc. VLDB*, 1994.

[17] Zloof, M. M. "Query-by-Example: a data base language", *IBM System J.*, 16(4):342-343, 1977.

[18] Zaniolo, C., Ceri, S., Faloutsos, C., Snodgrass, R. R., Subrahmanian, V.S., Zicari, R. "Advanced Database Systems", *Morgan Kaufmann*, 1997.

Decomposition by Pivoting and Path Cardinality Constraints

Sven Hartmann

FB Mathematik, Universität Rostock, 18051 Rostock, Germany

Abstract. In the relational data model, the problem of data redundancy has been successfully tackled via decomposition. In advanced data models, decomposition by pivoting provides a similar concept. Pivoting has been introduced by Biskup et al. [5], and used for decomposing relationship types according to a unary nonkey functional dependency. Our objective is to study pivoting in the presence of cardinality constraints which are commonly used in semantic data models.

In order to ensure the equivalence of the given schema and its image under pivoting, the original application-dependent constraints have to be preserved. We discuss this problem for sets of participation and co-occurrence constraints. In particular, we prove the necessity of path cardinality constraints, and give an appropriate foundation for this concept.

1   Introduction

Database modeling aims on designing efficient and appropriate databases. It is widely accepted now that databases are best designed first on a conceptual level. The result of this process is a conceptual schema which describes the requirements that the desired database must achieve. Usually, conceptual design does not consist of a single design step, but of a step-by-step process. Each step refines the schema, e.g. by adding information, changing the structure of the schema, refining types or generating subtypes. Composition, extension and decomposition are well-known operations to refine types during the design process.

In this paper, we are almost exclusively concerned with decomposition. In the relational model, this transformation has been successfully used e.g. for reducing redundancy, splitting overloaded concepts or making constraint enforcement more efficient. There is a fairly extensive literature on this strategy, to which the reader may wish to turn for details, e.g. [2,18,23]. The same idea may be used in semantic data models which are frequently used in conceptual design. Decomposition primitives in the entity-relationship model have been discussed e.g. in [2,3,22].

An interesting, new approach towards decomposition in semantic data models, was suggested by Biskup et al. [5,6]. They introduced the operation of pivoting for decomposing relationship types. Pivoting separates apart some components of a relationship type and assembles them in a newly generated type. Afterwards, the new type is linked to the old one via a pivot component.

A.H.F. Laender, S.W. Liddle, V.C. Storey (Eds.): ER2000 Conference, LNCS 1920, pp. 126–139, 2000.
© Springer-Verlag Berlin Heidelberg 2000

In the relational case, decomposition is usually motivated by functional dependency. In particular, normalization theory up to BCNF is based on this idea. In semantic data models, however, this concept falls short. During the last few decades, the area of integrity constraints has attracted considerable research interest. A large amount of different constraint classes has been discussed in the literature and actually used in database design.

Cardinality constraints are among the most popular classes of integrity constraints. They impose lower and upper bounds on the number of relationships an entity of a given type may be involved in. For cardinality constraints and generalizations, see [16,22]. The objective of this paper is to study pivoting in the presence of cardinality constraints. Our investigation is not restricted to ordinary cardinality constraints, but we also consider co-occurrence constraints, which among others include functional and numerical dependencies.

The paper is organized as follows. Section 2 describes the data model to be used. Though cardinality constraints have also been studied in object-oriented context, they are most popular in semantic data models. Therefore, all our considerations are carried out in an extended entity-relationship context. In Section 3, we briefly review decomposition under functional dependencies and highlight the idea of pivoting. Section 4 gives a formal definition of cardinality constraints. In Section 5 we give a generalized definition of pivoting, and show how to use pivoting under cardinality constraints. In Section 6, we discuss the problem of preserving constraints. We point out that the usual concept of cardinality constraints is not sufficient to cope with this problem. Finally, in Section 7, we identify conditions for a schema being equivalent to its pivoted version, namely the expressive power of path cardinality constraints.

2 The Data Model to Be Used

The entity-relationship approach to conceptual design provides a simple way of representing data in the form of entities and relationships among them. In this section, we shall briefly review basic concepts of this approach.

Let $\underline{E}$ be a finite set, whose elements are called *entity types*. With each entity type $\underline{e}$ we associate a finite set $\underline{e}^t$ called the population of the type $\underline{e}$ (at moment t). The members of $\underline{e}^t$ are entity instances or entities, for short. Intuitively, entities may be seen as real-world objects which are of interest for the application under consideration. By classifying them and specifying their significant properties (attributes), we obtain entity types which are frequently used to model the objects in their domains.

A *(first-order) relationship type* $\underline{r}$ is a collection $\{\underline{e}_1, \ldots, \underline{e}_n\}$ of elements from $\underline{E}$. Relationship types are used to model associations between real-world objects, i.e. entities. A relationship or instance of type $\underline{r}$ is a tuple from the cartesian product $\underline{e}_1^t \times \ldots \times \underline{e}_n^t$. A finite set $\underline{r}^t$ of such relationships forms the population of $\underline{r}$ at moment t. Analogously, higher-order relationship types may be defined hierarchically, cf. [21]. Given some entity and/or relationship types, a collection $\{\underline{u}_1, \ldots, \underline{u}_n\}$ of them forms a *(higher-order) relationship type*. As above, we define relationships of type $\underline{r}$ as tuples from the cartesian product $\underline{u}_1^t \times \ldots \times \underline{u}_n^t$ at moment t.

In a relationship type $\underline{r} = \{\underline{u}_1, \ldots, \underline{u}_n\}$ each of the entity or relationship types $\underline{u}_1, \ldots, \underline{u}_n$ is said to be a *component type* of $\underline{r}$. Additionally, each of the pairs $(\underline{r}, \underline{u}_j)$ is called a *link*.

A finite set $\underline{D} = \{\underline{r}_1, \ldots, \underline{r}_k\}$ of entity and relationship types used to represent information is called a *database schema*. Of course, we assume that with each relationship type $\underline{r}$, all its component types belong to $\underline{D}$, too. When replacing each type $\underline{r}$ in $\underline{D}$ by its population $\underline{r}^t$ at moment t, we obtain a database or instance $\underline{D}^t$ of $\underline{D}$.

One of the strengths of the entity-relationship approach is its ability to provide a simple graphical representation of information. Consider the digraph $\mathfrak{ERD}$ with vertex set $\underline{D}$ whose arcs are just the links of the database schema. This digraph is known as the *entity-relationship diagram* of $\underline{D}$. It is usual to represent entity types graphically by rectangles and relationship types by diamonds.

Example. Below we give a small motivating example to illustrate the issues we are going to tackle. Consider a part of a university database schema involving the entity types PROFESSOR, TUTOR, COURSE, BUILDING, HALL and DATE. On this set we define a relationship type LECTURE reflecting assignments of professors and their tutors to courses in combination with the dates and locations they take place at.

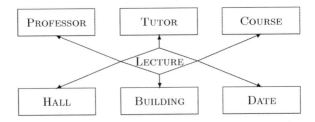

Fig. 1. Entity-relationship diagram for our example.

Some designers would perhaps tend to use smaller relationships. This is an issue where the designer must rely one his personal judgement and experience. However, in conceptual design, a designer should be supported to concentrate on the essential parts of the application: he has to determine the objects of interest and to specify their associations. In early design steps, he should not be confronted with restrictions, or to take into account the features of future design steps, such as breaking up overloaded relationships.

3   Decomposition by Pivoting

In database design, great attention is devoted to the modeling of semantics. Central to this idea is the notion of integrity constraints. Defining and enforcing integrity constraints helps to guarantee that the database correctly reflects the underlying domain of interest. Functional dependencies are commonly recognized as the most fundamental constraints. They have been introduced in the relational model by Codd [9], and later extended to other data models, cf. [21,24]. Among functional dependencies, key dependencies are of utmost interest, as they help to identify the tuples in a population.

Consider an instance r of the relationship type $\underline{r}$. By $r[X]$ we denote the restriction of r to a subset $X \subseteq \underline{r}$. Moreover, for a population $\underline{r}^t$, let $\pi_X(\underline{r}^t) = \{r[X] : r \in \underline{r}^t\}$ denote its projection to X. A subset $X \subseteq \underline{r}$ is a *key* for $\underline{r}$ if the restrictions $r[X]$, $r \in \underline{r}^t$, are pairwise distinct. A database instance $\underline{D}^t$ satisfies the *key dependency* $\underline{r} : X \to \underline{r}$, when X is a key for $\underline{r}$.

A *functional dependency* is an expression $\underline{r} : X \to Y$, where X and Y are subsets of $\underline{r}$. A database instance $\underline{D}^t$ satisfies this constraint, if we have $r_1[Y] = r_2[Y]$ whenever $r_1[X] = r_2[X]$ holds for any two relationships r_1 and r_2 in $\underline{r}^t$. A functional dependency is said to be *unary*, if X contains a single component type from $\underline{r}$.

Example. In the university schema, constraints are defined to meet the requirements of the university schedule. Let us consider some examples. If every professor has only one tutor per course, then the functional dependency LECTURE: {PROFESSOR, COURSE} → {TUTOR} holds. Since every lecture hall is situated in only one building, we also have the unary functional dependency LECTURE: {HALL} → {BUILDING}.

Constraints are to be considered as an integral part of a database schema. We use the notion $(\underline{D}, \Sigma)$ to refer to a schema $\underline{D}$ together with a set Σ of constraints defined on $\underline{D}$. A database instance $\underline{D}^t$, satisfying each constraint in Σ, is said to be *legal* for Σ.

If not all instances are possible, it stands to reason that there appears some kind of redundancy in legal instances. This problem and its consequences have been widely studied in literature, for a survey see [18,23]. In the relational model, redundancy has been successfully tackled via decompositions. To be brief, a decomposition of a relation type R is its replacement by two subsets X and Y such that $X \cup Y = R$ holds. In general, we only consider lossless decompositions: every legal population of R must be recoverable, i.e. the natural join of its projections $\pi_X(R^t)$ and $\pi_Y(R^t)$. It is well-known that this property holds if and only if one of the functional dependencies $R : X \cap Y \to X$ or $R : X \cap Y \to Y$ holds for R.

It is natural to ask whether this idea can be extended to advanced data models, such as semantic or object-oriented ones. In [5], Biskup et al. gave an interesting answer to this question. They suggest the transformation of pivoting to decompose relationship types. Assume, we are given a unary functional dependency $\underline{r} : \{\underline{u}\} \to Y$. Then, pivoting separates apart the component types in Y into a newly generated relationship type. The new type is declared to be a subtype of the so-called pivot component $\underline{u}$. Moreover the pivot component provides the link between the original relationship type and the newly generated one. For further details on this approach, we refer to [4,5,6].

The above discussion gives only an informal account of decomposition in database design. For a rigorous treatment, the reader should consult e.g. [18]. It is noteworthy, that eliminating redundancy is by far not the only motivation for performing decompositions: a schema where the relationship types are too large makes it necessary to use a single tuple for storing several facts. Such a schema is said to suffer from update anomalies. Moreover, decompositions are of considerable interest when translating conceptual schemas to target models such as object-oriented models. For a discussion of this issue, we refer to [5].

4   Cardinality Constraints

Cardinality constraints are often regarded as one of the basic constituents of the entity-relationship model. They are already present in Chen's seminal paper [8], and have been frequently used in conceptual design since then. A *cardinality constraint* is an expression $card(\underline{r}, \underline{u}) = (a, b)$, where $\underline{r}$ is a relationship type, and $\underline{u}$ is one of its component types. A database instance $\underline{D}^t$ satisfies this constraint if every instance of $\underline{u}$ appears in at least a and at most b relationships $r \in \underline{r}^t$. Often, cardinality constraints are reflected graphically in the entity-relationship diagram as (a, b)-labels associated to the corresponding links, see Figure 2.

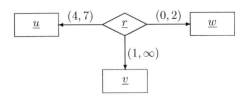

Fig. 2. An entity-relationship diagram with cardinality constraints.

Cardinality constraints are available in most semantic data models, but also in object-oriented models [7,10]. For a survey on cardinality constraints, the interested reader is referred to [16,22]. These monographs also furnish a great deal of information on generalized versions of cardinality constraints. In [10], ordinary cardinality constraints as defined above are known as *participation constraints*. In addition, [10] proposes a more general version of cardinality constraints, so-called co-occurrence constraints.

Formally, a *co-occurrence constraint* is an expression $\underline{r} : X \xrightarrow{(a,b)} Y$, where both X and Y are subsets of $\underline{r}$. A database $\underline{D}^t$ satisfies this constraint if

$$a \leq |\{z \in \pi_{X \cup Y}(\underline{r}^t) : z[X] = x\}| \leq b$$

holds for every tuple $x \in \pi_X(\underline{r}^t)$. Note that we do not claim X and Y to be disjoint, as done in [16]. It should be emphasized that co-occurrence constraints are also available in the Higher-order Entity-Relationship Model [21]. Co-occurrence constraints happen to be very powerful: for $(a, \overset{\frown}{b}) = (0, 1)$ they correspond to functional dependencies, for $(a, b) = (0, b)$ to numerical dependencies [11]. For results on these constraints and useful references, we refer to [13,16,20,22].

Figure 2 shows a relationship type $\underline{r}$ with given participation constraints. They tell us, that every entity of type $\underline{u}$ participates in between 4 and 7 instances of $\underline{r}$, each entity of type $\underline{v}$ participates at least once, and every entity of type $\underline{w}$ at most twice. Suppose, we are given a co-occurrence constraint $\underline{r} : \{\underline{u}, \underline{w}\} \overset{(3,5)}{\longrightarrow} \{\underline{v}\}$, too. Then, pairs of entities of types $\underline{u}$ and $\underline{w}$, that appear in some instance of $\underline{r}$, must co-occur (appear together) with at least 3 and at most 5 entities of type $\underline{v}$.

Example.   In the university schema, cardinality constraints provide useful information: a participation constraint $card(\text{LECTURE}, \text{PROFESSOR}) = (3, 6)$, for example, states that every professor will give 3 to 6 lectures per term. The university of Rostock frequently offers courses for external students. Lectures for these courses are usually scheduled for weekends or during the holidays. For external students it is rather inconvenient to come to Rostock just for a single lecture. Hence, there should always be between 3 and 5 lectures per course and date. This requirement can be modeled by the co-occurrence constraint LECTURE: $\{\text{COURSE}, \text{DATE}\} \overset{(3,5)}{\longrightarrow}$ LECTURE.

5   Pivoting under Cardinality Constraints

In this section, we shall discuss pivoting in the presence of cardinality constraints. To begin with, we have to define an appropriate notion of pivoting. Suppose we are given a database schema $\underline{D}$ containing a relationship type $\underline{r} = \{\underline{u}_1, \dots, \underline{u}_n\}$.

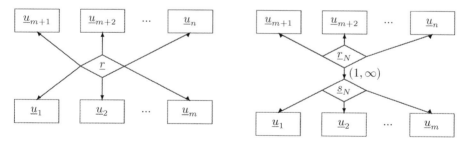

Fig. 3. ER diagrams for a database schema and its pivoted version.

Let S be a subset of $\underline{r}$, say $S = \{\underline{u}_1, \dots, \underline{u}_m\}$, where $2 \leq m < n$. We define a new relationship type $\underline{s}_N$, whose component types are just the elements of S. Next, these elements shall be removed from $\underline{r}$. Therefore, we replace $\underline{r}$ by a higher-order relationship type $\underline{r}_N$ obtained from $\underline{r}$ by deleting all elements of S, and by adding the newly generated relationship type $\underline{s}_N$ as a component type. Figure 3 shows the corresponding diagrams before and after pivoting. The elements of S

are said to be the *pivoted* components of r. Roughly speaking, they are shifted from r to s_N.

Throughout, let D be a fixed schema, and D_N the *pivoted schema* obtained from D by adding the new relationship type s_N, and replacing r by the modified relationship type r_N.

Usually, we want D_N to capture the application-dependent semantics of the original schema D. This natural requirement has been discussed in [2,19,22]. Informally, most of the proposals consider two schemas as being equivalent, if there exists a bijection between the instances of the schemas. We shall use transformation rules to show how instances of the pivoted schema are obtained from instances of the original schema:

- For each instance $r = (u_1, \ldots, u_n)$ of type r, the restriction $s_N := r[S] = (u_1, \ldots, u_m)$ becomes an instance of type s_N, and the tuple $r_N := (s_N, u_{m+1}, \ldots, u_n)$ becomes an instance of r_N.

Vice versa, we use the following rules to (re-)transform the instances of the pivoted schema to those of the original schema:

- Let $r_N = (s_N, u_{m+1}, \ldots, u_n)$ be an instance of r_N, where s_N denotes an instance $(u_1, \ldots, u_m)$ of s_N. Then $r := (u_1, \ldots, u_m, u_{m+1}, \ldots, u_m)$ becomes an instance of the original relationship type r.

We call two schemas *equivalent* if the transformation rules suggested above provide a bijection mapping legal instances of one schema onto legal instances of the other. Applying these transformation rules, we immediately obtain the following result.

Lemma 1 *Let Σ and Σ_N denote constraint sets defined on a schema D and its pivoted schema D_N, respectively. If (D, Σ) and (D_N, Σ_N) are equivalent, then Σ_N implies the cardinality constraint $card(r_N, s_N) = (1, \infty)$. On the other hand, (D, Σ) and (D_N, Σ_N) are equivalent whenever $\Sigma = \emptyset$ and $\Sigma_N = \{card(r_N, s_N) = (1, \infty)\}$.*

Note, that the participation constraint $card(r_N, s_N) = (1, \infty)$ forces the transformation presented above to be onto.

Usually, Σ is not empty, as most applications require integrity constraints. This points out a crucial problem of pivoting. All original constraints (given in Σ) have to be adjusted in the pivoted schema. Thus the question arises whether we are able to find a set of new constraints Σ_N defined on the pivoted schema, that may be used to express the original constraints. We shall consider this question in the following section. In this sense, Lemma 1 forms the basis for further investigations.

To continue with, we present a number of cases when pivoting should be taken into consideration:

- *First case:* Assume, we are given a co-occurrence constraint $r : X \xrightarrow{(a,b)} r$ with $b \geq 2$. Here, we suggest to choose S equal to X. In the pivoted schema, the original co-occurrence constraint may be expressed by the participation constraint $card(r_N, s_N) = (a, b)$.

Example. If our university schema is used to schedule the lectures for external students, we have a constraint LECTURE: $\{$COURSE, DATE$\} \xrightarrow{(3,5)}$ LECTURE. Following the proposal, we generate a new relationship type OFFER $=$ $\{$COURSE, DATE$\}$, and replace the old LECTURE-type by a new one involving the types PROFESSOR, TUTOR, HALL, BUILDING as well as the newly generated type OFFER. For the sake of simplicity, we reused the old type name rather than introducing an artificial new one. In the new schema, we have the participation constraint $card($LECTURE, OFFER$) = (3,5)$.

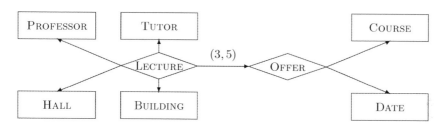

Fig. 4. The ER diagram for the pivoted schema of the university example.

In practice, it often happens that mentors have a prior engagement, or dates do not fit into the schedule of various students. Then dates have to be cancelled and replaced by new appointments. Before pivoting this information has to be updated in 3 to 5 tuples, after pivoting only in a single instance of type OFFER.

It should be mentioned, that mostly participation constraints are easier to understand and to handle than co-occurrence constraints. In particular, there exists a sound and complete system of inference rules for participation constraints, see [14,15].

- *Second case:* Even, if we have only a co-occurrence constraint $\underline{r} : X \xrightarrow{(a,b)} Y$ with $Y \subset \underline{r}$ and $b \geq 2$, we may perform pivoting with $S = X$. This time the old co-occurrence constraint is expressed by a new one, namely $\underline{r}_N : \{\underline{s}_N\} \xrightarrow{(a,b)} Y \backslash X$.

- *Third case:* A different (and perhaps better) way to use the co-occurrence constraint $\underline{r} : X \xrightarrow{(a,b)} Y$ is the following. We choose $S = X \cup Y$, and express the old constraint by a generalized key dependency $\underline{s}_N : X \xrightarrow{(a,b)} \underline{s}_N$.

Similar to ordinary keys, these generalized key dependencies can be enforced by keeping an index and specifying this index to be such that each value appears between a and b times. Moreover, they may be used for further investigation, like horizontal decomposition as studied in [11].

Example. External students may subscribe for magazines offered by the university or its departments. Let this be modeled by the relationship type SUBSCRIPTION containing the entity types STUDENT, ADDRESS and MAGAZINE.

External students usually have two addresses, a home address, and one in Rostock, where they live during their courses. Hence, we have a co-occurrence constraint SUBSCRIPTION: {STUDENT} $\overset{(1,2)}{\longrightarrow}$ {ADDRESS}. When pivoting as proposed above, we introduce a new relationship type SUBSCRIBER containing STUDENT and ADDRESS, and the old SUBSCRIPTION-type is replaced by a new one involving SUBSCRIBER and MAGAZINE. In the pivoted schema, we have the co-occurrence constraint SUBSCRIBER: {STUDENT} $\overset{(1,2)}{\longrightarrow}$ SUBSCRIBER.

Due to our definition of pivoting in the entity-relationship context, we are now able to use both, unary and non-unary functional dependencies, for pivoting.

- *Fourth case:* If we are given a nonkey functional dependency $\underline{r} : X \to Y$, then we again choose $S = X \cup Y$. In the pivoted schema, the old functional dependency may be expressed by the key dependency $\underline{s}_N : X \to \underline{s}_N$.

If the old functional dependency is unary, i.e. X consists of a single element $\underline{u}$, then it can also be expressed by the participation constraint $card(\underline{s}_N, \underline{u}) = (0, 1)$. In data models supporting subtype or ISA-hierarchies, $\underline{s}_N$ may be considered as a subtype of $\underline{u}$. This case has been studied extensively in [5,6].

In all four cases, pivoting removes (potential) redundancy and makes constraint enforcement more efficient. Usually, pivoting pays off only if instances of $\underline{s}_N$ are involved in more than just a single instance of $\underline{r}$. Clearly, this cannot happen if S is a key for $\underline{r}$. In this case, we do not recommend pivoting.

Before concluding this section, we have to mention that in all four cases the equivalence of the original schema and the pivoted schema can be ensured similarly to Lemma 1.

6   Preserving Cardinality Constraints

Due to the discussion above, the original schema and its pivoted version should be equivalent. For the designer, this requirement becomes something of a problem. We defined equivalence in terms of the instances of schemas, but these instances are not explicitly available during the design process. The designer however has to decide whether pivoting is useful or not. It goes without saying that the designer has to be supported in coping with this task.

The following theorems help us to decide whether the application-dependent constraints given in the original schema may be adjusted in the pivoted schema, or not. Our results use the idea of implication of constraints. A constraint set Σ *implies* a constraint σ if every legal database instance $\underline{D}^t$ for Σ is also legal for σ. At design time, inference rules may be used to investigate the implications of integrity constraints. Applying such rules, the designer must 'merely' focus on the purely syntactic items given in the schema.

Clearly, cardinality constraints not defined on $\underline{r}$ do not cause any problem. They may be reused in the pivoted schema, either immediately or, if $\underline{r}$ appears in the declaration of the constraints, after replacing it by $\underline{r}_N$. For cardinality constraints defined on $\underline{r}$ the problem is more involved.

Theorem 2 *Let Σ be a set of co-occurrence constraints defined on $\underline{r}$. There is a set Σ' of co-occurrence constraints such that $(\underline{D}, \Sigma)$ and $(\underline{D}_N, \Sigma_N)$ are equivalent and $\Sigma_N = \Sigma' \cup \{card(\underline{r}_N, \underline{s}_N) = (1, \infty)\}$ if and only if every constraint in Σ is trivial or implied by those constraints in Σ that are of the form*

$$\underline{r} : X \xrightarrow{(a,b)} Y, \text{ where } X, Y \subseteq S, \tag{1}$$

$$\underline{r} : X \xrightarrow{(a,b)} Y, \text{ where } X, Y \subseteq \underline{r} \backslash S, \tag{2}$$

$$\underline{r} : X \cup S \xrightarrow{(a,b)} Y, \text{ where } X, Y \subseteq \underline{r} \backslash S, \text{ or} \tag{3}$$

$$\underline{r} : X \xrightarrow{(a,b)} S, \text{ where } X \subseteq \underline{r} \backslash S. \tag{4}$$

Inference rules for co-occurrence constraints are given e.g. in [11,13,20,22]. Unfortunately there does not exist a finite complete system of inference rules for co-occurrence constraints as pointed out by Grant and Minker [11]. Hence, Theorem 2 may in general be used to prove the equivalence of schemas, but not to prove their non-equivalence. However, there are subclasses of cardinality constraints that do have finite sound and complete systems of inference rules, such as the Armstrong system [1] for functional dependencies.

Given functional dependencies Σ defined on $\underline{r}$, and any subset $X \subseteq \underline{r}$, let $cl_\Sigma(X) = \{\underline{u} \in \underline{r} : \Sigma \text{ implies } (\underline{r} : X \rightarrow \{\underline{u}\})\}$ denote the closure of X under functional dependencies.

Theorem 3 *Let Σ be a set of functional dependencies defined on $\underline{r}$. Then there is a set Σ' of functional dependencies such that $(\underline{D}, \Sigma)$ and $(\underline{D}_N, \Sigma_N)$ are equivalent and $\Sigma_N = \Sigma' \cup \{card(\underline{r}_N, \underline{s}_N) = (1, \infty)\}$ if and only if for every set $X \subseteq \underline{r}$ one of the following conditions hold:*

$$S \subseteq cl_\Sigma(X \cap S), \tag{5}$$

$$S \subseteq cl_\Sigma(X \backslash S), \text{ or} \tag{6}$$

$$cl_\Sigma(X) = cl_\Sigma(X \cap S) \cup cl_\Sigma(X \backslash S), \tag{7}$$

$$\text{where } cl_\Sigma(X \cap S) \subset S \text{ and } cl_\Sigma(X \backslash S) \subseteq \underline{r} \backslash S.$$

Theorem 3 in particular implies Theorem 6 of [5], which applies for pivoting guided by a unary functional dependency $\underline{r} : \{\underline{u}\} \rightarrow S$.

Finally, we give a similar result for participation constraints. A sound and complete system of inference rules is available in [14,15].

Theorem 4 *Let Σ be a set of participation constraints defined on $\underline{r}$. Then there is a set Σ_N of participation constraints such that $(\underline{D}, \Sigma)$ and $(\underline{D}_N, \Sigma_N)$ are equivalent iff every participation constraint $card(\underline{r}, \underline{u}) = (a, b)$ satisfies one of the following conditions*

$$\underline{u} \in \underline{r} \backslash S, \tag{8}$$

$$\underline{u} \in S, \text{ and } a \leq 1 \text{ and } b = \infty, \tag{9}$$

or there is some component type $\underline{w} \in S$ such that $card(\underline{r}, \underline{w}) = (0, 1)$ or $(1, 1)$ belongs to Σ.

7   Path Cardinality Constraints

Due to our definition of pivoting, this transformation is always lossless, but not necessarily constraint-preserving. Theorem 4, for example, states that most participation constraints declared for component types in S cannot be adjusted, unless there is a type $\underline{w} \in S$, which forms a key for the original relationship type $\underline{r}$. But in this case, as argued before, pivoting should not be applied. The problem, in general, is that we do not know exactly (neither in theory, nor in practice) how often instances of $\underline{s}_N$ are involved in instances of type $\underline{r}_N$.

Example. Remember our university schema. Motivated by the co-occurrence constraint LECTURE: $\{$COURSE, DATE$\} \xrightarrow{(3,5)}$ LECTURE, we performed pivoting and obtained new relationship types OFFER and LECTURE together with the participation constraint $card$(LECTURE, OFFER) $= (3, 5)$.

Now assume, the university decides to organize between 6 and 12 lectures per course. In the original schema, this corresponds to the participation constraint $card$(LECTURE, COURSE) $= (6, 12)$. How can we adjust this condition in the pivoted schema? It is easy to check, that $card$(OFFER, COURSE) $= (2, 3)$ is too sharp, while $card$(OFFER, COURSE) $= (2, 4)$ is not sharp enough.

In order to cope with this problem, we have to strengthen the concept of cardinality constraint. A (directed) *path* $P = (\underline{v}_0, \underline{v}_1, \ldots, \underline{v}_k)$ is a sequence of links $(\underline{v}_0, \underline{v}_1), (\underline{v}_1, \underline{v}_2), \ldots, (\underline{v}_{k-1}, \underline{v}_k)$ in the entity-relationship diagram of a schema. Here, $\underline{v}_0$ is the initial vertex of the path, and $\underline{v}_k$ is its terminal vertex. The integer k denotes the length of the path. Thus, paths of length 1 are just links. For standard graph-theoretic terminology, we refer to [12].

So far, when discussing cardinality constraints, we only considered component types of $\underline{r}$, i.e. types connected to $\underline{r}$ via a link. Now, we shall extend this concept to all types reachable from $\underline{r}$ via a path in the entity-relationship diagram. In the object-oriented context, this idea is well-known and formalized by path functions.

A *path function* ϕ_P maps instances of the initial vertex to instances of the terminal vertex. Path functions can be defined recursively as follows: To begin with, let $L = (\underline{r}, \underline{u})$ be a link. For every instance r of $\underline{r}$, we put $\phi_L(r) := u$ where u is the restriction of r to the component type $\underline{u}$. Next, consider any path $P = (\underline{r}, \underline{v}_1, \ldots, \underline{v}_{k-1}, \underline{v}_k)$ of length k. Let $Q = (\underline{r}, \underline{v}_1, \ldots, \underline{v}_{k-1})$ be its subpath of length $k - 1$, and $L = (\underline{v}_{k-1}, \underline{v}_k)$ be the remaining link. Then we put $\phi_P(r) := \phi_L(\phi_Q(r))$ for an instance r of type $\underline{r}$.

Now, we are ready to redefine the concept of cardinality constraints. Consider a path P with initial vertex $\underline{r}$ and terminal vertex $\underline{u}$. A *path participation constraint* is an expression $card\ P = (a, b)$. A database instance $\underline{D}^t$ satisfies this constraint if for every instance $u \in \underline{u}^t$ there are at least a and at most b instances $r \in \underline{r}^t$ such that $\phi_P(r) = u$ holds. If the path consists of a single link, this definition recaptures the original one in Section 4.

Example. In our university example, the pivoted schema contains a path $P=$(LECTURE, OFFER, COURSE). Using this path, the participation constraint $card$(LECTURE, COURSE) $= (6, 12)$ in the original schema may be replaced by the path participation constraint $card(\text{LECTURE, OFFER, COURSE}) = (6, 12)$ in the pivoted schema.

In a similar way, we proceed for co-occurrence constraints: While ordinary co-occurrence constraints relate some of the components of tuples r in a population $\underline{r}^t$, path co-occurrence constraints relate some of instances reachable by path functions from tuples $r \in \underline{r}^t$.

Let $X = \{P_1, \dots, P_l\}$ be a set of paths with common initial vertex $\underline{r}$, and terminal vertices $\underline{v}_1, \dots, \underline{v}_l$, respectively. For an instance r of type $\underline{r}$, we define $r[X] := (v_1, \dots, v_l) \in \underline{v}_1^t \times \dots \times \underline{v}_l^t$ by claiming $\phi_{P_i}(r) = v_i$, for $i = 1, \dots, l$.

A *path co-occurrence constraint* is an expression $\underline{r} : X \xrightarrow{(a,b)} Y$, where both, X and Y are sets of paths with common initial vertex $\underline{r}$. A database instance $\underline{D}^t$ satisfies this constraint if

$$a \leq |\{r[X \cup Y] : r \in \underline{r}^t \text{ and } r[X] = (v_1, \dots, v_l)\}| \leq b$$

holds for every $(v_1, \dots, v_l) \in \underline{v}_1^t \times \dots \times \underline{v}_l^t$, or there does not exist any $r \in \underline{r}^t$ satisfying $r[X] = (v_1, \dots, v_l)$ at all.

Again, path co-occurrence constraints of the form $\underline{r} : X \xrightarrow{(0,1)} Y$ correspond to path functional dependencies, which have been first studied by Weddell [24], and in a slightly stronger setting by Thalheim [21]. In both papers, you will also find a sound and complete system of inference rules for path functional dependencies.

In the sequel, we use the notion of a *path cardinality constraint* if we refer to a path participation or path co-occurrence constraint. Theorem 5 proves pivoting to be constraint-preserving, due to the expressive power of path cardinality constraints. In the special case, where pivoting is guided by a unary functional dependency, a similar result for path functional dependencies is given in [6].

Theorem 5 *Let Σ be a set of path cardinality constraints defined on $\underline{D}$. Then there is a set Σ_N of path cardinality constraints such that $(\underline{D}, \Sigma)$ and $(\underline{D}_N, \Sigma_N)$ are equivalent.*

The theorem may be verified using the transformation rules presented in Section 5. Consider a path P of the original schema used in a constraint in Σ. If P does not contain the fixed relationship type $\underline{r}$, then P is also a path of the pivoted schema. Otherwise, P has to be replaced by a modified path P_N. If P contains a link $(\underline{r}, \underline{u})$ with $\underline{u} \in S$, we replace this link by the new links $(\underline{r}_N, \underline{s}_N)$ and $(\underline{s}_N, \underline{u})$ to obtain P_N. If not, we simply replace $\underline{r}$ by $\underline{r}_N$ to get P_N. Clearly, P_N is a path of the pivoted schema. Now, for every constraint in Σ using P in its declaration, we replace P by P_N. As a result, we obtain Σ_N. Due to Lemma 1, it remains to add the (path) participation constraint $card(\underline{r}_N, \underline{s}_N) = (1, \infty)$.

Finally, a brief remark on the path manipulations is called for. As seen above, pivoting might increase the length of paths used in a cardinality constraint by 1. In a path co-occurrence constraint $\underline{r}_N : X \xrightarrow{(a,b)} Y$ where all paths in X and in Y start with the link $(\underline{r}_N, \underline{s}_N)$, we may delete the initial vertex $\underline{r}_N$ from each path, and redeclare this constraint on the relationship type $\underline{s}_N$. In practice, this often happens. The simplification has been applied in Theorems 2,3. For participation constraints however, Theorem 4 states that this simplification is very rare.

8   Concluding Remarks

Pivoting was suggested in [5] for breaking-up a relationship type on the semantic level. It generalizes the well-known decomposition of a relation type guided by a functional dependency. We continued this study by investigating pivoting in the presence of cardinality constraints. In Section 5, we defined a suitable notion of pivoting in an extended entity-relationship model. This is of interest as the entity-relationship model is not only popular as an intuitive tool for conceptual modeling, but becomes more and more prevalent as a basis upon which real database management systems are built.

It is natural to ask, whether all application-dependent constraints may be adjusted in the pivoted schema. Examples served to illustrate that this is not always possible, at least not if we insist on the usual concept of cardinality constraint. To overcome this limitation, we defined path cardinality constraints and proved that this generalization is sufficient to ensure the equivalence of the original schema and its pivoted version.

Our discussion showed that pivoting does not necessarily result in a schema which is 'better' than the original one. On the one side, pivoting may be used to increase the efficiency of constraint enforcement by reducing redundancy. On the other side, we have to cope with paths in the constraints. The decision depends on the pay-off expected from decomposition during the life-time of the designed database.

In Section 5, we assembled a number of cases when pivoting is reasonable. They should be considered as rules of thumb that can assist a designer in the design process. In Section 6, we characterized the well-behaving cases when path constraints do not come into play.

We believe that, path participation constraints shall not be seen as an obstacle that prevents a designer from pivoting a relationship type whenever this is promising. These constraints are easy to understand, and to enforce in a database system. It would be interesting to have a complete set of inference rules for path participation constraints. Unfortunately, reasoning about these constraints turns out to be different from the original case. While for ordinary participation constraints, the consistency merely depends on the existence of critical cycles [17] in the entity-relationship diagram, this does hold not for path participation constraints. However this issue is out of the scope of the paper, but suggested for future investigation.

References

1. W.W. Armstrong, Dependency structures of database relationship, Information Processing 74 (1974) 580-583.
2. P. Assenova and P. Johannesson, Improving quality in conceptual modelling by the use of schema transformations, in: B. Thalheim (ed.), Conceptual Modeling (Springer, Berlin, 1996) 277-291.
3. C. Batini, S. Ceri and S.B. Navathe, Database design with the ER model, (Benjamin/Cummings, Menlo Park, 1991).

4. J. Biskup, R. Menzel and T. Polle, Transforming an entity-relationship schema into object-oriented database schemas, in: Proc. ADBIS'95 (Moscow, 1995) 67-78.
5. J. Biskup, R. Menzel, T. Polle and Y. Sagiv, Decomposition of relationships through pivoting, in: B. Thalheim (ed.), Conceptual Modeling (Springer, Berlin, 1996) 28-41.
6. J. Biskup and T. Polle, Decomposition of database classes under path functional dependencies and onto constraints, in: B. Thalheim and K.-D. Schewe (eds.), Foundations of Information and Knowledge systems (Springer, Berlin, 2000) 31-49.
7. D. Calvanese and M. Lenzerini, Making object-oriented schemas more expressive, Proceedings of the Thirteenth ACM SIGACT-SIGMOD-SIGART Symposium on Principles of Database Systems, (ACM Press, Minneapolis, 1994) 243-254.
8. P.P. Chen, The Entity-Relationship Model: Towards a unified view of data, ACM Trans. Database Syst. 1 (1976) 9-36.
9. E.F. Codd, A relation model of data for large shared data banks, Commun. ACM 13 (1970) 377-387.
10. D.W. Embley, B.D. Kurtz and S.N. Woodfield, Object oriented systems analysis: a model-driven approach, (Yourdon Press Series, Prentice Hall, 1992).
11. J. Grant and J. Minker, Inferences for numerical dependencies, Theoretical Comput. Sci. 41 (1985) 271-287.
12. J. Gross and J. Yellen, Graph theory, (CRC press, Boca Raton, 1999).
13. S. Hartmann, Über die Charakterisierung und Konstruktion von ER-Datenbanken mit Kardinalitätsbedingungen, Ph.D. thesis (Rostock, 1996).
14. S. Hartmann, On the consistency of int-cardinality constraints, in: T.W. Ling, S. Ram and M.L. Lee (eds.), Conceptual Modeling, LNCS 1507 (Springer, Berlin, 1998) 150-163.
15. S. Hartmann, On interactions of cardinality constraints, key and functional dependencies, in: B. Thalheim and K.-D. Schewe (eds.), Foundations of Information and Knowledge systems (Springer, Berlin, 2000) 31-49.
16. S.W. Liddle, D.W. Embley and S.N. Woodfield, Cardinality constraints in semantic data models, Data Knowl. Eng. 11 (1993) 235-270.
17. M. Lenzerini and P. Nobili, On the satisfiability of dependency constraints in Entity-Relationship schemata, Information Systems 15 (1990) 453-461.
18. H. Mannila and K. Räihä, The design of relational databases, (Addison-Wesley, Reading, 1992).
19. J.A. Makowsky and E.V. Ravve, Dependency preserving refinements and the fundamental problem of database design, Data Knowl. Eng. 24 (1998) 277-312.
20. A. McAllister, Complete rules for n-ary relationship cardinality constraints, Data Knowl. Eng. 27 (1998) 255-288.
21. B. Thalheim, Foundations of Entity-Relationship Modeling, Ann. Math. Artif. Intell. 6 (1992) 197-256.
22. B. Thalheim, Entity-relationship modeling (Springer, Berlin, 2000).
23. J.D. Ullman, Principles of database and knowledge-base systems, Vol. I (Computer Science Press, Rockville, 1988).
24. G.E. Weddell, Reasoning about functional dependencies generalized for semantic data models, ACM Trans. Database Syst. 17 (1992) 32-64.

IS=DBS+Interaction:
Towards Principles of
Information System Design

Dina Goldin[3,1], Srinath Srinivasa[4,2], and Bernhard Thalheim[2]

[1] University of Massachusetts, Boston
100 Morrissey Blvd., Boston, MA 02125
U.S.A
dqg@cs.umb.edu
[2] Brandenburgische Technische Universität
Postfach 101344, D-03013 Cottbus
Germany
{srinath,thalheim}@informatik.tu-cottbus.de

Abstract. Even with the presence of active research communities that study information system design, the term *information system* (IS) still lacks precise formal underpinnings. Unlike for databases, there is no agreement on what constitutes "IS principles." Any significantly advanced IS contains some kind of a database system. On the other hand, any useful database system is actually an IS, providing additional services beyond simply maintaining data and running queries and updates. As a result, the distinction between issues related to databases and to ISs tends to get blurred, and it is not clear that the principles underlining the study of ISs should be different than those for databases. In this paper we argue that the interactive aspect of ISs necessitates a fundamentally different set of IS design principles, as compared to conventional database design. We provide some promising directions for a formal study of IS models, based on the observation that interactive behaviors cannot be reduced to algorithmic behaviors.
Keywords: Information System, Interaction, Database design, Algorithmic behavior, Interactive behavior, Interaction Machines.

1 Introduction

An *information system* (IS) is a generic term referring to software systems that provide information-based services. The notion of an IS has been addressed from various perspectives such as managerial, technical, organizational, etc. [3,30].

However, despite the presence of an active research community that studies conceptual models for ISs, this term still lacks precise formal underpinnings. Unlike for databases, there is no agreement on what constitutes *IS principles*. It

---
[3] Supported by NSF Grant #IRI-9733678.
[4] Supported by the German Research Society, Berlin-Brandenburg Graduate School in Distributed Information Systems (DFG grant no. GRK 316).

A.H.F. Laender, S.W. Liddle, V.C. Storey (Eds.): ER2000 Conference, LNCS 1920, pp. 140–153, 2000.

would hence be desirable to search for precise underpinnings for an IS and how can they be positioned with respect to database design.

Any significantly advanced IS contains some kind of a database system. As a consequence, IS design addresses a number of database issues like conceptual modeling, metadata management, etc. On the other hand, any large contemporary database system is actually an IS, providing additional services beyond simply storing data and running queries and updates. As a result, the distinction between a database and an IS tends to be blurred, and in common discourse it is not clear that the principles underlining the study of ISs should be different than those for databases.

The distinction between a database and an IS is best appreciated when we consider their function, or "job." The *"job of a database"* is to store data and answer queries. This entails addressing issues like data models, schema design, handling distributed data, maintaining data consistency, query evaluation, etc.

Given a query and a particular state of the system, the behavior of a database system, i.e. answering queries, is *algorithmic* in nature. *Algorithms* are Turing computable transformations from a predefined finite input (*question* state) to a finite output (*answer*). In particular, a database management system implements an algorithmic mapping of the form:

(DB contents, user query) → *(DB contents', query answer).*

On the other hand, the *job of an IS* is to provide a *service*, which entails considerations that span the life cycle of the larger system. Services over time are *interactive* in nature, involving user sessions with one or more users, and are specified by models of interaction not reducible to or expressible by algorithms or Turing machines [11,37,39,40]. Since models of interaction are a new area of research, it is not surprising that there have been no "IS principles" until now.

The distinction between a database that manages data and answers queries and an IS that provides services, is schematically depicted in Figure 1. 1(a) shows ISs as wrappers above database systems; while 1(b) shows issues of concern from a database and from an IS perspective.

In this paper, we argue that the interactive aspect of ISs necessitates a fundamentally different set of design principles for ISs, as compared to conventional database design. We provide some promising directions for a formal study of IS models, in particular based on concepts like Interaction Machines [37], Co-design [34] and Interaction Spaces [33]. Section 2 introduces some of the issues that contrast between the concerns of ISs and databases. Section 3 addresses these differences at the formal level – contrasting interactive behaviors with algorithmic behaviors. Section 4 surveys a few of the promising approaches to formalizing the notion of an IS.

2   Databases, Information Systems, and Their Users

This section contrasts between concerns that have motivated IS design versus those that have been the motivating factors for databases.

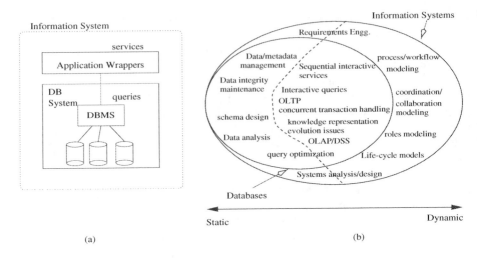

Fig. 1. Information Systems vs. Database Systems

2.1 Historical Perspective

IS design began from a managerial and organizational perspective, and addressed issues such as systems analysis and design, requirements engineering, data flow design, systems development life cycles, etc. Over time, many of these ideas became more formalized. Concepts like life cycle models of ISs obtained support in the form of various representational paradigms and tools. In addition, issues like organizational structure, process structure, coordination and collaboration came to be treated in mathematical terms [23,21,17].

In a parallel vein, the field of Database Systems (DBs) which began from issues like managing flat file systems of data, has progressed greatly to include a number of related issues. Contemporary database systems are no longer just sophisticated file systems; they are actually ISs, whose computation requires a richer model than just a mapping from one database state to another, and whose issues of concern extend beyond questions of data modeling and query evaluation.

The expansion of concerns in both ISs and DBs have resulted in the blurring of the domains. For example, Loucopoulos [22] defines ISs as "systems which are data-intensive, transaction-oriented, with a substantial element of human-computer interaction." It could well be argued that contemporary database research defines a database system similarly.

At the level of foundations, the outlook is somewhat different. Research on *principles of database systems* is well-defined, at least since the middle '80s [35]. It includes issues such as schema design, data modeling, query equivalence and expressiveness, and dependency preservation. On the other hand, there is no consensus on what are *IS principles*, and there are no accepted formalisms for studying them. Formulating the principles that define this area of research will guide its development, just as the principles of set-theoretic relations and first-order logic were useful in relational database research [19], or as object-oriented principles have been instrumental in defining what is meant by OODBs [4].

2.2   Statics vs. Dynamics

An IS functions within an organization or a system, affecting the flow of its activities through the management of information relevant to data gathering and decision making. The dynamics of an IS refers to the *services* that represent semantic information management processes pertinent to the organization.

IS dynamics are usually *interactive* in nature. An interactive service is the transduction of one or more autonomous *input streams* of user requests into *output streams* of system feedback, accompanied by an *evolving system state* [11].

Note that the input consists of all the user requests, throughout the (unbounded) lifetime of the system, and *not* of single commands that constitute the low-level tokens of the input. The system *evolves* over time, during its computation over its input, allowing its behavior to be *history dependent* and its services to be *individualized*.

A database, on the other hand, is fundamentally concerned with maintaining the *static* aspects of the system. Its computational episodes are the query and modification directives to the elements of the system structure. Database dynamics are carried out over individual strings representing single user requests (queries or updates), executed over the current database *instance*. The instance and the request can both be represented as finite strings over a finite alphabet. Together, they constitute the input for the database computation, which maps to a new database instance and/or to a query answer.

Whereas database computations can be modeled by algorithms or Turing machines, capturing IS dynamics requires interactive models of computation that are not directly reducible to algorithmic computation. Section 3 contrasts DB and IS dynamics in a formal manner.

2.3   Individualization of Interactive Services

Users of services provided by ISs play much more of a role than just supplying queries. They expect individualized services that continue meeting their needs as their relationship with the IS evolves over time. Individualization can be viewed as a *projection* of the IS onto the user's space, tailoring the available services to the user preferences and characteristics. Examples are: (a) *List the current valuation for all houses on my street* and, (b) *Supply purchase recommendations for me based on my purchase history.*

Individualization requires awareness of *user characteristics* and *user preferences*, as well as of the *history* of user interaction with the system. To individualize information services, an IS must be able to create and maintain *user profiles* with the relevant information. Some user characteristics are fixed (such as gender) while others may be reset by the user (such as their nickname). Some characteristics are never set by the user explicitly; these include simple facts (such as the user's software version number) as well as facts that are derived directly from the interaction history (such as the user's interests and patterns of behavior). The latter category of user preferences is the most interesting, from a research point of view.

Individualization of interactive services can be considered an extension of the concept of views in database design. Database views are static individualized structures, while IS individualization involves the dynamics as well. Database views contain data from the central database, but with possibly a different schematic structure. Individualized IS views may contain information about user preferences and interaction history that are not part of the larger IS, and also constraints which affect how IS services are used by the user. A database view is a subset of the larger database; but an individualized service may have completely different process structures that are not part of the larger IS.

2.4   Modeling the User

Since users generate the input streams for the IS, they constitute the external *environment* within which the IS performs its computation [11]. Representing the environment of an interactive system is necessary in order to have a complete model that can predict the system's behavior [38]. It is not surprising then that user modeling is an integral part of most IS design processes.

By contrast, it is an axiom of database theory that all queries are expressions over a non-ambiguous query language whose semantics depend only on the database, and not on the user who generated the request [35]. Though the user is implicitly present, it is assumed that his/her wishes are completely expressed by the query itself. Since user profiles are a part of the IS rather than a part of the database that underlies the IS, it follows that individualization (section 2.3) is strictly an IS issue.

The argument that a conceptual model for ISs (*hearers*) must also represent the *agents* (*actors, speakers*) in the environment of the system [18,9] underscores the importance of interaction as a part of that model: interaction involves at least two parties, so both must be modeled to at least some extent. Johannesson [18] points out that this is not (yet) the current practice in IS conceptual modeling.

The emergence of the *Unified Modeling Language* (UML) [8] as a new notation for modeling OODBs [20] is another acknowledgement that interaction must be modeled as a first-class entity. In an object-oriented setting, UML provides a new, interactive, modeling paradigm, by enabling the direct modeling of a system's environment and its interaction with the environment [12]. The blurring between DB and IS fields, discussed earlier, intensifies for OODBs because operations on objects are a part of the schema. Interaction modeling provides a useful metric for separating the two.

3   Formal Models of DBs and ISs

This section considers differences between DBs and ISs in a more formal manner, and provides motivations regarding why IS principles may need to be determined separately.

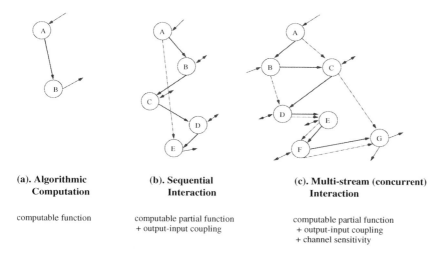

**(a). Algorithmic
Computation**

**(b). Sequential
Interaction**

**(c). Multi-stream (concurrent)
Interaction**

computable function

computable partial function
+ output-input coupling

computable partial function
+ output-input coupling
+ channel sensitivity

Fig. 2. Algorithmic and Interactive Processes

3.1   Algorithmic vs. Interactive Dynamics

With the onset of the object oriented paradigm, many researchers believe that
the distinction between high-level conceptual modeling and the lower-level data
organization and management gets bridged. This is because data and their associated processing may now be represented in terms of problem domain objects
and interactions among them. Many approaches towards database design hence
proceed from an application perspective – that is, progressively coming down
to low-level data modeling from higher-level conceptual modeling [7]. However,
paradigms for modeling the dynamics of object-oriented databases are unclear
regarding whether services provided by objects are part of the object database
or are outside it [25].

In a similar vein, the principles of IS design (as contended by Alter [3]) are
said to boil down to management of the following three issues: (a) information
content; (b) logical organization of the information; and (c) mechanisms to enable
users to obtain whatever information they need. This seems to suggest that
database design and IS design converge using the object paradigm. However,
as contended by Wegner and Goldin [37], the paradigm shift from the earlier
procedural (algorithmic) to object-oriented (interactive) paradigms involves a
qualitative shift in design principles, rather than a simple shift in the granularity
of programming.

Figure 2 contrasts between database and IS dynamics. Figure 2(a) depicts an
algorithmic mapping of a *problem state* to a *solution state*. Such closed-function
behavior is characteristic of read-only databases that answer queries by mapping
$(query, database)$ to $(answer)$. Figure 2(b) depicts a sequential interactive process, characteristic of a single-user read-write database. Here, the query answer
depends on the present database state, which is a result of past interactions between the database and the user. Figure 2(c) depicts multi-stream interaction,
characteristic of the dynamics of distributed ISs, where interactive processes oc-

cur concurrently over multiple I/O streams. The query answer on any stream depends on the current system state, and which in turn depends on past interactions on the current stream as well as other streams.

In the figure, while the system state changes according to the sequence $ABCDEFG$, the interaction stream on the right perceives the state change sequence as ACG and the one on the left perceives it as $BDEFG$. Multi-stream interactions also model true concurrency and activities like coordination and collaboration between two or more interacting actors.

A Turing Machine (TM) is a formal model for string-based algorithmic computation such as that of Figure 2(a). Sequential interaction, depicted in Figure 2(b), corresponds to stream-based computation, where the state is persistent over the computation. This is modeled by Sequential (single-stream) Interaction Machines (SIMs) [38]. Persistent Turing Machines (PTMs) are one example of SIMs [11], providing a minimal extension to Turing Machines by having an internal worktape whose contents are maintained intact from one algorithmic computation to the next. Sequential interaction is also expressed by other abstract models like *labeled transition systems* [24], *coalgebras* [16] and *transducers* [35].

Multi-stream interaction, on the other hand, is still largely unexplored. Many areas of computation have dealt with multi-stream interaction. Some examples are collaboration and coordination models in distributed computing [5,26], process management and resource sharing. However, a generic mathematical model of multi-stream interaction is still to be agreed upon. Wegner and Goldin [39, 40] propose Multi-stream Interaction Machines (MIMs) for this purpose, as an extension to the SIM model.

A MIM can be considered to be an abstract model for ISs, in the same way as a TM and SIM can represent read-only and read-write database management systems. We expect that a mathematical model for MIM computation will someday emerge from the theory community, that can be used to provide underpinnings for activities like collaboration, coordination and ISs in general.

3.2   Algorithmic vs. Interactive Solution Spaces

Definition: A *problem solving process* is a sequence of transitions in an abstract *state space* that consists of the different states of the system. A *problem solution space* is the set of all problem solving processes that reach a state in the *solution domain* of the problem.

The solution space of an *algorithmic problem* is represented by a relation over $I \times O$, where I is the *problem domain* and O is the *solution domain*. For example, if the problem is to compute the square of a given number, and the problem domain is the set of all integers, then the solution domain is the set of all positive integers that are perfect squares. The solution space is the set of all integer pairs (p, q), such that $q = p^2$. When represented by strings, this set of pairs is also called the *language* of the algorithm.

An interactive problem solving process may involve multiple interactions before the final state is reached. For example, in Figure 2(b), it is not possible to determine the end state E at the start of the computation, when in state A. The notion of a *language* consisting of a set of strings is no longer appropriate when the term "problem" denotes a task or a job rather than a single question. Interactive problem solving is described in terms of *streams*, which are possibly infinite sequences of computational mappings.

Example: Consider an *answering machine* which stores and plays back messages from the user. It supports three operations: `record`, `playback`, and `erase` [11]. Consider an interactive user session with the following input: `record(Y)`, `playback()`, `record(Z)`, `playback()`. The response `YZ` from the last `playback` depends on both of the previous `record` operations. This is an example of *history-dependent behavior*, characteristic of interactive computing.

Computation of a SIM such as an answering machine can be represented as a *partial function* that consists of mappings of the form $(s, i) \rightarrow (s', o)$, where s is the system state before the computation, i is the input, s' is the system state after the computation, and o is the output. A sequential interactive problem solving process that consists of n interactions is the *transitive closure* of n partial function mappings like the above.

In multi-stream interaction, the system state may be altered by interactions taking place on any stream. In such a case, the partial-function mappings can appear non-deterministic on any given interaction stream, even when the behavior of the process as a whole is deterministic.

The *problem solving process* of algorithmic, sequential interactive and multi-agent interactive computation may be represented in terms of their solution spaces, as follows.

Definition: An *algorithmic problem solving process* (APSP) is represented as a tuple $\langle I, O, \delta \rangle$, where I and O are input and output domains respectively, and δ is a computable function $\delta : I \rightarrow O$ that maps an element of an input (problem) domain to an element of an output (solution) domain.

Definition: A *sequential interactive process* (SIP) is a tuple $\langle S, I, O, \delta, E \rangle$ where S is the *system state*, I and O are *input and output domains* respectively, and δ is a partial function of the form $\delta : S \times I \rightarrow S \times O$ that maps between input and output domains based on the value of the system state, and possibly changes the system state as a result.

E is the environment of the process, generating the input tokens for the process and consuming the output tokens. In an interactive problem solving process, the input values may depend on earlier outputs. This is represented as a possibly non-computable function: $E : O \rightarrow I$

Definition: A *multi-stream interactive process* (MIP) that interacts with n interaction streams or *actors* in its environment, is a tuple $\langle S, I, O, \mathcal{E}, \delta \rangle$, where S is the *state space* of the system; I and O are *input and output domains* respectively, $\mathcal{E}$ is the environment consisting of n actors $\mathcal{E} = \{E_1, E_2, \dots E_n\}$ that are simultaneously interacting with the MIP via autonomous streams, and δ is a computable partial function of the form $S \times \mathcal{E} \times I \rightarrow S \times \mathcal{E} \times O$.

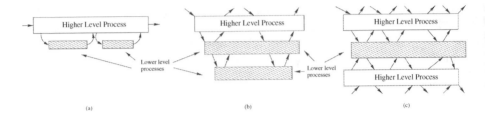

Fig. 3. Algorithmic and Interactive Spaces

Given an input from a particular actor, and the present state of the process, δ maps to an output on possibly another output stream and possibly changes the process state: $(o_k^q, s_k) = \delta(i_k^p, s_{k-1})$; where, at the kth interaction step, the output to the qth actor is a function of the input received from the pth actor. The input and output actors p and q may or may not be the same. The input from an actor in any instance may be coupled to the output received by it from a previous interaction: $i_k^p = E_p(s_j, o_j^p), j < k$.

Multi-stream interactive spaces are more complicated than algorithmic or sequential ones. The solution space of an algorithm or of a sequential interactive process can be broken down into well formed hierarchies, where each level in the hierarchy is at a strictly lower level than those which invoke it. Hence, database transactions that read or write elements of data can be broken down in a hierarchical fashion to deal with multiple data sources. However, multi-stream interactive spaces do not render themselves into well-formed hierarchies.

An interactive process maintains a *coroutine* relationship with its environment, rather than a *subroutine* relationship maintained by algorithmic processes. And when an interactive process interacts over multiple streams, the coroutine nature of an interactive process breaks down hierarchical structures defined by process granularities.

Figure 3 illustrates this distinction. 3(a) represents a database engine executing a transaction over a database, broken down into subtransactions within the scope of the larger transaction. 3(b) represents a read-write database executing transactions. The behavioral space can still be broken down in a hierarchical fashion, as long as the transactions being executed are serializable – that is, as long as the interactive process is sequential. The concept of a hierarchy here is established by the fact that any operation executed by a process at a lower level is within the scope of the higher level process. This criterion breaks down when processes interact over multiple streams.

Figure 3(c) represents multi-stream interaction, where two interactive processes are executing simultaneously on two streams. The processes interact with the IS "core" during their lifetime. However, because of the existence of the other interactive process, the behavior of any single process cannot be reduced to a hierarchical functional mapping. This situation is analogous to non-serializable transactions in groupware databases. Situations involving coordination, collaboration, workflows, etc. are common in IS design. Note that without the existence of multiple processes, 3(c) reduces to a hierarchical space of 3(b).

4   Towards Principles of Information Systems

Next, we consider promising directions for formalizing IS principles. We survey some paradigms that might be explored for formal underpinnings of multi-stream interaction and ISs in general. The motivating factors are the existence of open interactive spaces, and the individualization of interactive services.

Modeling Complex Systems. An IS is a *complex system* [31] whose *raison d'etre* is the set of interactions that it carries out with its environment. Interactive services that a system provides to its users result in evolution of the system behavior in time. A *snapshot* of the system captures its *static* aspects, such as the data schema, its contents, the current interaction history, and user account information. Some of these are fixed for a given system, while others change (*evolve*) over time; we refer to the latter as *system state*.

A snapshot model of a system, by capturing only its static aspects, is inadequate for expressing its *dynamics*. These are not just processes that manipulate the system state, but services that give emergent properties to the IS. IS dynamics can either be algorithmic, sequential interactive, or concurrent interactive – represented by a TM, SIM or MIM respectively. TMs, SIMs and MIMs also represent an increasing gradient of specification complexity. Specification of MIM spaces is harder (requires a more sophisticated notation) than specification of a SIM space, which in turn is harder than TM space specification.

Since MIMs are a generic model for ISs, the complexity of modeling an IS comes from the complex nature of MIM solution spaces. While there have been many paradigms that represent dynamic processes, the question of interactive services operating concurrently on multiple streams has not received adequate attention. In the list below, we survey some approaches towards this end which could provide promising directions towards principles of IS design.

Design Patterns. In order to model and address the complexities of multi-stream interactions, the first step is usually to look for behavioral archetypes or "patterns." Well-known patterns of behavior in ISs include the Model/View/Controller framework, the Publisher-Subscriber pattern, and the Observer pattern [10,27]. However, the notion of behavioral patterns still lacks a precise definition and a notational semantics, which impedes the formalization and automatization of their usage.

Actors and roles. Actors [1] are autonomous objects of the IS, whose behavior at any given instant is determined by the role that they have adopted. This paradigm has also been used by the *use case* modeling community [8] and the Artificial Intelligence (AI) community [2]. The evolution in AI from logic and search to agent-oriented models is not a tactical change, but a strategic paradigm shift to more expressive interactive models [36]. IS modeling based on actors has extended to different related paradigms like mobile agents and collaborative agents [41,15]. Actors also have been formalized using underpinnings from process calculi.

While the actor and agent paradigms provide an intuitive way of modeling a dynamic system, they are "entity centric" in nature. That is, domain entities or actors form the building blocks of the model, and the system dynamics are

represented on top of the actors. To address the complexities of interactive solution spaces, interactions between entities need to be modeled and managed
explicitly.

Coalgebras. Jacobs and Rutten [16,28] propose coalgebras as a general paradigm for modeling dynamic systems. A *dynamic system* is represented as a
tuple $\langle S, \alpha_S \rangle$, where S represents the state space of the system and α_S is called
the system "dynamics." α_S is of the form $S \rightarrow F(S)$ which maps each state
in S to a set of dynamics determined by $F(S)$. Such a system is also called an
F-coalgebra. Coalgebras present a powerful mechanism for representing many
different kinds of dynamic systems using a common formal framework. However,
channel sensitivity of multi-stream interactions cannot be modeled using the
above paradigm.

Categories and Abstract State Machines. A category K is a collection
of *objects obj(K)* and *morphisms mor(K)* such that:

1. Each morphism f has a *typing* on a pair of objects A, B written $f : A \rightarrow B$.
2. There is a partial function on morphisms called *composition* denoted by $\circ$. The
   composition $g \circ f : A \rightarrow C$ is *well-formed* if $g : B \rightarrow C$ and $f : A \rightarrow B \in mor(K)$.
3. Composition is associative: $h \circ (g \circ f) = (h \circ g) \circ f$.

If we take objects of a category to be the system state and morphisms to be the
dynamics between system states, then a category represents a dynamic entity.
If each object represents a subsystem, we obtain a model of an Abstract State
Machine (ASM) [13]. But, as with coalgebras, categories are not channel sensitive
and cannot model MIM solution spaces.

Speech-Act Formalisms. IS modeling based on speech-act formalisms [18]
considers IS dynamics to be made up of interrelated events. The events are
formalized using speech-act theory [29], classifying speech into different types
such as *assertive, commissive, directive declarative and expressive*, that affect
actions. The IS in [18] is made up of *discourses* based on speech acts that connect
events that comprise the IS dynamics. Discourses also connect events with *objects*
that comprise the IS statics.

It is possible that discourses can model multi-stream interaction by depicting
how interactions (*cause events*) on one stream affect interactions (*effect events*)
on an other stream. The logical formalisms for such a model are based on *deontic
logic*, a type of modal logic used to represent normative system behavior.

Stocks and Flows. *Stocks and flows* is a formal model reflecting managerial and organizational perspectives on IS design [14]. In this model, a system
consists of static elements called *stocks* (the "nouns" of the system) and dynamic elements called *flows* (the "verbs"). Stocks are represented by variables and
flows are represented by difference equations over stocks and environments. A
configuration of stocks and flows forms a system which interacts with one or
more "environments."

Stocks and flows modeling provides an intuitive abstraction for IS design.
However, it lacks mechanisms for abstraction, reasoning, and integrity checks on
the system that are usually found in conventional IS modeling paradigms.

Codesign and Interaction Spaces. IS *codesign* [6] addresses the design of both statics and dynamics simultaneously. Global and local (individualized) views are also defined for the dynamics, analogous to corresponding paradigms for the statics. Codesign classifies IS design issues into four categories: *individual-static, individual-dynamic, global-static and global-dynamic*. The *individual-static* aspects concern database views and their computation; while *global-static* concerns design of global static issues like conceptual modeling, logical design, etc. The *global-dynamic* aspect relates to the design issues for the application software of the IS; and *local-dynamic* aspects are individualized dynamic elements of the IS.

Interaction Spaces [32,33], which forms a part of the Codesign approach, proposes a formalism for representing the local-dynamic and global-dynamic aspects. In this approach, the IS concerns are the static *entity space* and the dynamic *interaction space*. The characterization of a space is called a *schema*. Thus the system model consists of the *entity schema* representing the system structure, and the *interaction schema* representing system dynamics. The interaction schema denotes the global-dynamic aspect of the IS, while entities that make up the interaction schema (*dialogs*) form the local-dynamic aspects of the IS.

5 Conclusions and Acknowledgements

This paper presented some promising directions for establishing *IS principles* as a well-defined area of research, unified with the *database principles* community by a common framework, but whose concerns and techniques are *interactive* and therefore distinct from those of databases. The remaining challenge seems to be the predominantly uncharted territory of modeling multi-stream interactive behaviors. Bridging this gap would provide strong foundations for areas of systems design and empirical computer science.

We would like to thank Peter Wegner, whose work on Interaction served as our inspiration, and David Keil, who spent many hours helping with the editing.

References

1. Agha, G., Mason, I.A., Smith, S., Talcott, C. Towards a Theory of Actor Computation. In Cleaveland, R. (ed.), *Proc. of CONCUR '92*, LNCS Vol. 630, Springer-Verlag, Berlin Heidelberg (1992), pp. 565–579.
2. Agre, P.E., Rosenschein, S.J. *Computational Theories of Interaction and Agency.* MIT Press (1996).
3. Alter, S. *Information systems: A Management Perspective.* Benjamin/Cummins (1996).
4. Atkinson, M., DeWitt, D., Maier, D., Bancilhon, F., Dittrich, K., Zdonik, S. The Object-Oriented Database System Manifesto. In Bancilhon, F. et al., (eds.) *Building an Object-Oriented Database System, The Story of O2.* Morgan Kaufmann (1992).
5. Carriero, N., Gelernter, D. Linda in Context. *CACM* **32**, 4 (1989), pp. 444–458.
6. Clauß, W., Thalheim, B. Abstraction Layered Structure Process Codesign. In Ram, J. (ed), *Proc. of COMAD 97* Narosa Publishers, Chennai, India (1997).

7. Embley, D. *Object Database Development: Concepts and Principles*. Addison-Wesley (1998).
8. Fowler, M.; with Scott K., *UML Distilled: Applying the Standard Object Modeling Language*. Reading, MA: Addison-Wesley (1997).
9. Falkenberg, Eckhard D., et al. A Framework of Information System Concepts (The FRISCO Report). IFIP WG 8.1 (1998).
10. Gamma, E., Helm, R., Johnson, R., Vlissides, J. *Design Patterns: Elements of Reusable Object-Oriented Software*. Addison-Wesley (1995).
11. Goldin, G. Persistent Turing Machines as a Model of Interactive Computation. *Proc. of FoIKS*. Burg (Spreewald), Germany (2000).
12. Goldin, D., Keil, D., Wegner, P. An Interactive Viewpoint on the Role of UML. Submitted for publication, June 2000.
13. Gurevich, Y. May 1997 Draft of the ASM Guide. Technical Report, Univ. of Michigan EECS Department, CSE-TR-336-97.
14. Hannon, B., Ruth, M. *Dynamic Modeling*. Springer-Verlag (1994).
15. Huhns, M.N., Singh, M.P., Les Gasser (eds.). *Readings in Agents*. Morgan Kaufmann Publishers (1998).
16. Jacobs, B., Rutten, J.J.M.M. A Tutorial on (Co)Algebras and (Co)Induction. *Bulletin of EATCS* **62**, (1997), pp. 222–259.
17. Jin, Y., Levitt, R.E. The Virtual Design Team: A Computational Model of Project Organizations. *Computational and Mathematical Organization Theory* **2** 3 (1996), pp. 171–196.
18. Johannesson, P. Representation and Communication - A Speech Act Based Approach to Information Systems Design. *Information Systems* **20** 4 (1995), pp. 291–303.
19. Kanellakis, P.C. Elements of Relational Database Theory. In van Leeuwen, J. (ed.) *Handbook of Theoretical Computer Science*, Vol. B, Elsevier, MIT Press (1990), pp. 1073–1158.
20. Kifer, M. Personal communication with D. Goldin, June 2000.
21. Lawrence, P. (ed.) *Workflow Handbook*. Workflow Management Coalition (1997).
22. Loucopoulos, P., Zicari, R. (eds.). *Conceptual Modeling, Databases and CASE*. John Wiley & Sons, (1992).
23. Malone, T.W., Crowston, K. The Interdisciplinary Study of Coordination. *ACM Computing Surveys* **26** 1 (1994), pp. 87–119.
24. Manna, Z., Pnueli, A. *The Temporal Logic of Reactive and Concurrent Systems*. Springer-Verlag (1992).
25. Nierstrasz, O. A Survey of Object-Oriented Concepts. In Kim, W., Lochovsky F.H. (eds.) *Object-Oriented Concepts, Databases, and Applications*. ACM Press (1989).
26. Papadopoulos, G., Arbab, F. Coordination Models and Languages. *Advances in Computers* **46** (1998), pp. 329–400.
27. Pree, W. *Design Patterns for Object-Oriented Software Development*. Addison-Wesley (1995).
28. Rutten, J.J.M.M. Universal Coalgebra: a Theory of Systems. Technical Report, CS-R9652, Centrum voor Wiskunde en Informatica, Amsterdam, Netherlands (1996).
29. Searle, J. *Speech Acts – An Essay in the Philosophy of Language*. Cambridge University Press (1969).
30. Senn, J., A. *Information Technology in Business: Principles, Practices, and Opportunities*. Prentice Hall (1997).
31. Simon, H. *The Sciences of the Artificial*. MIT Press (1996).

32. Srinivasa, S., Thalheim, B. Dialogs and Interaction Schema: Characterizing the Interaction Space of Information Systems. Technical Report 13/99, BTU-Cottbus (1999).

33. Srinivasa, S. The Notion of the Interaction Space of an Information System. *Proc. of CAiSE'00 Doctoral Consortium.* Stockholm, Sweden, (2000).

34. Thalheim, B. (ed.) Readings in Fundamentals of Interaction in Information Systems. BTU-Cottbus (2000).

35. Ullman, J.D. *Principles of Database and Knowledge-Base Systems.* W. H. Freeman & Co. (1988).

36. Wegner, P. Interactive Software Technology. *CRC Handbook of Computer Science and Engineering* (1996).

37. Wegner, P. Why Interaction is More Powerful than Algorithms. *CACM.* (May 1997).

38. Wegner, P., Goldin, D. Interaction as a Framework for Modeling. In Chen, et al (Eds.) *Conceptual Modeling: Current Issues and Future Directions*, LNCS Vol. 1565 (1999).

39. Wegner, P., Goldin, D. Coinductive Models of Finite Computing Agents. *Electronic Notes in Theoretical Computer Science* **19**, Elsevier (1999).

40. Wegner, P., Goldin, D. Interaction, Computability, and Church's Thesis. To appear in *British Computer Journal.*

41. Weiss, G. *Multiagent Systems: A Modern Approach to Distributed AI.* MIT Press (1999).

A Viewpoint-Based Framework for Discussing the Use of Multiple Modelling Representations

Nigel Stanger

University of Otago, Department of Information Science,
PO Box 56, Dunedin, New Zealand
nstanger@infoscience.otago.ac.nz

Abstract. When modelling a real-world phenomenon, it can often be useful to have multiple descriptions of the phenomenon, each expressed using a different modelling approach or *representation*. Different representations such as entity-relationship modelling, data flow modelling and use case modelling allow analysts to describe different aspects of real-world phenomena, thus providing a more thorough understanding than if a single representation were used. Researchers working with multiple representations have approached the problem from different directions, resulting in a diverse and potentially confusing set of terminologies. In this paper is described a viewpoint-based framework for discussing the use of multiple modelling representations to describe real-world phenomena. This framework provides a consistent and integrated terminology for researchers working with multiple representations. An abstract notation is also defined for expressing concepts within the framework.

1 Introduction

In this paper is described a framework for discussing the use of multiple modelling approaches or *representations* to describe a real-world phenomenon. This framework is derived from work on *viewpoint-oriented design*, and provides a consistent and integrated terminology for researchers working with multiple modelling representations. An abstract notation for expressing various concepts within the framework is also defined.

There are several reasons for using multiple modelling representations to describe the same phenomenon, such the ability to provide a more complete description of the phenomenon in question, and because some representations are better suited to particular problems than others. Use of multiple representations and associated issues are discussed in Sect. 2.

The framework described in this paper is primarily derived from earlier work in the area of viewpoint-oriented design methods. The basic concepts of viewpoint-oriented design methods are introduced in Sect. 3, and a lack of clarity is identified with respect to the definitions of some fundamental concepts (in particular, the meaning of the term 'representation').

The author's framework arose out of research into translating models among different modelling representations. There was a need to clarify the definitions

A.H.F. Laender, S.W. Liddle, V.C. Storey (Eds.): ER2000 Conference, LNCS 1920, pp. 154–167, 2000.

of various terms and also to integrate and simplify the potentially confusing range of pre-existing terminologies. The framework is discussed in Sect. 4. In particular, the terms 'representation', 'technique' and 'scheme' are clarified, and the new terms 'description', 'construct' and 'element' are defined.

The author has also developed an abstract notation for expressing various framework concepts in a concise manner. This notation is defined in Sect. 5, and the paper is concluded in Sect. 6.

2 Using Multiple Representations to Describe a Phenomenon

There are many different types of information to be considered when designing an information system, and a wide variety of modelling approaches and notations have been developed to capture these different types of information: entity-relationship diagrams (ERDs), data flow diagrams (DFDs), use case diagrams, the relational model, formal methods and so on. Problems can arise when useful information is omitted from a design. Consider an information system whose data structures are designed using entity-relationship diagrams and are implemented in a relational database. Data entry forms derived from these models are then built. Good design practices are followed throughout, yet the finished application is difficult to use. Some of the commonly used data entry forms have multiple states, but the transitions between these states are unclear to users because state information was not included in the system design.

While this is a purely theoretical example, it serves to illustrate an important point. Information systems are typically built to handle the data processing requirements of some real-world phenomenon. Such real-world phenomena may often be too complex to describe using a single modelling approach, or *representation*. This is supported by the plethora of different representations in existence [15,30], including those that model data structures (such as entity-relationship modelling), and those that model data movements (such as data flow diagrams). This implies that in order to completely model a phenomenon, multiple descriptions of the phenomenon are required, expressed using different representations.

Using multiple representations to describe a phenomenon is also important in other ways:

- if multiple developers are working on a project, each may prefer or be required to use a different representation to describe their particular part [1];
- particular subproblems may be better described using some representations than others [9]; and
- multiple representations are important when integrating heterogeneous data sources to form a federated or distributed system [1], as each data source may potentially use a different logical data model.

The idea of using multiple representations to model a phenomenon is not new. Grundy et al. examined the issues associated with building multi-view editing systems and integrated software development environments [12,13]. Their work was derived from earlier work on multi-view editing systems for software engineering [16,18].

Atzeni and Torlone [2] suggested the idea of translating between different representations as a means of facilitating the use of multiple representations. They proposed a formal model based on lattice theory that allowed them to express many different data modelling approaches using primitive constructs of a single underlying representation.

Su et al. [28] used multiple representations to assist in integrating heterogeneous data sources to build federated and distributed databases. Their approach is similar in many respects to that taken by Atzeni and Torlone, except that the underlying representation is object-oriented rather than mathematically based.

The Unified Modeling Language [21] also supports the use of multiple representations, and is oriented toward object-oriented modelling and development.

All of these approaches were developed independently of each other during the 1990's, and approached the use of multiple modelling representations from different starting points, resulting in a diverse and potentially confusing set of terminologies (see Table 1).

In addition, viewpoint researchers first suggested using multiple representations to describe viewpoints over a decade ago [10]. A viewpoint is effectively a formalisation of the perceptions of stakeholders with respect to some real-world phenomenon. Since we are dealing with the use of multiple representations to describe real-world phenomena, viewpoints provide a useful framework within which to discuss such use [27]. The author's framework will be described in Sect. 4, but first basic viewpoint concepts must be defined.

3   Viewpoint Concepts

A *viewpoint* can be thought of as a formalisation of the perceptions of a stakeholder group with respect to some real-world phenomenon that is being modelled. The first viewpoint-oriented approach was introduced in 1979 [19], but the concept of a viewpoint was not formalised until ten years later [10].

Viewpoint-oriented methods were originally developed to assist with requirements definition in a software engineering environment [19], and subsequent research has followed a similar direction [9,17,20]. The focus of the author's research has been on how to facilitate the use of multiple representations to describe a single viewpoint, in particular by performing translations among descriptions expressed using different representations [25,26,27].

In Fig. 1 are shown the relationships between the concepts of viewpoint-oriented methods. This initial framework was derived by the author [27] from the work of Finkelstein et al. [10], Easterbrook [9] and Darke and Shanks [7]. Their terminologies are also summarised in Table 1.

3.1   Perspectives and Viewpoints

Easterbrook [9] defines a *perspective* as "a description of an area of knowledge which has internal consistency and an identifiable focus of attention". During the requirements definition phase of systems analysis, developers may encounter many different perspectives on the problem being modelled. Perspectives may overlap and/or conflict with each other in various ways.

Table 1. Comparison of viewpoint/representation terminologies

Term used by the author	Meaning	Example	Corresponding term used by:						
			Finkelstein [10]	Easterbrook [9]	Darke & Shanks [7]	Atzeni & Torlone [2]	Su et al. [29]	UML [21]	Grundy et al. [13]
perspective	A description of a real-world phenomenon that has internal consistency and an identifiable focus.	–	–	perspective	perspective	–	–	–	–
viewpoint	The formatted description of a perspective.	–	ViewPoint	viewpoint	viewpoint	–	–	(use case)[a]	–
technique	A collection of abstract constructs that form a modelling 'method'.	relational model	style ⎫	style ⎫	technique	model ⎫	data model ⎫	meta-model ⎫	–
scheme	A collection of concrete constructs that form a modelling 'notation'.	SQL/92	style ⎭	style ⎭	scheme	model ⎭	data model ⎭	meta-model ⎭	–
representation	The combination of a particular technique and scheme.	relational model + SQL/92	–	–	representation	–	–	–	representation
description	An instantiation of a representation.	SQL/92 schema	specification	description	–	scheme	schema	model	view
construct	The basic unit of a representation.	a relation	–	–	–	construct	class[b]	metaobject[b]	–
element	An instantiation of a construct within a particular description.	Staff table	–	–	–	(varies)[c]	object	object	component

Notes on Table 1:

[a] This could be interpreted in one of three ways: first, a single use case could represent the complete viewpoint of a stakeholder with respect to a particular role; second, the complete viewpoint of a stakeholder could comprise the union of all their associated use cases; or third, a use case could merely be another representation used *within* a viewpoint.

[b] Also 'construct'.

[c] Terms used include 'component', 'element' and 'concept'.

'–' indicates that a term is not used by that author.

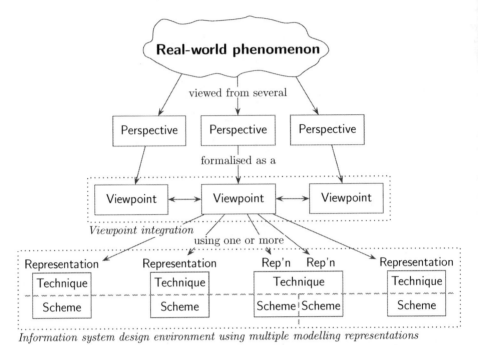

Information system design environment using multiple modelling representations

Fig. 1. Relationship between perspectives, viewpoints and representations

Finkelstein et al. [10] describe a *viewpoint* as comprising the following parts:

- "a *style*, the representation scheme in which the ViewPoint [*sic*] expresses what it can see (examples of styles are data flow analysis, entity-relationship-attribute modelling, Petri nets, equational logic, and so on);
- a *domain* defines which part of the 'world' delineated in the style (given that the style defines a structured representation) can be seen by the ViewPoint (for example, a lift-control system would include domains such as user, lift and controller);
- a *specification*, the statements expressed in the ViewPoint's style describing particular domains;
- a *work plan*, how and in what circumstances the contents of the specification can be changed; [and]
- a *work record*, an account of the current state of the development."

Easterbrook [9] simplifies this description by defining a viewpoint as "the formatted representation of a perspective", noting that a perspective is a "more abstract version of a viewpoint". In effect, a viewpoint is the formalisation of a particular perspective, so there is a one-to-one correspondence between a viewpoint and the perspective it formalises, as illustrated in Fig. 1.

The term 'viewpoint' is very similar to the term 'view' as used in multi-view editing systems [12,16,18]. These terms refer to different concepts, however: a 'view' is more akin to the concept of a *description*, which will be introduced in

Sect. 4. The similarity of the two terms has led to some confusion: the terms 'viewpoint' and 'view' have been used interchangeably in the past [16].

Darke and Shanks [7] define two main types of viewpoint:

1. *user viewpoints* that capture "the perceptions and domain knowledge of a particular user group, reflecting the particular portion of the application domain relevant to that group"; and
2. *developer viewpoints* that capture "the perceptions, domain knowledge and modelling perspective relevant to a systems analyst or other developer responsible for producing some component of the requirements specification".

Since a viewpoint is the formalisation of a perspective, some form of model is required to provide the formalised structure. The concept of a *representation* provides this.

3.2   Representations

Darke and Shanks [7] note that viewpoints may be described using different representation *techniques*, within each of which there may be available a number of representation *schemes*. Neither Darke and Shanks nor Finkelstein et al. [10] clearly define the terms 'representation', 'technique' or 'scheme'; rather, they introduce each term by example. This has led to some confusion in the use of this terminology. Darke and Shanks use the terms 'representation' and 'representation technique' interchangeably, while Finkelstein et al., as can be seen in their definition of a 'style', use the term 'representation scheme' in a similar way.

The intent appears to be that a *representation* should be thought of as a structured modelling approach that can be used to describe the content of a viewpoint. In order to clarify the confusion in terminology, the author has refined this informal definition and defined a representation as the combination of a particular technique and scheme to describe a viewpoint (see Sect. 4).

Finkelstein et al. [10] first mooted the idea of using multiple representations to describe a viewpoint in 1989, but there seems to have been little work in this area since. Darke and Shanks [8] reviewed twelve different viewpoint development approaches and found only two that supported multiple representations to describe a single viewpoint: the Soft Systems methodology [4] and Scenario Analysis [14], which are both user viewpoint approaches.

The author's own research [25,27] has followed the approach of using multiple representations to describe a single *developer* viewpoint, and uses an integrated terminology framework derived from viewpoint-oriented methods.

4   The Extended Viewpoint Framework

Looking at Table 1, there is fairly clear separation between the three viewpoint-oriented terminologies on the left of of the table and the four multiple-view-oriented terminologies on the right. The viewpoint-oriented terminologies deal more with 'high level' concepts and tend to ignore how representations are internally structured; conversely, the multiple-view-oriented terminologies deal more with constructs within representations and tend to ignore higher-level structures. The two sets of terminology are clearly related, yet the only real overlap between them is at the representation level.

The UML comes closest to providing a full set of terms. It was designed from the outset to support multiple representations, and includes a viewpoint-like formalism (use cases). It is, however, deliberately restricted to nine specific representations, whereas the author's framework could be applied to any combination of representations (including those of the UML).

There are also several synonyms in Table 1. For example, the terms 'style', 'representation', 'model' and 'data model' all refer to similar concepts. Conversely, the terms 'scheme' and 'model' are both used to refer to completely different concepts. Such variation can lead to confusion, so there is a definite need to develop a consistent and integrated terminology framework.

The initial framework shown in Fig. 1 needs to be refined. The author has addressed the confusion in terminology by clearly defining the terms 'representation', 'technique' and 'scheme', and has extended the original framework with the concepts of *descriptions*, *constructs* of representations and *elements* of descriptions (see Fig. 2).

4.1 Representations

Informally, a data modelling representation can be thought of as comprising two main parts:

1. a *generic* part that specifies the generic constructs that may be used to describe a viewpoint, such as entities, relations, and so on; which determines
2. a *specialised* part that specifies the constructs peculiar to the representation, along with their visual appearance or notation, such as boxes for entities, SQL `CREATE TABLE` statements for relations, and so on.

Finkelstein et al.'s [10] concept of a 'style' does not clearly distinguish between these two parts, as opposed to the 'techniques' and 'schemes' of Darke and Shanks [7]. It is therefore proposed to use the term *technique* to refer to the generic part of a representation, and the term *scheme* to refer to the specialised part of a representation. In practical terms, a technique can be thought of as a modelling 'approach', such as the entity-relationship approach or the relational model, and a scheme can be thought of as a particular 'notation' within that approach, such as Martin E-R notation or relational calculus. Another way to think of this is that a scheme is an 'instantiation' of a particular technique.

A *representation* can thus be defined as the combination of a particular technique with a particular scheme. In general, a technique may have one or more associated schemes, but each combination of a technique and a scheme forms a distinct representation. For example, the relational model is a technique, with SQL and $QUEL$ as two possible schemes, but the combinations $(Relational, SQL)$[1] and $(Relational, QUEL)$ form two distinct representations. Similarly, the entity-relationship approach (E-R) is a technique, with ERD_{Martin} and ERD_{Chen} as

---

[1] In practice, the many dialects of SQL will form many different representations. This has been ignored here in the interests of clarity.

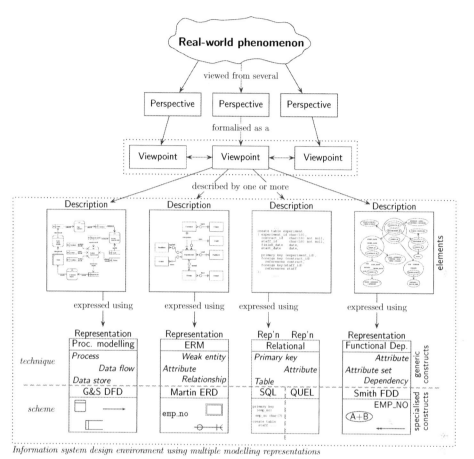

Fig. 2. The extended terminology framework

two possible schemes. The combinations $(E\text{-}R, ERD_{Martin})$ and $(E\text{-}R, ERD_{Chen})$ again form two distinct representations.

It is expected that a technique will not attempt to specify all possible concepts for all possible schemes within that technique. Rather, a technique defines the 'base' model, which is then specialised and extended by schemes to form a representation. This implies that a scheme may provide new constructs to a representation that have no direct analogue in the technique. For example, the relational technique [6] does not include general constraints, but they are an important feature of the relational scheme SQL/92. Similarly, type hierarchies are not part of the base E-R technique [5], but they do appear in some E-R schemes.

4.2   Descriptions

Representations are an abstract concept, so they must be instantiated in some way in order to describe the content of a viewpoint. One way to view the instantiation of a representation is as a set of 'statements' that describe a viewpoint or some subset thereof. Finkelstein et al. [10] refer to this as a 'specification' or 'description'; Easterbrook [9] also refers to this concept as a description. The author has adopted the term 'description' as it emphasises the idea that they are used to *describe* a viewpoint.

A viewpoint is thus specified by a set of *descriptions*, each expressed using some representation, as shown in Fig. 2. Each description may describe either all or some subset of the viewpoint; this is analogous to the concept of a 'view' in multi-view editing systems [12,16,18] or a 'model' in the UML. For example, a developer viewpoint might be specified by the union of these four descriptions:

1. an object class description expressed using UML class diagram notation [21];
2. a functional dependency description expressed using Smith functional dependency diagram notation [23,24];
3. a relational description expressed using SQL/92; and
4. a data flow description expressed using Gane & Sarson data flow diagram notation [11].

Descriptions may be distinct from each other, or they may overlap in a manner similar to viewpoints. Such redundancy can be useful in exposing conflicts both between descriptions and between viewpoints [9].

4.3   Constructs and Elements

Every representation comprises a collection of *constructs*. These may be divided into generic constructs associated with the technique and specialised constructs associated with the scheme, as shown in Fig. 2. The nature of a construct is defined by its *properties*, which include both its *relationships* with other constructs, and its *attributes*, such as name, domain or cardinality.[2] For instance, as illustrated in Fig. 3, a data store in a data flow diagram might have the attributes *name* (the name of the data store), *label* and *fields* (a list of data fields in the data store). The *flows* relationship specifies an association between the data store construct and a list of data flow constructs.

In the same way that a description is an instantiation of a representation, an *element* is an instantiation of a construct; elements are combined to build descriptions. Examples of constructs include object classes, processes and attributes; elements corresponding to these constructs could be Order, Generate invoice and address.

---

[2] The terms 'property', 'attribute' and 'relationship' come from the Object Data Management Group's object model [3].

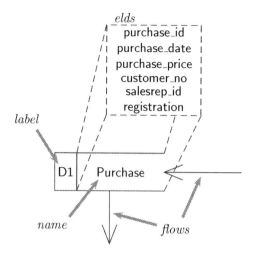

Fig. 3. Properties of a construct

5   A Notation to Express Representations, Descriptions, Constructs, and Elements

It can be cumbersome to discuss representations and descriptions using natural language, for example, 'the Staff regular entity element of the description D_1 (expressed using Martin entity-relationship notation) of the managers' viewpoint'. The author has therefore developed a concise abstract notation (modelled in part on the data transfer notation of Pascoe and Penny [22]) for expressing representations, descriptions, constructs of representations and elements of descriptions. Using the notation, the statement above could be expressed as:

$$D_1(V_{mgrs}, \text{E-R}, ERD_{Martin})\,[\mathit{staff} : \textsc{RegularEntity}]\,.$$

The notation is summarised in Table 2. The author has defined additional notations for expressing translations of descriptions from one representation to another [25]. For example, the expression:

$$D_1(V_t, FuncDep, FDD_{Smith}) \rightarrow D_2(V_t, \text{E-R}, ERD_{Martin})$$

denotes the translation within viewpoint V_t of a functional dependency description D_1 into an entity-relationship description D_2. Such additions are beyond the scope of this paper, however.

5.1   Description and Representation Notation

The notation $D_i(V, T, S)$ denotes that description D_i of viewpoint V is expressed using constructs of technique T and scheme S (this may be abbreviated to D_i when V, T and S are clear). Thus, $D_1(V_p, \text{E-R}, ERD_{Martin})$ denotes a description D_1 of the viewpoint V_p that is expressed using constructs of the entity-relationship technique (E-R) and the Martin ERD scheme (ERD_{Martin}).

Table 2. Summary of the abstract notation

Notation	Associated term	Definition
V	Viewpoint	A formatted expression of a perspective on a real-world phenomenon
T	Technique	A collection of generic constructs that form a modelling 'method', for example, the relational model or object modelling
S	Scheme	A collection of specialised constructs that form a modelling 'notation', for example, SQL/92 or UML class diagram notation
$R(T, S)$ or R	Representation	Representation R comprises constructs defined by the combination of technique T and scheme S
$D_i(V, T, S)$ or D_i	Description	Description D_i of viewpoint V is expressed using constructs of technique T and scheme S
$R(T, S)$ [CON], R [CON], or CON	Construct of a representation	CON specifies a construct of representation $R(T, S)$
$D_i(V, T, S)$ [e : CON], D_i [e : CON], or D_i [e]	Element of a description	e specifies an element (instantiated from construct CON) of description $D_i(V, T, S)$

The notation $\mathfrak{R}(T, S)$ denotes a representation $\mathfrak{R}$ that comprises a collection of constructs defined by the combination of technique T and scheme S (this may be abbreviated to $\mathfrak{R}$ when T and S are clear). Thus, $\mathfrak{R}_e(E\text{-}R, ERD_{Martin})$ denotes the representation $\mathfrak{R}_e$ formed by combining the constructs of the entity-relationship technique ($E\text{-}R$) with the Martin ERD scheme (ERD_{Martin}).

Representations may differ in both the technique and scheme used, or they may share the same technique and differ only in the scheme. For example, consider a viewpoint V_q that has three descriptions D_1, D_2 and D_3. D_1 is expressed using the entity-relationship technique and the Martin ERD scheme, and is denoted by $D_1(V_q, E\text{-}R, ERD_{Martin})$. D_2 is expressed using the object modelling technique and the UML class diagram scheme, and is denoted by $D_2(V_q, Object, Class_{UML})$. D_1 and D_2 differ in both the technique and the scheme used. D_3 is expressed using the entity-relationship technique and the Chen ERD scheme, and is denoted by $D_3(V_q, E\text{-}R, ERD_{Chen})$. D_3 differs from D_1 only in the scheme used.

If the viewpoint, technique or scheme are unspecified, they may be omitted from the notation. Thus, the notation $\mathfrak{R}_r(Relational,)$ denotes any relational representation, and $D_1(, Object, Class_{UML})$ denotes a UML class diagram in an unspecified viewpoint.

5.2   Construct and Element Notation

Constructs are the fundamental components of a representation, whereas elements are the fundamental components of a description. Given a representation $\Re(T, S)$, a construct CON of $\Re$ is denoted by $\Re(T, S)$ [CON], or, if T and S are clear, simply $\Re$ [CON]. Often $\Re$ may also be clear from the context, allowing the $\Re$ [] notation to also be omitted, leaving just CON. The name of the construct itself is denoted by SMALL CAPS.

The construct CON can be thought of as analogous to the concept of a relational domain in that it specifies a pool of possible 'values' from which an element e may be drawn. The notation $e :$ CON denotes that e is a member of the set of all possible elements corresponding to the construct CON.

Now consider a description $D_i(V, T, S)$. An element e of D_i (instantiated from construct $\Re$ [CON]) is denoted by $D_i(V, T, S)$ [$e :$ CON], or, if V, T and S are clear, simply D_i [$e :$ CON]. The construct may also be omitted if it is clear from the context, that is, D_i [e]. The representation $\Re$ is omitted from the construct CON because $\Re$ is implied by T and S in the description and would therefore be redundant.

Examples of construct and element expressions include:

- $\Re_e(E\text{-}R, ERD_{Martin})$ [ENTITYTYPE], denoting the generic entity type construct of the E-R/Martin representation $\Re_e$;
- $D_1(V, FuncDep, FDD_{Smith})$ [$s :$ SINGLEVALUED], denoting a functional dependency element in the Smith notation functional dependency description D_1;
- $D_2(V, Relational, SQL/92)$ [$c_1, \ldots, c_n :$ COLUMN], denoting a collection of column elements in the SQL/92 description D_2; and
- $\Re_s(Object, Statechart_{UML})$[STATE, EVENT], denoting the state and event constructs of the object modelling/UML statechart diagram representation $\Re_s$.

6   Conclusion

In this paper has been described a framework for discussing the use of multiple modelling representations to describe a viewpoint. Earlier work on the use of multiple representations has produced a diverse and potentially confusing set of terminologies, none of which provides a complete set of terms covering all concepts in the area (with the possible exception of the UML). Viewpoint concepts provide a useful framework within which to discuss the use of multiple representations, but there is a lack of clarity over the definitions of the terms 'representation', 'technique' and 'scheme'.

To remedy these issues, the author has clarified the definitions of 'representation', 'technique' and 'scheme', and extended the viewpoint framework with the concepts of *description, construct* of a representation and *element* of a description. Also described was an abstract notation for writing representation, description, construct and element expressions.

The framework described in this paper provides a consistent, integrated terminology and notation for researchers working on the use of multiple representations to describe a viewpoint.

References

1. Paolo Atzeni and Riccardo Torlone. Management of multiple models in an extensible database design tool. In P. Apers, M. Bouzeghoub, and G. Gardarin, editors, *Proceedings of the Fifth International Conference on Extending Database Technology (EDBT'96)*, volume 1057 of *Lecture Notes in Computer Science*, pages 79–95, Avignon, France, March 25–29 1996. Springer-Verlag.
2. Paolo Atzeni and Riccardo Torlone. MDM: A multiple-data-model tool for the management of heterogeneous database schemes. In Joan M. Peckman, editor, *Proceedings of the SIGMOD 1997 International Conference on the Management of Data*, pages 528–531, Tucson, Arizona, May 13–15 1997. Association for Computing Machinery, ACM Press.
3. R.G.G. Cattell, Douglas K. Barry, Mark Berler, Jeff Eastman, David Jordan, Craig Russell, Olaf Schadow, Torsten Stanienda, and Fernando Velez. *The Object Data Standard: ODMG 3.0*. Morgan Kaufmann, San Francisco, California, 2000.
4. P.B. Checkland. *Systems Thinking, Systems Practice*. John Wiley & Sons, Chichester, England, 1981.
5. Peter Pin-Shan Chen. The entity-relationship model — Toward a unified view of data. *ACM Transactions on Database Systems*, 1(1):9–36, 1976.
6. E.F. Codd. A relational model of data for large shared data banks. *Communications of the ACM*, 13(6), 1970.
7. Peta Darke and Graeme Shanks. Viewpoint development for requirements definition: Towards a conceptual framework. In *Proceedings of the Sixth Australasian Conference on Information Systems (ACIS'95)*, pages 277–288, Perth, Australia, September 26–29 1995.
8. Peta Darke and Graeme Shanks. Stakeholder viewpoints in requirements definition: A framework for understanding viewpoint development approaches. *Requirements Engineering*, 1:88–105, 1996.
9. Steve M. Easterbrook. *Elicitation of Requirements from Multiple Perspectives*. PhD thesis, Imperial College of Science Technology and Medicine, University of London, London, 1991.
10. A.C.W. Finkelstein, M. Goedicke, J. Kramer, and C. Niskier. ViewPoint oriented software development: Methods and viewpoints in requirements engineering. In J.A. Bergstra and L.M.G. Feijs, editors, *Proceedings of the Second Meteor Workshop on Methods for Formal Specification*, volume 490 of *Lecture Notes in Computer Science*, pages 29–54, Mierlo, The Netherlands, September 1989. Springer-Verlag.
11. C. Gane and T. Sarson. *Structured Systems Analysis: Tools and Techniques*. Prentice-Hall Software Series. Prentice-Hall, Englewood Cliffs, New Jersey, 1979.
12. John C. Grundy. *Multiple Textual and Graphical Views for Interactive Software Development Environments*. PhD thesis, Department of Computer Science, University of Auckland, Auckland, New Zealand, June 1993.
13. John C. Grundy and John G. Hosking. Constructing integrated software development environments with MViews. *International Journal of Applied Software Technology*, 2(3/4):133–160, 1997.
14. P. Hsia, J. Samuel, J. Gao, D. Kung, Y. Toyoshima, and C. Chen. Formal approach to scenario analysis. *IEEE Software*, 11(2):33–41, March 1994.
15. Richard Hull and Roger King. Semantic database modeling: Survey, applications, and research issues. *ACM Computing Surveys*, 19(3):201–260, 1987.
16. D.A. Jacobs and C.D. Marlin. Software process representation to support multiple views. *International Journal of Software Engineering and Knowledge Engineering*, 5(4), December 1995.

17. Gerald Kotonya and Ian Sommerville. Requirements engineering with viewpoints. *Software Engineering Journal*, 11(1):5–18, 1996.
18. S. Meyers. Difficulties in integrating multiview environments. *IEEE Software*, 8(1):49–57, January 1991.
19. G. Mullery. CORE — A method for controlled requirements specification. In *Proceedings of the Fourth International Conference on Software Engineering*, pages 126–135, Munich, Germany, September 17–19 1979. IEEE Computer Society Press.
20. B. Nuseibeh, J. Kramer, and A.C.W. Finkelstein. A framework for expressing the relationships between multiple views in requirements specification. *IEEE Transactions on Software Engineering*, 20(10):760–773, 1994.
21. Object Management Group. *OMG Unified Modeling Language Specification*. Object Management Group, Inc., 1.3 edition, June 1999.
22. Richard T. Pascoe and John P. Penny. Constructing interfaces between (and within) geographical information systems. *International Journal of Geographical Information Systems*, 9(3):275–291, 1995.
23. Henry C. Smith. Database design: Composing fully normalized tables from a rigorous dependency diagram. *Communications of the ACM*, 28(8):826–838, 1985.
24. Nigel Stanger. Modifications to Smith's method for deriving normalised relations from a functional dependency diagram. Discussion Paper 99/23, Department of Information Science, University of Otago, Dunedin, New Zealand, December 1999.
25. Nigel Stanger. *Using Multiple Representations Within a Viewpoint*. PhD thesis, University of Otago, Dunedin, New Zealand, November 1999.
26. Nigel Stanger. Translating descriptions of a viewpoint among different representations. Discussion Paper 2000/11, Department of Information Science, University of Otago, Dunedin, New Zealand, May 2000.
27. Nigel Stanger and Richard Pascoe. Environments for viewpoint representations. In Robert Galliers, Sven Carlsson, Claudia Loebbecke, Ciaran Murphy, Hans Hansen, and Ramon O'Callaghan, editors, *Proceedings of the Fifth European Conference on Information Systems (ECIS'97)*, volume I, pages 367–382, Cork, Ireland, June 19–21 1997. Cork Publishing.
28. S.Y.W. Su and S.C. Fang. A neutral semantic representation for data model and schema translation. Technical report TR-93-023, University of Florida, Gainesville, Florida, July 1993.
29. S.Y.W. Su, S.C. Fang, and H. Lam. An object-oriented rule-based approach to data model and schema translation. Technical report TR-92-015, University of Florida, Gainesville, Florida, 1992.
30. D. Tsichritzis and F. Lochovsky. *Data Models*. Prentice-Hall, 1982.

Practical Approach to Selecting Data Warehouse Views Using Data Dependencies

Gillian Dobbie and Tok Wang Ling

Department of Computer Science, National University of Singapore, Singapore
{dobbie,lingtw}@comp.nus.edu.sg

Abstract. Materialized views in data warehouses are typically complicated, making the maintenance of such views difficult. However, they are
also very important for improving the speed of access to the information
in the data warehouse. So, the selection of materialized views is crucial
to the operation of the data warehouse both with respect to maintenance
and speed of access. Most research to date has treated the selection of
materialized views as an optimization problem with respect to the cost
of view maintenance and/or with respect to the cost of queries. In this
paper, we consider practical aspects of data warehousing. We identify
problems with the star and snowflake schema and suggest solutions. We
also identify practical problems that may arise during view selection and
suggest heuristics based on data dependencies and access patterns that
can be used to measure if one set of views is better than another set of
views, or used to improve a set of views.

1 Introduction

A data warehouse stores huge volumes of data that has been gathered from one
or more sources for the purpose of efficiently processing decision support or on-
line analytic processing (OLAP) queries. Like in traditional database systems,
frequently asked queries or subparts of frequently asked queries may be precom-
puted and stored as materialized views, providing faster access. Obviously there
are many possible views that could be materialized, and the selection of which
views to materialize is a trade-off between the cost of the view maintenance and
the speed of access. Views in data warehouses are usually more complicated than
in traditional database systems, typically based on many tables and including
aggregation or summarization of the underlying data in the data warehouse.

Most research to date has treated the selection of materialized views as an
optimization problem with respect to the cost of view maintenance and/or with
respect to the cost of queries [1,4,8,9,10,11]. Each paper proposes an algorithm
designed within the framework of general query and maintenance cost models
without considering the physical properties of the actual data. In this paper,
we identify practical problems that may arise during view selection and suggest
heuristics based on data dependencies and access patterns that can be used to
measure if one view is better than another or used to improve a set of views.
Our work is related to physical database design and materialized view design in
the relational databases.

A.H.F. Laender, S.W. Liddle, V.C. Storey (Eds.): ER2000 Conference, LNCS 1920, pp. 168–182, 2000.
© Springer-Verlag Berlin Heidelberg 2000

The paper is organized as follows. Section 2 provides background information relevant to the rest of the paper, highlights problems with the widely accepted star and snowflake schema, and presents the enhanced star and snowflake schema. There is a sample data warehouse in Section 3 that is used in the following sections. Section 4 compares different views suggesting why one is better than another. Heuristics for good design are outlined in Section 5, and demonstrated in Section 6. We conclude in Section 7.

2 Background

In this section, we introduce the star and snowflake schema, describe inadequacies of the star and snowflake schema, present the enhanced star and snowflake schema, and introduce strong and weak functional dependencies.

The schema of a data warehouse that is built on top of a relational database is typically organized as a *star* or *snowflake* schema [6]. A *star schema* consists of a fact table, and a table for each dimension in the fact table. A star schema can be represented using an entity relationship diagram as shown in Figure 1. The fact table represents a relationship set between two or more dimension entities and has some measurement attributes (*salesQty* in Figure 1). Each dimension table represents an entity with an identifier and other single valued attributes. A *snowflake schema* is like a star schema except it represents the dimensional hierarchies directly, normalizing the dimension tables.

Example 1. Consider a data warehouse that stores information about employees, products and the quantity of each product sold by each employee. There would be a dimension table for employee and another for product storing the employee and product details respectively. The fact table would contain the identifier of employee, the identifier of product and the quantity of a product sold by an employee. A dimension table can contain a dimension hierarchy, e.g. if each product belongs to a category and each category has many products, then we say there is a product dimension hierarchy.   □

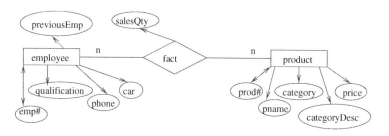

Fig. 1. ER diagram of typical star schema

While the star and snowflake schemas described above are convenient, they are overly simplistic. In practice, not all attributes in a dimension table are single valued, not all keys in dimension tables consist of just one attribute, and the relationships between the levels in a dimension hierarchy may be m-to-n (rather than 1-to-n). Consider the schema in Figure 2. Each employee may have more than one qualification, more than one previous employer, more than one phone and more than one car. That is the attributes in the employee dimension table are more likely to be multi-valued attributes and then the key is a composite key. Although we don't represent it in Figure 2, a product could also belong to more than one category. For example, a *health food drink* may belong to category *health food* and category *beverage*. The following functional dependencies hold on the schema in Figure 2:

$\{emp\#, prod\#\} \rightarrow salesQty$ $emp\# \twoheadrightarrow car$

$emp\# \twoheadrightarrow qualification$ $prod\# \rightarrow \{pname, category, price\}$

$emp\# \twoheadrightarrow previousEmp$ $category \rightarrow categoryDesc.$

$emp\# \twoheadrightarrow phone$

Fact Table	Employee Dimension Table	Product Dimension Table
emp#	emp#	prod#
prod#	qualification	pname
salesQty	previousEmp	category
	phone	categoryDesc
	car	price

Fig. 2. Sample enhanced star schema

While the snowflake schema eliminates the redundancy in the dimension hierarchy, the problems we have described are not alleviated. An alternative organization would be to store the composite key of the employee dimension table in the fact table: *Fact Table(emp#, qualification, previousEmp, phone, car, prod#, salesQty)*. This organization is worse than the schema in Figure 2. Not only are we duplicating all the employee information, we are also confusing the relationship between *emp#*, *prod#*, and *salesQty*.

In summary, the star and snowflake schemas that are proposed in many papers and books are overly simplistic and not practical for real world applications. Similarly, any view design heuristics that are based on the same assumptions will be overly simplistic.

Strong and *weak* functional dependencies [7] extend classical functional dependencies and when used in relational database design can provide better schemas than those produced using classical functional dependencies.

Strong functional dependency: Let $X \rightarrow Y$ be a functional dependency such that for each $z \in Y$, $X \rightarrow z$ is a full functional dependency. Then $X \rightarrow Y$ is a *strong functional dependency* if the values of all the attributes in Y will not be updated, or if the update need not be performed at real-time or on-line and such updates seldom occur.

Example 2. A person's name seldom changes, so we can say there is a strong functional dependency between *employee_number* and *employee_name*. We write this as $employee\_number \overset{S}{\to} employee\_name$.   □

Weak functional dependency: Let X and Y be subsets of a table R, such that there is a non-trivial multi-valued dependency $X \twoheadrightarrow Y$ in R. If most of the X values are associated with a unique Y-value in R, except for the occasional X-value that may be associated with more than one Y-value, we say Y is *weakly functionally dependent* on X, and write $X \overset{W}{\to} Y$.

Example 3. Assume that typically an employee lists only one phone number, and very occasionally an employee lists more than one phone number, then we can say that $employee\_number \overset{W}{\to} employee\_phone$.   □

3   Motivating Example

In this section we introduce a populated sample data warehouse that we use in later sections of this paper. While the volume of data in this sample data warehouse (see Figure 3) is unrealistic, the relationships between the fields are realistic and are used to motivate our work. The *Fact table* stores the quantity of each product sold by each employee. The *Employee* table contains, for each employee, their employee number, qualifications, previous employers, phone numbers and car registration. It is uncommon for an employee to have more than one telephone number or more than one car, but quite common for an employee to have more than one qualification and previous employer. The *Product* table contains the product number, name, the category the product belongs to, a text description of the category, and the price of the product. Each product belongs to one category. We expect the fact table to be modified often, the price to be modified sometimes and other attributes to change less frequently. The following dependencies hold on the data:

$$emp\# \overset{W}{\to} phone \qquad emp\# \twoheadrightarrow qualification \qquad prod\# \to price$$
$$emp\# \overset{W}{\to} car \qquad \{emp\#, prod\#\} \to salesQty \qquad prod\# \overset{S}{\to} pname$$
$$emp\# \twoheadrightarrow previousEmp \quad category \overset{S}{\to} categoryDesc \qquad prod\# \overset{S}{\to} category.$$

The key of the *Fact table* is $\{emp\#, prod\#\}$. The key of the *Employee* table is $\{emp\#, qualification, previousEmp, phone, car\}$. The key of the *Product* table is $\{prod\#\}$. Access patterns are derived from typical queries. The monthly reports include:

Q1. total quantity of items sold by each employee,
Q2. total quantity of items sold by product, listing both product number and product name,
Q3. total quantity of items sold by category, listing both category and category description,

Fact table

emp#	prod#	salesQty
1	10	10
2	10	15
3	10	35
1	20	8
2	20	14
3	20	30
1	30	6
2	30	10
3	30	25

Employee

emp#	qualification	previousEmp	phone	car
1	HSC	Bob's fruit shop	8725911	DX6195
1	BSc	Bob's fruit shop	8725911	DX6195
1	MSc	Bob's fruit shop	8725911	DX6195
2	HSC	Quality Groceries	8741834	LX5255
2	BSc	Quality Groceries	8741834	LX5255
2	HSC	Shop and Save	8741834	LX5255
2	BSc	Shop and Save	8741834	LX5255
3	HSC	Shop and Save	7794580	MM5735
3	HSC	Shop and Save	6741856	MM5735

Product

prod#	pname	category	categoryDesc	price
10	weetbix	cereal	A cereal is typically eaten for breakfast. Cereals can be stored on shelves and have a shelf life of 3 months. ...	2
20	vegemite	spread	A spread is eaten with bread. Speads can be stored on shelves and have a shelf life of 6 months. ...	4
30	apple	fruit	Fruit (excluding bananas) must be stored in a partially refrigerated part of the shop. Fruit has a shelf life of 1 week to 1 month. ...	1
40	orange	fruit	Fruit (excluding bananas) must be stored in a partially refrigerated part of the shop. Fruit has a shelf life of 1 week to 1 month. ...	2

Fig. 3. Example data warehouse

Q4. total sales of items sold by employee,

Q5. total sales of items sold by product, listing both product number and product name,

Q6. total sales of items sold by category, listing both category and category description,

Q7. average number of items for all products sold by each employee,

Q8. average number of items for all employees by each product, and

Q9. details of each product, and employee number and phone number of the top performing employee for each product.

Less frequently the following information is required:

Q10. A manager is interested in the correlation between the average number of products sold by employees and the average number of qualifications of the employees.

Q11. To get a feel for what to look for when hiring people, a manager may want to know who the previous employers are of his three top sales staff.

4   Problems

In this section, we discuss possible problems in view selection, briefly compare different views and give reasons why one set of views is better than another. A fuller comparison can be found in [2].

If there is an attribute that changes frequently, how should the materialized views be selected? In traditional databases, if there is a relation $R(A, B, C)$ where $A \rightarrow B$ and $B \rightarrow C$ then it is suggested that the relation is decomposed into $R_1(A, B)$ and $R_2(B, C)$. The reason being that when C is updated, it will need to be updated in many places if the schema with R is used, but only once if the schema with R_1 and R_2 is used. Consider the scenario where the values of C are very unlikely to change, (i.e. $B \xrightarrow{S} C$) then there is no need to decompose R. The same reasoning can be followed when selecting data warehouse views.

If a multi-valued attribute is usually single-valued, how should the views be selected? In traditional databases, if there is a relation $R(A, B, C)$ and $A \twoheadrightarrow B$, then it is suggested that the relation be decomposed into $R_1(A, B)$ and $R_2(A, C)$. However, if the multi-valued attribute usually has only one value, like an employee usually having only one phone number, then this information can be stored as though it is single valued with an extra (overflow) table for the occasional extra phone number. The same reasoning can be followed when selecting data warehouse views. Methods for maintaining the overflow tables are described in [7].

If a multi-valued attribute is usually multi-valued, how should the views be selected? When designing traditional database systems, one of the overriding aims is to reduce the number of joins when answering queries, however there are occasions when selecting views for data warehouses where this is not the best approach. A view should be decomposed into two views if two attributes in the view are independent and frequently have multiple values.

Example 4. Consider view $V(emp\#, previousEmp, prod\#, pname, salesQty)$ where an employee has on average 10 previous employers and sells 100 products. To find the previous employer of an employee, it is necessary to access 1000 tuples on average. If instead, the view is decomposed into $V_1(emp\#, previousEmp)$ and $V_2(emp\#, prod\#, pname, sales)$, it'd be necessary to access only 10 tuples on average to find the previous employers of an employee. On the other hand, view V would be preferable to views V_1 and V_2 if very few employees have more than one previous employer, or *previousEmp* and *pname* are frequently accessed together.   □

Is there a possibility of anomalous data being introduced when views are selected? If there is an attribute in a table that you will aggregate on, then this table can only be joined with other tables with the same key, otherwise anomalous data will be introduced and the following aggregation will be incorrect. This usually arises when you are joining a fact table (or a summarization of a fact table) and a dimension table, and the fact table contains a measurement attribute that is likely to be aggregated in the future.

Example 5. Consider creating a view $V(emp\#, prod\#, qualification, salesQty)$ with employee number 1, and product number 10. Using the sample data warehouse in Section 3, the view would contain the following information:

emp#	prod#	qualification	salesQty
1	10	HSC	10
1	10	BSc	10
1	10	MSc	10

The total sales calculated from view V is 30 whereas the total sales for $emp\#$ 1, $prod\#$ 10 calculated from the Fact table is 10. The problem is that the $salesQty$ has been replicated for each $qualification$ of the employee.   □

If the size of attributes varies greatly or views are large, how should the views be selected? There are two situations in which vertical partitioning is advantageous. If there are some attributes that are small and some that are very large, then when the smaller attributes are accessed, the whole tuple including the very large attributes must be read into temporary storage. A better design is to vertically partition the large attributes from the other attributes. Another related situation is where a view has many attributes and those attributes are never accessed together. The access speed is negatively affected by the extra attributes in the view.

If a view is very large and queries are often based on particular values for an attribute, the table can be horizontally partitioned on that attribute to improve performance. For example, consider a large view with an attribute $city\#$, where access frequently involves individual cities, then performance will be improved if the view is horizontally partitioned based on the $city\#$.

If a key attribute is large, how should the views be selected? If there is a set of views where one table has a large key and that key is used as a foreign key in another view then a surrogate key can be introduced and the surrogate key can replace the foreign key e.g. where the attribute $categoryDesc$ is a key in one view and a foreign key in another view.

When should two views be joined? Consider joining (a subset of the attributes of) a fact table and (a subset of the attributes of) a dimension table to form a view. If on average the key of the dimension table occurs infrequently in the fact table, then (the subset of the attributes of) the dimension table can be joined to (the subset of the attributes of) the fact table to form a view, otherwise it isn't worthwhile forming a view.

Example 6. Consider the view $V(emp\#, prod\#, pname, category, price)$ formed by joining a subset of the attributes of the fact table with a subset of the attributes of the product table. View V can be formed if on average each $prod\#$ value occurs infrequently in the fact table. If on the other hand, each $prod\#$ value occurs frequently on average then forming view V wastes a lot of space and the product information is harder to maintain, so no such view should be formed.

How should views with aggregates be selected? The way aggregates are stored for a dimension hierarchy is dependent on the frequency of updates, and the access pattern as demonstrated in Example 7. Recall that when you are decomposing a table while normalizing a traditional database, it is important that no information is lost. When you are selecting views for data warehouses, this constraint no longer applies because the underlying tables remain.

Example 7. Assume we require both the sum of *salesQty* grouped by *prod#*, $\Sigma salesQty_{prod\#}$, and the sum of *salesQty* by category, $\Sigma salesQty_{category}$.[1] The following are possible sets of views:

- $V(prod\#,\ pname,\ category,\ \Sigma salesQty_{prod\#})$,
- $V_1(prod\#,\ pname,\ \Sigma salesQty_{prod\#})$, $V_2(category,\ \Sigma salesQty_{category})$,
- $V_3(prod\#,\ pname,\ category,\ \Sigma salesQty_{prod\#})$,
  $V_4(category,\ \Sigma salesQty_{category})$.

The view V is preferable over the other sets of views under the following conditions:

- View V is preferable if the number of sales is updated often. If either of the alternative sets of views are used, when the sales quantity is updated, both aggregations of the sales quantity must also be recalculated.
- View V is preferable if there are very few queries asking for sales by category, because only then is it worthwhile computing the aggregation at the time of the query.
- Otherwise, one of the other sets of views is preferable.

The set of views with V_1 and V_2 is preferable if there are few queries involving the relationship between *pname* and *category* otherwise the set of views with V_3 and V_4 is preferable.   □

5 Design Heuristics

The following heuristics take into account the problems described in Section 4. The heuristics can be used for two related but different purposes, to judge if one set of views is better than another, and to improve a set of views. The aims are to make maintenance easier, and to reduce the access cost, without introducing anomalous data. The process of selecting materialized views involves the following three steps:

S_1. Create a temporary view for each query by selecting the attributes that are required to answer the query.

S_2. Join temporary views from Step S_1 grouping attributes that belong to the same entities together. This minimizes the space that will be needed to store the materialized views and makes updating the views less error prone.

S_3. Finally select the best set of views by decomposing the temporary views found in Step S_2, using the following heuristics.

---

[1] $\Sigma salesQty_{category}$ can also be calculated by grouping $\Sigma salesQty_{prod\#}$ by *category*.

5.1   Reduce Access Cost Using Data Dependencies

The heuristics in this section are based on the weak and strong functional dependencies that were introduced in Section 2.

OK Rule

a1. The view $V(A, B, C, D)$ is OK if
   - $A \to \{B, C, D\}$, or
   - $A \to \{B, D\}$ and $B \xrightarrow{S} C$. [2]
a2. The view $V(A, B, C)$ is OK, if there is a non-trivial multi-valued dependency $A \twoheadrightarrow B$, on average there are not many values for B and C, and the attributes B and C are frequently accessed together.

Not OK Rule

b1. If $A \to \{B, D\}$ and $B \to C$ but $B \xrightarrow{S} \hspace{-1.3em}/\;\; C$,[3] then the view $V(A, B, C, D)$ should be decomposed into $V_1(A, B, D)$ and $V_2(B, C)$.
b2. If there is a non-trivial multi-valued dependency $A \twoheadrightarrow B$ and $A \xrightarrow{W} B$ and $A \xrightarrow{W} C$, the view $V(A, B, C)$ is replaced by $V_1(A, B, C)$, $V_{bOverflow}(A, B)$ and $V_{cOverflow}(A, C)$.
b3. Let there be a fact table, R(A,E,F) where $\{A, E\} \to F$, and $\{A, E\} \xrightarrow{S} \hspace{-1.3em}/\;\; F$. Let B be the aggregation of F grouped on A (i.e. ΣF_A) and C be an aggregate that is computed using B (e.g. sum of B, average of B), then view $V(A, B, C)$ should be replaced by $V_1(A, B, D)$, where D is another attribute can be used to compute C from B. [4]
b4. If there is a non trivial multi-valued dependency $A \twoheadrightarrow B$ and $A \xrightarrow{W} \hspace{-1.3em}/\;\; B$, on average there are many values for B and C, and the attributes B and C are not frequently accessed together the view $V(A, B, C)$ should be decomposed into $V_1(A, B)$ and $V_2(A, C)$.

5.2   Reduce Access Cost Using Physical Database Principles

Traditionally in database systems the query cost is reduced by choices made during physical database design. The heuristics in this section are based on physical database design principles, like those in [3], and adapted for views in data warehouses.

OK Rule

c1. A view $V(A, B, C)$ is OK if $A \to \{B, C\}$ and each of the attributes is a "reasonable" size. It is up to the designer to judge what a reasonable size is.
c2. A view is OK, if it has a large key but the key is not used as a foreign key in another view.

---

[2] The values of the attributes in C are not updated frequently.
[3] The values of the attributes in C are updated frequently.
[4] If B is ΣF_A, and C is **av** F_A, then D may be **count** F_A.

c3. A view that is created by joining (a subset of the attributes of) a fact table with (a subset of the attributes of) a dimension table is OK if
  − the key of the dimension table is a subset of the key of the fact table, and
  − each value of the key of the dimension table occurs infrequently, on average, in the fact table.
c4. Consider a view $V(A, B, C)$ (with key A) where A and B are two levels in a dimension hierarchy and C is a measurement for A. For example, A could be a product number, B a category and C the quantity of the product sold. The view V is OK if
  − the attribute C is updated often, or
  − there are few queries that aggregate C grouping by B.

Not OK Rule

d1. If there is a view $V(A, B, C)$ where $A \rightarrow \{B, C\}$ and attribute C is very large then decompose the view into $V_1(A, B)$ and $V_2(A, C)$.
d2. If there is a view $V(A, B, C)$ where $A \rightarrow \{B, C\}$ and queries are frequently asked for particular values of attribute B then V can be horizontally partitioned on B.
d3. If the key in a view V_1 is large and the key is being used as a foreign key in another view V_2 then introduce a surrogate key into V_1 and replace the foreign key in V_2 by the surrogate key.
d4. A view $V(A, B, C, E, F, G)$ that is created by joining (a subset of the attributes of) a fact table $R(A, B, C)$ (with key $\{A, B\}$) with (a subset of the attributes of) a dimension table $R_1(B, E, F, G)$ (with key B) should not be formed if each value of the key of the dimension table occurs frequently, on average, in the fact table.
d5. Consider a view $V(A, B, C)$ (with key A) where A and B are two levels in a dimension hierarchy and C is a measurement of A. The view V should be decomposed into $V_1(A, C)$ and $V_2(B, \Sigma C_B)$ if
  − the attribute C is not updated often,
  − there are few queries that include the relationship between A and B,
  − there are frequent queries that aggregate C grouping by B.
  If the above properties hold but there are frequent queries that include the relationship between A and B then replace V with $V_3(A, B, C)$ and $V_4(B, \Sigma C_B)$.
d6. If there is a view $V(A, B, C, D, E, F)$ with key A and there are frequent queries that access A, B, C and frequent queries that access D, E, F then vertically partition V into $V_1(A, B, C)$ and $V_2(A, D, E, F)$.

5.3   Do Not Introduce Anomalous Data
The heuristics in this section guard against anomalous data being introduced when views are selected. The anomalous data problem arises where there is aggregation or summarization of the underlying data. Views are built from underlying tables using select, project, join, and aggregation operations. As we

demonstrated in Example 5 anomalous data may be introduced if there is a measurement that is likely to be aggregated or a measurement that is aggregated, and the keys of the underlying tables are not the same.

Let there be a table $R_1(A, B, C)$ and another table $R_2(A, D, E)$. We write $\mathcal{F}C_A$ to denote that function $\mathcal{F}$ is likely to be or has been applied to attribute C after grouping on A.

OK Rule

e1. It is OK to create a view as the result of joining a temporary table $R_T(A, \mathcal{F}C_A)$ to any table that has key A.

Not OK Rule

f1. If there is a table (or temporary table) $R_T(A, \mathcal{F}C_A)$ and a table $R_2(A, D, E)$ with a composite key e.g. $\{A, D\}$ then the tables should not be joined.

Example 8. Consider a fact table $F(A, B, C)$ with key $\{A, B\}$ and a dimension table $D(A, E, F)$ with key $\{A, E\}$. Anomalous data will be introduced if the view $V(A, E, \Sigma C_A)$ is created. The view must be replaced by $V_1(A, E)$ and $V_2(A, \Sigma C_A)$. □

6   Demonstrating the Heuristics

In this section we demonstrate how views are selected for the example data warehouse in Section 3 based on the procedure outlined in Section 5.

Step S_1, we create a view for each of the queries presented in Section 3.[5] Both $Q9$ and $Q11$ involve selecting top performing employees.[6]

$Q1(\underline{emp\#},\ \Sigma salesQty_{emp\#})$
$Q2(\underline{prod\#},\ pname,\ \Sigma salesQty_{prod\#})$
$Q3(\underline{category},\ categoryDesc,\ \Sigma salesQty_{category})$
$Q4(\underline{emp\#},\ \Sigma(price \times salesQty)_{emp\#})$
$Q5(\underline{prod\#},\ pname,\ \Sigma(price \times salesQty)_{prod\#})$
$Q6(\underline{category},\ categoryDesc,\ \Sigma(price \times salesQty)_{category})$
$Q7(\underline{emp\#},\ \mathbf{av}salesQty_{emp\#})$
$Q8(\underline{prod\#},\ pname,\ \mathbf{av}salesQty_{prod\#})$
$Q9(\underline{prod\#},\ emp\#,\ phone,\ pname,\ category,\ categoryDesc,\ price)$
$Q10(\underline{emp\#},\ \Sigma salesQty_{emp\#},\ \mathbf{count}\ qualification_{emp\#})$
$Q11(\underline{emp\#},\ previousEmp)$

---

[5] For simplicity we use the query number (from Section 3) as the name of the matching view.

[6] Horizontal partitioning could be appropriate but because the tuples in the partition change often, we do not perform the partitioning.

Step S_2, new temporary views are formed by grouping all attributes that belong to an entity together. For example, we group all attributes that belong to the *employee* entity in one view.[7] The resulting views are:

- $VT_{emp\#}(\underline{emp\#},\ previousEmp,\ phone,\ \Sigma salesQty_{emp\#},\ \mathbf{av}salesQty_{emp\#},$
  $\Sigma(price\ \times\ salesQty)_{emp\#},\ \mathbf{count}\ qualification_{emp\#})$,[8]
- $VT_{prod\#}(\underline{prod\#},\ pname,\ category,\ \Sigma salesQty_{prod\#},\ \mathbf{av}salesQty_{prod\#},$
  $\Sigma(price\ \times\ salesQty)_{prod\#},\ price)$, [9]
- $VT_{category}(\underline{category},\ \Sigma salesQty_{category},\ \Sigma(price\ \times\ salesQty)_{category},$
  $categoryDesc)$.[10]

Step S_3, we use the heuristics together with the dependencies described in Section 3 to improve the views produced in the previous step as illustrated in Figures 4 and 5. We consider $VT_{emp\#}$, $VT_{prod\#}$ and $VT_{category}$, and then the resulting views together. Based on the heuristics, the set of views, $V_{overflow}$, V_3, V_4, V_7 and V_9, are the best for this schema, with the given access pattern.

7 Conclusions

The selection of materialized views is crucial to data warehouse performance. To date, most of the research in the area uses general cost models to find the optimal or near optimal set of views, without considering the physical properties of the data.[11] In this paper, we have shown that the widely recognized star and snowflake schema are based on overly simplistic assumptions, so any view design heuristics based on these schema are also overly simplistic. We discussed problems that may arise when selecting which views to materialize for a data warehouse. Our main contributions are the enhanced star and snowflake schema and the set of heuristics that can be used to improve a set of materialized views, or judge if one set of views is better than another set of views. The heuristics are based on physical properties of the data such as how likely it is that the value of an attribute will change, how often a multi-valued attribute will have more than one value, access patterns, and size of attributes. We have provided a preliminary investigation on which further practical work, involving the selection of materialized views in data warehouses, can be based.

This is preliminary work and there are many directions we could take from here. One direction is to perform experiments to verify our claims of improved performance and another is to formalize the heuristics presented and further

---

[7] The attributes in $Q9$ are split between the three entity views.

[8] Formed by joining the attributes from $Q1$, $Q4$, $Q7$, $Q10$, $Q11$ and the attributes *emp#* and *phone* from $Q9$.

[9] Formed by joining the attributes from $Q2$, $Q5$, $Q8$, and the attributes *prod#*, *pname*, *category* and *price* from $Q9$.

[10] Formed by joining the attributes from $Q3$, $Q6$, and the attributes *category* and *categoryDesc* from $Q9$.

[11] See [2] for a more thorough discussion of the related work.

$VT_{emp\#}(\underline{emp\#}, phone, previousEmp, \Sigma salesQty_{emp\#}, \Sigma(price \times salesQty)_{emp\#},$
$\qquad \mathbf{av}salesQty_{emp\#}, \mathbf{count}qualification_{emp\#})$

$\Downarrow b_2 \hookleftarrow emp\# \xrightarrow{w} phone \hookleftarrow$
$VT_{emp\#} isreplacedby V_{overflow} and VT_{emp\#}$

$V_{overflow}(\underline{emp\#}, phone)$
$VT_{emp\#}(\underline{emp\#}, previousEmp, phone, \Sigma salesQty_{emp\#}, \Sigma(price \times salesQty)_{emp\#},$
$\qquad \mathbf{av}salesQty_{emp\#}, \mathbf{count}qualification_{emp\#})$

$\Downarrow b_3 \hookleftarrow aggregatesbasedon \Sigma salesQty_{emp\#} \hookleftarrow$
$VT_{emp\#} isreplacedby VT_2$

$V_{overflow}(\underline{emp\#}, phone)$
$VT_2(\underline{emp\#}, previousEmp, phone, \Sigma salesQty_{emp\#}, \mathbf{count}qualification_{emp\#})$

$\Downarrow b_4 \hookleftarrow emp\# \twoheadrightarrow previousEmp \hookleftarrow$
$VT_2 isreplacedby V_3 and V_4$

$V_{overflow}(\underline{emp\#}, phone)$
$V_3(\underline{emp\#}, previousEmp)$
$V_4(\underline{emp\#}, phone, \Sigma salesQty_{emp\#}, \mathbf{count}qualification_{emp\#})$

$VT_{prod\#}(\underline{prod\#}, pname, category, \Sigma salesQty_{prod\#}, \Sigma(price \times salesQty)_{prod\#},$
$\qquad \mathbf{av}salesQty_{prod\#}, price)$

$\Downarrow b_3 \hookleftarrow aggregatesbasedon \Sigma salesQty_{prod\#} \hookleftarrow$
$VT_{prod\#} isreplacedby VT_5$

$VT_5(\underline{prod\#}, pname, category, \Sigma salesQty_{prod\#}, price)$

$VT_{category}(\underline{category}, \Sigma salesQty_{category}, \Sigma(price \times salesQty)_{category},$
$\qquad categoryDesc)$

$\Downarrow b_3 \hookleftarrow aggregatesbasedon \Sigma salesQty_{category} \hookleftarrow$
$VT_{category} isreplacedby VT_6$

$VT_6(\underline{category}, categoryDesc, \Sigma salesQty_{category})$

$\Downarrow d_1 \hookleftarrow categoryDescislargefield \hookleftarrow$
$VT_6 isreplacedby V_7 and VT_8$

$V_7(\underline{category}, categoryDesc)$
$VT_8(\underline{category}, \Sigma salesQty_{category}) \triangleright$

Fig. 4. Views after heuristics applied in Step S_3 to 3 entity views from Step S_2

$VT_5(\underline{prod\#},\ pname,\ category,\ \Sigma salesQty_{prod\#},\ price)$
$VT_8(\underline{category},\ \Sigma salesQty_{category})$

$\Downarrow$ c4 ($salesQty$ updated often)
VT_5 and VT_8 are replaced by V_9

$V_9(\underline{prod\#},\ category,\ pname,\ \Sigma salesQty_{prod\#},\ price)$.

Fig. 5. Resulting views in Step S_3 from VT_5 and VT_8

investigate the heuristics. We need to find answers to the following questions: Does one heuristic make something that was OK, not OK or vice versa? Do any of the rules conflict? Is the result different if the rules are applied in a different order? Could some of the heuristics be replaced by first ensuring that the views are in relax-replicated 3NF (see [7])? Following this line of investigation would clarify the distinction between traditional database design and data warehouse view design and could lead into investigating the kinds of indexes that are best, from a practical perspective, for materialized views in data warehouses.

References

1. Elena Baralis, Stefano Paraboschi and Ernest Teniente. Materialized Views Selection in a Multidimensional Database. In *Proceedings of 23rd International Conference on Very Large Data Bases (VLDB'97)*, 1997.
2. Gillian Dobbie and Tok Wang Ling. Practical Approach to Selecting Data Warehouse Views Using Data Dependencies. *Technical Report from School of Computing, National University of Singapore, No. TRA7/00.*
3. Rob Gillette, Dean Muench and Jean Tabaka. *Physical Database Design for SYBASE SQL Server.* Prentice Hall PTR, 1995.
4. Himanshu Gupta and Inderpal Singh Mumick. Selection of Views to Materialize Under a Maintenance Cost Constraint. In *Database Theory - ICDT '99, 7th International Conference on Database Theory (ICDT)*, 1999, pages 453–470, Springer-Verlag LNCS 1540.
5. Ashish Gupta and Inderpal Singh Mumick. Maintenance of Materialized Views: Problems, Techniques, and Applications. In *Data Engineering Bulletin*, 18(2), pages 3–18, 1995.
6. R. Kimball. *The data warehouse toolkit.* John Wiley and Sons, 1996.
7. Tok Wang Ling, Cheng Hian Goh and Mong Li Lee. Extending Classical Functional Dependencies for Physical Database Design. In *Information and Software Technology*, pages 601-608, vol 38 (1996), Elsevier Science.
8. Kenneth A. Ross, Divesh Srivastava and S. Sudarshan. Materialized View Maintenance and Integrity Constraint Checking: Trading Space for Time. In *Proceedings of the 1996 ACM SIGMOD International Conference on Management of Data*, pages 447-458.
9. Amit Shukla, Prasad Deshpande and Jeffrey F. Naughton. Materialized View Selection for Multidimensional Datasets. In *Proceedings of 24th International Conference on Very Large Data Bases, (VLDB'98)*, 1998, pages 488-499.

10. Dimitri Theodoratos, Spyros Ligoudistianos, and Timos Sellis. Designing the Global Data Warehouse with SPJ Views. In *Proceedings of 11th International Conference on Advanced Information Systems Engineering, (CAiSE'99)*, 1999, Springer-Verlag LNCS 1626.
11. Jian Yang, Kamalakar Karlapalem and Qing Li. Algorithms for Materialized View Design in Data Warehousing Environment. In *Proceedings of 23rd International Conference on Very Large Data Bases, (VLDB'97)*, 1997, pages 136-145.

Semantic Analysis Patterns

Eduardo B. Fernandez and Xiaohong Yuan

Department of Computer Science and Engineering
Florida Atlantic University
Boca Raton, FL 33431
{ed, xhyuan}@cse.fau.edu

Abstract. The development of object-oriented software starts from requirements expressed commonly as Use Cases. The requirements are then converted into a conceptual or analysis model. Analysis is a fundamental stage because the conceptual model can be shown to satisfy the requirements and becomes the skeleton on which the complete system is built. Most of the use of software patterns until now has been at the design stage and they are applied to provide extensibility and flexibility. However, design patterns don't help avoid analysis errors or make analysis easier. Analysis patterns can contribute more to reusability and software quality than the other varieties. Also, their use contributes to simplifying the development of the analysis model. In particular, a new type of analysis pattern is proposed, called a Semantic Analysis Pattern (SAP), which is in essence a miniapplication, realizing a few Use Cases or a small set of requirements. Using SAPs, a methodology is developed to build the conceptual model in a systematic way.

1 Introduction

The development of object-oriented software starts from requirements expressed normally as Use Cases [17]. The requirements are then converted into a conceptual or analysis model. Analysis is a fundamental stage because the conceptual model can be shown to satisfy the requirements and becomes the skeleton on which the complete system is built. No good design or correct implementation is possible without good analysis, the best C++ or Java programmers cannot make up for conceptual errors. The correction of analysis errors becomes very expensive when these errors are caught in the code. It is therefore surprising how poorly understood is this stage and how current industrial practice and publications show a large number of analysis errors [6]. We have found that industrial software developers usually have trouble with analysis. What is worse, even serious journals and conferences publish papers or tutorials that contain clear analysis errors.

A possible improvement to this situation may come from the use of patterns. A pattern is a recurring combination of meaningful units that occurs in some context. Patterns have been used in building construction, enterprise management, and in several other fields. Their use in software is becoming very important because of their value for reusability and quality; they distill the knowledge and experience of many designers.

A.H.F. Laender, S.W. Liddle, V.C. Storey (Eds.): ER2000 Conference, LNCS 1920, pp. 183-195, 2000.

Most of the use of patterns until now has been at the design stage. However, design patterns don't help to avoid analysis errors or to make analysis easier. We believe that we need analysis patterns to improve the quality of analysis and they can contribute more to reusability and software quality than the other varieties. We also intend to show that their use contributes to simplifying the development of the application analysis model. In particular, we propose a new type of analysis pattern, called a Semantic Analysis Pattern (SAP), which is in essence a miniapplication, realizing a few Use Cases or a small set of requirements [8]. Using SAPs we develop a methodology to build the conceptual model in a systematic way. We use UML (Unified Modeling Language) [2], as a language to describe our examples.

Section 2 introduces SAPs and how they are obtained. We show analogy and generalization as ways to develop SAPs. Section 3 describes how SAPs are used in producing conceptual models from Use Cases. Section 4 compares SAPs to other varieties of analysis patterns and evaluates their use. A last section presents conclusions and suggestions for future work.

2   Analysis Patterns and Their Use

2.1 Semantic Analysis Patterns

The value of analysis is played down in practice. The majority of the papers published about object-oriented design as well as the majority of textbooks concentrate on implementation. Books on Java, C++, and other languages outnumber by far the books on object-oriented analysis/design (OOA /OOD). On top of that, most books on OOA/OOD present very simple examples. To make things worse, professional programmers need to implement as soon as possible, there is pressure to show running code and they may skip the analysis stage altogether. What is deceiving is that software may appear to work properly but may have errors, not be reusable or extensible, be unnecessarily complex. In fact, most of the software built without some model exhibits some or all of these defects. Most schools emphasize algorithms, not the development of software systems. There is a large literature on methods of system development that although oriented to other disciplines [24], is very applicable to software, but rarely used (In fact, design patterns originated from ideas about buildings). Some people believe that with components we don't need to understand what is inside each component. The result of all this is that analysis is skipped or done poorly.

We need to look for ways to make analysis more precise and easier for developers. The use of patterns is a promising avenue.

A Semantic Analysis Pattern is a pattern that describes a small set of coherent Use Cases that together describe a basic generic application. The Use Cases are selected in such a way that the application can fit a variety of situations.

Semantic Analysis Patterns differ from design patterns in the following ways:

- Design patterns are closer to implementation, they focus on typical design aspects, e.g., user interfaces, creation of objects, basic structural properties.
- Design patterns apply to any application; for example, all applications have user interfaces, they all need to create objects.
- Design patterns intend to increase the flexibility of a model by decoupling some aspects of a class.

An instance of a SAP is produced in the usual way: Use Cases, class and dynamic diagrams, etc. We select the Use Cases in such a way that they leave out aspects which may not be transportable to other applications. We can then generalize the original pattern by abstracting its components and later we derive new patterns from the abstract pattern by specializing it (Figure 1). We can also use analogy to directly apply the original pattern to a different situation.

We illustrate these two approaches in the next sections. We develop first a pattern from some basic use cases. We then use analogy to apply it to a different situation, then we generalize it and finally we produce another pattern for another application specializing the abstract pattern. We then show how to use these patterns in building conceptual models.

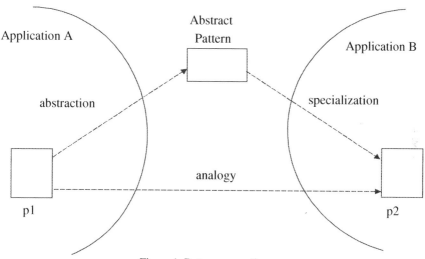

Figure 1. Pattern generation

2.2 An Example

In a project we developed a design for a computer repair shop. The specifications for this application are: A computer repair shop fixes broken computers. The shop is part of a chain of similar shops. Customers bring computers to the shop for repair and a reception technician makes an estimate. If the customer agrees, the computer is assigned for repair to some repair technician, who keeps a Repair Event document. All the Repair Event documents for a computer are collected in its repair log. A repair event may be suspended because of a lack of parts or other reasons.

These requirements correspond to two basic Use Cases:
- Get an estimate for a repair
- Repair a computer

A class diagram for this system is shown in Figure 2, while Figure 3 shows a state diagram for Repair Event. Figure 4 shows a sequence diagram for assigning the repair of some computer to a technician. The class diagram reflects the facts that a computer can be estimated at different shops in the chain and that one of these estimates may become an actual repair. A computer that has been repaired at least once has a repair log that collects all its repair events. The collection of repair shops is described by the repair shops chain.

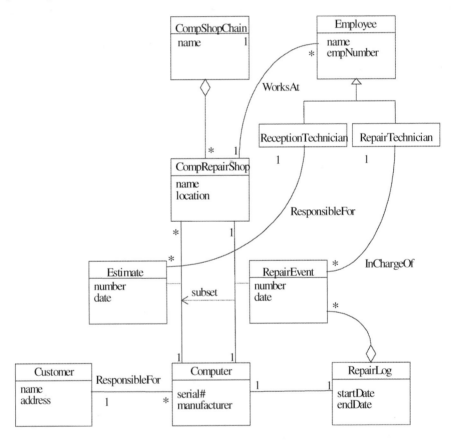

Figure 2. Class diagram for the computer repair shop.

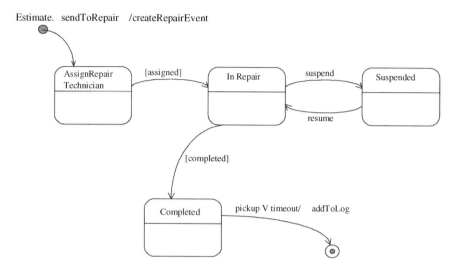

Figure 3. State diagram for Repair Event

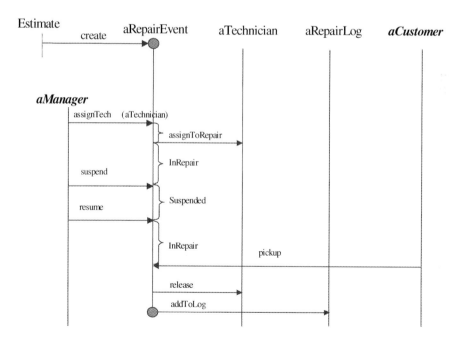

Figure 4. Sequence diagram for assigning repair jobs to technicians

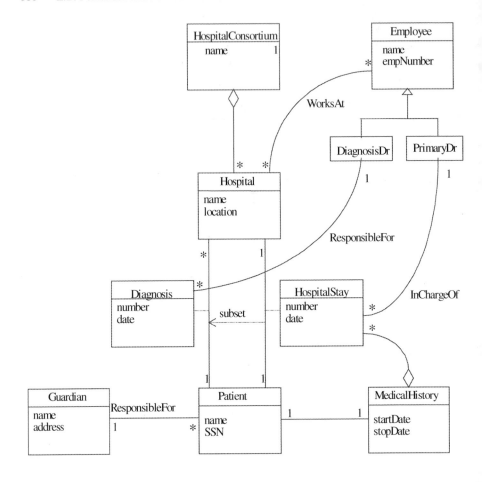

Figure 5. Class diagram for hospital registration.

2.3 Looking for Analogies

In a later project we needed to design a system to handle patient registration in a hospital. Noticing that a hospital, instead of broken computers, fixes sick people, we arrived at the class diagram of Figure 5. We just needed to reinterpret each class in the repair shop as a corresponding class for a hospital. For example, the computer becomes a patient, the estimate becomes the diagnosis, the repair event a hospital stay, etc. Similarly, sequence diagrams and state diagrams are developed in analogy with those used in computer repair [8].

2.4 Pattern Generalization

We can generalize the patterns of Figures 2 and 5 by noticing that their essential actions are:

- Application to a collection of places
- Selection of one place to stay
- Keep a Stay record for each stay.
- Keep History of stays
- Personnel is assigned to the evaluation of applications and to do something during the stays.

With these concepts we can define the abstract pattern of Figure 6, that includes the two previous patterns. Here the specific institutions have become generic institutions, patients have become applicants, etc. This pattern fits a wide range of applications and we can use it in a new application, student admissions in a university (Figure 7). Here, we particularize institution into university, applicant into student, etc.

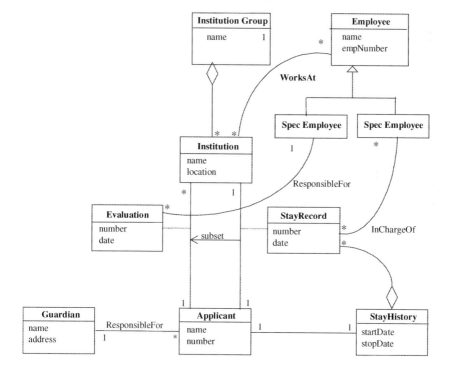

Figure 6 Admissions pattern

One can take portions of these patterns and find simpler patterns. For example, the set of personnel-related classes in Figures 2 and 5 can define a Personnel pattern, the collection pattern appears twice, as a collection of shops and as a collection of repair events. Also, some design patterns, e.g., the composite pattern [13], are useful in analysis.

3   Analysis Method Using SAPs

To use the methodology it is necessary to have first a good collection of patterns. We have developed four analysis patterns [8], [9], [10], [11]. We are also collecting the most interesting analysis patterns that have appeared in the literature [1], [3], [4], [18], [21], [27].

A possible analysis method using SAPs is described now. We assume we have a catalog of concrete and abstract patterns as well as catalogs of subpatterns, Fowler-style patterns[1], and design patterns. We examine the Use Cases and/or other requirements and:

♦   Look for SAPs. We look first for concrete patterns that match exactly or closely the requirements. Then we try to specialize analogous or abstract patterns that may apply. This stage is shown in Figure 8, where patterns p3, p5,...,have been identified and cover some of the requirements.

♦   Look for smaller patterns, such as the subpatterns of Figures 2 and 5.

♦   See if there are appropriate design or architectural patterns. As indicated earlier, some design or architectural patterns may be useful in analysis.

♦   Add Fowler-style patterns for flexibility and extensibility. This involves examining classes for possible breakup.

This procedure results in a skeleton, where some parts of the model are fairly complete while other portions are partially covered or not covered at all. We still need to cover the rest of the model in an ad hoc way but we already have a starting model. Naturally, we can still add design patterns in the design stage.

As an example, consider the following requirements: We need a system to handle the Soccer World Cup. Teams represent countries and are made up of 22 players. Countries qualify from zones, where each zone is either a country or a group of countries. Each team plays a given number of games in a specific city. Referees from different countries are assigned to games. Hotel reservations are made in the city where the teams play.

Figure 9 shows that this model was almost completely covered with the following patterns:

1)   An instance of the composite pattern [13]
2)   An instance of the collection pattern (a subpattern of Figure 2)
3)   An instance of the reservation pattern [9]
4)   Another instance of the collection pattern
5)   Another instance of the reservation pattern

In addition to these patterns one needs several associations to connect them. These associations correspond to specific requirements , e.g., a referee represents a country, a game is played in a given city. Of course, this is a simple example, larger examples are not so easily covered. However, the example exposes the flavor of the methodology.

---

[1] Fowler tries to increase flexibility by decoupling some aspects of a class into a separate class, e.g., the physical characteristics of a person would be separated from class Person [12]. His patterns in general are small, two or three classes.

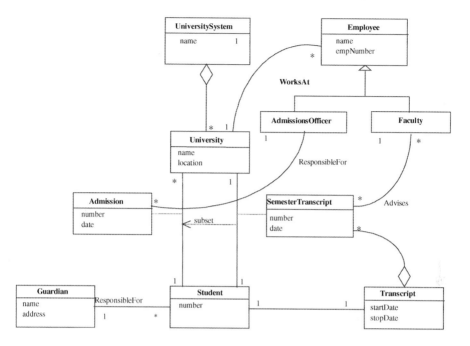

Figure 7.   Student admissions

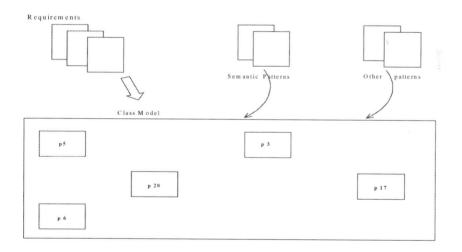

Figure 8.   Use of semantic APs

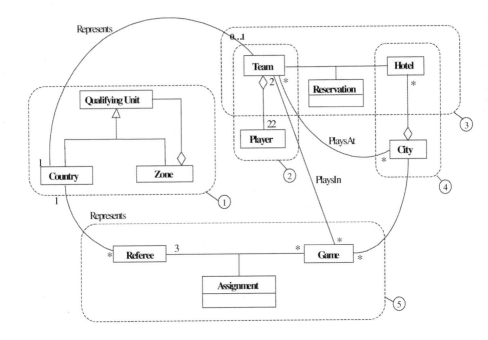

Figure 9. Analysis model for World Cup example

4 Discussion

The first work on object-oriented analysis patterns is due to P. Coad [5]. More influential has been the work of M. Fowler, who produced the first book on this topic [12]. His patterns (and Coad's) emphasize similar objectives as design patterns, i.e., to decouple parts of a class to increase the model flexibility and extensibility. These authors consider some dynamic aspects in the form of sequence diagrams but do not use statecharts. R. Johnson and his group at the University of Illinois have developed several analysis patterns [19], including banking and security; in general, they follow Fowler's style. Nature's project in the UK intends to classify application requirements into problem domains [22]. Their emphasis is not on modeling although they describe requirements using class models. Note that our approach results in much fewer patterns since we can abstract patterns. Another related work is the book of D.Hay [15], where he describes data patterns; however, he doesn't consider dynamic aspects.

All these projects have different objectives; in particular they do not emphasize synthesis of complex models. They are certainly a source of possible patterns and they are being mined to build the SAP catalog. The design patterns group shows in its book [13] and in a paper [14], another approach to synthesis: start from a class or two and keep adding patterns until the requirements are satisfied[2]. This approach doesn't

---

[2] A similar approach is used by Johnson's group [28].

appear very feasible when designing systems with long and complex specifications. However, it could be useful in the latter stages of our approach. Synthesis of complex systems is also the objective of a project at Georgia State University [23], and some of their ideas could be useful in this work.

Methods for synthesis of models using SAPs could result in considerable improvement to the quality of the software produced in industry. Design patterns have had a strong effect in design, we need now to do the same for analysis. Other than improving the analysis modeling there are more benefits:

◆ Test cases are developed from Use Cases but a good conceptual model helps define the needed preconditions and postconditions. Traceability can also be improved.

◆ Testing of object-oriented systems involves inspection of the models developed at each stage. In particular, domain models require careful validation [20]. The use of SAPs can help make these inspections faster and more accurate.

The actors of the use cases in the minimal application must be given the required rights to perform their functions. We can consider actors as roles and if we assign rights accordingly we have a Role-Based Access Control model of security [7]. We can define in this way "authorized SAPs', that include the needed authorizations.

The Object Constraint Language can be used to define precise constraints in the models [26]. Using OCL the minimal application is defined more precisely using business rules. Authorizations can also be expressed more precisely.

5 Conclusions

A good analysis model for a portion of a complex system can be abstracted and become an analysis pattern that can be used in other applications[1]. Analogy and abstraction play an important role in reusing an analysis pattern. Subsets of a pattern may also have their own application in other situations. All this can save time and improve the quality of a system. One of the most difficult steps in practice is to get an initial model; this approach makes it easier. In subsequent steps the initial model can be modified to suit better the needs of the application. An analysis model using patterns is easier to understand and to extend. It should also result in a higher quality design. A software architecture constructed this way is more reusable and extensible than an architecture defined directly from the requirements or where patterns are applied later. Note that SAP-based development is different from domain analysis, SAPs cut through several domains.

There are several aspects that we are developing or intend to develop:

◆ The design methodology has been applied to relatively small examples and it appears useful, but we need larger examples. We are collecting large-system specifications that we intend to model.

◆ Related SAPs can be combined into frameworks. For example, Order Entry, Inventory, and Reservations could make up a manufacturing framework. We

---

[1] We can think also that a SAP is a pattern recurrent in several frameworks.

have used this approach with security patterns [16], but we have not applied it to SAPs.

♦ Pattern languages. The SAPs by their nature leave out many aspects. A collection of related patterns, a pattern language, is needed to cover a domain or a significant part of it. We are developing a pattern language for reservation and use of entities, to complement the basic reservation SAP [9].

♦ SAPs as composite design patterns. A composite design pattern [25] is a pattern composed of other patterns where the composite is more than the sum of its parts. SAPs can be studied as special cases of composite design patterns.

References

1. Arsanjani, A.: Service provider: A domain pattern and its business framework implementation, *Procs. of PloP'99.* http://st-www.cs.uiuc.edu/~plop/plop99
2. Booch, G., Rumbaugh, J., Jacobson, I.: *The Unified Modeling Language User Guide*, Addison-Wesley 1998.
3. Braga, R.T.V., Germano, F.S.R., Masiero, P.C.: A confederation of patterns for resource management, *Procs. of PLoP'98*, http://jerry.cs.uiuc.edu/~plop/plop98
4. Braga, R.T.V., Germano, F.S.R., Masiero, P.C.: A pattern language for business resource management, *Procs. of PloP'99*, http://st-www.cs.uiuc.edu/~plop/plop99
5. Coad, P.: *Object models – Strategies, patterns, and applications* (2nd. Edition), Prentice-Hall 1997
6. Fernandez, E. B.: Good analysis as the basis for good design and implementation, Report TR-CSE-97-45, *Dept. of Computer Science and Eng., Florida Atlantic University*, September 1997. Presented at OOPSLA'97
7. Fernandez, E.B., Hawkins, J.: Determining role rights from use cases, *Procs. 2nd ACM Workshop on Role-Based Access Control*, 1997, 121-125
8. Fernandez, E.B.:Building systems using analysis patterns, *Procs. 3rd Int. Soft. Architecture Workshop (ISAW3)*, Orlando, FL , November 1998 , 37-40
9. Fernandez, E.B, Yuan, X.: An analysis pattern for reservation and use of entities, *Procs.of PLoP99* , http://st-www.cs.uiuc.edu/~plop/plop99
10. Fernandez, E. B.: Stock manager: An analysis pattern for inventories, to appear in *Procs. of PLoP 2000.*
11. Fernandez, E. B., Yuan, X., Brey, S.: Analysis Patterns for the Order and Shipment of a Product, to appear in *Procs. of PLoP 2000.*
12. Fowler, M.: *Analysis patterns -- Reusable object models*, Addison- Wesley, 1997
13. Gamma, E., Helm, R., Johnson, R. and Vlissides, J.: *Design patterns –Elements of reusable object-oriented software*, Addison-Wesley 1995
14. Gamma, E., Beck, K.: JUnit: A cook's tour, *Java Report*, May 1999, 27-38
15. Hay, D.: *Data model patterns-- Conventions of thought*, Dorset House Publ., 1996
16. Hays, V., Loutrel, M., Fernandez, E.B.: The Object Filter and Access Control Framework, to appear in *Procs. of PLoP 2000*
17. Jacobson, I., Booch, G., Rumbaugh, J.: *The Unified Software Development Process*, Addison-Wesley 1999
18. Johnson, R., Woolf, B.: Type Object, Chapter 4 in *Pattern Languages of Program Design 3*, Addison-Wesley, 1998
19. Johnson, R.: http://st-www.cs.uiuc.edu/users/Johnson
20. McGregor, J. D.: Validating domain models, *JOOP*, July-August 1999, 12-17
21. Mellor, S.J.: Graphical analysis patterns, *Procs. Software Development West98*, February 1998. http://www.projtech.com

22. Nature Project. http://www.city.ac.uk/~az533/main.html
23. Purao, S. and Storey, V.: A methodology for building a repository of object-oriented design fragments. *Procs. of 18ʰ International Conference on Conceptual Modeling (ER'99)*, 203-217.
24. Rechtin, E.: The synthesis of complex systems, *IEEE Spectrum*, July 1997, 51-55
25. Riehle, D.: Composite design patterns, *Procs. of OOPSLA'97*, 218-228.
26. Wanner, J., Kloppe, A.: *The OCL: Precise modeling with UML*, Addison-Wesley 1998
27. Yoder, J. and Johnson, R.: *Inventory and Accounting patterns*, http://www.joeyyoder.com/marsura/banking
28. Yoder, J., Balaguer, F.: Using metadata and active object-models to implement Fowler's analysis Patterns, Tutorial Notes, OOPSLA'99

Tool Support for Reuse of Analysis Patterns -
A Case Study

Petia Wohed

Department of Computer and Systems Sciences
Stockholm University/Royal Institute of Technology
Electrum 230, 164 40 Kista, Sweden
petia@dsv.su.se

Abstract: The size and complexity of modern information systems together with requirements for short development time increase the demands for reuse of already existing solutions. The idea of reuse itself is not novel and the ability of reuse is even a part of the learning process. However, not much support for reuse can be found for the analysis phase of information systems design. Collecting reusable solutions, called patterns, in a library and supporting the search for an appropriate pattern within such a library is one approach addressed in this area. A tool for this purpose has been partly implemented and the results from a case study testing this tool are reported here.

1   Introduction

The demands put on the new software systems increases continuously with the massive introduction of IT in our society. This line of development influences in turn the requirements during the development of these systems, which often have both comprehensive functionality as well as manage large volumes of data. To meet the needs of the market, short development time has become a necessity and also one of the most competitive instruments for software vendors. This development together with the human ability of reusing old, well tested solutions over and over again has forced modern development strategies into the approaches of object-orientation and component based development: approaches advocating the formalisation of diverse solutions in the shape of objects and components aimed to facilitate reuse.

However, there is a considerable difference on how long the adoption of these approaches has come within the different phases of information systems development. The contrast is clear between the implementation phase, supported by modern environments for standard programming languages offering large libraries of reusable objects, and the analysis phase where no such libraries have yet been implemented. The development of a tool supporting particularly the analysis phase of information systems development and more precisely supporting the conceptual modelling has therefore been the subject of our work.

The terminology of patterns is adopted. The concept was first established by Christopher Alexander. In the beginning of his book "A Pattern Language" [1] he

A.H.F. Laender, S.W. Liddle, V.C. Storey (Eds.): ER2000 Conference, LNCS 1920, pp. 196-209, 2000.
© Springer-Verlag Berlin Heidelberg 2000

writes: "Each pattern describes a problem which occurs over and over again in our environment, and then describes the core of the solution to that problem, in such a way that you can use this solution a million times over, without ever doing it the same way twice". Even if Alexander worked in, and developed his patterns for the field of architecture, his concept of pattern was applicable for, and inspired people working in the area of object-oriented information systems development.

The aims of our work is to collect a library of patterns and automate the support for finding the most appropriate one for a certain situation, hoping that it is a suitable way for supporting the reuse of the huge amount of knowledge saved in the patterns. As a first step a pattern library including patterns from one domain only, the booking domain, has been collected and a modelling wizard tool for the same domain has been constructed and implemented [10]. As a second step a case study evaluating the usability of the tool has been performed and reported in this paper.

The next section summarises the work on analysis patterns related to our work. Section 3 describes the tool. The goal and the design of the study are described in Sections 4 and 5. Sections 6 and 7 document the outcome of the study. Finally, section 8 summarises the paper and gives directions for further work.

2   Related Work on Analyses Patterns

One of the earliest contributions on patterns in information systems analysis was provided by Coad, who in 1995 published "Object Models: Strategies, Patterns, and Applications" [2]. The book presents 148 strategy steps for building systems and 31 object-model patterns for building object models, and demonstrates their applicability by using them for building five different systems. Even if Coad dissociates from Alexander's definition of a pattern claming that "A pattern is a template, not a specific solution", his main idea is still reuse.

The characteristic for Coad's 31 patterns is that they are very small and generic, usually consisting of no more than two object types and a relationship between them. Coad divides 30 of his patterns in the following four categories: transaction patterns; aggregate patterns; plan patterns; and interaction patterns, and all of these patterns follow a template, the first pattern called the fundamental patterns (redrawn on the right). Coad's patterns do not address a particular problem, but are rather general and may be considered as building blocks of a schema.

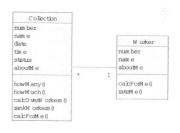

Later, in 1996 and 1997 Hay and Fowler, respectively, independently published two more books on analysis patterns ([5] and [4]). Both these books collect a number of patterns addressing different problem domains. They do not pretend to be complete solutions, but rather consider parts of solutions, which can be reused in different situations. The main difference from Coad is that Hay's and Fowler's patterns could represent directly or with small modification at least some domains, whereas to model any domain with Coad's patterns a number of patterns need to be synthesized.

However, neither Coad, Hay, nor Fowler provides any automated support for extracting the most suitable pattern for a particular situation, but all these approaches assume that the systems analyser selects the necessary pattern by herself. A step in the direction for supporting the search of a pattern is provided by Purao and Storey. Their initial work in this area [8],[9], is based on the patterns defined by Coad, but it focuses on automating the systems design by intelligent pattern retrieval and synthesis. In order to provide such pattern retrieval and synthesis a sentence, of maximum of 24 words, describing the system is required. Based on this system description, patterns (from the library suggested by Coad) are retrieved and synthesised by the usage of techniques for natural language processing, automated reasoning and learning heuristics. When applying some of these techniques extra knowledge is used. The required system description and the retrieval of the solution are the most important differences between Purao's and Storey's approach and the approach in our work.

The modelling wizard tool described in the next section does not require any description of the system as a starting point. Instead it gathers information from the user and suggests the most suitable pattern from the pattern library. Besides the information received from the user and the pattern library no other information is used.

3   The Modelling Wizard Tool

The conceptual modelling wizard is based, as can be seen from its name, on the concept of a wizard tool, i.e., a tool gathering information from the user by asking him a number of questions and suggesting a solution tailored to the set of answers. Applied to the area of conceptual modelling, this wizard poses questions about a domain and suggests a conceptual schema for this domain (a detailed description of the tool can be found in [10]). So far the wizard supports the booking domain only.

Six questions were collected and implemented in the tool (see Figure 1). When collecting the questions, different booking solutions - both such found in Fowler [4] and Hay [5], and such constructed to capture as different booking situation as possible - were analysed. The questions were aimed to cover the differences between distinct booking situations so that a pattern, which provides a satisfactory solution, shall be identified. A brief description of the questions is given below.

The first question is about the cardinality of a booking, i.e., where a booking may consist of several booking lines (e.g. when booking a trip, both tickets and hotel rooms may be reserved), or not (usually when booking a rental car). The second question is about whether a booking concerns a (number of) concrete objects (like when booking a time to dentist) or not (e.g. when booking a book in the library, not a particular exemplar from the book, but rather the title is booked). The third question investigates whether the bookings made within a system have the same character (as is the case for cinema tickets bookings) or whether they may differ. (In the trip booking example above, tickets bookings require departure and destination places and time for departure, whereas hotel room bookings require arrival and departure times and number of beds.) The fourth question is aimed to clarify whether the request for a booking is necessary to keep information about. (For instance, when scheduling room

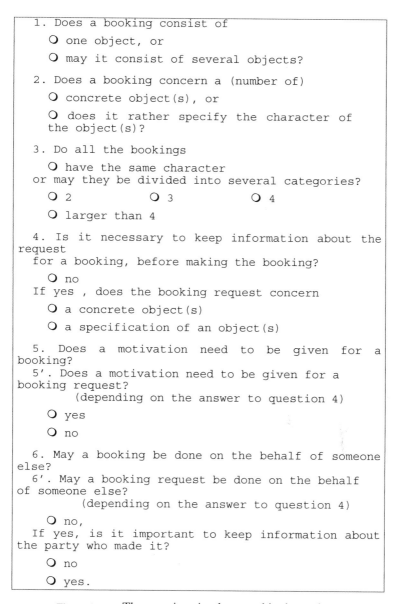

1. Does a booking consist of
  O one object, or
  O may it consist of several objects?

2. Does a booking concern a (number of)
  O concrete object(s), or
  O does it rather specify the character of the object(s)?

3. Do all the bookings
  O have the same character
  or may they be divided into several categories?
  O 2   O 3   O 4
  O larger than 4

4. Is it necessary to keep information about the request
  for a booking, before making the booking?
  O no
  If yes , does the booking request concern
    O a concrete object(s)
    O a specification of an object(s)

5. Does a motivation need to be given for a booking?
  5'. Does a motivation need to be given for a booking request?
        (depending on the answer to question 4)
  O yes
  O no

6. May a booking be done on the behalf of someone else?
  6'. May a booking request be done on the behalf of someone else?
        (depending on the answer to question 4)
  O no,
  If yes, is it important to keep information about the party who made it?
  O no
  O yes.

Figure 1   The questions implemented in the tool

bookings for university courses, booking requests are collected from the head teachers for each course and used as input in the scheduling work.) The fifth and sixth questions depend on the answer to the fourth question. If booking requests are necessary to keep information about these questions are posed for the booking requests otherwise they are posed for the bookings. The fifth question asks if a motivation (a purpose) for a booking/booking request is necessary. (For the university scheduling work each head teacher gives the course he/she will make the bookings for.) The sixth question asks whether a booking/booking request may be done on behalf of someone else (e.g. a secretary may book the business trips for her/his boss).

For the sake of user-friendliness the questions are placed in sequence by showing a new question first when the previous one has been answered. A conceptual schema solution is gradually built and refined, and graphically presented after each new answer. For instance the first alternative in the first questions should result in the schema in Figure 2a) whereas the second alternative should result in the solution presented in Figure 2b). Choosing further the first alternative from the second question should result in expanding the schema into the one shown in Figure 3a) or 3c), depending on the point of the departure, which is the result from the first question. Choosing the second alternative should give one of the solutions in Figures 3b) and 3d).

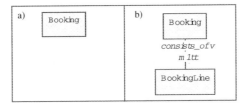

Figure 2   The alternative solutions after the first question[1]

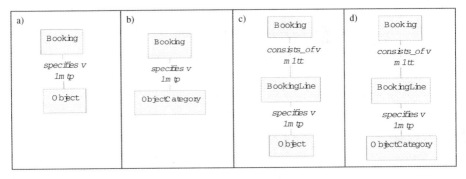

Figure 3   The alternative solutions after the second question

In this way, the wizard interviews the user and piece by piece build a schema depending on the answers. A very basic help function for clarifying the last unanswered question is implemented. Facilities for changing the schemas will also be implemented with the development of the tool. In Figure 4, two different solutions, a minimal contrasted to a more complex solutions, are shown to illustrate the tool.

---

[1] The notation is UML similar, except the direction and the cardinality constraints for a relation, which follows Johannesson's notation [6] and are represented by an error in the name of the relation and a quadruple <1m,1m,tp,tp> under the name of the relation, respectively. The first position in the quadruple shows whether a relation is single (1)- or multy (m)- valued. The third position shows where a relation is total (t) or partial (p). The second and the fourth positions show the same properties for the inverse of a relation.

4   The Motivation and the Goal of the Study

When reasoning about a tool, two aspects are necessary to investigate namely, the quality and the usability of the tool. Furthermore, two necessary criteria for the quality of this kind of tools are the quality of the proposed solutions and the complete-

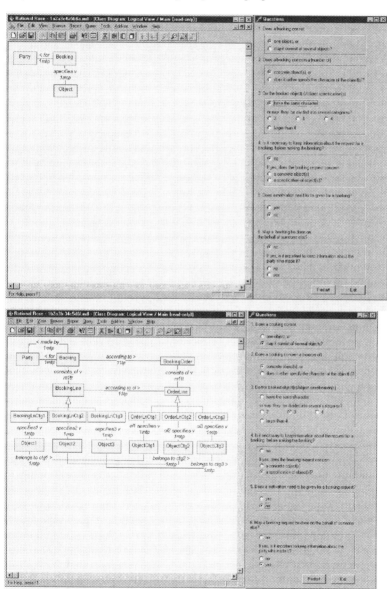

Figure 4   Two answer sets and the contrasting solutions corresponding to them

ness of the questions, where the fulfilment of the second one is much more difficult to guarantee. In the works provided in [10] and [11] the completeness of the questions is discussed based on Fillmore's case grammar theory [3] and Coad's library of modelling patterns [2], respectively. However no one of these approaches can guarantee the completeness of the questions. In order to do that a huge work has to be done empirically, testing the tool on real cases.

According to the usability, the tool may be used for at least two different purposes: by expert software engineers as a starting point in data analysis and reuse processes; and, in the education during the self-study process for improving students' modelling skills. Different case studies testing the usability for these two purposes have to be provided. Due to the easy access to students, our work has initially been concentrated on the usability of the tool as a pedagogical aid.

A first study has been provided. The general hypothesis was that the wizard might successfully be used in the education process. The main expectation from the tool as a teaching aid was that it should stimulate the students to reflect on the different solutions depending on the differences in describing the corresponding situations to be modelled. In this way the students should be pushed to catch different modelling patterns with an underlying understanding about them. The intention was to compare the learning process between a test and a reference group, as well as to analyse the usage of the modelling wizard by the test group. Due to the small number of the subjects in the test group no quantitative, but only qualitative analysis has been done for the second matter.

5   Method and Design of the Study

37 subjects participated and worked in groups of two or three. Only one subject worked alone. The groups were divided into one test group and one reference group. The test group included nine groups organising 20 subjects and the reference group included seven groups of 17 subjects. All subjects were computer and systems sciences students attending an advanced course on logical and physical database design. The study was performed as an optional assignment of the course and gave a bonus point on the final exam to the students who participated in it. Approximately 80% of the class attended it.

The study consisted of three parts. Firstly, after the usual questions about the age and gender the students were given an initial modelling task for a hospital booking system. The time for this task was limited to 20 minutes and the task was deliberately constructed to be difficult.

Secondly, the students were given four small tasks to be done in a sequence. The domain was chosen to be a university booking system in order to ensure the students' knowledge about it and reduce the uncertainty about the domain. The test groups were asked to use the modelling wizard when performing the tasks, whereas the reference groups were asked to model by pen on paper. This sequence of tasks was constructed to exemplify four different booking situations starting with small and uncomplicated modelling cases and moving into more complex situations. The intention was to push the students to reflect on the similarities and differences between the four different situations. For the test groups, this task also intended to exemplify the capacity of the tool. During this task the test groups ware expected to learn the tool, as well as to reflect on the solutions they were proposed by it. The only introduction to the tool was the information that it consisted of a number of questions and about the help facilities it offered. For the sake of learning the tool, the groups were offered an example solution for each task, as a feed back so the subjects should be able to verify whether they correctly understood the questions placed by the tool or not. The groups with deviating solutions from the example solutions were asked to identify the differences between both solutions in order to force them to further discuss the meaning of each question. To ensure that the reference group had access to the same kind of information as the test group, the example solution were even offered to them as well as a book on conceptual modelling which they were allowed to use during the whole study.

During the third part of the study, the students were asked to reconsider their solutions from the initial task and document improvements, where they had any. The time for this task was limited to 15 minutes. Finally the subjects were asked to individually fill in a questionnaire which for the test group included questions about the reflection on the tool.

The time for the study was limited to two hours. The subjects were asked to evaluate their modelling skills within a scale of one to five[2] three times: in the beginning of the study, after the first part of the study, and finally before the final questionnaire. The study was documented by recording all groups on video.

6   Results

The schemas produced by the groups during the initial task as well as those produced in the final phase were evaluated by two independent judges. The judges used the same scale as the one used by the students (1-5) for evaluating their modelling skills. The results from these evaluations are shown in Figure 5. The minus in some cells

		T1	T2	T3	T4	T5	T6	T7	T8	T9	R1	R2	R3	R4	R5	R6	R7
judge1	Initial	2	3	3	3	3	3	3	2	3	2	2	2	2	2	3	2
	Final	3	3	3	2	4	-	-	3	-	2	2.	2	(3)	-	-	(2.5)
judge2	Initial	1	2	4	2	3	4	3	3	2	2	3	3.5	2	2.5	3	1
	Final	2.5	2.5	4.5	3	3	-	-	4	-	2	3	4	(3.5)	-	-	(2)

Figure 5   The marks for the initial and the final models produced by the test (T1,... ,T9) and the reference (R1,...,R7) groups given by the two judges[3].

stands for an unchanged schema. Two of the groups did not modify their schemas because they did not thought they gained any new knowledge during the study, whereas three of the groups reconsidered, but ware satisfied with their initial schemas. Generally, the changes improved the schemes. In the cases where it was not noticed

---

[2] The scale 1 to 5, where 1 stands for bad and 5 for excellent, was until some years ago the national scale within high school in Sweden. The evaluation within high schools followed the normal dispersion. This scale was chosen in the study, due to the population's familiarity with it.

[3] Tow groups, R4 and R7, documented the changes they wanted to perform in text only. They did not had time to draw the changes graphically. Therefore, the judgements of their final solutions as well as the differences from the initial solution are presented in parentheses.

with the mark is due to the changes was not big enough or due to in changing parts of the schema some other parts failed outside consideration and was forgotten in the new schema. However, no big difference between the test and the reference group can be noticed on the modelling skills progress during the study.

However, a behaviour difference is that the test groups generally suggested much more comprehensive changes when improving their original models, whereas all the reference groups, except one, performed small changes only. The changes are measured by the number of object types introduced or reduced, the number of relations introduced, reduced or changed, and the number of generalization constraints (Isa) introduced to the initial solution (see Figure 6). The introduction/ reduction of an object type is considered as a bigger change compared to the changes of a relation, since they have a bigger influence on the stability of a schema [7].

		T1	T2	T3	T4	T5	T6	T7	T8	T9	R1	R2	R3	R4	R5	R6	R7
Object	Introduced	2	1	2	2	2	-	-	2	-	6	-	-	(-)	-	-	(2)
	Reduced	1	-	3		3	-	-	-	-	9	-	-	(-)	-	-	(-)
Relation	Introduced	5	2	1	3	5	-	-	1	-	8	-	-	(1)	-	-	(1)
	Reduced	-	1	5	6	10	-	-	-	-	8	-	-	(-)	-	-	(-)
	Changed	1	5	2	3	2	-	-	3	-	-	1	1	(1)	-	-	(-)
Isa	Introduced	-	-	1	2	1	-	-	-	-	-	-	-	(-)	-	-	(-)

Figure 6   The differences between the initial and the final schemas[3]

Beside the evaluation of the schemas done by the judges, the students' own evaluations on their modelling skills were considered. The subjects were asked to evaluate their modelling skills three times during the study:

- firstly in the beginning of the study;
- secondly after the difficult initial modelling task; and
- thirdly at the end of the study after the four small tasks were performed and the reconsideration of the initial schema with possible changes was done.

The reason with the second evaluation was to confirm the first one, whereas the reason with the third evaluation was to measure the subjects' own impression of their progress during the study. The results are presented in Figure 7 and Figure 8. Most of the subjects kept their evaluations unchanged (shown by minuses in the tables) and no significant differences can be seen between the test and the reference groups. One reason for which the subjects did not increase their marks in the third evaluation may lie in the limited time for which the study was performed. One hour, which the subject approximately had for performing the sequence of the four modelling tasks, was not long enough to give the students confidence to evaluate their own skills with a higher mark. Instead some subjects decreased their marks. Notable, even if not surprisingly, is that the five subjects who decreased in their evaluation were all female. Notable is also that one subject, subject number 34, evaluated his skills with five plus, a mark outside the given scale.

gr	T1		T2		T3			T4		T5	T6		T7			T8			T9	
nr	1	2	3	4	5	6	7	8	9	10	11	12	13	14	15	16	17	18	19	20
gen	f	f	m	m	f	m	f	m	m	f	m	m	f	f	f	f	m	f	m	m
ev1	2	2	3	3	2	3	3	3	1	4	3	4	4	2.5	3	4	4	4	3	3
ev2	-	-	-	2	-	-	-	-	-	-	-	-	-	-	-	-	-	-	-	-
ev3	-	-	-	3	-	-	-	-	-	3	-	-	-	-	-	-	-	-	-	-

Figure 7   The test subjects own evaluations of his/her modelling skills (gr = the number of the group, nr = the number of the subject, gen = the gender of the subject, ev = evaluation)

gr	R1		R2		R3			R4		R5		R6			R7		
nr	21	22	23	24	25	26	27	28	29	30	31	32	33	34	35	35	37
gen	f	f	f	f	m	f	m	m	f	m	m	f	f	m	f	f	f
ev1	2	2	3	3	3	3	3	4	3	3	3	2.5	2.5	5+	3	3	3
ev2	-	-	-	-	-	-	-	-	-	-	-	-	3	-	2	-	-
ev3	1	-	-	-	-	-	-	-	2.5	-	-	-	2.5	-	-	-	-

Figure 8   The references subjects own evaluations of his/her modelling skills

Finally, an interesting observation is that no one from the reference groups used the book they were offered. Most of them just forgot, immediately after reading the instructions, that a book was available and they did not even note the presence of it beside other necessary material like paper and pens.

7   Evaluation

In this section some reflection made from the test groups' work with the modelling tool are presented. Since the number of the test subjects is only 20 no reliable quantitative measures may be done. Instead qualitative analysis has been performed. It is based on the questions posed to the subjects after their work with the tool. The analysis was supported by the video records from the study. In the questionnaire, the subjects were asked to answer about the questions, the examples, and the solutions from the tool as well as whether they reflected on the schemas, and finally how they assessed the tool for teaching aims. Six alternatives: 'not at all', 'very little', 'little', 'much', 'very much', and 'very very much' and place for comments were given to each question.

7.1 Were the Questions Implemented in the Wizard Clear?

The questions: 1) "Did you understood the questions which the tool placed?" and 2) "How difficult did you find the questions placed by the tool?" were placed to gather information for the formulations of the questions and their impact on the domain analysis process. The subjects' impressions from the question were mixed. Some of them found the questions clear where other found them unclear. Some of the subjects found the questions difficult whereas others did not. However, generally the meaning of the second question in the tool "Does a booking concern a (number of) a) concrete object(s), or b) does it rather specify the character of the objects(s)?" was difficult to grasp by the subjects. Even the fourth question, which includes the same distinction, i.e., the distinction between an object and an object specification, caused problems. This distinction needs therefore to be better clarified.

Furthermore, the fifth question, asking whether a motivation is needed for a booking, was usually misinterpreted, even if it did not confuse the subjects, since they did believe they correctly understood it. Unfortunately, the choice of the word motivation was the failure since it did not get the subjects to associate with the action, for which a booking is done.

7.2 Did the Students Use the Help Examples?

Most of the subjects did use the help examples. Most of them thought that the examples were clearly written. Still, some subjects thought that the examples were difficult to transfer to the problems they had to solve and had therefore difficulties in using them immediately. Even the opinion that all the examples should follow one and the same case was uttered. The reason for not giving the examples from the same case now was the intention to find the most suitable examples rather than to pick them up from the same case.

Furthermore, more sophisticated help was required. Some subjects wanted to see in advance what the result of choosing different answer alternatives should be. This effect is for the moment obtained by picking different answer alternatives and considering the suggested solutions. However, in this way only one schema is shown at the time, which does not for instance support comparison of schemas resulting from different answers. Comparing two or three solutions may, therefore, be facilitated by extending the help function in the way proposed by the subjects.

7.3 Did the Tool Support Reflection?

On the question "Did the tool help you to reflect on your schemas?" answered only one subject 'not at all'. This was the same subject who worked alone. Ten subjects (of 20) answered very positively by choosing one of the alternatives 'much', 'very much' or 'very very much'. Almost all of these ten subjects have also answered positively on the first question whether they understood the questions or not. Therefore, a possible relation may be that students who understood the questions were also able to understand and reflect on the solutions suggested by the tool.

7.4 Shall Such a Tool Be Used for Teaching Aims?

Fifteen (of 20) subjects answered positively on this question. One subject did not have any opinion on this subject. Most of those who judged the tool positively as a teaching aim thought they had learned 'a little' or 'much' from it and/or that it was 'motivating' or 'very motivating' to use it. Also all the subjects who thought that the tool supported reflection answered positively on this question.

The four subjects who were negative for using the tool within the education motivated it in the following way: "The general object names were not good. Explanation on why the tool models in a certain way is missing", "The answer alternatives were too limited", "The tool does not give chance to experiment, which often teaches more", "I better like to work for myself instead of getting questions". The last two comments are difficult to do anything about and they are therefore left aside. The generality, both of the questions and of the suggested solutions, is necessarily for such a tool. It is indeed desirable to be able to change the names in the suggested solutions with application specific words, which is not yet supported by the prototype, but is planned to be added to the tool's functionality. According to the limitation of the answer alternatives it is important to guarantee the completeness of the alternatives for each question, which we consider they are. However, much more difficult is to guarantee the completeness of the questions and a lot of empirical studies are necessary before doing this. Finally, since the comment on missing explanations from the tool was even left from other subjects we are discussing it in the following subsection.

7.5 General Observations

Some subjects did not get the connection between the questions and the presented solutions. One suggestion was to improve the help function by explaining how a particular answer influences the development of a schema and why. Another possible solution is to implement verbal feedback explaining impact of the answers on the suggested solutions. Beside, a different reflection that may have a big influence on the understanding of a schema, is that all compulsory relations should be drawn from the very start. For instance, a booking is always made for a party. Introducing an object type Party connected to the object type Booking already in the beginning should help on understanding the intermediate solutions considerably. For the moment the Party object type is introduced very late in the intermediate solutions and it turns out from the study that this is inappropriate.

Surprisingly, some subjects found difficulties with the language. Two of those who were negative for using the tool as a teaching aim pointed out this as a problem. The language in the tool is English. However, the instructions for the study and the exercises were given in Swedish. The language difficulty was explicitly noted by some subjects when they had to answer on how well they understood the questions and the examples. Other had difficulties with interpreting some parts of the suggested solution due to problems with understanding some words. It was surprising since a large part of the literature for the university students is in English and the subjects were not expected to have problems with the language. However, the language used in a conceptual schema is really limited and not recognising a single word may cause problems for understanding a whole schema.

Furthermore, a limitation of the tool was identified when preparing the study. This limitation appears in the case when different booking categories are distinguished (identified by question 4) and they differ in matters identified by the rest of the questions. For the moment, the rest of the questions are posed for all the different bookings together, instead of posing them for each one category separately. The subjects were asked to note any limitations they discovered. The initial modelling task was deliberately constructed to include the situation described above. However, no one of the subjects found this limitation. Only one of the test groups had a suggestion about extension of the tool namely, that it should even ask whether a withdrawn booking should be taken care of, or not.

8 Conclusions and Further Work

The subject for this work has been reuse in the information systems analysis process. It is one in a series of works on documentation and collection of analysis patterns in order to expose them and make them easily available for the systems engineers. The approach is however slightly different from the proceeding approaches [2],[4], and [5] by its attempt for automating the search for the most suitable pattern when a library has been collected. It also differs from [9] by its attempt of gathering the necessary information directly from the user, instead of using meta–knowledge for the retrieval of an appropriate solution. The goal is to realize these properties by a modelling wizard tool. Such a tool should be useful in some of the following situations:

- To support novices in the process of domain analysis and conceptual schema construction;
- To be used by expert software engineers as a starting point in data analysis and reuse processes;
- With some extensions to be tailored for novice database developers with no formal knowledge, of database design, developing their own databases for use in the home;
- In the education during the self-study process for improving students' modelling skills.

As a necessary step of the development of a tool, a prototype developed for one domain has been implemented [10]. Before any work on generalising the prototype was provided [11], a case study investigating the usability of the tool has been performed. The study aimed to evaluate the tool as a pedagogical aid, as well as to give input about the opinion about such a tool. It was performed on students. The hypothesis of the study was that the tool could successfully be used in the education process. The outcome of the study is that, for the short time the subjects had, there was not a considerable difference in the learning capacity between subjects who used it and those who did not. However, a positive result from the study was that the subjects who used the tool differed from the subjects who did not, in the number of changes they wanted to perform in a schema constructed by them earlier in the study. Furthermore, the subjects who used the tool were positively for using it within the education. Some of them explicitly noted that the usage of such tools should be on voluntary bases and as a complement to, rather as a reduction of, the teachers.

The future work will go in three main directions carried out in parallel.

- The generalisation of the tool into domain independent wizard, continuing the work provided in [11]. It is necessary to provide research on how a general tool shall be constructed in order to support the modelling of complex UoD ranging over, not only one but, several different domains. One possibility should be to equip the modelling wizard with schema integration facilities. Still a number of questions should need to be considered in order to co-ordinate the behaviour of the tool with the behaviour of a user. One such question is to investigate the human's techniques when modelling complex domains and her way of dividing complex problems into smaller parts, either to make them easily to grasp and solve, or to distribute the responsibilities between different parties;
- The evaluation of the completeness of the wizard, which have to be done by empirical work either by using the wizard in a number of real cases or by letting a number of experts use and evaluate the tool; and

- The evaluation of the usability of the tool for the different categories of users mentioned above. The work on evaluating the wizard as a pedagogical aid can be extended to also consider the usage of the tool for a longer time in the education process, i.e., the impact of the tool when using it during a whole course, rather than using it for one hour only.

Acknowledgements
I would like to thank Paul Johannesson, Maria Bergholtz, Eva Fåhræus, Sari Hakkarainen, Klas Karlgren, Robert Ramberg and Jakob Tholander, for their help with the case study. I would also like to thank the anonymous reviewers for the valuable comments on an earlier draft of the paper.

References

[1]   C. Alexander et al. *A Pattern Language*, NY: Oxford University Press, 1977.

[2]   P. Coad et al. *Object Models: Strategies, Patterns, & Applications*, Yourdon Press, Prentice Hall, 1995.

[3]   C.H. Fillmore, "The case for case", in Bach and Harms (eds.) *Universals in Linguistic Theory*, Holt, Rinehart and Winston, New York, 1968.

[4]   M. Fowler, *Analysis Patterns: Reusable Object Models*, Addison-Wesley, 1997.

[5]   D.C. Hay, Data Model Patterns: Conventions of Thought, Dorset House Publishing, 1996.

[6]   P. Johannesson, *Schema Integration, Schema Translation, and Interoperability in Federated Information Systems*, Dissertation at the Dept. of Computer and Systems Sciences, Stockholm University/KTH, Sweden, 1993.

[7]   P. Johannesson and P. Wohed, "Improving Quality in Conceptual Modelling by the Use of Schema Transformations", in Thalheim, B. (Ed.), Proc. of 15th Int. Conference on Conceptual Modeling ER96, LNCS 1157, Springer-Verlag, 1996

[8]   S. Purao, "APSARA: A Tool to Automate Systems design via Intelligent Pattern Retrieval and Synthesis", The Data Base for Advances in Information Systems – Fall, vol. 29, no. 4, 1998.

[9]   S. Purao and V.C. Storey, "Intelligent Support for Retrieval and Synthesis of Patterns for Object-Oriented Design", in D.W. Embley and R.C. Goldstein (eds.) *Conceptual Modeling – ER'97*, LNCS 1331, Springer, pp. 30-42, 1997.

[10] P. Wohed, "Conceptual Patterns for Reuse of Information Systems Design", in B. Wangler and L. Bergman (Eds.), Proc. of 12th Int. Conf. on Advanced Information Systems Engineering – CAiSE 2000, LNCS 1789, pp 157-175, Springer-Verlag, 2000.

[11] P. Wohed, "Conceptual Patterns – a Consolidation of Coad's and Wohed's Approaches", to appear in M. Bouzeghoub, Z. Jedad and E. Métais (eds.), Proc. of the 5th International Conference on Application of Natural Language to Information Systems – NLDB'2000, LNCS, Springer-Verlag, 2000.

Ontological Analysis of Taxonomic Relationships

Nicola Guarino and Christopher Welty[†]

LADSEB/CNR
Padova, Italy
{guarino,welty}@ladseb.pd.cnr.it
http://www.ladseb.pd.cnr.it/infor/ontology/ontology.html
† on sabbatical from Vassar College, Poughkeepsie, NY

Abstract. Taxonomies based on a partial-ordering relation commonly known as is-a, class inclusion or subsumption have become an important tool in conceptual modeling. A well-formed taxonomy has significant implications for understanding, reuse, and integration, however the intuitive simplicity of taxonomic relations has led to widespread misuse, making clear the need for rigorous analysis techniques. Where previous work has focused largely on the semantics of the *is-a* relation itself, we concentrate here on the ontological nature of the *arguments* of this relation, in order to be able to tell whether a single *is-a* link is ontologically well-founded. For this purpose, we discuss techniques based on the philosophical notions of *identity, unity, essence,* and *dependence*, which have been adapted to the needs of information systems design. We demonstrate the effectiveness of these techniques by taking real examples of poorly structured taxonomies, and revealing cases of invalid generalization. The result of the analysis is a cleaner taxonomy that clarifies the modeler's ontological commitments.

1 Introduction

Taxonomies are an important tool in conceptual modeling, and this has been especially true since the introduction of the extended ER model [6,26]. Properly structured taxonomies help bring substantial order to elements of a model, are particularly useful in presenting limited views of a model for human interpretation, and play a critical role in reuse and integration tasks. Improperly structured taxonomies have the opposite effect, making models confusing and difficult to reuse or integrate.

Many previous efforts at providing some clarity in organizing taxonomies have focused on the semantics of the taxonomic relationship (also called is-a, class inclusion, subsumption, etc.) [3], on different kinds of relations (generalization, specialization, subset hierarchy) according to the constraints involved in multiple taxonomic relationships (covering, partition, etc.) [23], on the taxonomic relationship in the more general framework of data abstractions [7], or on structural similarities between descriptions [2,5]. Our approach differs in that we focus on the arguments (i.e. the properties) involved in the taxonomic relationship, rather than on the semantics of the relationship itself. The latter is taken for granted, as we take the statement "φ subsumes ϕ" to mean simply:

$$\forall x \; \phi(x) \to \varphi(x) \qquad (1)$$

A.H.F. Laender, S.W. Liddle, V.C. Storey (Eds.): ER2000 Conference, LNCS 1920, pp. 210–224, 2000.
© Springer-Verlag Berlin Heidelberg 2000

Our focus here will be on verifying the plausibility and the well-foundedness of single statements like (1) on the basis of the *ontological nature* of the two properties ϕ and φ.

In this paper we present and formalize four fundamental notions of so-called *Formal Ontology* [11]: *identity, unity, essence, and dependence*, and then show how they can be used as the foundation of a methodology for conceptual modeling. Implicitly, we also assume a fifth fundamental notion, *parthood,* whose role for conceptual analysis has been extensively discussed elsewhere [1,22]. Finally, we demonstrate the effectiveness of our methodology by going through a real example of a poorly structured taxonomy, and revealing cases of invalid generalization. The result of the analysis is a cleaner taxonomy that clarifies the modeler's ontological assumptions.

2   Background

The notions upon which our methodology is based are subtle, and before describing them with formal rigor we discuss the basic intuitions behind them and how they are related to some existing notions in conceptual modeling.

2.1   Basic Notions

Before presenting our formal framework let us informally introduce the most important philosophical notions: *identity, unity, essence,* and *dependence*. The notion of identity adopted here is based on intuitions about how we, as cognitive agents, in general interact with (and in particular recognize) individual entities in the world around us. Despite its fundamental importance in Philosophy, it has been slow in making its way into the practice of conceptual modeling for information systems, where the goals of analyzing and describing the world are ostensibly the same.

The first step in understanding the intuitions behind identity requires considering the distinctions and similarities between *identity* and *unity*. These notions are different, albeit closely related and often confused under a generic notion of identity. Strictly speaking, identity is related to the problem of distinguishing a specific instance of a certain class from other instances of that class by means of a *characteristic property,* which is unique for *it* (that *whole* instance). Unity, on the other hand, is related to the problem of distinguishing the *parts* of an instance from the rest of the world by means of a *unifying relation* that binds the parts together, and nothing else. For example, asking, "Is that my dog?" would be a problem of identity, whereas asking, "Is the collar part of my dog?" would be a problem of unity.

Both notions encounter problems when time is involved. The classical one is that of *identity through change*: in order to account for common sense, we need to admit that an individual may remain *the same* while exhibiting different properties at different times. But which properties can change, and which must not? And how can we reidentify an instance of a certain property after some time? The former issue leads to the notion of an *essential property*, on which we base the definition of *rigidity,* discussed below, while the latter is related to the distinction between *synchronic* and *diachronic* identity. An extensive analysis of these issues in the context of conceptual modeling has been made elsewhere [14].

The fourth notion, *ontological dependence,* may involve many different relations such as those existing between persons and their parents, holes in pieces of cheese and the cheese, and so on [22]. We focus here on a notion of dependence as applied to properties. We distinguish between *extrinsic* and *intrinsic* properties, according to whether they depend or not on other objects besides their own instances. An intrinsic property is typically something inherent to an individual, not dependent on other individuals, such as having a heart or having a fingerprint. Extrinsic properties are not inherent, and they have a relational nature, like "being a friend of John". Among these, there are some that are typically assigned by external agents or agencies, such as having a specific social security number, having a specific customer i.d., even having a specific name.

It is important to note that our ontological assumptions related to these notions ultimately depend on our *conceptualization* of the world [12]. This means that, while we shall use examples to clarify the notions central to our analysis, *the examples themselves will not be the point of this paper*. For example, the decision as to whether a cat remains the same cat after it loses its tail, or whether a statue is identical with the marble it is made of, are ultimately the result of our sensory system, our culture, etc. The aim of the present analysis is to clarify the formal tools that can both make such assumptions explicit, and reveal the logical consequences of them. When we say, e.g. that "having the same fingerprint" may be considered an identity criterion for *PERSON*, we do *not* mean to claim this is the universal identity criterion for *PERSONs*, but that *if this were* to be taken as an identity criterion in some conceptualization, what would that mean for the property, for its instances, and its relationships to other properties?

2.2   Related Notions

Identity has many analogies in conceptual modeling for databases, knowledge bases, object-oriented, and classical information systems, however none of them completely captures the notion we present here. We discuss some of these cases below.

Membership conditions. In description logics, conceptual models usually focus on the sufficient and necessary criteria for class *membership*, that is, recognizing instances of certain classes [4]. This is not identity, however, as it does not describe how instances of the same class are to be told apart. This is a common confusion that is important to keep clear: membership conditions determine when an entity is an instance of a class, i.e. they can be used to answer the question, "Is that *a* dog?" but not, "Is that *my* dog?"

Globally Unique IDs. In object-oriented systems, uniquely identifying an object (as a collection of data) is critical, in particular when data is persistent or can be distributed [28]. In databases, *globally unique id's* have been introduced into most commercial systems to address this issue. These solutions provide a notion of identity for the descriptions, for the units of data (objects or records), but not for the entities they describe. It still leaves open the possibility that two (or more) descriptions may refer to the same *entity*, and it is this entity that our notion of identity is concerned with. In other words, globally unique IDs can be used to answer, "Is this the same description of a dog?" but not, "Is this my dog."

Primary Keys. Some object-oriented languages provide a facility for overloading or locally defining the equality predicate for a class. In standard database analysis, introducing new tables requires finding unique keys either as single fields or combinations of fields in a record. These two similar notions very closely approach our notion of identity as they do offer evidence towards determining when two descriptions refer to the same entity. There is a very subtle difference, however, which we will attempt to briefly describe here and which should become more clear with the examples at the end of the paper.

Primary (and candidate) keys and overloaded equality operators are typically based on *extrinsic properties* (see Section 2.1) that are required by a system to be unique. In many cases, information systems designers add these extrinsic properties simply as an escape from solving (often very difficult) identity problems. Our notion of identity is based mainly on *intrinsic properties*—we are interested in analyzing the inherent nature of entities and believe this is important for understanding a domain.

This is not to say that the former type of analysis never uses intrinsic properties, nor that the latter never uses extrinsic ones – it is merely a question of emphasis. Furthermore, our analysis is often based on information which *may not be represented in the implemented system*, whereas the primary key notion can never use such information. For example, we may claim as part of our analysis that people are uniquely identified by their brain, but this information would not appear in the final system we are designing. Our notion of identity and the notion of primary keys are not incompatible, nor are they disjoint, and in practice conceptual modelers will often need both.

3   The Formal Tools of Ontological Analysis

In this section we shall present a formal analysis of the basic notions discussed above, and we shall introduce a set of *meta-properties* that represent the behaviour of a property with respect to these notions. Our goal is to show how these meta-properties impose some constraints on the way subsumption is used to model a domain.

Our analysis relies on certain fairly standard conventions and notations in logic and modal logic, which are described in more detail in [16]. We shall denote primitive meta-properties by bold letters preceded by the sign "+", "-" or "~" which will be described for each meta-property. We use the notation ϕ^M to indicate that the property ϕ has the meta-property **M**.

3.1   Rigidity

The notion of rigidity was defined previously in [10] as follows:

Definition 1 A *rigid property* is a property that is essential to *all* its instances, i.e. a property ϕ such that: $\forall x \, \phi(x) \rightarrow \Box \, \phi(x)$.

Definition 2 A *non-rigid property* is a property that is not essential to *some* of its instances, i.e. $\exists x \, \phi(x) \wedge \neg \Box \, \phi(x)$.

Definition 3 An *anti-rigid property* is a property that is not essential to *all* its instances, i.e. $\forall x \, \phi(x) \rightarrow \neg \Box \, \phi(x)$.

For example, we normally think of *PERSON* as rigid; if x is an instance of *PERSON*, it must be an instance of *PERSON* in every possible world. The *STUDENT* property, on the other hand, is normally not rigid; we can easily imagine an entity moving in and out of the *STUDENT* property while being the same individual.

Anti-rigidity was added as a further restriction to non-rigidity. The former constrains all instances of a property and the latter, as the simple negation of rigidity, constrains at least one instance. Anti-rigidity attempts to capture the intuition that all instances of certain properties must possibly not be instances of that property. Consider the property *STUDENT*, for example: in its normal usage, every instance of *STUDENT* is not necessarily so.

Rigid properties are marked with the meta-property **+R**, non-rigid properties are marked with **-R**, and anti-rigid properties with **~R**. Note that rigidity as a meta-property is not "inherited" by sub-properties of properties that carry it, e.g. if we have *PER-SON*$^{+R}$ and $\forall x\, STUDENT(x) \rightarrow PERSON(x)$ then we know that all instances of *STUDENT* are necessarily instances of *PERSON*, but not *necessarily* (in the modal sense) instances of *STUDENT*, and we furthermore may assert *STUDENT*$^{-R}$. In simpler terms, an instance of *STUDENT* can cease to be a student but may not cease to be a person.

3.2   Identity

In the philosophical literature, an *identity condition* (IC) for an arbitrary property ϕ is usually defined as a suitable relation ρ satisfying the following formula:

$$\phi(x) \wedge \phi(y) \rightarrow (\rho(x, y) \leftrightarrow x = y) \tag{2}$$

For example, the property *PERSON* can be seen as carrying an IC if relations like *having-the-same-SSN* or *having-the-same-fingerprints* are assumed to satisfy (2).

As discussed in more detail elsewhere [14], the above formulation has some problems, in our opinion. The first problem is related to the need for distinguishing between *supplying* an IC and simply *carrying* an IC: it seems that non-rigid properties like *STUDENT* can only carry their ICs, inheriting those supplied by their subsuming rigid properties like *PERSON*. The intuition behind this is that, since the same person can be a student at different times in different schools, an IC allegedly supplied by *STUDENT* (say, having the same registration number) may be only local, within a certain studenthood experience.

The second problem regards the nature of the ρ relation: what makes it an IC, and how can we index it with respect to time to account for the difference between *synchronic* and *diachronic* identity?

Finally, deciding whether a property carries an IC may be difficult, since finding a ρ that is both necessary *and* sufficient for identity is often hard, especially for natural kinds and artifacts.

For these reasons, we have refined (2) as follows:

Definition 4 An *identity condition* is a formula Γ that satisfies either (3) or (4) below, excluding trivial cases and assuming a predicate E for *actual existence* (see [15]):

$$E(x,t) \wedge \phi(x,t) \wedge E(y,t') \wedge \phi(y,t') \wedge x=y \rightarrow \Gamma(x,y,t,t') \tag{3}$$
$$E(x,t) \wedge \phi(x,t) \wedge E(y,t') \wedge \phi(y,t') \wedge \Gamma(x,y,t,t') \rightarrow x=y \tag{4}$$

An IC is necessary if it satisfies (3) and sufficient if it satisfies (4). Based on this, we define two meta-properties:

Definition 5 Any property *carries* an IC iff it is subsumed by a property supplying that IC (including the case where it supplies the IC itself).

Definition 6 A property ϕ *supplies* an IC iff i) it is rigid; ii) there is a necessary or sufficient IC for it; and iii) The same IC is not carried by *all* the properties subsuming ϕ. This means that, if ϕ inherits different (but compatible) ICs from multiple properties, it still counts as supplying an IC.

Definition 7 Any property carrying an IC is called a *sortal* [25].

Any property carrying an IC is marked with the meta-property **+I** (**-I** otherwise). Any property supplying an IC is marked with the meta-property **+O** (**-O** otherwise). The letter "O" is a mnemonic for "own identity". From the above definitions, it is obvious that **+O** implies **+I** and **+R**. For example, both *PERSON* and *STUDENT* do carry identity (they are therefore **+I**), but only the former *supplies* it (**+O**).

3.3   Unity

In previous work we have extensively discussed and formalized the notion of unity, which is itself based upon the notion of part [14]. This formalization is based on the intuition that a whole is something all of whose parts are connected in such a way that each part of the whole is connected to all the other parts of that whole and nothing else. Briefly, we define:

Definition 8 An object *x is a whole under* ω iff ω is an equivalence relation such that all the parts of *x* are linked by ω, and nothing else is linked by ω.

Definition 9 A property φ *carries a unity condition* iff there exists a single equivalence relation ω such that each instance of φ is a whole under ω.

Depending on the ontological nature of the ω relation, which can be understood as a "generalized connection", we may distinguish three main kinds of unity for concrete entities (i.e., those having a spatio-temporal location). Briefly, these are:

- *Topological unity*: based on some kind of topological or physical connection, such as the relationship between the parts of a piece of coal or an apple.

- *Morphological unity*: based on some combination of topological unity and shape, such as a ball, or a morphological relation *between wholes* such as for a constellation.

- *Functional unity*: based on a combination of other kinds of unity with some notion of purpose as with artifacts such as hammers, or a functional relation between wholes as with artifacts such as a bikini.

As the examples show, nothing prevents a whole from having parts that are themselves wholes (with a different UC). This can be the foundation of a theory of *pluralities*, which is however out of this paper's scope.

As with rigidity, in some situations it may be important to distinguish properties that do not carry a *common* UC for all their instances, from properties all of whose instances are not wholes. As we shall see, an example of the former kind may be *LEGAL AGENT*, all of whose instances are wholes, although with different UCs (some legal agents may be people, some companies). *AMOUNT OF MATTER* is usually an example of the latter kind, since none of its instances can be wholes. Therefore we define:

Definition 10 A property has *anti-unity* if every instance of the property is not a whole.

Any property carrying a UC is marked with the meta-property **+U** (**-U** otherwise). Any property that has anti-unity is marked with the meta-property **~U**, and of course **~U** implies **-U**.

3.4   Dependence

The final meta-property we employ as a formal ontological tool is based on the notion of dependence. As mentioned in Section 2.1, we focus here on ontological dependence as applied to properties. The formalization adopted below is refined from previous work [9], and is based on Simons' definition of *notional dependence* [22]. We are aware that this is only an approximation of the more general notion of extrinsic (or relational) property, and that further work may be needed (see for instance [19]).

Definition 11 A property ϕ is *externally dependent* on a property ψ if, for all its instances x, necessarily some instance of ψ must exist, which is not a part nor a constituent of x:

$$\forall x \,\Box\, (\phi(x) \rightarrow \exists y\, \psi(\dot{y}) \wedge \neg P(y, x) \wedge \neg C(y, x)) \tag{5}$$

The part and constituent relations are discussed further in [16]. In addition to excluding parts and constituents, a more rigorous definition must exclude qualities (such as colors), things which necessarily exist (such as the universe), and cases where ψ is subsumed by ϕ (since this would make ϕ dependent on itself). Intuitively, we say that, for example, *PARENT* is externally dependent on *CHILD* (one can not be a parent without having a child), but *PERSON* is not externally dependent on heart nor on body (because any person has a heart as a part and is constituted of a body).

An externally dependent property is marked with the meta-property **+D** (**-D** otherwise).

3.5   Constraints and Assumptions

Our meta-properties impose several constraints on taxonomic relationships, and to these we add several methodological points that help to reveal modeling problems in taxonomies.

A first observation descending immediately from our definitions regards some *subsumption constraints*. If ϕ and ψ are two properties then the following constraints hold:

$$\phi^{-R} \text{ can't subsume } \psi^{+R} \tag{6}$$
$$\phi^{+I} \text{ can't subsume } \psi^{-I} \tag{7}$$
$$\phi^{+U} \text{ can't subsume } \psi^{-U} \tag{8}$$
$$\phi^{-U} \text{ can't subsume } \psi^{+U} \tag{9}$$
$$\phi^{+D} \text{ can't subsume } \psi^{-D} \tag{10}$$
$$\text{Properties with incompatible ICs/UCs are disjoint.} \tag{11}$$

Constraints (6-10) follow directly from our meta-property definitions (see [15] for more discussion and examples), and (11) should be obvious from the above discussion of identity and unity, but it is largely overlooked in many practical cases [13,15]. Concrete examples will be discussed at the end of this paper.

Finally, we make the following assumptions regarding identity (adapted from [20]):

* *Sortal Individuation.* Every domain element must instantiate some property carrying an IC (+I). In this way we satisfy Quine's dicto "No entity without identity" [21].

* *Sortal Expandability.* If two entities (instances of different properties) are the same, they must be instances of a property carrying a condition for their identity.

4   Methodology

We are developing a methodology for conceptual analysis whose specific goal is to *make modeling assumptions clear*, and to produce *well-founded taxonomies*. The analysis tools that make up our methodology can be grouped into four distinct layers, such that the notions and techniques within each layer are based on the notions and techniques in the layers below.

4.1   First Layer: Foundations

In the lowest, foundational, layer of the methodology are the meta-properties described in Section 3, and in more detail in [16].

4.2   Second Layer: Useful Property Kinds

The second layer in the methodology contains an ontology of useful property kinds. This is an extension of the formal ontology of *basic* property kinds presented in [15], which includes further specializations of *sortal* properties (Def. 7), each one corresponding to an identity or unity condition commonly found in practice. This ontology can be seen as a library of reference cases useful to characterize the meta-properties of a given property, and to check for constraints violations. We sketch below the basic property kinds as well as some further specializations of sortal properties.

Basic Property Kinds. The formal ontology of properties discussed in [15] distinguishes eight different kinds of properties based on the valid and most useful combinations of the meta-properties discussed in Section 3. In this paper we mention only three of these combinations: *categories* (**+R-I**), *types* (**+R+O**), and *quasi-types* (**+R-O+I**). These and the other five property kinds add to a modeler's ability to specify the meaning of properties in an ontology, since the definition of each property kind includes an intuitive and domain-independent description of what part that kind of property should play in an ontology.

CO. Countable Properties. This is an important specialization of sortals. In many cases, besides carrying identity (**+I**), countable properties also carry unity (**+U**). All subsumed properties must also be countable. Note that we appeal to a strict defintion of countability provided in [14], which may not be immediately intuitive in the case of collections, such as a group of people. One can count possible groups of people in a combinatoric sense, but by our current definition of countability, a group of people is not countable because it does not have unity.

ME. Properties carrying a mereologically extensional IC. Certain properties concerning masses or plural entities, such as *LUMP-OF-CLAY* or *GROUP-OF-PEOPLE*, have as a necessary identity condition that the parts of their instances must be the same (instances cannot change their parts). They cannot subsume properties with **-ME**.

UT. Properties carrying topological unity. See Section 3.3. Properties with **+UT** have unity (**+U**), and can not subsume properties with **-UT**.

UM. Properties carrying morphological unity. See Section 3.3. Properties with **+UM** have unity (**+U**), and can not subsume properties with **-UM**.

UF. Properties carrying functional unity. See Section 3.3. Properties with **+UF** have unity (**+U**), and can not subsume properties with **-UF**.

4.3   Third Layer: Ontology-Based Modeling Principles

The third layer in the methodology contains the notions of *backbone property* and *stratification*.

The backbone taxonomy. One of the principal roles of taxonomies is to impart structure on an ontology, to facilitate human understanding, and to enable integration. We have found that a natural result of our analysis is the identification of special properties in a taxonomy that best fill this role. We call these properties *backbone properties*, which constitute the *backbone taxonomy* [15].

The backbone taxonomy consists only of rigid properties, which are divided into three kinds (as discussed above): *categories*, *types*, and *quasi-types*. Categories can not be subsumed by any other kinds of properties, and therefore represent the highest level (most general) properties in a taxonomy. They are usually taken as primitive properties because defining them is too difficult (e.g. entity or thing).

Types are critical in our analysis because according to the assumptions presented in Section 3.5, *every instance instantiates at least one of these properties*. Therefore considering only the elements of the backbone gives someone a survey of the entire universe of possible instances.

Stratification. A very important result of our analysis is the recognition of multiple entities, based on different identity or unity criteria, where usually only one entity is conceived. The classical example is the statue and the clay it is made of, which count as different objects in our analysis. As discussed further in [13], this view results in a *stratified ontology*, where entities belong to different levels, depending on their identity and unity assumptions: we may distinguish for instance the physical level, the functional level, the intentional level, the social level. Entities at the higher levels are *constituted* (and co-located with) entities at the lower levels. The advantage of this view is a better semantic account of the taxonomic relation, a better account of the hidden ontological assumptions, and in general better ontologies. The costs are: i) a moderate proliferation (by a constant factor corresponding to the number of levels) of the number of entities in the domain; ii) the necessity to take into account different relations besides *is-a*, such as dependence, spatio-temporal colocalization, and constitution.

4.4   Fourth Layer: Top Level Ontology

The highest layer of our methodology is a top-level ontology designed using the notions and techniques of the layers below. This layer of the methodology is not yet complete, however first steps towards this have been discussed in [13].

4.5   Question/answer system

Finally, we are capturing the notions and techniques from these four layers in a knowledge-based question/answer system that guides conceptual modelers through the analysis process. This approach is similar to that of [24], and seemed necessary for two principal reasons:

- While rigidity and dependence tend to be simpler concepts to grasp, identity and unity are not. In addition to determining if a property carries identity and unity conditions, it is also useful to know, when possible, what those criteria are, because they must be consistent with subsuming and subsumed properties.

- Most of our analysis tools impose fairly strict constraints on the taxonomic links between properties, but verifying that these constraints are not violated can be tedious and difficult in human hands.

The Q/A system is implemented in CLASSIC [4], a description logic that provides simple rules and can perform subsumption testing between descriptions. Each meta-property, for example, is a concept in CLASSIC and properties in an ontology are individuals which are either derived or asserted to be instances of the concepts representing the meta-properties. All the principles, meta-properties, and constraints described in this paper have been represented in this system.

A demo of the system and a more detailed description is available on the web [27].

5   Example

We now discuss a more in-depth example. Figure 1 shows a messy taxonomy, which has mostly been drawn from existing ontologies such as WordNet, Pangloss, and Mikrokosmos. An initial scan of the taxonomy looks reasonable. To save space, we concentrate on just a few taxonomic pairs to explore how the system works. A more detailed discussion of the same example can be found in [16].

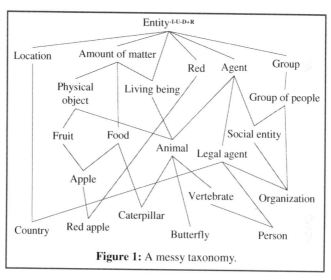

Figure 1: A messy taxonomy.

Assuming *ENTITY* has been already defined as shown, a dialog with the modeler might be:

```
What is the property name? amount-of-matter
What is the subsuming property? Entity
Is amount-of-matter rigid? (Y or N) Y
Does amount-of-matter supply identity? (Y,N,U) U
Does amount-of-matter carry identity? (Y,N,U) U
If an instance of amount-of-matter changes its parts, may it remain the same
instance? (Y,N,U) N
amount-of-matter carries identity.
amount-of-matter supplies identity.
Does amount-of-matter carry unity? (Y,N,U) U
Can instances of amount-of-matter be counted? (Y,N,U) N
amount-of-matter does not carry unity.
Is amount-of-matter dependent on any other properties? (Y,N) N
RESULT: amount-of-matter is +O+I+R-D-U
```

```
What is the property name? physical-object
What is the subsuming property? amount-of-matter
Is physical-object rigid? (Y or N) Y
Does physical-object supply identity? (Y,N,U) U
Does physical-object carry identity? (Y,N,U) U
If an instance of physical-object changes its parts, may it remain the same
instance? (Y,N,U) Y
VIOLATION: Non-mereologically extensional properties (physical-object) can not
be subsumed by mereologically extensional ones (amount-of-matter).
```

The modeler here has uncovered an inconsistency. A physical object, such as a car, can change some of its parts without becoming a different thing - we can change the tires of a car without making it a different car. An amount of matter, such as a lump of clay, is completely identified by its parts - if we remove some clay it is a different lump. A physical object is not, then, an amount of matter in this conceptualization, but it is *constituted* of matter. The relationship should therefore be one of constitution, not subsumption. The choice at this point is either to change the conceptualization of one of the two properties, or to change the taxonomic link.

The modeler chooses to put *PHYSICAL-OBJECT* below *ENTITY*, and continues, changing only the answer to the second question this time. We pick up the dialog from the last question:

```
If an instance of physical-object changes its parts, may it remain the same
instance? (Y,N,U) Y
Does physical-object have a characterizing feature that is unique to each
instance? (Y,N) U
Does physical-object carry unity? (Y,N,U) U
Can instances of physical-object be counted? (Y,N,U) Y
physical-object carries identity
physical-object supplies identity
physical-object carries unity
Is physical-object dependent on any other properties? (Y,N) N
RESULT: physical-object is +O+I+R-D+U
```

We now skip to the point where the modeler examines the class *ANIMAL*:

```
What is the property name? animal
What is the subsuming property? physical-object
Is animal rigid? (Y or N) Y
Does animal supply identity? (Y,N,U) U
Does animal carry identity? (Y,N,U) U
If an instance of animal changes its parts, may it remain the same instance?
(Y,N,U) Y
Does animal have a characterizing feature that is unique to each instance? (Y,N)
Y
What is the feature? brain
animal carries identity
animal supplies identity
Does animal carry unity? (Y,N,U) U
Can instances of animal be counted? (Y,N,U) Y
animal carries unity.
Is animal dependent on any other properties? (Y,N) N
```

```
Rigidity check: If an instance of animal ceases to be an instance of animal, does
it cease to be an instance of physical-object? (Y,N) N
VIOLATION: Rigidity check with physical-object failed.
```

The rigidity check question is only asked between rigid subsuming *sortals* (Def. 7). This question forced the modeler to think about the nature of the rigidity of this property. When an animal ceases to exist, i.e. when a person dies, their physical body remains. This indicates that, at least according to one conceptualization, *ANIMAL* is not subsumed by *PHYSICAL-OBJECT*, but as in the previous example, perhaps constituted of one.

To save space, we briefly describe the remaining changes instead of providing sample dialogs.

The analysis proceeds until all rigid properties in the taxonomy have been specified.   *GROUP-OF-PEOPLE* carries the meta-property **+ME**, however *ORGANIZATION* does not, since people in organizations change, therefore this taxonomic link is removed.   *SOCIAL-ENTITY* is also found not to have **+ME**, and therefore the taxonomic link to *GROUP-OF-PEOPLE* is removed from it as well. For similar reasons we remove the taxonomic link between *LIVING-BEING* and *AMOUNT-OF-MAT-TER*.

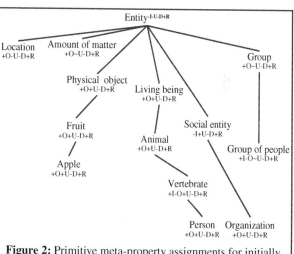

Figure 2: Primitive meta-property assignments for initially rigid properties.

The analysis of rigid properties continues through *LOCATION, GROUP, FRUIT, APPLE, VERTEBRATE,* and *PERSON*. The result of the meta-property analysis for the rigid properties is shown in Figure 2.

Once the rigid properties have been specified, we begin adding the non-rigid properties one at a time.

The property *AGENT* immediately causes problems because it is anti-rigid and subsumes two rigid properties (*ANIMAL* and *SOCIAL-ENTITY*). These taxonomic links are removed, which forces the modeler to consider why they were there. In this case, it is likely that a common misuse of subsumption as a sort of type restriction, such as "all agents must be animals or social entities," was meant. It should be clear that logically, subsumption is not the same as disjunction, and another representation mechanism should be used to maintain this restriction.

The link between *FOOD* and *APPLE* must also disappear for the same reason (apple is not necessarily food), and *LEGAL-AGENT* is anti-rigid and can not subsume *PERSON* or *ORGANIZATION*. In this case, the link was being used as a type restriction, it was not the case that all *PERSON*s are *LEGAL-AGENT*s.

Analyzing *COUNTRY* brings us to an interesting case. *COUNTRY* may be, upon first analysis, anti-rigid, because a country, e.g. Prussia, may no longer exist, yet still be a place someone can go. Our deeper analysis reveals however, that two senses were being collapsed into one property: the sense of a geographical region, which is rigid, and a political or social entity, a country, which is also rigid (Prussia the country no longer exists, Prussia the region does). Again, countries are constituted of regions.

Analysis of *CATERPILLAR* and *BUTTERFLY* yields interesting examples. Closer inspection of these two properties reveals a special type of property, known as a *phased sortal* [29]. A phased sortal is a property whose instances can change from one sortal to another and still remain the same thing, i.e. a caterpillar becomes a butterfly, or in some systems, we can imagine that a student becomes an alumnus. Our methodology requires that phased sortals be identified along with all the corresponding phases, and grouped under a rigid property that subsumes *only them* [15]. We add, therefore, *LEP-IDOPTERAN*.

Finally, the property *RED* is non-rigid because, although most things are not necessarily red, there may be some things that are. Non-rigid properties make very little commitment, and are often confusing parts of a taxonomy. In this case, it is used to distinguish *RED-APPLE* from presumably other color apples.

With the analysis of all properties complete, our technique identifies the backbone taxonomy and the final cleaned taxonomy, shown in Figure 3. Note that one result of this "cleaning" process is the removal of many occurrences of multiple inheritance. This is not necessarily a specific goal, however it naturally follows from the fact that, as discussed in [16], multiple inheritance is often used as a tool to represent more than simply subsumption – as we found in this

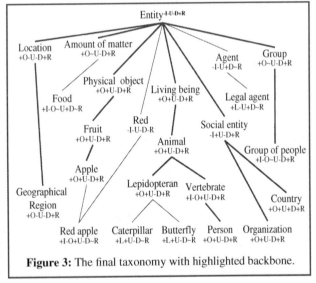

Figure 3: The final taxonomy with highlighted backbone.

example. We believe that these cases make taxonomies confusing; if the purpose of an ontology is to make the meaning clear, then the meaning should not be clouded by using the same mechanism to signify more than one thing, since there is no way to disambiguate the usage. Furthermore, there is at least some empirical evidence derived from studies of programmers who maintain object-oriented programs that multiple inheritance is confusing and makes taxonomies difficult to understand [18].

6   Conclusion

We have discussed several notions of Formal Ontology used for ontological analysis in Philosophy: identity, unity, essence, and dependence. We have formalized these notions in a way that makes them useful for conceptual modeling, and introduced a methodology for ontological analysis founded on these formalizations.

Our methodology is supported by a question/answer system that helps the conceptual modeler study the deep ontological issues surrounding the representation of properties in a conceptual model, and we have shown how this methodology can be used to analyze individual taxonomic links and make the taxonomy more understandable. In particular, we have also shown how to identify the backbone taxonomy, which represents the most important properties in an ontology that subsume every instance.

Unlike previous efforts to clarify taxonomies, our methodology differs in that:

- It focuses on the nature of the properties involved in subsumption relationships, not on the nature of the subsumption relation itself (which we take for granted).

- It is founded on formal notions drawn from Ontology (a discipline centuries older than database design), and augmented with practical conceptual design experience, as opposed to being founded solely on the former or latter.

- It focuses on the validation of single subsumption relationships based on the *intended meaning* of their arguments in terms of the meta-properties defined here, as opposed to focusing on structural similarities between property descriptions.

Finally, it is important to note again that in the examples we have given, we are providing a way to make the *meaning* of properties in a certain conceptualization clear. We do not, for example, mean to claim that "Person is-a Legal-Agent" is wrong. We are trying to point out that *in a particular conceptualization* where *LEGAL-AGENT* has certain meta-properties (such as being anti-rigid) and *PERSON* certain others (such as being rigid), it is inconsistent to have person subsumed by legal-agent.

References

1. Artale, A., Franconi, E., Guarino, N., and Pazzi, L. 1996. Part-Whole Relations in Object-Centered Systems: an Overview. *Data and Knowledge Engineering,* **20**(3): 347-383.

2. Bergamaschi, S. and Sartori, C. 1992. On Taxonomic Reasoning in Conceptual Design. *ACM Transactions on Database Systems,* **17**(3): 285-422.

3. Brachman, R. 1983. What IS-A Is and Isn't: An Analysis of Taxonomic Links in Semantic Networks. *IEEE Computer,* **16**(10): 30-36.

4. Brachman, R. J., McGuinness, D. L., Patel-Schneider, P. F., Resnick, L., and Borgida, A. 1990. Living with CLASSIC: When and How to Use a KL-ONE-like Language. In J. Sowa (ed.) *Principles of Semantic Networks.* Morgan Kaufmann: 401-456.

5. Calvanese, D., Lenzerini, M., and Nardi, D. 1998. Description Logics for Conceptual Data Modeling. In J. Chomicki and G. Saake (eds.), *Logics for Databases and Information Systems.* Kluwer: 229-264.

6. Elmasri, R., Weeldreyer, J., and Hevner, A. 1985. The category concept: An extension to the Entity-Relationship model. *Data and Knowledge Engineering,* **1**(1): 75-116.

7. Goldstein, R. C. and Storey, V. C. 1999. Data abstractions: Why and how? *Data and Knowledge Engineering,* **29**: 293-311.

8. Gruber, T. R. 1993. Model Formulation as a Problem-Solving Task: Computer-Assisted Engineering Modeling. *International Journal of Intelligent Systems,* **8**: 105-127.

9. Guarino, N. 1992. Concepts, Attributes and Arbitrary Relations: Some Linguistic and Ontological Criteria for Structuring Knowledge Bases. *Data & Knowledge Engineering,* **8**(2): 249-261.

10.   Guarino, N., Carrara, M., and Giaretta, P. 1994. An Ontology of Meta-Level Categories. In D. J., E. Sandewall and P. Torasso (eds.), *Principles of Knowledge Representation and Reasoning: Proceedings of the Fourth International Conference (KR94)*. Morgan Kaufmann, San Mateo, CA: 270-280.

11.   Guarino, N. 1995. Formal Ontology, Conceptual Analysis and Knowledge Representation. *International Journal of Human and Computer Studies*, **43**(5/6): 625-640.

12.   Guarino, N. 1998a. Formal Ontology in Information Systems. In N. Guarino (ed.) *Formal Ontology in Information Systems*. Proceedings of FOIS'98, Trento, Italy, 6-8 June 1998. IOS Press, Amsterdam: 3-15.

13.   Guarino, N. 1999. The Role of Identity Conditions in Ontology Design. In *Proceedings of IJCAI-99 workshop on Ontologies and Problem-Solving Methods: Lessons Learned and Future Trends*. Stockholm, Sweden, IJCAI, Inc.

14.   Guarino, N. and Welty, C. 2000a. Identity, Unity, and Individuality: Towards a Formal Toolkit for Ontological Analysis. In *Proceedings of ECAI-2000: The European Conference on Artificial Intelligence*. Berlin, Germany, IOS Press. Available from http://www.ladseb.pd.cnr.it/infor/ontology/Papers/OntologyPapers.html

15.   Guarino, N. and Welty, C. 2000b. A Formal Ontology of Properties. In Rose Dieng (ed.), *Proceedings of 12th Int. Conf. on Knowledge Engineering and Knowledge Management*, Springer Verlag. Available from http://www.ladseb.pd.cnr.it/infor/ontology/Papers/OntologyPapers.html

16.   Guarino, N. and Welty, C. 2000c. Towards a methodology for ontology-based model engineering. In *Proceedings of ECOOP-2000 Workshop on Model Engineering*. Cannes, France. Available from http://www.ladseb.pd.cnr.it/infor/ontology/Papers/OntologyPapers.html

17.   Hirst, G. 1991. Existence Assumptions in Knowledge Representation. *Artificial Intelligence*, **49**: 199-242.

18.   Huitt, R. and Wilde, N. 1992. Maintenance Support for Object-Oriented Programs. *IEEE Trans. on Software Engineering*, **18**(12).

19.   Humberstone, I. L. 1996. Intrinsic/Extrinsic. *Synthese*, **108**: 205-267.

20.   Lowe, E. J. 1989. *Kinds of Being. A Study of Individuation, Identity and the Logic of Sortal Terms*. Basil Blackwell, Oxford.

21.   Quine, W. V. O. 1969. *Ontological Relativity and Other Essays*. Columbia University Press, New York, London.

22.   Simons, P. 1987. *Parts: a Study in Ontology*. Clarendon Press, Oxford.

23.   Storey, V. C. 1993. Understanding Semantic Relationships. *Very Large Databases Journal*, **2**: 455-488.

24.   Storey, V., Dey, D., Ullrich, H., and Sundaresan, S. 1998. An ontology-based expert system for database design. *Data and Knowledge Engineering*, **28**: 31-46.

25.   Strawson, P. F. 1959. *Individuals. An Essay in Descriptive Metaphysics*. Routledge, London and New York.

26.   Teorey, T. J., Yang, D., and Fry, J. P. 1986. A Logical Design Methodology for Relational Databases Using the Extended Entity-Relationship Model. *ACM Computing Surveys*, **18**(2): 197-222.

27.   Welty, C. *A description-logic based system for ontology-driven conceptual analysis*. System demo available at http://untangle.cs.vassar.edu/odca/.

28.   Wieringa, R., De Jonge, W., and Spruit, P. 1994. Roles and dynamic subclasses: a modal logic approach. In *Proceedings of European Conference on Object-Oriented Programming*. Bologna.

29.   Wiggins, D. 1980. *Sameness and Substance*. Blackwell, Oxford.

A Conceptual Model for the Web

Mengchi Liu[1] and Tok Wang Ling[2]

[1] Department of Computer Science, University of Regina
Regina, Saskatchewan, Canada S4S 0A2
mliu@cs.uregina.ca
[2] School of Computing, National University of Singapore
Lower Kent Ridge Road, Singapore 119260
lingtw@comp.nus.edu.sg

Abstract. Most documents available over the web conform to the HTML specification. Such documents are hierarchically structured in nature. The existing graph-based or tree-based data models for the web only provide a very low level representation of such hierarchical structure. In this paper, we introduce a conceptual model for the web that is able to represent the complex hierarchical structure within the web documents at a high level that is close to human conceptualization/visualization of the documents. We also describe how to convert HTML documents based on this conceptual model. Using the conceptual model and conversion method, we can capture the essence (i.e., semistructure) of HTML documents in a natural and simple way.

1 Introduction

Most documents available over the web conform to the HTML specification. They are intended to be human readable through a browser and thus are constructed following some common conventions and often exhibit some hierarchical structure. How to represent such documents at a conceptual level is an important issue.

In the past few years, a number of data models have been developed in the database community to retrieve data from the web, such as UnQL [2], OEM [12], Strudel [4], etc. For a survey, see [6]. These proposals mainly use relational, graph-based or tree-based data models to represent the web data. They focus on inter-document structures, with little attention to intra-document structures and thus can only represent the web data at a very low level. For example, none of the existing data models and languages can represent Michael Ley's DBLP bibliography web document at http://www.informatik.uni-trier.de/~ley/db in a natural and simple way.

Because of the difficulties with HTML, XML [1] has been proposed to uniformly represent data over the Internet. It provides natural support for describing the hierarchical structure in web documents and allows specific markup to be created for specific data. It may become as important and as widely used as HTML in the future.

A.H.F. Laender, S.W. Liddle, V.C. Storey (Eds.): ER2000 Conference, LNCS 1920, pp. 225–238, 2000.
© Springer-Verlag Berlin Heidelberg 2000

The World Wide Web Consortium (W3C) has recently recommended the Document Object Model (DOM) [14] as an application programming interface for HTML and XML documents. DOM defines the logical structure of documents and the way a document is accessed and manipulated. It represents documents as a hierarchy of various node objects. With DOM, programmers can access, change, delete, add or build HTML or XML documents, and navigate their structure. Nevertheless, DOM is a very low level data model for HTML/XML documents as it is intended for programmers to write programs to access and manipulate HTML and XML documents, rather than for the user to find the information within the documents.

W3C has also recommended XPath [3] as a language for operating on the abstract, logical structure of an XML document, rather than its surface syntax. It models an XML document as a tree of various nodes at a level higher than DOM, such as element nodes, attribute nodes, text nodes, etc.

Before XML becomes widely used as HTML, a powerful data model to describe the abstract, logical structure of HTML documents and a flexible language to query large HTML documents is still in high demand.

In this paper, we propose a conceptual model for HTML that is able to represent any complex hierarchical structure within the web documents at a high level that is close to human conceptualization/visualization of the documents. We also describe how to convert HTML documents based on this conceptual model.

The rest of the paper is organized as follows. Section 2 introduces our conceptual model for the web. Section 3 presents a set of rules to automatically structure HTML documents. Section 4 shows how to use the rules to convert HTML documents. Section 5 summarizes and points out further research issues.

2   Conceptual Model for the Web

In this section, we introduce our conceptual model for the web. We note that this conceptual model is very simple with only a few constructs but powerful enough to represent any HTML documents. First, we assume the existence of two kinds of symbols: a set $\mathcal{U}$ of urls, and a set $\mathcal{S}$ of strings. Note that $\mathcal{U}$ is a subset of $\mathcal{S}$.

Each web document in HTML is considered structured in our conceptual model. They have a title and a body. Our purpose is to extract conceptual structures within web documents. To this end, we ignore the features that are used to enhance the visual aspects of the web documents, such as fonts, colors, style, etc.

The notion of *objects* in our conceptual model is defined recursively as follows.

(1) A string $s \in \mathcal{S}$ is a *lexical* object.
(2) Let o be an object and $u \in \mathcal{U}$. Then $o\langle u \rangle$ is a *linking* object, and o is called the *label* and u is called the *anchor* of the linking object.
(3) Let a, o be objects. Then $a \Rightarrow o$ is an *attributed* object, and a is called the *attribute* and o is called the *value* of the attributed object.

(4) If $o_1, ..., o_n$ are distinct objects with $n > 1$, then $o_1|...|o_n$ is a *or-value* object.
(5) If $o_1, ..., o_n$ are objects with $n > 1$, then $\{o_1, ..., o_n\}$ is a *list* object.

In an object-relational database, we can have homogeneous tuples and sets. Tuple elements can be accessed using attribute names while set elements are directly accessed. In a web document, we may have attributed objects together with other objects. Thus, it is sometimes impossible to distinguish tuples from sets. In addition, web documents conceptually supports lists instead of sets as duplication is allowed and the order of elements exists. Thus, we use list objects for tuples, sets and lists in our conceptual model. The attribute names can be used to access components of list objects that matches them.

An or-value object is used to indicate that the intended value is one of them. It was first used in [10] and then extended in [9] to record conflicting information when manipulating semistructured objects.

The following are examples of objects:

Lexical objects:	*Computer Science, Database Systems*
Linking objects:	$Faculty\langle faculty.html\rangle$, *Jim Carter*$\langle/faculty/carter/\rangle$
Attributed objects:	*Title* $\Rightarrow$ *CS Department, Location* $\Rightarrow$ *Europe*
Or-value objects:	*Female*|*Male*, 15|20|23
List object:	$\{Title \Rightarrow CS\ Dept, Research\langle research.html\rangle\}$

In our conceptual model, a web document is modeled as a list object. As we access a web document using its URL, we treat the URL together with its web document as a web object. We formalize it as follows.

Let u be a URL and l a list object. Then $u : l$ is a *web object*.

The following is an example of web object:

$csdept.html$: {
    $Title \Rightarrow CSDept$,
    $People\langle people.html\rangle \Rightarrow \{$
        $Faculty\langle faculty.html\rangle$,
        $Staff\langle staff.html\rangle$,
        $Students\langle students.html \rangle\}$,
    $Programs \Rightarrow \{$
        $Ph.D\ Program\langle phd.html\rangle$,
        $M.Sc\ Program\langle msc.html\rangle$,
        $B.Sc\ Program\langle bsc.html\rangle\}$,
    $Research\langle research.html\rangle$
}

Note that a linking object is different from a web object even though it is somewhat similar to a web object. A web object corresponds to a web document while a linking object is simply part of a web object.

Example 1. Part of Michael Ley's latest DBLP bibliography server web document at `http://www.informatik.uni-trier.de/~ley/db` can be represented in our conceptual model with simplified URLs such as a_1, b_1 etc. to fit in the paper as follows:

$http://www.informatik.uni-trier.de/\sim ley/db$: {

    Title $\Rightarrow DBLP$ *Bibliography,*

    Body $\Rightarrow$ {

        Search $\Rightarrow \{Author\langle a_1\rangle,\ Title\langle a_2\rangle,\ Advanced\langle a_3\rangle\}$, *Home Page Search*$\langle a_4\rangle\}$,

        Bibliographies $\Rightarrow$ {

            Conferences$\langle b_1\rangle \Rightarrow \{SIGMOD\langle b_{11}\rangle,\ VLDB\langle b_{12}\rangle,\ PODS\langle b_{13}\rangle,\ EDBT\langle b_{14}\ \rangle\}$,

            Journals$\langle b_2\rangle$ $\Rightarrow \{CACM\langle b_{21}\rangle,\ TODS\langle b_{22}\rangle,\ TOIS\langle b_{23}\rangle,\ TOPLAS\langle b_{24}\ \rangle\}$,

            Series$\langle b_3\rangle$ $\Rightarrow \{LNCS/LNAI\langle b_{41}\rangle,\ DISDBIS\langle b_{42}\rangle\}$,

            Books $\Rightarrow \{Collections\langle b_{51}\rangle,\ DB\ Textbooks\langle b_{52}\rangle\}$,

            By Subjects$\langle b_4\rangle \Rightarrow \{Database\ Systems\langle b_{61}\rangle,\ Logic\ Prog\langle b_{62}\rangle,\ IR\langle b_{63}\rangle\}\}$,

        Full Text $\Rightarrow ACM\ SIGMOD\ Antholog\langle c_1\rangle$,

        Reviews $\Rightarrow ACM\ SIGMOD\ Digital\ Review\langle c_2\rangle$,

        Links $\Rightarrow$ {

            Research Groups $\Rightarrow \{Database\ Systems\langle d1\rangle,\ Logic\ Programming\langle d2\rangle\}$,

            Computer Science Organization$\langle e_1\rangle \Rightarrow$ {

                $ACM\langle e_{11}\rangle\ (DL\langle e_{12}\ \rangle,\ SIGMOD\langle e_{13}\ \rangle,\ SIGIR\ \langle e_{14}\rangle)$,

                IEEE Comptuer Society$\langle e_{15}\rangle \Rightarrow DL\langle e_{16}\rangle\}$

            Related Services $\langle f_1\rangle \Rightarrow$ {

                $CoRR\langle f_{11}\rangle,\ ResearchIndex\langle f_{12}\rangle,\ NZ\text{-}DL\langle f_{13}\rangle$,

                $CS\ BibTex\langle f_{14}\rangle,\ HBP\langle f_{15}\rangle,\ Virtual\ Library\ \langle f_{16}\rangle\}\}$

    }

As the above examples show, our conceptual model provides an intuitive representation of the complex hierarchical structure within a web document at a high level that is close to human conceptualization/visualization of the documents. None of the existing data models can do so in such a simple way.

3   Converting HTML Documents

Web documents are intended to be human readable though a browser. They are constructed following some common conventions and often exhibit some hierarchical structure. By examining a large number of web documents, we have discovered a set of general rules that can be used to convert web documents into web objects in our conceptual model. We now introduce our conversion method. In particular, we focus on HTML 4.01 [13], the latest version of HTML.

    Web documents are not parsed by browsers and may have syntactic errors. To make our presentation simple, we assume that web documents are syntactically correct. Our conversion method is mainly based on the contents shown by a web browser and therefore we ignore the features that are used to enhance the visual aspects of the web documents, such as fonts, colors, style, etc. We assume that there is an HTML interpreter that can be used to parse the HTML files, remove optional tags and irrelevant components, and automatically identify the corresponding components for our definition below in the web document based on various tags, such as <p>, <hn>, <br>, etc.

    A web document in HTML starts with <html> tag and ends with </html> tag. It normally consists of two parts: *head* and *body*. The head mainly specifies the *title* of the document while the body specifies the *contents* of the web document. Web documents can be classified into two kinds based on the kind of the contents: *Frame-based web documents* whose body starts with `<frameset>` and ends with `</frameset>`, and *Regular web documents* whose body starts with `<body>` and ends with `</body>`.

The purpose of frame-based web documents is to allow the visitor to see more than one page at a time, without completely cluttering up their screen. Frame-based web documents have a number of frames in their body and each frame contains its own web document. A frame-based web document has the following form, where F_i is a frame or a nested frameset:

$$\texttt{<html><head><title>} T \texttt{</title></head>}$$
$$\texttt{<frameset ...>} F_1, ..., F_n \texttt{</frameset>}$$
$$\texttt{</html>}$$

Let u be the URL of a web document of the above form, C a converter that converts components of a web document, and $\uplus$ list concatenation operator. Then we can convert a frame-based web document into a web object as follows:

$$u : \{Title \Rightarrow T, Frameset \Rightarrow C(F_1) \uplus ... \uplus C(F_n)\}$$

For each frame $F =$ `<frame name = "`N`" src = "`U`">` where N is the name and U is the url of the frame, we convert it into $C(F) = \{N\langle U\rangle\}$. For a frameset $FS =$ `<frameset ...>`$F_1, ..., F_n$`</frameset>`, we convert it recursively into $C(FS) = C(F_1) \uplus ... \uplus C(F_n)$.

In other words, we convert a possibly nested frameset into a list object and each frame in a frameset into a linking object in the list object.

Regular web documents have a number of sections in their body with the following form, where $S_1, ..., S_n$ are sections:

$$\texttt{<html><head><title>} T \texttt{</title></head>}$$
$$\texttt{<body>} S_1 ... S_n \texttt{</body>}$$
$$\texttt{</html>}$$

Let u be the URL of a web document of the above form and C a converter that converts components of a web document. Then we can convert a regular web document into a web object as follows:

$$u: \{Title \Rightarrow T,$$
$$Body \Rightarrow \{C(S_1), ..., C(S_n)\}\}$$

3.1 Converting Sections

In HTML documents, there can be two kinds of sections: with a heading and without a heading. If it has a heading, then we can convert it into an attributed objects with the heading as its attribute and the rest as its value. Formally, Let $S =$ `<h`n`>`H`</h`n`>`T be a section with a heading H and contents T. Then

$$C(S) = C(H) \Rightarrow C(T)$$

If a section has no heading $S = T$, then $C(S) = C(T)$.

Example 2. Consider the HTML document with URL *http://www.cs.uregina.ca* as follows:

```
<html>
<head><title>Computer Science Department</title></head>
<body>
<h2>History</h2>
The department was founded in 1970.
<h2>Programs</h2>
The department offers B.Sc. M.Sc. and Ph.D degrees
<h2>Facilities</h2>
The department has up to date equipment and software
</body>
</html>
```

This HTML document has 3 sections with headings. We can convert it into a web object as follows:

http://www.cs.uregina.ca: {
   Title ⇒ *Computer Science Department,*
   Body ⇒{
      History ⇒ *The department was founded in 1970,*
      Programs ⇒ *The department offers B.Sc. M.Sc. and Ph.D degrees,*
      Facilities ⇒ *The department has up to date equipment and software*}
   }

Within a section, there may be a sequence of paragraphs, lists, tables, etc. In what follows, we discuss how to translate each of them.

3.2   Converting Paragraphs

A section in HTML may have a sequence of paragraphs. Let $S = P_1, ..., P_n$ be a section where $P_1, ..., P_n$ are paragraphs. Then we convert it as follows:

$$C(S) = \{C(P_1), ..., C(P_n)\} \quad \text{when } n > 1$$
$$C(S) = C(P_1) \qquad\qquad\qquad \text{when } n = 1$$

A paragraph in a section starts with a tag `<p>`. Some paragraphs may have a emphasized beginning that is usually bold, italic, etc. or is followed by a colon ':'. Let $P = $ `<t>`B`</t>` $: R$ or $P = $ `<t>`$B : $`</t>`$R$ be such a paragraph, where t is either `b`, `i`, `em`, or `strong`. Then we convert it as follows:

$$C(P) = C(B) \Rightarrow C(R)$$

If a paragraph has no emphasized beginning $P = R$, then $C(P) = C(R)$.

Let R be the part of a paragraph other than the emphasized heading. If R has logical parts $R_1, ..., R_n$ with $n \geq 1$, then

$$C(R) = \{C(R_1), ..., C(R_n)\} \quad \text{if } n > 1$$
$$C(R) = C(R_1) \qquad\qquad\qquad \text{if } n = 1$$

Each logical part is converted as a paragraph recursively.

Example 3. The following is a section of CIA World Factbook page about Canada at `http://www.odci.gov/cia/publications/factbook/`. It consists of a sequence of paragraphs.

```
<p>
<b>Location:</b> Northern North America
<p>
<b>Geographic coordinates:</b> 60 00 N, 95 00 W
<p>
<b>Map references:</b> North America
<p>
<b>Area:</b>
<br><i>total:</i> 9,976,140 sq km
<br><i>land:</i> 9,220,970 sq km
```

We can convert it into the following list object:

$\{ Location \Rightarrow Northern\ North\ America,$
$\quad Geographic\ coordinates \Rightarrow 60\ 00\ N,\ 95\ 00\ W$
$\quad Map\ references \Rightarrow North\ America,$
$\quad Area \Rightarrow \{ total \Rightarrow 9,976,140\ sq\ km,\ land \Rightarrow 9,220,970\ sq\ km \}\}$

3.3 Converting Multimedia Features and Hypertexts

In HTML documents, we can include multimedia features such as images, applets, video clips, sound clips, etc. In our conceptual model, we convert them into attributed objects using keywords *Image, Applet, Video, Sound*, etc. as their attributes.

For an image link $I = $ `<img src = "U" alt="T">`, where U is a URL and T is a string, we convert it as follows:

$$C(I) = Image \Rightarrow T\langle U \rangle.$$

The cases for other multimedia features are handled in the similar way. Due to space limitation, we omit them here.

For a hypertext link $H = $ `<a href = "U">T</a>`, where U is a URL and T is a string, we convert it into a linking object:

$$C(H) = C(T)\langle U \rangle.$$

For each character string S, $C(S) = S$.

Consider the following paragraph without emphasized heading:

```
<a href="dick.html">Dick</a> likes <a href="jane.html">Jane</a>
```

It has three logical units and thus can be converted into the following object:

$\{ Dick\langle dick.html \rangle,\ likes,\ Jane\langle jane.html \rangle \}$

3.4 Converting Lists

HTML supports three kinds of lists: ordered lists, unordered lists and definition lists. One can also nest one kind of list inside another.

For an unordered list $L = $ `<ul><li>`L_1`...<li>`L_n`</ul>` or an ordered list $L = $ `<ol><li>`L_1`...<li>`L_n`</ol>`, we convert it into a list object as follows:

$$C(L) = \{C(L_1), ..., C(L_n)\} \quad \text{when } n > 1$$
$$C(L) = C(L_1) \qquad\qquad\qquad \text{when } n = 1$$

For a definition list $L = $ `<dl><dt>`N_1`<dd>`$L_1 ...$`<dt>`N_m`<dd>`L_m`</dl>`, we convert it into a list object as follows:

$$C(L) = \{C(N_1) \Rightarrow C(L_1), ..., C(N_m) \Rightarrow C(L_m)\} \quad \text{when } n > 1$$
$$C(L) = C(N_1) \Rightarrow C(L_1) \qquad\qquad\qquad\qquad \text{when } n = 1$$

For a nested ordered or unordered list item $I = T$`<`l`>``<li>`$T_1 ...$`<li>`$T_m$`</`$l$`>`, where $l$ is either `ol` or `ul`, we convert it into a nested list objects;

$$C(I) = C(T) \Rightarrow \{C(T_1), ..., C(T_m)\}$$

Example 4. Consider the following HTML document containing nested lists with URL `research.html`:

```
<html>
<head><title>CS Department Research</title></head>
<body>
<h2>Research Areas</a></h2>
<ol>
<li>Artificial Intelligence
<ul>
<li>Cognitive Science <li>Linguistics <li>Reasoning
</ul>
<li>Database Systems
<ul>
<li>Query Processing  <li>Data Models <li>Active Databases
</ul></ol>
<h2>Research Groups and Labs</a></h2>
<ol>
<li>Programming Languages
<li>Intelligent Systems
<li>Natural Language Lab
</ol>
</body>
</html>
```

We can convert it into a web object in our conceptual model as follows:

research.html : {
    Title $\Rightarrow$ *CS Department Research*,
    Body $\Rightarrow$ {
        Research Areas $\Rightarrow$ {
            Artificial Intelligence $\Rightarrow$ {*Cognitive Science, Linguistics, Reasoning*},
            Database Systems $\Rightarrow$ {*Query Processing, Data Models, Active Databa-*
ses}},
        Research Groups and Labs $\Rightarrow$ {
        Programming Languages,
        Intelligent Systems,
        Natural Language Lab}}
    }

Example 5. Consider the following HTML document containing definition lists with URL `index.html`.

```
<html>
<head><title>Graduate Studies in CS</title></head>
<body>
<h2></h2>
<dl>
<dt>General Information
<dd>The department was established in 1970
<dt>Programs of Study
<dd>It offers M.Sc, and Ph.D in Computer Science
<dt>Financial Support
<dd>A variety of scholarships are available
<dt>Facilities
<dd>The research labs have all kinds of state-of-the-art equipment
</dl>
</body>
</html>
```

We convert it as follows:

index.html : {
    Title ⇒ *Graduate Studies in CS,*
    Body ⇒{
        General Information ⇒ *The department was established in 1970,*
        Programs of Study ⇒ *It offers M.Sc, and Ph.D in Computer Science,*
        Financial Support ⇒ *A variety of scholarships are available,*
        Facilities ⇒ *The research labs have all kinds of state-of-the-art equipment*}
    }

3.5 Converting Tables

Tables in HTML are used to arrange data into rows and columns of cells. Tables may have caption and column/row headings. Tables can be used in two different ways: to present a list with a better layout; to present real tabular information. We convert tables as follows.

For a table with a caption T = `<table><caption>`H`</caption>`TC `</table>` where H is the caption and TC is the table contents

$$C(T) = C(H) \Rightarrow C(TC)$$

For a table without a caption T = `<table>`TC`</table>`, $C(T) = C(TC)$

Let $R_1, ..., R_n$ be rows other than the row for column headings in the table contents TC. Then

$$C(TC) = \{C(R_1), ..., C(R_n)\}$$

Each row R_i is converted as follows:

1. If the table has column headings $H_1, ..., H_n$ and each row R has a row heading
   H, R = `<tr><th>`H`<td>`$C_1 ...$`<td>`C_n, then
   $C(R) = C(H) \Rightarrow \{C(H_1) \Rightarrow C(C_1), ..., C(H_n) \Rightarrow C(C_n)\}$
2. If the table has column headings $H_1, ..., H_n$, but each row R has no row
   heading R = `<tr><td>`$C_1 ...$`<td>`C_n, then
   $C(R) = \{C(H_1) \Rightarrow C(C_1), ..., C(H_n) \Rightarrow C(C_n)\}$
3. If the table does not have column headings but each row R has a row heading
   R = `<tr><th>`H`<td>`$C_1 ...$`<td>`C_n,
   $C(R) = C(H) \Rightarrow \{C(C_1), ..., C(C_n)\}$
4. If the table has no column and row headings, then for each row R =
   `<tr>`$C_1 ...$`<td>`C_n,
   $C(R) = \{C(C_1), ..., C(C_n)\}$

Example 6. Consider the following table in an HTML document:

```
<table>
<caption align = top>Bear Sightings</caption>
<tr>
<td><br><th>Babies<th>Adults<th>Total
<tr>
    <th>Northampton<td>2<td>4<td>6
<tr>
    <th>Becket<td>5<td>22<td>27
<tr>
    <th>Worthington<td>7<td>5<td>12
</table>
```

We can convert it into the following object based on Rule (13) Case 1:

Bear Sightings $\Rightarrow \{$
    Northampton $\Rightarrow \{$*Babies* $\Rightarrow 2$, *Adults* $\Rightarrow 4$, *Total* $\Rightarrow 6\}$,
    Becket $\Rightarrow$ $\{$*Babies* $\Rightarrow 5$, *Adults* $\Rightarrow 22$, *Total* $\Rightarrow 27\}$,
    Worthington $\Rightarrow \{$*Babies* $\Rightarrow 7$, *Adults* $\Rightarrow 5$, *Total* $\Rightarrow 12\}\}$

In HTML documents, tables without column headings and row headings are
often used to arrange items for visual effects. Consider the following portion of
an HTML document:

```
<h2 align="center"> Faculty Profiles</h2>
<table border="0" cellpadding=3 cellspacing=3 align=center>
<tr>
<td colspan=50% align=left>
<a href="/faculty/bunt/">Rick Bunt</A>
<td colspan=50% align=left>
<a href="/faculty/carter/">Jim Carter</A>
<tr>
<td colspan=50% align=left>
<a href="/faculty/cheston/">Grant Cheston</A>
...
</table>
```

We can convert it into the following object:

Faculty Profiles ⇒ {
   Rick Bunt⟨*/faculty/bunt/*⟩,
   Jim Carter⟨*/faculty/carter/*⟩,
   Grant Cheston⟨*/faculty/cheston/*⟩,
   ... }

Using our conversion method, we can convert part of Michael Ley's latest DBLP web document at `http://www.informatik.uni-trier.de/~ley/db` into our conceptual model as shown in Example 1.

3.6 Converting Forms

HTML forms enable visitors to communicate with the web server. There are two basic parts of a form: the structure that consists of fields, labels and buttons that the visitor sees and fills out on a page, and the processing part that process the information the visitor fills out using CGI script typically written in Perl or some other programming language.

Conceptually, we are interested in the structure part; that is, what kind of information is presented and can be communicated with the web server. Each element on a form has a name and a value/type associated with it. The name identifies the data that is being sent, the value is the data that is built-in the HTML document, the type specifies that value that comes from the visitor.

HTML supports two ways to send information to the web server: *GET* and *POST*. The GET method appends the name-value pairs to the end of the URL and is for small amount of data. The POST method sends a data file with the name-value pairs to the server's standard input without size limit.

Using either GET or POST is mainly a physical level concern and does not have much conceptual value.

A form normally has two special buttons: *submit* and *reset*. The *submit* button is used to send the information and the *reset* button is used to reset the form.

Let F be a form as follows:

```
F = <form ...>
    S
    <input type = "submit" ...>
    <input type = "reset"...>
    </form>
```

We convert it as follows:

$$C(F) = FORM \Rightarrow C(S)$$

Note that we don't keep information about the *submit* and *reset* buttons in the result as we make *FORM* a reserved attribute in our conceptual model to indicate that its value are used for communication with the web server and those two buttons are implied.

For a text field of the form $TF = L$:`<input type = "`T`" name = "`N`">`, we convert it into

$$C(TF) = L \Rightarrow T$$

Note that T is a build-in type and is used to indicate that the form will communicate value of type T to the web server for the label L. Also note that the name N does not occur in our conceptual model as it is the internal representation of L and the visitor cannot see it. In practice. N and L are the same quite often.

Text areas in HTML allows the user to enter more text than can fit in the a text field. Let $TA =$ `<textarea name = "`N`" ... >T</textarea>` be a text area, we convert it into

$$C(TA) = N \Rightarrow T$$

Radio buttons on forms are used to allow the user to make only one choice. In our conceptual model, we use an or-value to indicate the choices. For a group of radio buttons of the forms:

$R =$ `<input type = "radio" name = "`N_1`" value = "`V_1`">`L_1

...

`<input type = "radio" name = "`N_m`" value = "`V_m`">`L_m

we convert it into the following or-value object:

$$C(R) = L_1 \mid ... \mid L_m$$

While radio buttons can accept only one answer per set, Check boxes allow the visitor to check as many check boxes as they like. For a group of check boxes of the form:

$B =$ `<input type = "checkbox" name = "`N_1 value = "V_1`">`L_1

...

`<input type = "checkbox" name = "`N_m value = "V_m`">`L_m

we convert it into the following list of or-value object to indicate that the value can be a list:

$$C(R) = \{L_1 \mid ... \mid L_m\}$$

Menus allow visitors of HTML document to enter information easily. For a menu of the form:

$N =$ `<select name = "`N`" ...>`
    `<option value = "`V_1`">`L_1

    ...

    `<option value = "`V_n`">`L_n
    `</select>`

We convert it into the following attributed or-value object:

$$C(N) = N \Rightarrow L_1 \mid ... \mid L_n$$

Example 7. Consider the following portion of an HTML document:

```
<form method=post action="http://site.com/cgi-bin/get_menu"><br>
<p>Name:<input type="text" name="Name">
<p><strong>Age</string><br>
<select name = "Age" size ="5">
<option value = "18-65">18-65
<option value = "66-10o">66-100
</select>
<p>Gender:
<input type = "radio" name = "Gender" value = "female">Female
<input type = "radio" name = "Gender" value = "male">Male <br>
<input type = "submit" value = "Send info">
<input type = "reset" value = "Start over">
</form>
```

We convert it into the following object:

$Form \Rightarrow \{$
    $Name \Rightarrow text,$
    $Age \Rightarrow 18\text{-}65|66\text{-}100,$
    $Gender \Rightarrow Female|Male\}$

4   Conclusion

The main contribution of the paper is the following. First, we have proposed a conceptual model for capturing, in a natural and simple way, the internal structure of HTML documents. Our conceptual model is close to human conceptualization/visualization of the documents. Unlike the data models that deal with the hierarchical structure of web documents at a very low level using trees and graphs such as WebOQL, OEM, UnQL, Strudel, our conceptual model only have a few simple but powerful high level constructs that can be used to best describe the contents in the web documents. Also, using trees and graphs cannot support ordering which is essential for various lists in HTML. Our conceptual model uses list objects which naturally correspond to various lists in HTML.

Second, we have presented a set of generic rules to automatically convert HTML documents into this conceptual model. Discovering rules to structure the HTML text has already been addressed in the literature such as [5,7,8,11]. However, the rules presented in this paper are more general and systematic. Because we use list objects to represent tuples, lists, tables, frames, forms, menus, etc. and ignore many features that are used to enhance the visual aspects of the web documents, it may be impossible to convert some objects in our data model back into HTML format. In other words, we lose some information during conversion. However, we can capture the essence of HTML documents in a natural and intuitive way.

A wrapper to convert HTML documents into our conceptual model based on the conversion method presented here has been implemented in Java and can be downloaded from the web page at http://www.cs.regina.ca/~mliu/webmodel.

We would like to extend our work by developing powerful query languages based on our conceptual model presented here to query the structure and contents of HTML documents. Also, we would like to extend the conceptual model to properly represent the features omitted here so that we can convert between our conceptual model and HTML as suggested by one of the referees.

References

1. T. Bray, J. Paoli, and C.M. Sperberg-McQueen. Extensible Markup Language (XML) 1.0. *W3C Recommendation.* See http://www.w3c.org/TR/1999/REC-xml-19980210, February 1998.
2. P. Buneman, S. Davidson, G. Hilebrand, and D. Suciu. A Query Language and Optimization Techniques for Unstructured Data. In *Proceedings of the ACM SIG-MOD International Conference on Management of Data*, pages 505–516, 1996.
3. J. Clark and S. DeRose. XML Path Language (XPath) Version 1.0. *W3C Recommendation.* See http://www.w3c.org/TR/1999/REC-xpath-19991116, November 1999.
4. M. Fernandez, D. Florescu, A. Levy, and D. Suciu. A Query Language for a Web-Site Management System. *SIGMOD Record*, pages 4–11, 1997.
5. M. Fernandez, D. Florescu, A. Levy, and D. Suciu. Reasoning About Web-Site Structure. In *Proceedings of AAAI'98 Workshop on AI and Information Integration*, 1998.
6. D. Florescu, A. Levy, and A. Mendelzon. Database Techniques for the World-Wide Web: A Survey. *SIGMOD Record*, 27(3):59–74, 1998.
7. J. Hammer, H. Garcia-Molina, J. Cho, A. Crespo, and R. Aranha. Extracting Semistructured Information from the Web. In *Proceedings of the Workshop on Management of Semistructured Data*, 1997.
8. C. A. Knoblock, S. Minton, J. L. Ambite, N. Ashish, P. J. Modi, I. Muslea, A. G. Philpot, and S. Tejada. Modeling Web Sources for Information Integration. In *Proceedings of the 15th National Conference on AI*, 1998.
9. M. Liu and T. W. Ling. A Data Model for Semistructured Data with Partial and Inconsistent Information. In *Proceedings of the International Conference on Advances in Database Technology (EDBT 2000)*, pages 317–331, Konstanz, Germany, March 27-31 2000. Springer-Verlag LNCS 1777.
10. M. Liu, T. W. Ling, and T. Guan. Integration of Semistructured Data with Partial and Inconsistent Information. In *Proceedings of the International Database Engineering and Application Symposium (IDEAS '99)*, pages 44–52, Montreal, Canada, August 2-4 1999. IEEE-CS Press.
11. I. Muslea, S. Minton, and C. A. Knoblock. Hierarchical Wrapper Induction for Semistructured Information Sources. *To appear in Journal of Autonomous Agents and Multi-Agent Systems.*
12. Y. Papakonstantinou, H. Garcia-Molina, and J. Widom. Object Exchange across Heterogeneous Information. In *Proceedings of the International Conference on Data Engineering*, pages 251–260. IEEE Computer Society, 1995.
13. D. Raggett, A. L. Hors, and I. Jacobs. HTML 4.01 Specification. *W3C Recommendation.* See http://www.w3c.org/TR/html401, December 1999.
14. L. Wood, A. L. Hors, et al. Document Object Model (DOM) Level 2 Specification. *W3C Recommendation.* See http://www.w3c.org/TR/2000/CR-DOM-Level-2-20000307, March 2000.

Adapting Materialized Views after Redefinition in Distributed Environments

Zohra Bellahsene

LIRMM
UMR 9928 CNRS - Montpellier II
161 rue ADA 34392 Montpellier Cedex 5, France
E-mail: bella@lirmm.fr

Abstract. In this paper, we show how view adaptation can be supported in order to keep views up to date and running after view redefinition. In particular in the area of integrated systems such as Web sites, data warehousing systems and mediator systems, efficient solutions to this task are of most importance. The proposed view adaptation is based on a novel view selection approach that consists in decomposing views into fragments and then merging them into a Multi View Materialization Graph, sharing materialized fragments where possible. The goal is to minimize both communication cost (i.e., accessing the sources) and the cost of adapting the view materialization. In related work, view adaptation is based solely on the old materialization of the same view. Our approach performs view adaptation regarding all the materialized views therefore it optimizes the communication cost.

1 Introduction

The view mechanism has been recognized as a good technology for providing integrated *views* over distributed and heterogeneous information. The resulting views are often materialized in order to quickly answer user queries independently of the availability of the data sources [9, 10]. Taking changes at the data sources and view redefinition into account, is an important feature in these integrated systems, such as data warehousing or mediator systems. In the literature, most of the related work deals with the problem of view maintenance, which consists in maintaining the materialized view in response to data modifications of the source relations [6, 15, 18]. The problem of view adaptation is recomputing a materialized view in response to changes in the view definition [7, 11, 12]. The main issue is to avoid the recomputing of views from scratch, especially when the views are defined over source relations from multiple sources. The key idea of adaptation techniques is the use of the old view's materialization and the source relations.

In this paper, we investigate the view adaptation issue, which consists in adapting a set of materialized views after view redefinition. This adaptation concerns both the view definition and the view extent. Furthermore, in our approach, the adaptation of a view is performed regarding all materialized views. On the contrary, in related work, adaptation technique is based solely on the old materialization of the same view. The

A.H.F. Laender, S.W. Liddle, V.C. Storey (Eds.): ER2000 Conference, LNCS 1920, pp. 239-252, 2000.
© Springer-Verlag Berlin Heidelberg 2000

goal of our adaptation strategy is to minimize both communication cost (i.e., accessing the sources) and the cost of adapting the materialized view.

View adaptation technique is needed in all integrated systems that are based on materialized views: Web sites, data warehousing systems, mediator systems. A new view can be added or the view definition can be changed. Consequently, the system must bring the materialized views up-to-date. Such dynamic views are needed in interactive applications such as dynamic queries, data visualization, biological databases, etc. Some approaches for adapting materialized views in response to view redefinition are given in [7, 11, 12]. A more detailed discussion about related work is given in Section 4. As it has been noted in [7], view adaptation differs from the problem of rewriting queries since in view adaptation setting the new view is not always equivalent to the old view. Besides, in structural view maintenance the view changes are assumed to be local, i.e., the new view schema is close to the old one.

The present paper focuses on adaptation algorithms by exploiting a central construct called Multi View Materialization Graph (MVMG), which provides a framework for the representation of a set of materialized views. Our view selection strategy consists in decomposing a view into fragments. Some fragments will be materialized and others will remain virtual. Materialized fragments will be reused for other views as much as possible in order to reduce storage space and maintenance cost. This means, shared fragments are materialized once.

We present a comprehensive study of different types of schema changes that can be performed on the view definition and the appropriate algorithms to adapt the view materialization in response to these changes.

This paper is organized as follows. Section 2 is devoted to the presentation of the view adaptation problem and our view selection approach. Section 3 describes the view adaptation algorithms. Section 4 gives an overview of related work. We conclude in Section 5.

2 Preliminaries

We consider SPJ (Selection-Projection-Join) views that may involve aggregate functions and *group by* clause as well. They can also include set operators like union or intersection. However, for reuse purpose, aggregate functions are not materialized except if an aggregate function is a sub-expression shared by several views.

A view is a derived relation defined by a query in terms of source relations. It is said to be materialized when its extent is computed and persistently stored. Otherwise, it is said to be virtual.

2.1 Principle of Our View Selection

2.1.1 Decomposing a View into Fragments. In our approach, a view is not fully materialized but decomposed into a set of materialized fragments. Intuitively, a fragment is a piece of a relation, which is computed by a Selection-Projection (i.e., SP_fragment), a Join (i.e., J_fragment) or an Aggregation (i.e., A_fragment).
Our adaptation strategy attempts to reuse, as much as possible, the already materialized fragments in the MVMG. To determine which views should be materialized, it is necessary to know if some views are contained in others. However, it is known that the problem of query containment is in general undecidable[1]. This is why we propose smaller containment granularity (i.e., a fragment) in order to detect overlapping between views.

Definition. (Virtual/Materialized fragment): A fragment is said to be materialized if the corresponding operation is computed and its result is stored. Otherwise, the fragment is virtual: its data are computed at the invocation of the view.

2.1.2 Principle of Our View Selection. The selection of views to materialize is a very important issue in data warehousing and mediator systems. We present the view selection strategy on which our view adaptation algorithms are based on. However, this topic is not the main subject of this paper. The problem is selecting a set of views to materialize that optimizes both the view maintenance cost and query processing cost. Indeed, to improve query response time, the number of views to be materialized should increase. Consequently, the maintenance cost of the materialized views will increase and vice-versa. Since these two costs are in conflict, the issue is finding a selection strategy that ensures a balance between maintenance and query processing costs. The most of the approaches proposed in related work lead to huge multiquery graphs [17, 2, 8]. Moreover, when a view or a query is added, the view selection should be reconsidered. This process entails some important overhead costs, such as the bookkeeping for determining which views are materialized, and the costs (both direct costs and the indirect costs of interrupted service) for reorganizing the data warehouse.

The first step of our view selection method consists in building incrementally the multi View Materialization Graph (MVMG) by using merging rules based on attributes and conditions involved in the view query. These same rules will be used in our adaptation algorithms that will be described in the present paper. In the second step, we apply materialization/dematerialization rules to attributes according to their update and query frequencies. Lastly, we make use of optimization rules to reduce maintenance and storage costs. Due to space limitations, the second and third steps will not be described in this paper. For more details, please refer to [4].

Our adaptation algorithms are based on the use of a central construct called Multi View Materialization Graph (MVMG), which provides a framework for the representation of a set of materialized views.

---

[1] However, for conjunctive queries, the problem is decidable.

Definition. (Multi View Materialization Graph). Let G(N, E) be a directed graph, where
- N is the set of nodes. Each node is a fragment associated to an operation of view query.
- E is a set of edges such that an edge (a, b) ∈ E if and only if a is an operand of the operation associated to the fragment b

We call G the Multi View Materialization Graph of a given views set.

In fact, the Multi View Materialization Graph (MVMG) is a DAG with two kinds of nodes: terminal fragments and intermediary fragments. Each leaf node is a materialized SP_fragment, called terminal fragment. An intermediary fragment is a fragment that is situated on a non-leaf node of the MVMG. It can be a J_fragment, an A_fragment or a SP_fragment that may be materialized or virtual.

2.1.3 The Adaptation Rules. We propose a set of rules to find the right position of a fragment in the MVMG. These rules are used both for building the MVMG and for performing view adaptation. To decide which parts of view to be computed from the sources, our method compares the fragments of the view with the already materialized fragments in order to reuse them as much as possible. For the comparison of attributes, we make use of the inclusion operator. Comparing conditions, a constraint relation is used:

$C_1 > C_2$ if and only if C_1 is less restrictive than C_2. Due space limitation, we cannot provide all the rules. We choose to present the Materializing/dematerializing rule that is used in most of the adaptation algorithms.

Let $\xi[V](R_i)$ be the set of attributes of relation Ri that is involved in view V.
Let $C[V](R_i)$ be the selection conditions on R_i in view V.
Let $\xi[Vp](R_i)$ be the set of attributes of relation R_i in view Vp.
Let $TF[Vp](R_i)$ be a terminal fragment in the MVMG belonging to view Vp.
Let $VF[V](R_i)$ be the virtual SP_fragment of view V over relation R_i that will be integrated in the MVMG.

Materializing/dematerializing rule. This situation happens when the set of attributes in a terminal fragment is included in the set of attributes in fragment $VF[V](R_i)$ and when the conditions of $TF[Vp](R_i)$, the terminal fragment on R_i, is more restrictive than the conditions in the view on the same relation R_i. We have to consider two situations:

(a) $\xi[Vp](R_i) \subset \xi[V](R_i)$ and $C[V](R_i) = C[Vp](R_i) = \varnothing$
(b) $\xi[Vp](R_i) \subseteq \xi[V](R_i)$ and $C[V](R_i) > C[Vp](R_i)$

If (a) or (b) occurs then the procedure consists of materializing the fragment of view V and in dematerializing the terminal fragment $TF[Vp](R_i)$. This means $TF[Vp](R_i)$ becomes a virtual fragment. Therefore, the tuples in the old terminal fragment will be transferred to the new terminal fragment. Besides, tuples that didn't satisfy the conditions of the old terminal fragment but do satisfy those of the new terminal fragment should be inserted into it. This operation needs to access the data sources.

2.2 Example of Multi View Materialization Graph

Figure 1 depicts a MVMG with two views defined over the TPC-D schema [16]. Let us consider a first view providing a list of the customers who have ever placed large quantity orders, say a quantity greater than 10,000, after 1999 25th December. The view query lists the customer name, customer key, and the order key, orderdate and total price.

```
Create View V₁ As
Select C.c_name, C.c_custkey, O.o_orderkey, O.o_orderdate,
       O.o_totalprice,
From customer C, order O, lineitem L
Where O.o_orderkey in(Select l_orderkey
               From lineitem
               Group by l_orderkey
               Having sum(l_quantity)>10,000)
    and C.c_custkey = O.o_custkey and O.o_orderkey = L.l_orderkey
    and O.o_orderdate > '1999/25/12';
```

The first step of our approach consists of decomposing each view into a set of fragments. For the view V_1, three SP_fragments are generated, one per involved relation. Each SP_fragment has four parameters: the view name, the relation name, the attributes of projection and the selection condition.

SP_fragment(V_1, customer, ξ, C), where ξ={c_custkey, c_name, c_nationkey} and C = $\varnothing$

SP_fragment (V_1, order, ξ, C), where ξ={o_custkey, o_orderkey, o_orderdate, o_totalprice} and C = (o_orderdate > '1999/25/12')

SP_fragment (V_1, lineitem, ξ, C) where ξ= {l_orderkey, l_quantity} and C= $\varnothing$

A_fragment (V_1, lineitem, ξ, C) where ξ= {l_orderkey} and C =(sum(l_quantity)>10,000)

Then, one J_fragment per join will be generated. Each J_fragment has four parameters: the two relations, the attributes needed later on for performing other operators, and the join condition. For instance, the join J_{11} includes the attributes needed by the join J_{12}.

J_fragment[V_1](customer, order, ξ, C) is noted J_{11}, where ξ={c_custkey, c_name, o_orderkey, o_orderdate, o_totalprice }and C= (c_custkey = o_custkey)

J_fragment[V_1](lineitem, order, ξ, C) is noted J_{12}, where ξ={c_custkey, c_name, o_orderkey, o_orderdate, o_totalprice} and C =(l_orderkey =o_orderkey)

Let us consider the second view named V_2 providing a special treatment for computer items sold in France. The view-defining query is as follows:

```
Create View V₂ As
Select    P.p_partkey,    P.p_name,    P.p_retailprice,    P.p_brand,
O.o_orderdate
From Customers C, Lineitem L, Nation N, Order O, Part P
Where P. p_type = 'computer' and
           N.n_name = 'France' and
           N.n_nationkey = C.c_nationkey and
           P.p_partkey = L.l_partkey and
           L.l_orderkey = O.o_orderkey and
           O.o_custkey = C.c_custkey;
```

For the view V_2, Five SP_fragments are generated, one per involved relation.

SP_fragment(V_2, part, ξ, C), where $\xi = \{$p_partkey, p_name, p_retailprice, p_brand$\}$ and

C= (p_type='computer')

SP_fragment (V_2, order, ξ, C), where $\xi = \{$o_custkey, o_orderkey, o_orderdate$\}$ and C=$\varnothing$

SP_fragment (V_2, lineitem, ξ, C), where $\xi = \{$l_orderkey, l_partkey$\}$and C=$\varnothing$

SP_fragment(V_2, customer, ξ, C), where $\xi = \{$c_custkey, c_nationkey$\}$ and C=$\varnothing$

SP_fragment(V_2, nation, ξ, C), where $\xi = \{$n_nationkey$\}$and

C= (n_name='France').

Then, one J_fragment per join involved in the view query will be generated.

J_fragment[V_2](lineitem, order, ξ, C) is noted J_{11}, where $\xi=\{$l_partkey, o_custkey, o_orderdate $\}$ and C=(l_orderkey = o_orderkey)

J_fragment[V_2](customer, nation, ξ, C) is noted J_{22}, where $\xi = \{$c_custkey$\}$ and C= (c_nationkey = n_nationkey)

J_fragment[V_2](order, custumer, ξ, C) is noted J_{23}, where $\xi=\{$o_orderdate$\}$ and C= (o_custkey = c_custkey)

J_fragment[V_2](lineitem, part, ξ, C) is noted J_{24} and $\{$p_partkey, p_name, p_retailprice, p_brand, o_orderdate$\}$ and C= (l_partkey = p_partkey)

The MVMG will be initialized with one of the two views. The order in which the view is integrated in the MVMG is not important. For instance, let us begin with the view V_1. Then, all SP_fragments of V_1 will become the terminal fragments of the MVMG, and then be materialized.

After integrating view V_1 into the MVMG, we have the following transformations:

SP_fragment(V_1, customer, ξ, C) will be materialized and becomes terminal fragment noted TF(customer).

SP_fragment (V_1, order, ξ, C) will be materialized and becomes terminal fragment noted TF(order).

SP_fragment(V_1, lineitem, ξ, C) will be materialized and becomes a terminal fragment noted TF(lineitem).

Next, integrating V_2 consists in comparing each of its SP_fragment with the already materialized fragment of the same relation in the MVMG.

SP_fragment(V_2, nation, ξ, C) will be materialized and becomes terminal fragment noted TF(nation) because this relation was not present in the MVMG.

SP_fragment(V_2, part, ξ, C) will be materialized and becomes terminal fragment noted TF(part) for the same reason.

Thus, as depicted in figure 1, the two views share three terminal fragments defined over source relation's customer, lineitem and order respectively. VF[V_1](order) corresponds to the SP_fragment with condition C=(o_orderdate >'1999/25/12'). This condition is more restrictive than the empty condition of the terminal fragment TF(order), which is shared by the view V_2. This is why VF(lineitem) is placed above TF(order).

The attributes set and the condition of SP_fragment(V_2, lineitem, ξ, C) are not comparable with those of terminal fragment TF(lineitem). Consequently, a new fragment computing the union of the attributes of both SP_fragments of V_1 and V_2, and the union of both conditions, will be created and materialized. This new fragment becomes TF(lineitem). Then, the old TF(lineitem) becomes virtual and noted VF[V₁](lineitem) that will placed above the new F(lineitem). Then, SP_fragment(V₂, customer, ξ, C), noted VF[V₁](lineitem), remains virtual and will be placed above TF(lineitem). It represents the virtual A_fragment with the condition (sum(l_quantity)>10,000).

Now, let us consider SP_fragment(V₂, customer, ξ, C). Its attributes and condition are included in those of TF(customer). Consequently, SP_fragment(V₂, customer, ξ, C) remains virtual and noted VF[V₂](customer).

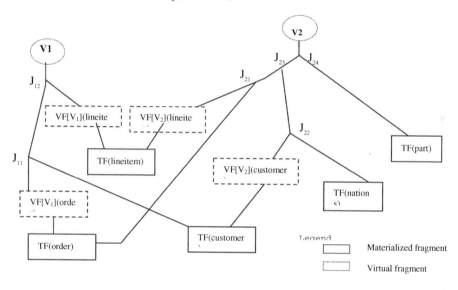

Fig. 1. The MVMG with two views

For simplification reason, in the sequel of this paper, a SP_fragment defined over the relation R₁ and belonging to the view V will be noted SP_fragment[V](R₁) and other parameters will be omitted. We will adopt the same simplification for the other types of fragments.

3 View Adaptation after Redefinition

In this section, we present the different view redefinition changes and their related adaptation algorithms. Our adaptation strategy is based on the reuse of existing materialized fragments of all materialized views. However, data independence is guaranteed. For instance, when an attribute is deleted from a view, those views, which do not involve this attribute, will not be affected. Our approach supports a set of comprehensive changes operations on view definition.

- Addition and deletion of relations,
- Addition and deletion of attributes,
- Addition and deletion of conditions.

The following algorithm integrates a given SP_fragment VF[V](R$_i$) into the MVMG. This algorithm will be called by some adaptation algorithms.

SP_Adaptation(VF[V](R$_i$))
Input : a MVMG, a SP_fragment VF[V](R$_i$) of view V over source relation R$_i$
Output : the new MVMG
If there is a terminal fragment TF(R$_i$) in the MVMG Then
      If (VF[V](R$_i$) = TF[Vp](R$_i$)) Then /*Equality of fragments */
      The terminal fragment TF(Ri) is reused by V

      Else If (TF[Vp](R$_i$) ` VF[V](R$_i$)) Then
      /*TF[Vp](R$_i$) contains more attributes and is less constrained than VF[V](R$_i$)
*/
            VF[V](R$_i$) is placed above TF[Vp](R$_i$)
            Else If (VF[V](R$_i$) ` TF[Vp](R$_i$)) then
            /* VF[V](R$_i$) contains more attributes and is less constrained than
                TF[Vp](Ri) */
                Dematerialize TF[Vp](R$_i$) ;
                TF[Vp](R$_i$) takes the place of VF[V](R$_i$)
                Materialize VF[V](R$_i$);
                VF[V](R$_i$) becomes TF[V](R$_i$)
      If VF[Vj](R$_i$) exists such that VF[Vj](R$_i$) was projected over TF[Vp](R$_i$)
      Then VF[Vj](R$_i$) is placed above TF[V](R$_i$)
      Else If (VF[V](R$_i$) and TF[Vp](R$_i$) are not comparable) Then
            Create a new fragment TF[V, Vp](R$_i$) AS the union of
            VF[V](R$_i$) and TF[Vp](R$_i$)
            Dematerialize TF[Vp](R$_i$) ; TF[Vp](R$_i$) becomes VF[Vp](R$_i$)
            VF[V](Ri) and VF[Vp](R$_i$) are placed above TF[V, Vp](R$_i$)).
            If VF[Vj](R$_i$) exists such that VF[Vj](R$_i$) was placed above
                  TF[Vp](R$_i$) Then VF[Vj](R$_i$) is placed above TF[V, Vp](R$_i$)
      Else Materialize VF[V](R$_i$); VF[V](R$_i$) becomes TF[V](R$_i$)
End

Fig. 2. The merge algorithm for a SP_fragment

3.1 Adding a Join Relation by Redefinition

Adding a join relation affects also the WHERE clause since it adds an equi-join condition in order to connect the new relation to the other relations included in the view query. Our algorithm attempts to reuse the materialized fragment of the new relation if it is already present in the MVMG.

Add(V, $R_i.A_1$, $S.B_1$, JC) /* *add a join relation S and its join condition with relation* R_i */

Input: MVMG, view V, attributes $R.A_1$, $S.B_1$ and a join condition JC
Output: MVMG
    Create a SP_fragment VF[V](S) AS
    $S.B_1$, S.key are in VF[V](S)
    If there is TF[Vp](S_i) Then If there is a selection condition on TF[Vp](S_i) Then
        VF[V](S) will be materialized and becomes TF[V](S)
        TF[Vp](S) becomes VF[Vp](S) and will be placed above TF[Vp,V](S)
        Else TF[Vp](S_i) will be used by view V
    Else /* VF[V](S) becomes a terminal fragment since S does not exist in the MVMG */
        VF[V](S) will be materialized and becomes TF[V](S)
    Create a SP_fragment VF[V](R_1) AS
    $R_1.A_1$, R_1.key are in VF[V](R_1)
    If there is another view Vp using the terminal fragment TF[Vp](R_1) Then
        If there is a selection condition on TF[Vp](R_1) Then
            VF[V](R_1) will be materialized and becomes TF[V](R_1)
            TF[Vp](R_1) becomes VF[Vp](R_1) and will be placed above TF[V](R_1)
    Create a J_fragment JF[V]($R_1.A_1$, $S.B_1$, JC) AS
    $R_1.A_1$, $S.B_1$ are in VF[V]
    JC is in VF[V]; /* JC is the join condition */
    Place JF[V]($R_1.A_1$, $S.B_1$, JC) above TF[V](R_1) and TF[V](S)
End

Fig. 3. Adding a join relation by redefinition

3.2 Adding an Attribute and/or a Condition by Redefinition

Now, we consider the case where adding a set of attributes with eventual conditions on these attributes. The algorithm for performing view adaptation in response to this change is depicted figure 4. Our strategy avoids accessing the sources if the added attribute is already present in a terminal fragment of the MVMG, while in the related work, adding an attribute in a view entails always accessing the sources [7, 11, 12]. Furthermore, this algorithm performs simultaneously several changes on a view.

Add(V, R$_i$, A, C) /* add a set of attributes A of a relation R$_i$ and a condition on these
                attributes */
Input: MVMG, attribute R$_i$.A, view V
Output: MVMG
If there exists TF[V](R$_i$) Then If it is not shared by other views Then
        Delete From TF[V](R$_i$) the tuples that do no match *C*
        Else /* *the terminal fragment is shared by other views* */
                Create a virtual SP_fragment VF[V](R$_i$) having as
                Attributes = {R$_i$.A}
                Condition = C
Else /* *there exists a virtual fragment VF[V](R$_i$); it will be placed in the MVMG
        according to its new condition and* attributes */
                Create a virtual SP_fragment VF[V](R$_i$) having as
                Attributes = {attributes of VF[V](R$_i$) $\cup$ R$_i$.A}
                Condition = (condition of VF[V](R$_i$) $\cup$ C)
SP_Adaptation(VF[V](R$_i$)) /* *call the SP_Adaptation algorithm*/
End

Fig. 4. Adaptation algorithm for adding attributes and/or a condition

3.3 Deleting an Attribute and/or a Condition by Redefinition

We consider three situations: (1) the attribute could be involved in the Select clause; (2) the attribute could be involved in a predicate of the Where clause; (3) both (1) and (2). Our algorithm is able to perform simultaneously these three situations in one pass. Deleting the concerned selection condition from the view definition may entail inserting tuples into the new SP_fragment. Once again, accessing the sources can be avoided by reusing tuples that are already present in materialized fragments of the MVMG.

Delete (V, A, C, R₁) /* *delete a set of attributes A and the selection condition on these attributes from view V* */

Input: MVMG, view V, a set of attributes A and a condition C

Output: MVMG

    If C <> ∅ Then

        If TF[V](Rᵢ) involves condition C Then

            If TF[V](Rᵢ) is shared by other views Then

                Create a virtual fragment VF[V](Rᵢ) that keeps condition C but projects

                out the unneeded attributes from TF[V](Rᵢ);

                /* *VF remains virtual and connected to TF[V](Rᵢ)* */

                Add to TF[V]](Rᵢ) the tuples discarded by the condition C

            Else /* *C belongs to a virtual fragment in the MVMG* */

            Delete the corresponding virtual fragment VF[V](Rᵢ);

        Else /* *C is empty* */

        If TF[V](Rᵢ) is shared by other views Then

            Create a virtual fragment VF[V](Rᵢ) that project out the unneeded attributes from

            TF[V](Rᵢ);

            /* *VF will remain virtual and connected to TF[V](Rᵢ)* */

        Else project out the unneeded attribute from TF[V](Rᵢ);

    End

Fig. 5. Adaptation algorithm after deleting an attribute and/or a condition.

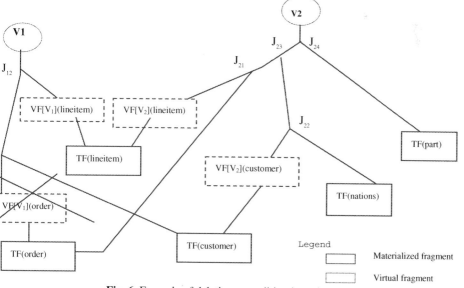

Fig. 6. Example of deleting a condition in a view.

Example. Let us consider deleting a condition from the view V_1. In related work, this deletion should in all cases entail accessing the sources in order to compute the tuples that have been discarded by the to-be-deleted condition. In our approach, accessing the sources is avoided because these tuples are already materialized by view V_2. Consequently, the adaptation consists in deleting the virtual fragment VF(order) from the MVMG (see figure 6).

4 Related Work

The view adaptation problem was first investigated in [7], in relational DBMS context. In this approach, the adaptation process is expressed as an additional query or update upon the old view and the source relations that will be executed to adapt the view. Another approach for the view adaptation problem in data warehousing systems was proposed in [11, 12]. The objective of this approach is minimizing the communication cost. A view query is represented as an expression tree. It has been shown that if changes are made at the root, it only entails the evaluation of right hand child.

The problem of the *view synchronization* caused by external environment changes was addressed in [13]. The proposed approach focuses on the replacement of invalidated views. For this purpose an extension to SQL was provided for expressing user preferences for view evolution. An interesting idea concerns the exploitation of meta-knowledge involving inter-site constraints existing between different information sources. This strategy is so-called POC (PrOject Containment).

The major differences between the related work and our approach are:

- The related work [7, 11, 12, 13] are based on the full materialization of views versus materialization of view fragments.
- Their adaptation strategies are based on one view versus on the all-existing views. Moreover, the view adaptation approach described in [7] was done in the context of a relational DBMS and did not take the communication cost into account.

5 Conclusion

This paper addresses an important issue in integrated systems such as Web sites, data warehousing systems and mediator systems: the view adaptation after redefinition. We have shown how view adaptation can be supported in order to keep views up to date and running after view redefinition.

We proposed a view adaptation strategy based on a novel view selection approach that consists of decomposing views into fragments and then merging them into a Multi View Materialization Graph, sharing nodes where possible. Our adaptation algorithms deal with a set of affected views and perform the adaptation regarding all materialized views. On the contrary, in related work, the adaptation of the view materialization is performed regarding solely the old materialization of the same view. Therefore, our adaptation strategy takes advantage by reusing materialized data. Thus, accesses to data sources are reduced. Moreover, data independence is guaranteed. For instance, when an attribute is deleted from a view, those views, which do not involve this attribute, will not be affected.

To summarize, the contributions of this paper are:

-   A novel approach for selecting a set of views to materialize,
-   Adaptation algorithms dealing with view redefinition,
-   The combination of multiple changes on a view-defining query.

The work presented in this paper is currently implemented on top of Oracle DBMS. There are several directions of future work. First, work on extending the adaptation algorithms for a largest class of views is currently investigating.

Although our approach encourages reuse of terminal fragments over relations that are included in different views, opportunities of sharing information should be extended to sub graphs of the Multi View Materialization Graph. We plan to investigate this issue. Another one of our ongoing work deals with view definition for XML [1] and warehousing XML data.

Acknowledgements

We would like to thank Marianne Huchard for our discussion about the graphs and for his comments on the earlier draft of this paper. We also thank Tigist Alemu for his help proof reading the English manuscript.

References

1. Abiteboul S., on Views and XML. invited talk in PODS'99, ACM SIGACT-SIGMOD-SIGART, Philadelphia, May 31-June2 (1999)
2. Akinde M.O., M.H. Böhlen. Constructing GPSJ View Graphs. In Proceedings of the International Workshop on Design and Management of Data Warehouses (DMDW'99) (1999)
3. Bellahsene Z., View Adaptation in Data Warehousing Systems, in Proc. of International Database and Expert Applications Conference, DEXA'98, Lectures Notes in Computer Science, Springer Verlag, Vienna (1998)
4. Bellahsene Z., Marot P., A dynamic approach for selecting views to Materialize, Technical Report, LIRMM N° 99074, June, Montpellier (1999)
5. Gupta, A., Harinarayan, V., Quass, D. Aggregate Query Processing in Data Warehousing Environments. in Proc of International Conference on Very large Databases, Zurich, Switzerland, september (1995)
6. Gupta, A., Mumick, I.S. Maintenance of materialized views: Problems, techniques, and applications. IEEE Data Eng. Bulletin, Special Issue on Materialized Views and Data Warehousing, 18(2) (1995)
7. Gupta, A., Mumick, I.S., Ross K.A., Adapting Materialized after Redefinitions, in Proc of ACM SIGMOD International Conference on Management of Data, San Jose, USA (1995)
8. Gupta H., Selection of Views to Materialize in a Data Warehouse. In Proceedings of the International Conference on Database Theory, Delphi, Greece, January (1997)
9. Hull R., Zhou G., A Framework for supporting Data Integration using the Materialized and Virtual Approaches. Proc. of SIGMOD'96 Conference, Montreal, Canada, (1996)

10. Hammer, J., Garcia-Molina, H., Widom, J., Labio, W., and. Zhuge, Y. The Stanford Data Warehouse Project. IEEE Data Engineering Bulletin, in Proc. Special Issue -48, June (1995)
11. Mohania M., Dong G., Algorithms for Adapting Materialized Views in Data Warehouses, In Proc. of International Symposium on Cooperative Database Systems for Advanced Applications, Kyoto, Japan (1996) 62-69
12. Mohania M., Avoiding Re-computation: Views Adaptation in data warehouses, Proc. of 8th. International Database Workshop, Springer Verlag, pp. 151-165, Hong Kong (1997)
13. Nica A., Lee A. J., Rundensteiner E. A., The CVS Algorithm for View synchronisation in Evolvable large-Scale information systems, in Proc. of International Conference on Extending database technology, EDBT'98, Lectures Notes in Computer Science, Springer Verlag (1998)
14. Nica A., Lee A. J., Rundensteiner E. A., View Maintenance after View Synchronization in Proc. of the International Conference of IDEAS'99, Montreal, Canada (1999)
15. Quass D., Gupta A., Mumick I.S., J. Widom, Making Views self-Maintainable for data Warehousing, in Proc. Proc. Of Parallel and Distributed Information Systems, Miami Beach, FL, Dec.ember (1996)
16. TPC-D Benchmark Standard Specification 2.01, http://www.tpc.org, January (1999)
17. Theodoratos D., Sellis T., Data warehouse configuration. In Proceedings of the 23rd International Conference Very Large Data Bases, Athens, (1997)
18. Zhuge, Y., Garcia-Molina, H., Wiener, J.: Consistency Algorithms for Multi-Source Warehouse View Maintenance. Journal of Distributed and Parallel Information Systems, Kluwer Academic Publishers, pp 1-36, December 1996.Engineering, April (1991) 146-182

On Warehousing Historical Web Information*

Yinyan Cao, Ee-Peng Lim, and Wee-Keong Ng

Centre for Advanced Information Systems, School of Applied Science, Nanyang
Technological University, Nanyang Avenue, Singapore 639798, SINGAPORE,
aseplim@ntu.edu.sg

Abstract. We present a temporal web data model designed for warehousing historical data from World Wide Web that changes with time.
As the Web is now populated with large volume of web information, it
has become necessary to capture some useful web information in a data
warehouse that supports further intelligent data analysis. Nevertheless,
due to the unstructured and dynamic nature of Web, the traditional relational model and its temporal variants could not be used to build such
a data warehouse. In this paper, we therefore propose a temporal web
data model that captures the connectivities of web documents and their
content in the form of temporal web tables. To support the analysis of
web data that evolve with time, valid time intervals are associated with
each web document. To manipulate temporal web tables, we define a variety of web operators and illustrate their usefulness using some realistic
motivating examples.

1 Introduction

Nowadays, World Wide Web (WWW) serves as a huge repository of data formatted as web pages hosted by large number of autonomous web sites. As web
sites update their web pages regularly, users can easily obtain from them the
most up-to-date information at very low cost. Nevertheless, once a web page
is updated, its previous content will be overwritten and can never be retrieved
again from the HTTP server that manages the web site unless the web site administrator implements some facility to archive web pages and make the archived
information available to the web users. To manage historical web pages and to
systematically derive useful knowledge from the Web, we would like to justify the
need for a powerful data warehousing system for web information by presenting
three motivating examples based on real web sites.

Example 1: The Business Times, a major newspaper in Singapore, operates
an online collection of regional market reports within its web site located at
http://business-times.asia1.com.sg. Since the reports contain daily updated useful business information, they are suitable candidates to be maintained by a

---

* This work was supported in part by the Nanyang Technological University, Ministry
of Education (Singapore) under Academic Research Fund #4-12034-5060, #4-12034-
3012, #4-12034-6022. Any opinions, findings, and recommendations in this paper are
those of the authors and do not reflect the views of the funding agencies.

A.H.F. Laender, S.W. Liddle, V.C. Storey (Eds.): ER2000 Conference, LNCS 1920, pp. 253–266, 2000.
© Springer-Verlag Berlin Heidelberg 2000

warehousing system. Figure 1 provides a schematic abstraction for capturing the regional market information from The Business Times site. Each node in the graph (in this example, a linear graph) denotes a web page, and each directed link denotes a hyperlink between two web pages. There are a few predicates that represent constraints to be satisfied by the web pages and links. Node a denotes the home page of The Business Times web site by the constraint that $a.url = $ "http://business-times.asia1.com.sg". Link j leads to a web page d containing a directory of hyperlinks including those for finance/commodity news. Link k leads to a web page e which serves as a directory for some regional market reports, one for each major financial center such as Hong Kong or Tokyo. The regional market report web page is represented by f, and it contains a hyperlink (represented by m) to the stock exchange table page (denoted by g) for that particular market. For example, the web page for Hong Kong market report always contains a hyperlink to the web page containing the Hong Kong stock exchange table. ■

a.url= "http://business-times.asia1.com.sg"
j.labelcontain "Finance"
k.labelcontain "BT Stocks"
f.title = "RegionalMarketReport"
m.labelcontain "Stock Exchange Table"

Fig. 1. Regional market reports

Example 2: The Business Times web site provides online company and market news updated daily. Such information has been of primary interest to the managers and investors. Figure 2 depicts the paths from the home page of The Business Times to web pages containing company & market news articles. Node a represents home page of The Business Times. Via hyperlink represented by link h labeled Markets & News, a web page (represented by node b) containing a daily updated list of news can be reached. Node c depicts a web page containing one market news article and the label of i contains the title of the news represented by c. ■

a.url= "http://business-times.asia1.com.sg"
h.labelcontain "Markets & News"

Fig. 2. Company and market news

Example 3: The web site http://www.fish.com.sg operated by Financial Interactive Services Hub provides frequently updated online stock information listed at the Stock Exchange of Singapore. Figure 3 describes the navigation paths that lead to the web pages containing price information of individual shares. Node x represents the home page of http://www.fish.com.sg. Hyperlink represented by link n leads to web page (captured by node y) of a directory of share categories.

x .url= "http://www .fish.com .sg "
n .label contain "Stock Info "
o .label contain "All"
w .text contain p .label
p .target_url contain "cbin/getCounterInfo "

Fig. 3. Stock information

Node z captures a share category (for example, SES Main). The label of hyperlink represented by link p contains the name of share, whose price information is contained in web page represented by node w. ∎

The above three examples highlight the strategic web information often accessed by business managers and investors. The conventional approach of accessing the Web is not sufficient to manage and analyze the web data. Firstly it is time consuming because the user has to manually categorize and browse the relevant web pages. Secondly, it is not scalable for historical web data, which is very large in volume. Thirdly, the user needs to have prior knowledge about the structure and content of their interested web sites because the conventional tools do not maintain such knowledge. To overcome the difficulties, it is crucial to devise a data warehousing system to capture and maintain the historical web data systematically and provide the appropriate support for more advanced data analysis applications.

At the Center for Advanced Information Systems, we are in the process of developing a web warehousing system known as **WHOWEDA** (WareHouse Of WEb DAta) for storing and managing web information extracted from selected web sites [15]. The extracted web information represents segments of the Web which users are interested in and would like to perform further queries and analysis on. The primary goal of WHOWEDA is to provide a comprehensive information infrastructure for (a) materializing and indexing web information; (b) maintaining the schema of stored web information; and (c) supporting expressive queries to manipulate the materialized web information.

In this paper, we propose a **temporal web data model** for WHOWEDA to equip the warehousing system with the capability to manage historical web information. While web pages at the selected web sites may be updated regularly, WHOWEDA can materialize a history of the web pages by timestamping each web page with its valid time duration. A set of **web operators** are further developed to query and manipulate the web pages in WHOWEDA using the temporal features of the pages.

The rest of the paper is organized as follows. Section 2 presents the temporal web data model. Sections 3 to 5 describe the core, supporting, and meta web operators respectively. Section 6 discusses the related work. We conclude the paper and provide future directions in Section 7.

2   Temporal Web Data Model

In this section, we present the temporal web data model adopted by WHO-WEDA, which essentially maintains a warehouse consisting of a set of **web tables**. The web information required by Examples 1, 2 and 3 in Section 1 has been captured in web tables *Reports*, *News*, and *Fish*. The last two tables are shown in Figures 4 and 5 respectively. In the following, we define essential elements of a web table. Each web table keeps of a set of **web tuples**, each of which can be seen as a directed graph. Each **node** in the graph denotes a web document and each **link** denotes a hyperlink. The web table is described by a **web schema** that defines constraints on the structure and content of its tuples.

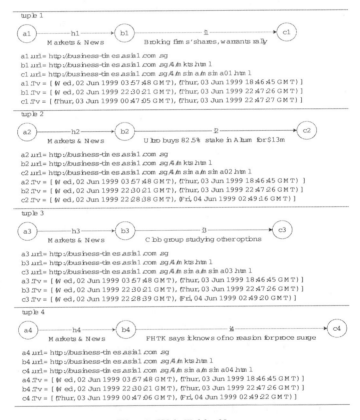

Fig. 4. Web Table *News*

In order to create and manipulate temporal web tables, we have developed several web operators. They can be categorized into **core operators**, **supporting operators**, and **meta operators**. Core operators provide the most basic operations on the creation and manipulation of web tables. Supporting operators provide additional facilities to adjust the valid time of tuples in a web table to prepare the web tables for further manipulation. A meta operator is built up on several core and supporting operators to maintain a web table over time.

There are two temporal data types in our data model: **time point** and **time interval**. A time point is represented in date/time format[1] defined by RFC822 (updated by RFC 1123). An example time point is (Sun, 06 Nov 1994 08:49:37 GMT). An example time interval is [(Sun, 06 Nov 1994 08:49:37 GMT), (Mon, 07 Nov 1994 10:10:10 GMT)] . This fine granular representation of time allows us to perform fine grain temporal manipulation of web information especially those from frequently updated web sites.

2.1   Node and Link

A node captures the information about a single web document with a set of **attributes** including *url, valid time, title, format, size*, and *text*. *Valid time* is called the **timestamp attribute**. The rest are **non-timestamp attributes**. Non-timestamp attributes are collectively denoted by A_n and *valid time* is denoted by T_v. A_n captures the URL, title, document format, size(in bytes), and textual content of the web document represented by the node. The valid time is an interval enclosed by two time points, namely *last-modified time* and *downloading time*. Let t_m denote the *last-modified time* and t_d denote the *downloading time*, $[t_m, t_d]$ indicates the period within which the web document captured by the node remains unchanged.

When a web document is accessed, we take the Last-modified field in the HTTP[8] header for *last-modified time*, and the Date field for *downloading time*. Last-modified field indicates the latest time a web object is changed. Date represents the date and time at which a web page departs the HTTP server. Given a web page returned by a HTTP server, the interval between Last-modified and Date represents the **valid duration** of the web document up to the time of retrieval. We have chosen to exclude temporal information that can be found within web documents as part of their HTML content. This is because this type of temporal information are ad hoc in nature and their semantics and validity are largely determined by their creators themselves.

Here we define the concept of **node equivalence**. Two nodes n_i and n_j are said to be equivalent, denoted by $n_i \equiv n_j$, if they share the same values for all the non-timestamp attributes and the same *last-modified time*. i.e. $(\forall a \in A_n,\ n_i.a = n_j.a) \wedge (n_i.t_m = n_j.t_m)$. Two equivalent nodes capture the same version of the same web document. Any change in the Last-modified of the web document results in a new node. In our web warehouse, a node is created when a web document is captured for the first time. Subsequently, a new node is created when it is updated. In this way, the history of a web document can be saved in the warehouse. A link captures the information of a hyperlink. Its attributes include *source_url, target_url, label*, and *link_type*, which correspond to the URL of the source document containing the hyperlink, the URL of the target document, the label of the link, and the type of the link respectively. There are three types of hyperlinks in the Web: **interior**, **local**, and **global**. A link is interior if its source and target documents are the same. A link is local if

---

[1] It is the one of the formats used used by HTTP/1.1

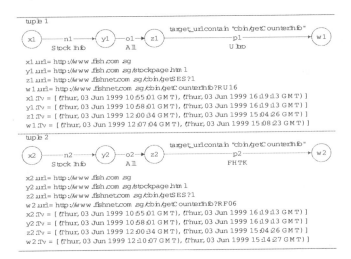

Fig. 5. Web Table *Fish*

its source and target documents are hosted in the same server machine. A link is global if its source and target documents are hosted in different server machines. Here we define the concept of **link equivalence**. Two link objects l_i and l_j are said to be equivalent, denoted by $l_i \equiv l_j$, if they share the same values for all the attributes.

2.2   Web Tuple and Web Schema

A web tuple captures a fragment of the WWW including a set of web documents and the hyperlinks among them. A tuple remains valid for a certain period, denoted by T_w, which is the intersection of the valid time values of all its nodes. If a node is updated in its last modification time, a new tuple is created. In this way, our warehouse can save the history of user-interested fragments from the WWW. The existence of tuple objects eliminates the necessity of introducing temporal information in link objects. Although a link links different versions of web pages at different times, a tuple can define which node (which version of a web page) is connected to which node by which link. We therefore do not need to store any temporal information in the link objects. A web schema, which describes the structure and content of web tuples of a web table, is formally defined as an ordered 4-tuple $M = \langle X_n, X_l, C, P \rangle$, where X_n, X_l, C, and P denote a set of **node variables**, a set of **link variables**, a set of **connectivities** in conjunctive normal form, and a set of **predicates** in conjunctive normal form respectively. The directed graphs shown in Figures 1, 2, and 3 are visualizations of the web schemas M_1, M_2 and M_3, which describe the web tables *Reports*, *News*, and *Fish* respectively. For example, M_1 is defined as:

- $X_{n1} = \{a, d, e, f, g\}$
- $X_{l1} = \{j, k, l, m\}$
- $C_1 = \{a\langle j \rangle d \wedge d\langle k \rangle e \wedge e\langle l \rangle f \wedge f\langle m \rangle g\}$
- $P_1 = \{a.url = \text{"http://business-times.asia1.com.sg"} \wedge$
     $f.title = \text{"Regional Market Report"} \wedge$

j.label contain "Finance" ∧ *k.label contain* "BT Stocks" ∧
m.label contain "Stock Exchange Table" }

The node variables, link variables, and the connectivities in a web schema determine how the node and link instances are connected in the tuples belonging to the web table. The node and link variables further allow us to specify predicates on the attributes of node and link instances. In our schema examples, the string comparator *contain* checks if a string is contained by another. A connectivity in the form of $a\langle e \rangle b$ defines the inter-linked relationship between node variables a and b via link variable e. For a tuple to be part of web table *Reports*, the tuple must consist of node and link instances that satisfy both the connectivities and predicates defined by M_1.

3   Core Operators

3.1   Global Couple

Web schema reflects the requirement of extracting a set of inter-related web documents. Since the conventional search engines cannot accept a web schema as input, neither can they return inter-linked documents as query results, a **global couple** operator, denoted by symbol Γ, has been defined. Given a web schema M, the global couple operation denoted by $\Gamma_M WWW$ creates a new temporal web table, whose tuple captures a set of inter-linked web documents matching M^2. To ensure that the global couple operation is tractable, we require the URL of every source node variable in the input web schema to be specified by some predicates. The operator searches the WWW for inter-linked web documents matching M and retrieves a collection of them into the result temporal web table.

3.2   Web Concatenate

When two web tables of the same schema are constructed at different points in time, they can be combined into one web table using the **web concatenate** operator, denoted by φ. Given two web tables W_i and W_j sharing the same schema M, the web concatenate operation, denoted by $W = W_i \varphi W_j$, creates a new web table W whose tuples are either from W_i or W_j, with pairs of equivalent tuples from both web tables merged together. The schema of the resultant web table is the same as that of operand web tables.

Two tuples t_i and t_j are said to be equivalent, denoted by $t_i \equiv t_j$, if t_i and t_j comply to the same schema $M = \langle X_n, X_l, C, P \rangle$, and the corresponding node and link instances in t_i and t_j are also equivalent. i.e. $(\forall n \in X_n, t_i.n \equiv t_j.n) \wedge (\forall l \in X_l, t_i.l \equiv t_j.l)$. As content equivalent tuples capture exactly the same fragment of the WWW, our data model does not allow content equivalent tuples existing in

---

[2] Since every global couple takes the WWW as the only operand, the operation can be simply written as Γ_M.

the same web table. To combine two equivalent tuples from different web tables, we perform a **Merge** operation on the tuples by unioning the valid time at the node level. Given two content equivalent web tuples t_i and t_j, the resultant tuple t of the merge operation satisfies $(t \equiv t_i) \wedge (\forall n \in X_n, t.n.T_v = t_i.n.T_v \cup t_j.n.T_v)$.

3.3   Web Select

Web select, denoted by σ, extracts from a web table a subset of tuples satisfying some given **selection condition**. Given an input web table W_i with schema M_i and a selection condition sc, a web select operation is denoted by $\sigma_{\langle sc \rangle} W_i$. The selection condition sc consists of a set of predicates over node and link variables of the schema of W_i. The resultant web table essentially retains the schema of the operand web table except that the new schema also includes predicates in the selection condition.

For example, a user needs to find news articles related to Ultro released after (Wed, 02 Jun 1999 22:00:00 GMT) from The Business Times. In this case, the user can specify a web select operation on *News*. In *News*, the link variable i represents the hyperlink from the Markets & News web document to the individual market news articles, and label of i captures the title of the corresponding article. Thus, *i.label contain* "Ultro" specifies that the title of news articles contain Ultro. The other constraint is that release time of the news should be later than (Wed, 02 Jun 1999 22:00:00 GMT). It can be expressed as $c.t_m > $ (Wed, 02 Jun 1999 22:00:00 GMT). Hence, the complete selection condition is $c.t_m > $ (Wed, 02 Jun 1999 22:00:00 GMT) $\wedge$ *i.label contain* "Ultro". Only tuple 2 in *News* satisfies the selection condition. Thus it is the only tuple in the result.

A temporal web table may contain very large number of tuples since it keeps historical data. Very often, one would like to extract tuples which are valid within a certain time period. Here we allow an optional temporal constraint to be added to the selection condition and write a web select operation as $\sigma_{(\langle sc \rangle, V_t)} W$, where V_t is a time interval denoting a **time slice window**. A tuple is selected if it satisfies the selection condition and it remains valid for certain time within V_t. For example, if a user wishes to extract historical new articles about Ultro published at The Business Times site from June 1999 to September 1999, he should specify: $\sigma_{((\ i.label\ contain\ "Ultro"\ ),\ [(Tue,\ 01\ Jun\ 1999\ 00:00:00\ GMT),\ (Thur,30\ Sep\ 1999\ 23:59:59\ GMT)])} News$.

3.4   Web Project

Web project, denoted by π, is a unary operator extracting fragments of each web tuple satisfying certain requirement. For example, *Reports* contains tuples about the regional market report from The Business Times. Each tuple also contains home page of The Business Times and other documents through which one can reach the Regional Market Report documents. To concentrate on the regional market report, a user may need a web table whose tuples only include web documents about Regional Market Report and Stock Exchange Table. A web project is required in this situation.

Given input web table W_i with schema $M_i = \langle X_{ni}, X_{li}, C_i, P_i \rangle$, a web project operation is written as $\pi_{\langle pr \rangle} W_i$, where pr denotes the project requirement satisfying $pr \subset X_{ni}$. The resultant web table contains a set of web tuples, each containing node instances of some tuple from W_i that correspond to node variables in pr.

3.5   Web Associate

Web associate, denoted by $\otimes$, correlates tuples from two different web tables. It allows users to discover new implicit relationship between different web tables. The relationship is defined by predicates that relate node or link variables from different web tables. For example, to find how the release of market news from The Business Times affects the share prices of Singapore Stock Market, one would like to associate *News* that contains market news from The Business Times, with *Fish* that contains share information from Singapore Stock Market. This is similar to the join operation in relational model.

Given two operand web tables W_i and W_j with schemas $M_i = \langle X_{ni}, X_{li}, C_i, P_i \rangle$ and $M_j = \langle X_{nj}, X_{lj}, C_j, P_j \rangle$ respectively, a web associate operation is denoted by $W_i \otimes_{\langle ac \rangle} W_j$, where ac is the associate condition. An associate condition determines how tuples from W_i are to be correlated with the tuples from W_j. An associate condition uses the format of $p_1 \wedge p_2 \wedge \cdots \wedge p_n$, where each p_i is a predicate involving node or link variables from the schemas of both web tables. Web associate returns a set of tuples each of which consists of two original tuples t_i and t_j obtained from W_i and W_j respectively such that t_i and t_j satisfy the associate condition defined in ac. The schema M of the resultant web table satisfies: (1) $X_n = X_{ni} \uplus X_{nj}$, (2) $X_l = X_{li} \uplus X_{lj}$, (3) $C = C_i' \wedge C_j'$, (4) $P = P_i' \wedge P_j' \wedge \langle ac' \rangle$. Note that a new symbol $\uplus$ is used instead of usual union symbol $\cup$. This is to disambiguate the identical names used for node and link variables in different schemas. To disambiguate two variables sharing the same name, say a, web associate prefixes each variable name with the operand table name. Thus in the resultant schema, the name of the two node variables become $W_i.a$ and $W_j.a$ respectively. The variable names in C_i, C_j, P_i, P_j, and ac are changed accordingly in M.

In our example of associating *News* and *Fish*, predicate *i.label contain p.label* specifies that the news article pointed by i is about the particular share whose name is indicated by *p.label*. However, the prices of the shares are affected by many factors such as the release of relevant news, the performance of other major stock markets, and technical corrections of the market. The release of a piece of news is likely to have its effect superseded by the other factors after some time. Thus, it is necessary to restrict the time difference between the release of the news and the share price movement. We assume the effect is most obvious within one day[3]. Hence, two more predicates $w.t_m \leq c.t_m + 1$ and $w.t_m > c.t_m$ have been included by the associate condition. The resultant web table W_a is shown in Figure 6.

---

[3] It is only a convenient assumption

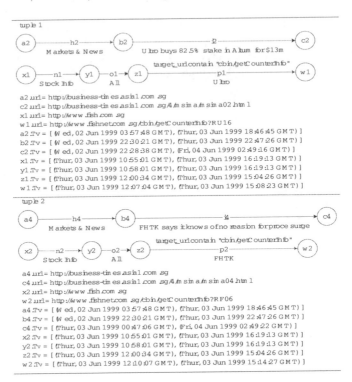

Fig. 6. Web Associate Result W_a

Note that each tuple returned by web associate consists of components from different web tables and these components could have non-overlapping valid times. We therefore provide an optional facility for web associate to ensure that each tuple in the resultant web table consists of components that share some common validity time. The web associate operation can therefore be written as $W_i \otimes_{(\langle ac \rangle,\ overlap)} W_j$.

4   Supporting Operators

4.1   Intervalize

Some HTTP servers do not give the value of Last-modified in their message headers. In this case, we use the *downloading time* as *last modification time*. The valid time of a node obtained from such servers or generated by CGI programs is thus a time interval which spans over a single time point. In other words, the valid time can be written as $[t_m, t_m]$.

Consider the following scenario. In web table W, nodes $a_1, a_2, \ldots, a_n$ are instances of node variable a, which models documentd from URL url_a. The documents represented by these nodes are identical in content. However, they have different Last-modified time. If the document is generated by CGI program, then $a_1.T_v = [t_1, t_1]$, $a_2.T_v = [t_2, t_2]$, ..., and $a_n.T_v = [t_n, t_n]$. It is very likely that the documents are identical throughout $[t_1, t_n]$. The temporal data model only manages to obtain a sequence of time points is due to insufficient information provided by the HTTP message header or the ad hoc regeneration of web documents by CGI programs. We therefore provide a supporting operator **intervalize**, denoted by ϑ, to consolidate the valid time of certain node variables. If intervalize is deployed for the above scenario, then the valid time of the instances of a is consolidated as $[t_1, t_n]$ in the resultant web table, thus overcoming the restrictive temporal information returned by the autonomous web servers.

Here we need to define the concept of **node content equivalence**. Two nodes n_i and n_j are content equivalent, denoted by $n_i \simeq n_j$, if and only if $\forall n \in A_n$ $n_i.n = n_j.n$. Note that content equivalence is less restrictive than pure node equivalence. Given input web table W_i with schema $M_i = \langle X_{ni}, X_{li}, C_i, P_i \rangle$, an intervalize operation is written as $\vartheta_{\langle ir \rangle} W_i$, where ir denotes the intervalize requirement, which is a subset of X_{ni}. The result of pure intervalize operation contains all web tuples as from W_i, except that all node instances of the node variables listed in ir have their valid time consolidated. The consolidation of the valid time of a node variable a results in a set of new instances S_n, such that $\forall a_i \in S_n$:

- The valid time of a_i is disjoint from the valid time of all other instances in S_n. In other words, $\forall a_j \in S_n$, $a_i.T_v$ *disjoint* $a_j.T_V$
- If there exists an $a_j \in S_n$, such that a_i is content equivalent to a_j, then there must exist an $a_k \in Sn$, such that $(a_i \not\simeq a_k) \wedge (a_i.t_m < a_k.t_m < a_j.t_m \vee a_i.t_m > a_k.t_m > a_j.t_m)$

4.2   Snap

A temporal web table consists of a set of historical tuples. A new tuple is appended to the table if the captured web documents or hyperlinks are updated. This results in large number of **url equivalent** tuples residing within the same web table. Two tuple objects are said to be url equivalent, denoted by $t_i \simeq t_j$, if (1) t_i and t_j comply to the same schema $M = \langle X_n, X_l, C, P \rangle$, and (2) the node instances in t_i and t_j corresponding to the same node variable have the same url value. i.e. $\forall n \in X_n$, $t_i.n.url = t_j.n.url$. Note that $t_i \simeq t_j$ is a less stringent equivalence than normal equivalence $t_i \equiv t_j$ and content equivalent. The url equivalent tuples capture the same fragment of the Web at different times. Given web table W_i, the unary operation **snap**, denoted by τW_i, extracts the latest versions of url equivalent tuples from the web table W_i. The latest version of a set of rul equivalent tuples is the tuple which carries the most recent last modification time or downloading time. The resultant web table contains the latest versions of all the Web segments captured by W_i.

5 Meta Operators

5.1 Create

Global couple retrieves from the web a collection of inter-related web documents matching the input web schema. However, its resultant web table only reflects the information of the WWW at the time when global couple is performed. To construct a temporal web table which captures the evolution of web information over time, we define a meta operator **create**, denoted by ε. It is known as a meta operator because it involves a procedure that builds a web table incrementally by using the existing core and supporting operators. To create a temporal web table W using schema M, the create operation is denoted by $W = \varepsilon_{(p_s,p_e,f,s,i,ir,M)} WWW$.

M is the schema of the temporal web table to be created. *Start time point* and *end time point*, p_s and p_e, determine when the create operator first accesses the World Wide Web and when the create operation should be terminated respectively. The *frequency of the periodic polling*, f, determines how often the World Wide Web should be accessed. Each poll updates the existing temporal web table. The *deployment of snap operator*, s, whose value is either true or false, determines whether snap operator is used in create operation. The use of snap operator ensures that only the latest versions of the WWW information are stored in the web table. The *deployment of intervalize operator*, i, whose value is either true or false, determines whether intervalize operator is used in create operation. The use of intervalize operator ensures that the instances of specified node variables always keep their valid time intervalized. The *intervalize requirement*, ir, determines the instances of which node variables are to be intervalized. It is a subset of the node variables defined in M. Note that this parameter is in use only when i is true. Note that snap is performed before intervalize if both s and i are true. In this case, the intervalize operation does not have any effect on the web table since snap returns only the lastest version of url equivalent tuples.

6 Related Work

Our temporal web data model is built upon concepts from diverse research areas, mainly existing temporal databases and web data models. The previously proposed temporal data models usually extend relational data models [6,14,16]. Our data model is designed to overcome limitations of relational model which is flat and unsuitable for Web. A detailed survey on database techniques applied on the Web is given in [9]. WebSQL [3,13] and W3QL [11] assume the internal structure of document is not known. WebLog [12] is a logic based language for querying as well as restructuring the Web. WebOQL [2], Lorel [1], UnQL [5], and Florid [10] are also designed to query semistructured data. Araneus [4] and Strudel [7] are both web site management systems, supporting restructuring and creation of Web sites. Our work follows the same line as WebSQL in assuming no knowledge on the internal structure of web documents. Different from all the above research, our data model captures historical web data, and web segments are modeled as web tables, which are treated as first-class objects. Similar to Florid, the captured information are stored locally. While the other research works focus on web restructuring besides querying, we concentrate on further manipulation and analysis of web data and to provide the DBMS support for web applications.

Several web search engines, such as Altavista, Hotbot, Yahoo, and Snap, are equipped with search capabilities in the time dimension. For example, the advanced search interface of Hotbot allows users to specify as part of a query the last modified time to be satisfied by the returned web pages. The search engines do not materialize historical web information systematically. They usually support very simple temporal search criteria, while our data model has been designed to support complex queries or data analysis on the web information.

7   Conclusion and Future Work

Due to the ever changing nature of the World Wide Web, the ability to store, query, and analyze historical web information is vital to the modern businesses. In this paper, we have described a data warehousing system known as WHO-WEDA for managing selected useful historical web information, which are interconnected and highly unstructured. To support temporal queries and data analysis in WHOWEDA, we have defined a temporal web data model which is able to capture web documents together with their updating history into temporal web tables. We attach a temporal attribute to each node variable in the web schema to denote the valid time interval of its web document instances. We have also developed a set of time-aware web operators to manipulate the web tables. We have used real life examples to illustrate our ideas on temporal web data model and temporal web operators. In the future, we will look into more detailed analysis of the web operators including their properties and expressiveness. We are also implementing the temporal web warehouse.

References

1. S. Abiteboul, D. Quass, J. McHugh, J. Widom, and J. L. Wiener. The Lorel query language for semistructured data. *International Journal on Digital Libraries*, 1(1):68–88, April 1997.
2. G. Arocena and A. Mendelzon. WebOQL: Restructuring documents, databases and webs. In *Proceedings of ICDE'98*, Orlando, Florida, February 1998.
3. G. Arocena, A. Mendelzon, and G. Mihaila. Applications of a web query language. In *Proceedings of the 6th International WWW Conference*, Santa Clara, April 1997.
4. P. Atzeni, G. Mecca, and P. Merialdo. To weave the web. In *Proceedings of the 23rd VLDB Conference*, Athens, Greece, 1997.
5. P. Buneman, S. Davidson, and G. Hillebrand. A querying language and optimization techniques for unstructured data. In *Proceedings of ACM SIGMOD Conference on Management of Data*, pages 505–516, Montreal, Canada, 1996.
6. J. Clifford and A. Croker. The historical relational data model (HRDM) and algebra based on lifespans. In *Proceedings of the International Conference on Data Engineering*, pages 528–537. IEEE Computer Society, February 1987.

7. M. Fernandez, D. Florescu, J. Kang, and A. Levy. Catching the boat with Strudel: Experiences with a web-site management system. In *Proceedings of ACM SIGMOD Conference on Management of Data*, Seattle, WA, 1998.
8. R. Fielding, J. Gettys, J. Mogul, H. Frystyk, and T. Berners-Lee. *Hypertext Transfer Protocol – HTTP/1.1*, Jan 1997.
9. D. Florescu, A. Levy, and A. Mendelzon. Database techniques for the world-wide web: A survey. *ACM SIGMOD Record*, 27(3):59–74, September 1998.
10. R. Himmeroder, G. Lausen, B. Ludascher, and C. Schlepphorst. On a declarative semantics for web queries. In *Proceedings of the 5th International Conference on Deductive and Object-Oriented Databases*, Montreux, Switzerland, December 1997.
11. D. Konopnicki and O. Shmueli. W3QS: A query system for the world wide web. In *Proceedings of the 21st VLDB Conference*, Zurich, Switzerland, 1995.
12. L. V. S. Lakshmanan, F. Sadri, and L. N. Subramanian. A declarative language for querying and restructuring the web. In *Proceedings of the 6th International Workshop on Research Issues in Data Engineering, RIDE '96*, New Orleans, February 1996.
13. A. Mendelzon, G. Mihaila, and T. Milo. Querying the world wide web. *International Journal on Digital Libraries*, 1(1):54–67, April 1997.
14. S.B. Navathe and R. Ahmed. A temporal relational model and a query language. *Information Sciences*, 49(1-3):147–175, 1989.
15. W.-K. Ng, E.-P. Lim, C.-T Huang, S.S. Bhowmick, and F.-Q. Qin. Web warehousing: An algebra for web information. In *Proceedings of IEEE International Conference on Advances in Digital Libraries (ADL'98)*, April 1998.
16. Richard Snodgrass. The temporal query language TQuel. *ACM Transactions on Database Systems*, 12(2):247–298, June 1987.

On Business Process Model Transformations [*]

Wasim Sadiq and Maria E. Orlowska

Distributed Systems Technology Centre
School of Computer Science & Electrical Engineering
The University of Queensland
Qld 4072, Australia
email: {wasim,maria}@dstc.edu.au

Abstract. A business process model represents the basic building block for a workflow-enabled enterprise information system. Generally, a process model evolves through numerous changes during its lifetime to meet dynamic and changing business requirements. It is essential that such changes are introduced systematically and their impact is clearly understood. Process model transformation is a suitable approach for this purpose. Applying pre-defined transformation operations can ensure that the modified process conforms to a given class of constraints specified in the original model. Using a generic process modelling language, we identify three classes of transformation principles – equivalent, imply, and subsume – to manage changes in process models. A simple algebraic notation for representing process graphs is also presented that can be used to reason about transformation operations.

1 Introduction

The use of workflow technology in enterprises has grown substantially during the past few years. It is now considered an appropriate platform for building and integrating component-oriented enterprise systems. The workflow management systems provide a flexible environment to manage process logic of business applications just like database management systems have been providing functionality for managing application data. Data modelling techniques like entity-relationship diagrams are applied to identify data management requirements of business applications. These entity-relationship models are mapped to relational schemas of target database management systems. Similarly, process modelling techniques and tools are applied to capture process-relevant requirements of the business applications in the form of business process models. Workflow management systems take up the responsibility for coordinating business processes on the basis of process models that define workflow tasks and associated coordination constraints.

There are several aspects of a business process model including structural / control flow, data flow, roles, application interfaces, temporal constraints, and others [8]. The structural modelling – also known as control flow modelling – of a process model

---

[*] The work reported in this paper has been funded in part by the Cooperative Research Centres Program through the Department of the Prime Minister and Cabinet of the Commonwealth Government of Australia.

A.H.F. Laender, S.W. Liddle, V.C. Storey (Eds.): ER2000 Conference, LNCS 1920, pp. 267–280, 2000.

defines the way a workflow management system would order and schedule workflow tasks. This is the primary and perhaps the most important aspect of a process model. It defines the coordination constraints between workflow tasks and builds a foundation for capturing other aspects of workflow requirements.

Generally, a process model goes through several evolutionary changes during its lifetime to satisfy dynamic and changing business requirements. There are generally two approaches for changing process models to meet such new requirements. We classify them as *constrained* and *unconstrained* approaches respectively.

The unconstrained approach, as the name suggests, does not put any restrictions on the way a process model may be modified. The process designer takes the original process model as a starting point for building a new process model that would incorporate required changes. As an outcome of the modification process, we get a new version of the process model with some arbitrary changes. However, it is generally not possible to sensibly relate the modified model with the original process model it is derived from. Unconstrained approach is useful in situations where there has been significant process evolution and adaptation to business requirements [7] [13].

In contrast, the constrained approach maintains certain relationships between the original and modified process models. Using this approach, the workflow designer introduces changes in the original process model in such a way that the modified process model conforms to a specific class of coordination constraints specified in the original process model. This approach supports a systematic way of introducing process model changes and provides means to reason about their impact.

A suitable approach for introducing constrained changes in the process models is through transformation operations. In this paper, we will identify transformation principles and associated operations for introducing constrained changes in process models. Such transformation operations may be applied to a process model G to transform it into G' such that G and G' still maintain underlying structural relationship with each other. We will be concentrating only on structural / control flow aspects of process models. We will identify three classes of structural relationships – equivalent, imply, and subsume – and associated transformation operations. An analogy between process model transformations and schema transformations in database research is useful here.

Before we present these transformation principles, let us consider a few scenarios where changing a process model through transformation operations rather than arbitrary modifications would be useful.

Given a set of tasks and coordination constraints, it is possible to build two equivalent process models that may look quite different graphically on the surface but express identical functionality. Let us assume that a process designer correctly captures the process requirements of a system in the form of a process model. One of his colleagues, while looking at the model, realizes that certain changes in the process model would simplify and clarify the design and make it simpler to comprehend. However, he would not want to introduce any changes in the process model that might violate any existing coordination constraints. In such a case, transformation operations that keep the modified model semantically equivalent with the original model can be applied. A transformed equivalent model may also allow improved ability to easily introduce and reason about other changes in the process model.

It is important to point out here transformation changes in the process model result in addition or deletion of modelling objects from the process models. We are not considering changes in graph layouts here. There are several graph-drawing algorithms available in the literature that can improve the visual layout of graphs. The graph layout algorithms just improve the visual representation of a process graph without making any semantic changes to the underlying modelling structures.

Another useful application of process transformations is in gradual development of a process model. Let us suppose that a process designer captures the necessary and required process tasks and associated coordination constraints in the form of a higher-level process model. This process model then may be used as a basis to define other specialized process models that would add additional tasks and coordination constraints without violating the constraints specified in the original model. In such a case, we would get several versions of the original process model with additional process modelling objects and structures.

Process transformations may also be used to effectively manage dynamic modification of process models at run-time. In workflow environments supporting dynamic changes, a process model template is used as a basis for creating new instances of business processes. Each of such process model instances could be modified at run-time to handle specialized requirements of the instance. A framework to introducing such changes through process transformations allows process designers to correctly understand the impact of their change on overall execution of the process model.

Rest of the paper is organized as follows. In section 2, we present a simple and generic process modelling language to build process models. This modelling language will be used to present three structural relationships and associated transformation operations in section 3 and 4. An algebraic notation for transformation principles will be introduced in section 5 to provide a means for formal considerations. Section 6 concludes the paper and introduces a process-modelling tool that has been developed to implement some of the ideas presented in this paper.

2 Process Modelling Structures

A process model contains tasks and associated coordination constraints to control the scheduling and execution of defined tasks. The tasks in a process model are performed to achieve some business objectives. Generally, these tasks are inter-related in such a way that the initiation of one task is dependent on the successful completion of a set of other tasks. Therefore, the order in which tasks are executed is very important. The structural / control flow modelling of a process defines the way a workflow management system would order and schedule workflow tasks.

In this section, we briefly describe the structural aspects of a typical workflow modelling language and its graphical representation. This language conforms to the generic workflow modelling concepts as described by Workflow Management Coalition [11]. The process models in this graphical language are modelled using two types of objects: node and transition. Node is classified into two subclasses: task and choice/merge coordinator. A *task*, graphically represented by a rectangle, represents the work to be done to achieve some objectives. It is also used to implicitly build sequence, fork, and synchronizer structures. It is the primary object in workflow specifications and could represent both automated and manual activities. A *choice/merge coordinator*, graphically represented by a circle, is used to build choice and merge structures. A transition links two nodes in the graph and is graphically represented by a directed edge. It shows the execution order and flow between its head and tail nodes. By connecting nodes with transitions through modelling structures, as shown in Figure 1, we build directed acyclic graphs (DAG) called workflow graphs where vertices represent nodes and directed edges represent transitions.

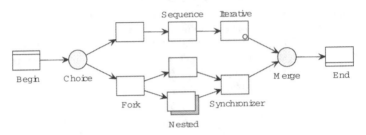

Fig. 1. Process modelling structures

Sequence is the most basic modelling structure and defines the ordering of task execution. It is constructed by connecting at most one incoming and one outgoing transition to a task. A *fork* structure is used to represent concurrent paths within a workflow graph and is modelled by connecting two or more outgoing transitions to a task. At certain points in workflows, it is essential to wait for the completion of more than one execution path to proceed further. A *synchronizer* structure, represented by attaching more than one incoming transition to a task, is applied to synchronize such concurrent paths. A task waits until all the incoming transitions have been triggered.

A *choice* structure is used to model mutually exclusive alternative paths and is constructed by attaching two or more transitions to a choice/merge coordinator object. At run-time, the workflow selects one of the alternative execution paths for a given instance of the business process by activating one of the transitions originating from the choice coordinator object. The choice structure is exclusive and complete. The exclusive characteristic ensures that only one of the alternative paths is selected. The completeness characteristic guarantees that, if a choice coordinator object is activated, one of its outgoing flows will always be triggered. A *merge* structure is "opposite" to the choice structure. It is applied to join mutually exclusive alternative paths into one path by attaching two or more incoming transitions to a merge coordinator object.

The fork and synchronizer structures are represented implicitly by directly connecting transitions to task objects. This approach keeps the resulting workflow model compact as well as graphically explicit. Nevertheless, in certain cases, it requires the use of *null* or dummy tasks to model proper coordination of flow and to conform to the syntactical correctness criteria of workflow structures.

Since a workflow model is represented by a directed acyclic graph (DAG), it has at least one node that has no incoming transitions (source) and at least one node that has no outgoing transitions (sink). We call these *begin* and *end* nodes respectively node. A workflow instance completes its execution after its end node has completed its execution. However, if a workflow graph contains more than one end node, then its instance completes its execution after executing a subset of these end nodes depending on whether choice or fork structures have been used in preceding paths of the end nodes. A bar at the top of a task or choice/merge coordinator represents a begin node. Similarly, a bar at the bottom represents an end node.

The *nesting* structure simplifies the workflow specifications through abstraction. Using this construct, we can encapsulate a workflow specification into a task and then use that nested task in other workflow specifications. For each execution of a nested task, the underlying workflow is executed. A nested task is graphically represented through a shaded rectangle under the task rectangle. The *iteration* structure is needed to model the repetition of a group of tasks within a workflow. One way to support iteration is through exit conditions. As long as a certain condition is not met, a particular task is repeatedly executed. The nesting structure could be used if there is a need to repeat a sub graph of the workflow model. This technique for iteration, however, can only support blocked iteration.

We have come across a variety of process modelling languages both in research papers and commercial products [12]. Most of the languages support the generic modelling structures introduced here. We have defined this simple language to provide a basis for introducing process transformation relationships and rules. However, the transformation principles are applicable to other forms of process modelling languages after appropriately modifying the transformation rules.

3   Structural Transformations

In this section, we will introduce concepts of three types of structural relationships – equivalent, imply, and subsume – that two process graphs G and G' may satisfy. These relationships are progressively relaxed and subsume the restrictive ones. This means if G and G' satisfy equivalent relationship, they also satisfy imply and subsume relationships. Similarly, if G and G' satisfy imply relationship then they also satisfy subsume relationship.

While defining relationships we will use the notion of *execution nodes*. The nodes in a process graph can be classified into two groups: execution nodes and coordination nodes. In the process modelling language described in the previous section, all tasks except null tasks represent execution nodes, i.e., the nodes that directly perform some business activity. Choice/merge nodes and null tasks represent coordination nodes that are used to explicitly capture the coordination constraints graphically.

3.1 Equivalent Relationship

A workflow graph G' is structurally equivalent to a graph G if sets of execution nodes in both G and G' are equal and each one of them preserves the structural / control flow constraints specified in the other. The structurally equivalent relationship is represented by notation $G' \leftrightarrow G$. Informally, if G and G' are structurally equivalent workflow graphs then they may be used interchangeably to accomplish identical functionality.

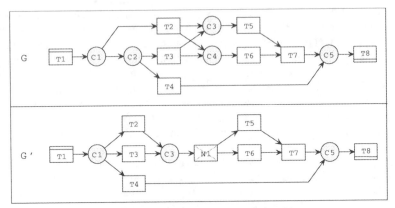

Fig. 2. Structurally equivalent relationship

In Figure 2, we show an example of two structurally equivalent workflow graphs G and G'. In G, after completing $T1$, a choice is made between $T2$, $T3$, and $T4$ through two choice structures $C1$ and $C2$. In G', identical functionality is achieved through a single choice construct $C1$. In G, completion of either $T2$ or $T3$ would trigger $T5$ and $T6$ in parallel. The graph G' achieves identical functionality through different modelling structures.

3.2 Imply Relationship

A workflow graph G' implies G if sets of execution nodes in both G and G' are equal and G' preserves the structural / control flow constraints specified in G. However, G may not satisfy all of the structural / control flow constraints specified in $G^{\cdot}$.

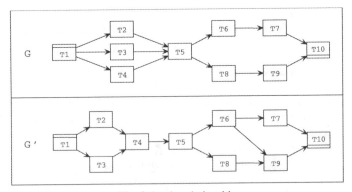

Fig. 3. Imply relationship

The imply relationship is represented by notation $G' \to G$. Informally, if a workflow graph G' implies G, then G' may be used to accomplish the functionality of G with less parallelism. However, the reverse does not hold. Additionally, if G and G' conform to equivalent relationship, then also conform to imply relationship. In graph G of Figure 3, $T2$, $T3$, and $T4$ may be performed in any order between $T1$ and $T5$. However, $T4$ in G' can only be performed after completing $T2$ and $T3$. That means,

G' is restrictive than G in specifying the order in which some of the tasks may be executed in parallel. Similarly, $T9$ in G' will be initiated only after both $T6$ and $T8$ have finished. In graph G, however, $T9$ does not have to wait for the completion of $T6$. This example shows that G' will perform all tasks of G without violating any control flow constraint specified in G. However, G' contains some additional control flow constraints that would result into less parallelism between workflow tasks.

3.3 Subsume Relationship

A workflow G' subsumes G if the set of execution nodes in G is a subset of the execution nodes in G' and G' preserves the structural / control flow constraints specified in G. However, G may not satisfy all of the structural / control flow constraints specified in G'. The subsume relationship is represented by notation $G' \supset G$. Again, informally, if a workflow graph G' subsumes G, then G' may be used to accomplish the functionality of G with less parallelism and additional tasks. However, the reverse does not hold. Additionally, if G and G' conform to either equivalent or imply relationship, then they also conform to subsume relationship.

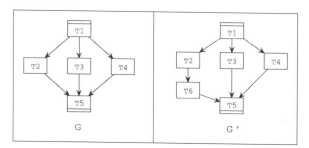

Fig. 4. Subsume relationship

Figure 4 shows an example of subsume relationship. In graph G, after $T2$, $T3$ and $T4$ are completed, $T5$ may be started. However, in G', $T6$ has to be completed after $T2$ as well before $T5$ can be performed. This means, G' is performing all the tasks in G without violating any control flow constraints and in addition is performing an additional task $T6$.

4   Transformation Operations

In this section, we will identify a set of transformation operations that can be applied to transform a workflow graph G into a workflow graph G' such that both G and G' conform to the three types of relationships introduced in previous section. Accordingly, three sets of transformation rules will be presented that maintain these relationships. The transformation rules are defined on the basis of following assumptions:

– Each execution node in the workflow graph is unique and it is not possible to identify if two distinct nodes perform exactly the same function.

- The workflow graph is free of errors and conforms to the correctness criteria. It does not contain deadlock and lack of synchronization structural conflicts [9].
- The transformation rules take into account only the semantics of adjacent modelling structures.

4.1 Structurally Equivalent Transformation

We will define a set of transformation rules that may be applied in any order to transform a graph *G* into *G'* such that both *G* and *G'* are structurally equivalent. It may seem that some of the transformation rules add or remove redundant modelling structures from workflow graphs. The addition or removal of such structures in workflow graphs is also used in applying the transformation rules for imply and subsume relationships.

SE1: Choice/merge child node of a single parent choice/merge node
A child choice/merge coordinator node of a single parent choice/merge node may be removed from the graph after moving outgoing transitions of the child choice/merge node to the parent choice/merge node. Inversely, a choice/merge node may be added as a child of another choice/merge node and a subset of the outgoing transitions of the parent choice/merge node may be moved as outgoing transitions of the newly added choice/merge node.

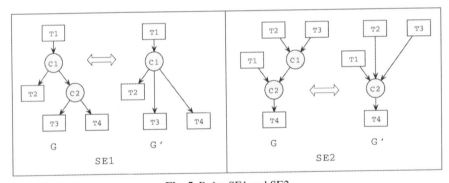

Fig. 5. Rules SE1 and SE2

In Figure 5 (SE1), *C2* has a single parent choice/merge node *C1*. By applying this transformation rule, we can move the two outgoing transitions from *C2* to *C1* and remove *C2* from the workflow graph. It is assumed that when this transformation is applied, the underlying transition conditions for the choice structure are also appropriately modified.

SE2: Choice/merge parent node of single child choice/merge node
A parent choice/merge node of a single child choice/merge node may be removed from the graph after moving incoming transitions of the parent choice/merge node to the child choice/merge node. Inversely, a choice/merge node may be added as a parent of another choice/merge node and a subset of the incoming transitions of the child

choice/merge node may be moved as incoming transitions of the newly added choice/merge node.

In Figure 5 (SE2), *C1* has a single child choice/merge node *C2*. By applying this transformation rule, we can move the two incoming transitions from *C1* to *C2* and remove *C1* from the graph.

SE3: Child null task node of a single parent task node

A child null task node of a single parent task node may be removed from the graph after moving outgoing transitions of the null task to the parent task. Inversely, a null task node may be added as a child of another task and a subset of the outgoing transitions of the parent task may be moved as outgoing transitions of the newly added null task.

In Figure 6 (SE3), *N1* is a null task that has a single parent task node *T1*. By applying this transformation rule, we can move the two outgoing transitions from *N1* to *T1* and remove *N1* from the graph. The use of a null task in following example may seem redundant. However, it could be used to clarify the design of a workflow model. This transformation rule is also quite useful when applied in conjunction with imply and subsume transformation operations that will be explained later.

SE4: Parent null task node of a single child task node

A parent null task node of a single child task node may be removed from the graph after moving incoming transitions of the null task to the child task. Inversely, a null task node may be added as a parent of another task and a subset of the incoming transitions of the child task may be moved as incoming transitions of the newly added null task.

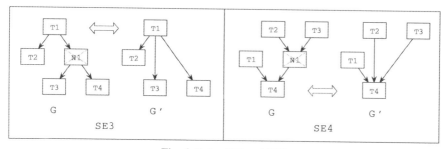

Fig. 6. Rules SE3 and SE4

In Figure 6 (SE4), *N1* is a null task that has a single child task node *T4*. By applying this transformation rule, we can move the two incoming transitions from *N1* to *T4* and remove *N1* from the graph.

SE5: Overlapping synchronizer and fork structures

Overlapping implicit synchronizer and fork structures are formed whenever two or more task nodes have identical sets of two or more parent task nodes. Such structures may be merged into explicit structures using a null task. Inversely, the null task may be removed to transform explicit synchronizer and fork structures into structurally equivalent implicit representation.

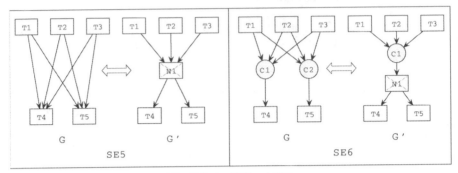

Fig. 7. Rules SE5 and SE6

In graph G of Figure 7 (SE5), $T4$ and $T5$ can be started in parallel as soon $T1$, $T2$, and $T3$ have been completed. This synchronization and fork behavior has been modelled implicitly in G. By applying SE5 transformation operation, we can introduce a null task $N1$ in G' that explicitly model the synchronization $T1$, $T2$, and $T3$ and parallel fork of $T4$ and $T5$.

SE6: Overlapping merge and fork structures

Overlapping implicit merge and fork structures are formed whenever two or more choice/merge nodes have identical sets of two or more parent task nodes. Such structures may be merged into explicit structures using a single explicit merge structure and a single explicit fork construct using a null task. Inversely, the null task and merge construct may be removed to transform explicit constructs into structurally equivalent implicit representation.

In graph G of Figure 7 (SE6), $C1$ and $C2$ are two merge nodes that have identical sets of parent task nodes $T1$, $T2$, and $T3$. By applying this transformation rule, we can remove $C2$, insert a null task $N1$ and move outgoing transitions of $C1$ and $C2$ to $N1$.

SE7: Redundant synchronizing transitions

A redundant synchronizing transition may be added between two workflow tasks only if it does not introduce a deadlock structural conflict. Inversely, a redundant synchronizing transition may be removed from the workflow graph.

In graph G of Figure 8, a transition is added between $T1$ and $T5$. This new transition adds a constraint that $T5$ cannot start until $T1$, $T3$, and $T4$ have finished. However, completion of $T3$ and $T4$ transitively ensures that $T1$ has also been completed. Therefore, a transition between $T1$ and $T5$ is redundant since it does not affect the execution order of the workflow tasks.

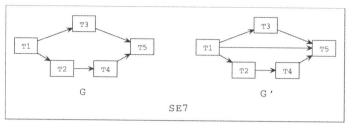

Fig. 8. Rule SE7

4.2 Imply Transformation

The transformation rules for structurally equivalent relationship can be applied since they maintain the imply relationship. In addition, the following transformation rule may also be applied.

IM1: Adding non-redundant synchronizing transitions
A non-redundant transition may be added between two workflow tasks only if it does not introduce a deadlock structural conflict. The addition of a non-redundant transition between two workflow tasks introduces some new restrictions on the possible parallel execution of one or more tasks in the workflow definition. It may also make some other transitions in the model redundant that may be removed by applying SE7.

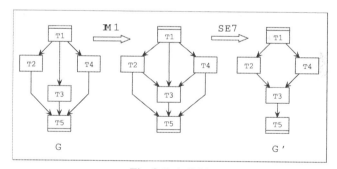

Fig. 9. Rule IM1

In graph *G* of Figure 9, IM1 is applied twice to add two new transitions, from *T2* to *T3* and from *T4* to *T3*. Then SE7 is applied to remove three transitions from *T1* to *T3*, *T2* to *T5* and *T4* to *T5* that have become redundant after applying IM1.

4.3 Subsume Transformation

The transformation rules for structurally equivalent relationship and imply relationship can be applied since they maintain the subsume relationship. In addition, the following transformation rule may also be applied.

SU1: Insert a subgraph in place of a transition
A transition in a workflow graph may be replaced by a workflow graph by adding the subgraph between the head and tail node of the transition. However, the added subgraph must conform to certain properties to ensure that it does not introduce errors in workflow graph and for all possible executions of the subgraph the tail node of the replaced transition is always triggered.

In graph *G* of Figure 10, SU1 is applied to replace the transition from *T3* to *T4* with a subgraph starting from *C1* and ending at *C2*. It is also applied to insert a task *T7* between *T2* and *T4*.

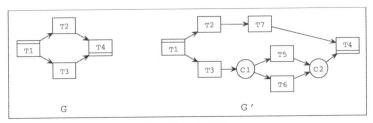

Fig. 10. Rule SU1

5   Formal Considerations

In this section, we will introduce an algebraic notation to represent process model structures. Using the notation, a process graph may be represented through a set of triggering constraints.

A triggering constraint (TC) is of the form $X \rightarrow Y$, implying that the completion of an instance of X triggers the start of an instance of Y. Here X and Y represent algebraic terms building upon workflow nodes and transitions.

Given a process graph, we can represent it through a set of simple triggering constraints of the form:

> *Simple triggering constraints:*
> Triggering Constraint: Node $\rightarrow$ Term | Term $\rightarrow$ Node
> Term: Node | (Node $\vee$ Node ...) | (Node $\wedge$ Node ...)

We can also transform a set of simple triggering constraints to a set of composite triggering constraints by removing non-executing nodes and merging associated triggering constraints. This transformation is performed using a set of triggering constraint transformation principles that will be presented later in this section.

> *Composite triggering constraints:*
> Triggering Constraint: Term $\rightarrow$ Term
> Term: Node | (Term $\vee$ Term ...) | (Term $\wedge$ Term ...)

Figure 11 shows a mapping of process modelling structures to simple triggering constraints. The notation used for triggering constraints is simple and self-explanatory.

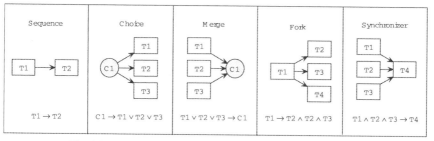

Fig. 11. Mapping modelling structures to triggering constraints

The underlying principles for transforming simple triggering constraints (STCs) to composite triggering constraints (CTCs) are shown in Table 1. In these principles, Rx represent triggering constraint terms, Cx represent choice / merge nodes, and Nx represent null task nodes. By applying these principles, we can reason about the transformation operations presented in section 4.

Table 1. Transformation principles for triggering constraints

$R1 \vee R2 \Leftrightarrow R2 \vee R1$	(P1)
$R1 \wedge R2 \Leftrightarrow R2 \wedge R1$	(P2)
$R1 \vee (R2 \vee R3) \Leftrightarrow R1 \vee R2 \vee R3 \Leftrightarrow (R1 \vee R2) \vee R3$	(P3)
$R1 \wedge (R2 \wedge R3) \Leftrightarrow R1 \wedge R2 \wedge R3 \Leftrightarrow (R1 \wedge R2) \wedge R3$	(P4)
$R1 \rightarrow R3, R2 \rightarrow R3 \Leftrightarrow R1 \vee R2 \rightarrow R3$	(P5)
$R1 \rightarrow R2, R1 \rightarrow R3 \Leftrightarrow R1 \rightarrow R2 \wedge R3$	(P6)
$R1 \rightarrow C1, C1 \rightarrow R2 \Leftrightarrow R1 \rightarrow R2$	(P7)
$R1 \rightarrow N1, N1 \rightarrow R2 \Leftrightarrow R1 \rightarrow R2$	(P8)
$R1 \rightarrow C1, C1 \vee R2 \rightarrow R3 \Leftrightarrow R1 \vee R2 \rightarrow R3$	(P9)
$R1 \rightarrow N1, N1 \wedge R2 \rightarrow R3 \Leftrightarrow R1 \wedge R2 \rightarrow R3$	(P10)
$R1 \rightarrow C1 \vee R2, C1 \rightarrow R3 \Leftrightarrow R1 \rightarrow R2 \vee R3$	(P11)
$R1 \rightarrow N1 \wedge R2, N1 \rightarrow R3 \Leftrightarrow R1 \rightarrow R2 \wedge R3$	(P12)

6 Concluding Remarks

The main contribution of the paper is to identify three useful classes of structural relationships for transforming process models and to introduce a framework for systematically introducing changes in process models. We have presented three structural relationships – equivalent, imply, and subsume – and associated transformation operations that may be used to systematically changing process models. We have also introduced a simple algebraic notation to represent and reason about process model transformations.

The structural relationships and transformation principles have been introduced using a generic process modelling language conforming to the process definition concepts defined by Workflow Management Coalition. The language makes use of core modelling structures – sequence, fork, synchronizer, choice, merge, iteration, and nesting – to build structural / control flow specifications of process models.

The concepts introduced in this paper have been implemented in a workflow modelling and verification tool called FlowMake. The tool allows both constrained and unconstrained approaches for introducing changes in the process models. It provides workflow analysts and designers a well-defined framework to model and reason about various aspects of workflows. It has been designed to augment workflow products with enhanced modelling capabilities. More information about FlowMake is available at http://www.dstc.edu.au/Research/Projects/FlowMake/.

References

1. Sadiq W and Orlowska ME (1997). On Correctness Issues in Conceptual Modeling of Workflows. In Proceedings of the 5th European Conference on Information Systems (ECIS '97), Cork, Ireland, June 19-21, 1997.
2. Carlsen S (1997). Conceptual Modeling and Composition of Flexible Workflow Models. PhD Thesis. Department of Computer Science and Information Science, Norwegian University of Science and Technology, Norway, 1997.
3. Casati F, Ceri S, Pernici B and Pozzi G (1995). Conceptual Modeling of Workflows. In M.P. Papazoglou, editor, Proceedings of the 14th In-ternational Object-Oriented and Entity-Relationship Modeling Conference, volume 1021 of Lecture Notes in Computer Science, pages 341-354. Springer-Verlag.
4. Georgakopoulos D, Hornick M and Sheth A (1995) An Overview of Workflow Management: From Process Modeling to Workflow Automation Infrastructure. Journal on Distributed and Parallel Databases, 3(2):119-153.
5. Hofstede, AHM ter, Orlowska ME and Rajapakse J (1998). Verification Problems in Conceptual Workflow Specifications. Data & Knowledge Engineering, 24(3):239-256, January 1998.
6. Rajapakse J (1996). On Conceptual Workflow Specification and Verification. MSc Thesis. Department of Computer Science, The University of Queensland, Australia, 1996.
7. Reichert M and Dadam P (1997). ADEPTflex - Supporting Dynamic Changes of Workflow without loosing control. Journal of Intelligent Information Systems (JIIS), Special Issue on Workflow and Process Management.
8. Sadiq W and Orlowska ME (1999). On Capturing Process Requirements of Workflow Based Information Systems. In Proceedings of the 3rd International Conference on Business Information Systems (BIS '99), Poznan, Poland, April 14-16, 1999.
9. Sadiq W and Orlowska ME (1999). Applying Graph Reduction Techniques for Identifying Structural Conflicts in Process Models. In Proceedings of the 11th International Conference on Advanced Information Systems Engineering (CAiSE '99), Heidelberg, Germany, 14-18 June 1999. Lecture Notes in Computer Science 1626. pp. 195-209. Springer-Verlag.
10. Workflow Management Coalition (1996) The Workflow Management Coalition Specifications - Terminology and Glossary. Issue 2.0, Document Number WFMC-TC-1011.
11. Workflow Management Coalition (1998). Interface 1: Process Definition Interchange, Process Model, Document Number WfMC TC-1016-P.
12. Butler Report (1996). Workflow: Integrating the Enterprise. The Butler Group, 1996.
13. Sadiq S (2000). Handling Dynamic Schema Change in Process Models. In Proceedings of Australian Database Conference, Canberra, Australia. January, 2000.

Towards Use Case and Conceptual Models through Business Modeling[1]

J. García Molina, M. José Ortín, Begoña Moros, Joaquín Nicolás, and
Ambrosio Toval

Software Engineering Research Group[2], Departamento de Informática y Sistemas
Facultad de Informática, Universidad de Murcia
Campus de Espinardo. C.P. 30071. Murcia, Spain
{Jmolina, Mjortin, Bmoros, Jnicolas, Atoval}@um.es

Abstract. A guide to requirements modeling is presented in this paper, in which use cases and the conceptual model are directly obtained from a business modeling based on UML activity diagrams. After determining the business processes of the organization, and describing their workflows by means of activity diagrams, use cases are elicited and structured starting from the activities of each process, while the concepts of the conceptual model are obtained from the data that flow between activities. Furthermore, business rules are identified and included in a glossary, as part of the data and activities specification. One notable aspect of our proposal is that use case and conceptual modeling are performed at the same time, thus making the identification and specification of suitable use cases easier. Both use case and conceptual modeling belong to the requirements analysis phase, which is part of a complete process model on whose definition we are currently working. This process is being experimented in a medium-sized organism of a Regional Public Administration.

1 Introduction

Since UML [1] was adopted as the OMG standard language for modeling, a large number of UML-based process models for object-oriented (OO) development have been proposed. These approaches are usually use-case driven, and therefore capture the functional system requirements as use cases, which provide the foundation for the rest of the development process: iteration planning, analysis, design, and testing.

Nowadays, a lot of research concerning use cases can be found in the literature, and there is an agreement on their usual misunderstanding and on the lack of precise guides to properly organize them. In this sense, several approaches have been published (cf. [3, 7, 8]) dealing with issues such as use case granularity, the level of detail in which use cases should be described, and the suitability of creating a use case hierarchy.

Based on the OOram *three-model architecture* [13] and the IDEA method [2], we are working on the definition of a UML-based process model for application in the

---

[1] Partially supported by the CICYT (Science and Technology Joint Committee), Spanish Ministry of Education and Ministry of Industry, project MENHIR TIC97-0593-C05-02.
[2] Member of RENOIR (European Requirements Engineering Network of Excellence).

information systems domain. This process includes a business modeling phase, aimed at describing the business processes of the organization, and which then allows the elicitation of the system use cases and the conceptual model in a simple, straightforward way. Inspired by the process view of the enterprise model of the three-model architecture, we describe each business process by a UML activity diagram with swimlanes. Next, we identify the system use cases from the activities, and the *concepts* (domain classes) from data (the information objects that flow among the activities).

In this paper, we describe an approach to business modeling, and how it can serve as a basis for requirements analysis (conceptual and use cases models). This approach is currently being experimented within the framework of a project which aims to provide a specific process model, based on requirements, for the development of information systems with intensive use of data. The scope of the research is the Regional Information Systems and Telecommunications Office (RISTO), of the Ministry of Finance (MF) in the Regional Government of Murcia (RGM)– Spain.

This paper is structured in the following way: some issues related to use cases are presented in section 2; next, a summarized version of our approach can be found in section 3; section 4 deals with business modeling, and our proposal for its realization; the rules that govern the transition from the business model to the use case and the conceptual models are presented in section 5; and finally, in section 6 we present our conclusions.

2   Use Cases in Practice

Most of the process models currently proposed for UML are defined as *use-case driven*. A use case can be defined as a sequence of actions, including variations, that the system can execute and that produce an observable result which has some value for an actor that interacts with the system [1].

Although the success of use cases is usually justified by the simplicity and intuitiveness of the technique, several authors (cf. [3, 7, 8]) have pointed out difficulties in discovering and specifying useful use cases, and finding consensus about how to organize and manage them. These are the reasons why we believe that it is necessary to establish a set of principles to guide the identification, description and organization of use cases.

Some interesting discussions about use case management have been made by T. Korson and A. Cockburn. Korson [7] claims that requirements (and therefore use cases) have to be hierarchically organized, in order to be able to understand them, reason about them, refine them and use them to validate the developed products. He establishes a set of recommendations, including the following statements: i) each level of use cases should conform to a complete set of requirements, i.e. each level does not add any new requirements, but it refines those in the previous level; and ii) the use case hierarchy should not be the result of a functional decomposition and it should be developed in an iterative and incremental way.

Cockburn [3], on the other hand, uses the concept of *goal* to organize use cases hierarchically. He basically distinguishes *strategic goals* (the business processes of the organization) and *user goals* (the system functions). Strategic goals are traced to a set of user goals, and, likewise, a user goal can be subsequently decomposed into a set of user goals. Thus, the concept of *summary goal* arises, which corresponds to either a composite user goal, or to a strategic goal.

Last but not least, another important issue is the allocation of use case modeling in the process model. Use case modeling is usually conceived as a previous step to conceptual modeling. However, Korson [8] claims that it is not possible to create adequate and useful use cases (nor correctly implement them) without understanding the domain, and therefore, use case and conceptual modeling have to be two parallel activities.

Usually, use cases are intuitively elicited from the system specification, an then the entities of the conceptual model are elicited based on the use cases specification. In the following sections, we present an approach to obtain the use cases and the conceptual models from a business model in a systematic way. Inspired in the *OOram Three Model Architecture* [12,13], business modeling is performed by means of UML activity diagrams. After determining the business processes of the organization, and describing their workflows by means of activity diagrams, use cases are elicited and structured starting from the activities of each process, while the concepts of the conceptual model are obtained from the data that flow between activities. Furthermore, business rules are identified and included in a glossary, as part of the data and activities specification. One notable aspect of our proposal is that use case and conceptual modeling are performed at the same time, thus making the identification and specification of suitable use cases easier.

3 Our Proposal in Brief

Our approach can be summarized in the following steps, which are not performed sequentially, but in an iterative and, at times, concurrent way:

1. The identifying and delimiting of the *business processes* within the organization under study, according to the enterprise strategic *goals*. A *business use case* is defined for each business process, and a *business use case diagram* is used to show the context and the boundary of the enterprise.
2. The discovery of the *roles* involved in the business processes, and their description in a *role model* that describes the interactions among the workers in the enterprise during the execution of a business use case. These interactions are represented by UML interaction diagrams (behavioral aspect) and a stereotyped class diagram (structural aspect).
3. The modeling of the *workflow* of every business process by means of activity diagrams, thus showing the interaction among roles to achieve the goal. The business rules that constrain the business processes are also elicited.
4. The extraction of the *system use cases* from the activities making up a business use case (which are included in the corresponding activity diagram).
5. The establishing of the *conceptual model* from the data (*information objects*) in the activity diagrams.

All the elements created during the modeling are specified in a glossary. The overall process outlined before is shown in Fig. 1.

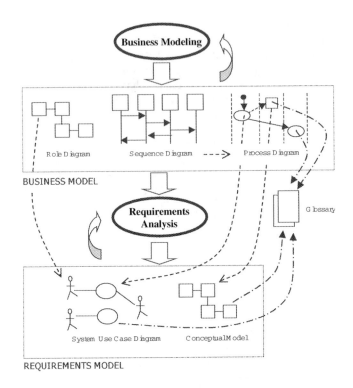

Fig. 1. Traceability relationships between Business and Requirements models

Moreover, use case are organized in two levels: firstly, each business process is associated with a *business use case*, which maps to the Cockburn's strategic goals; secondly, from these business use cases, a collection of *system use cases* is defined, once the activities involved in each business process have been considered.

4 Business Modeling

To achieve its goals, an enterprise organizes its activity through a set of business processes. Each business process is characterized by a collection of data which are produced and manipulated by means of a collection of *tasks* in which certain *agents* (for instance, workers or departments) participate according to a *workflow*. In addition, these business processes are constrained by *business rules*, which determine the policies and structure of the information of the enterprise.

The purpose of business modeling is to describe every business process, specifying the corresponding data, activities (tasks), roles (agents) and business rules. At this stage, our purpose is to understand the activity of the organization related to the system to build, considering "what" the system is supposed to do, instead of "how" it will support its goals.

The first step of business modeling is to capture the business processes of the organization under study. The elicitation of an adequate set of business processes is a crucial task, since it establishes the boundaries of the later modeling process. Following the concept of strategic goal given by Cockburn [3], we capture the business processes from the main goals of the enterprise. Firstly, we consider the strategic goals of the organization. Since these objectives are extremely complex, they are decomposed into a set of a few subgoals, which are more specific and which have to be accomplished to achieve the strategic goal. These subgoals can be subsequently divided into some more subgoals, and therefore a hierarchy of goals arises. In our research, we have experienced that two (or a maximum of three) levels of decomposition are enough. For every one of those subgoals we define a business process, whose purpose is to achieve that goal. We represent every business process as a business use case, which is initially specified by using a textual description.

We will use as running example the case study of a company that manufactures products by demand (following a *just in time* scheme). The strategic goals of that company might include *Satisfy a customer order, Increment sales by 25%*, or *Improve the manufacturing time by 15%*. Thus, the goal *Satisfy an order* can be divided in the following subgoals: *Register order, Manufacture product, Stock management* and *Generate orders to providers*. These are the subgoals that we use to discover the business processes.

4.1   Role Identification in the Business Context

Once business use cases are identified, we must discover the agents that are involved in their realization. Every agent plays a certain role when it collaborates with other agents to carry out the activities making up that business use case. In fact, we identify roles which are played by enterprise agents (including workers, departments, and devices) or external agents (as customers or other systems). For the moment, we only pay attention to those roles with which the organization interacts to carry out its business use cases. In our example, we have two roles which are clearly external to the organization: *Customer* and *Provider*.

To have a general view of the collection of business processes of the organization, we can create a business use case diagram, in which every business process is represented as a use case (see Fig. 2).

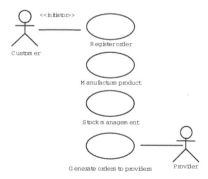

Fig. 2. Business use case diagram for the *Just in Time Manufacturing System*

The business use case diagram allows us to show the boundary and environment of the organization under study. This is the reason why only the business actors that correspond to external roles are shown in this diagram. In this way, the business use cases only involving roles which are internal to the organization are not connected to any actor. The business use case diagram shown in Fig. 2 is a UML use case diagram that consists of business use cases and actors. The diagram also specifies that the agent *Customer* initiates the realization of its related use case, while *Provider* is an actor that just participates in the associated use case.

4.2   Describing the Business Use Cases

The following step consists of describing in detail every business use case previously identified. We will focus on one of the business use cases of our example, namely *Register Order*, whose description is shown in Fig. 3. This description can be easily validated by the users.

1.   A customer submits an order, which has to include the order date, the customer data and the desired products. A clerk of the sales department might also introduce the order on request of a customer who has placed their order by phone, or has sent it by fax or ordinary mail to the sales department of the company.
2.   The clerk revises the order (and completes it, if necessary), and begins its processing by sending it to the catalog manager, who is in charge of its analysis.
3.   The catalog manager analyses the viability of each product of the order separately:
   • if the ordered product is in the catalog, its manufacturing is accepted.
   • otherwise, it is considered as a *special product*, and the catalog manager studies its manufacturing:
      - if it is viable, the manufacturing of the special product is accepted;
      - if it is not viable, the product is not going to be manufactured.
4.   Once the whole order has been studied, the catalog manager...
   • informs the sales department if every ordered product is accepted or rejected;
   • in the case that all the products of an order have been accepted, a work order for every product is created, starting from a manufacturing template (the standard one, if the product was in the catalog, or a new one, specifically designed for the product, if it was not present in the catalog). Every work order is sent to the manufacturing manager, and its launching is considered pending.
5.   The clerk informs the customer about the final result of the analysis of his or her order.

Fig. 3. Description of the *Register Order* business use case

Now we have to determine the internal agents that play a role in each business use case. We have identified the roles that belong to the business environment and now, we have to study the description of each business use case, and observe the complete set of involved roles, both external as well as internal to the organization. For instance, the roles in the business use case example are *Customer, Clerk, Catalog manager* and *Manufacturing manager* (where the last three are internal to the system).

The static (or structural) aspect of the collaboration among the roles to perform the business process, can be represented in a *role diagram*, in which each role (a stereotyped UML class) appears linked to the roles with which it can collaborate (see Fig. 4). Therefore, this diagram allows us to express the knowledge that some roles have about the others, as well as the characteristics of the relationships between roles (such as multiplicity). In addition, this diagram can serve to define some characteristics of the identified roles, such as their attributes and responsibilities. Ortín and García Molina [11] discuss role modeling in UML in more detail.

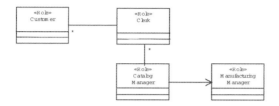

Fig. 4. Role diagram for the *Register Order* business process

Next, we create *scenarios* to show the behavioral aspect of the role collaboration. Here, we use UML sequence diagrams (see Fig. 5), in which the objects denote the instances of the roles that participate in the interaction.

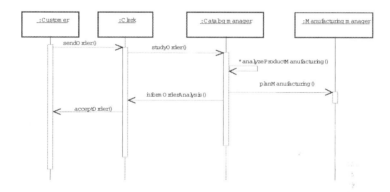

Fig. 5. Sequence diagram for the *Register Order* business use case

In every business use case we must distinguish the basic path of the interaction (in our example, request of an order that is finally accepted), and the existing alternative paths (for instance, canceling or rejecting an order). To improve legibility, it is convenient to associate several scenarios to the same business use case, instead of including all the possibilities in the same sequence.

The OOram three-model architecture [13] includes a business model represented through *process views* based on the standard IDEF0 [5], showing the workflow performed to obtain some goal of the organization, indicating the roles that are in charge of each activity, and the data required and produced by each activity. We consider these kind of diagrams very suitable for model business use cases, since they are very expressive and simple, thus facilitating discussions with users. These diagrams can be easily adapted to UML, by using activity diagrams with swimlanes. Thus, we use this type of UML diagram, which we have called *process diagrams*, to show the workflow that makes up the business use case in more detail.

A process diagram that includes the scenario of Fig. 5 is shown in Fig. 6. There exists a swimlane for every role participating in the scenario, including the activities performed by that role. The diagram also shows the data needed and produced by

288   J. García Molina et al.

each activity, as well as the synchronization required between different activities. Data appear as objects that flow between activities and can have a state. For instance, the activity *Pass on* order receives a proposed order and initiates its revision (see Fig. 6). We refer to these objects as *information objects*.

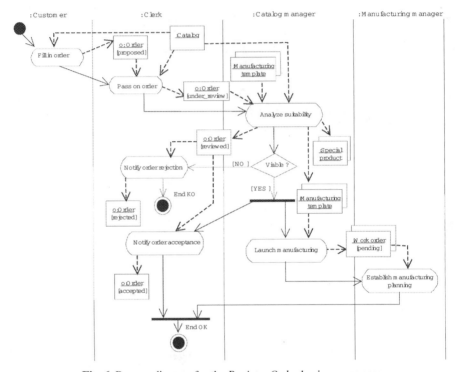

Fig. 6. Process diagram for the *Register Order* business use case

During the description of a business use case by means of a process diagram, it is possible to find out an activity which is complex enough to be described in another activity diagram. Thus, this new activity diagram will describe a subgoal in relation to the goal related to the original business process. In this fashion, business processes can be hierarchically organized.

4.3   Business Rules Specification

In an organization, both business processes and the data that they manage are constrained by business rules. As Whitenack [14] states, business rules are seldom explicitly captured during product development, regardless of the fact that they are often important constraints on system behavior. Because there is no well-defined framework in which to plug rules, and because there is a variety of rules types that are not well understood, rules are often ignored until the implementation phase.

With the aim of understanding the various types of rules that we have to take into account in a requirement specification, we use the taxonomy described by Odell [9]. This classification is simple but comprehensive and it covers every kind of business rule. Business rules are divided into two categories: constraint and derivation rules.

- *Constraint rules* specify policies or conditions that constrain the structure and behavior of the objects. Moreover, these rules can be subdivided into *stimulus-response rules* (they constrain the behavior and specify the conditions that must hold to activate an operation), *operation constraints rules* (they specify conditions that must be true before and after an operation is performed) and *structural rules* (they specify constraints about object types and associations, and these rules must always hold true).
- *Derivation rules* specify policies and conditions to infer or calculate facts from other facts in the business.

According to this classification, we explicitly collect each type of rule in the business model by means of the specification of the activities and information objects shown in the process diagrams. These specifications are gathered in a glossary.

Each information object is described by a set of attributes and their integrity constraints (if any exist). Therefore, we explicitly state structural and derivation rules. On the other hand, the semantics of each activity is described by its *source* (that is, the previous activities), *agent* (who is the responsible for doing the activity), *pre* and *post conditions* (stating what has to hold before and after the activity). The latter establish the operation rules, whereas stimulus-response rules are represented in the source part, where we express the order between the activities. Fig. 7 shows how the information object *Order* and the activities *Launch Manufacturing* and *Notify Order Acceptance* could be specified.

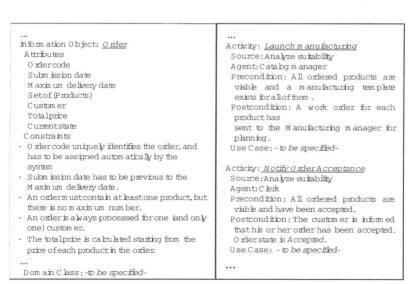

Fig. 7. Glossary: information objects (left) and activities (right)

The glossary will have a hypertext (cross-references) structure, in order to maintain the traceability relationships from the business processes to the classes and use cases that specify the functionality of the system. During the development life cycle, suitable links will be established. For example, each activity will be connected to the system use case where it is performed, and each information object will be linked to its related domain class, as we will see in the next section.

5   Requirements Analysis: Use Case and Conceptual Models

Starting from the business model described in the previous section, it is possible to obtain both the initial collection of system use cases and the earliest conceptual model in a systematic and straightforward way. Next, we are going to describe separately how to obtain each model.

The requirements which are elicited and specified in this phase will be included in a Software Requirements Specification (SRS) document. We recommend the use of a SRS standard template, such as the IEEE 830-1998.

5.1   Transition to the Initial System Use Case Model

We believe that the activities in a process diagram have the appropriate level of granularity to be associated to a single system use case. In this manner, we create a use case for each activity of a process diagram that will be supported by the software system. Thus, the role performing the activity will be the primary actor of the use case. Note that, according to the use case definition, not all the activities in a process diagram will be considered as use cases, but only those which have some value for an actor.

For instance, consider that the *Customer* role could not fill in the order on their own (through a web form, for example). Then, he or she would have to send all the data by fax or by telephone or some other way, as the result of the activity *Fill in Order*. As this activity would be performed outside the software system, neither the *Customer* role (since he or she will not interact with the software system) nor the activity *Fill in Order* (see Fig. 6) would be created in the system use case diagram. Fig. 8 shows the *system use cases diagram* for the business process *Register order*, whose process diagram was illustrated in Fig. 6, with the consideration that all the activities will be supported by software.

The use case diagram shown in Fig. 8 contains the architecturally most important use cases. We have to remark that some use cases could not be directly obtained from the process diagrams, but would be detected when the elicited use cases were described, thus gaining more knowledge about the requirements to be supported. These new use cases represent functions that the system must perform in order to achieve the goal related to some existing system use case. For instance, in our running example, to *Analyze suitability* it is necessary to look up in the products catalog whether an ordered product exists and so this catalog must be up to date. Thus, we would have to add the use case *Maintain product catalog*. Another example of a new use case could be *Maintain manufacturing templates*.

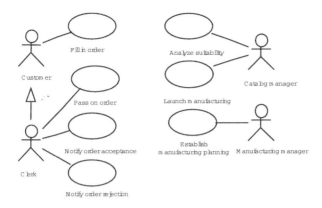

Fig. 8. Initial system use cases diagram

Moreover, the use cases could be organized into levels (two o three levels as maximum) according to the hierarchical decomposition proposed in the business modeling.

Each use case will be described by means of a template which can be filled in starting from the description of the associated activity, which is specified in the glossary as we saw before. We have chosen the template proposed by Coleman [4] because it combines simplicity and completeness, as shown in Fig. 9.

Use Case	Launch manufacturing.
Description	Work orders for every ordered product will be created, and will be sent to the Manufacturing manager so as to be planned.
Actors	Catalog manager.
Assumptions	– All ordered products are viable. – There exist manufacturing templates for all of them.
Steps	1. REPEAT 1.1. Obtain a product from the order. 1.2. Look for the manufacturing template of this product. 1.3. Create the work order. 1.4. Store the work order with *pending* state.
Variations	—
Non-Functional	—
Issues	—

Fig. 9. *Launch manufacturing* use case description

Once we have described the use case, it will be linked to the related activity in the glossary, with the aim of keeping traceability between business use cases and system use cases.

Relationships between uses cases could also be found, such as *include*, if common aspects to various use cases are found, and *extend*, to express an optional or alternative path in a use case. Nevertheless, we agree with the recommendations about not overusing these relationships and not showing them in the use case diagrams.

In order to complete this phase, non-functional requirements should be stated. If they are related to a specific use case, they will be specified in the proposed use case template [4]. If they are global to the system, they will be gathered in a section of the chosen SRS.

5.2   Transition to the Initial Conceptual Model

The information objects that flow between the activities of a business use case represent domain data and therefore they are a good base to create the initial conceptual model. This conceptual model will include the concepts and their relationships, and will be represented by a UML class diagram, where concepts are represented by classes (domain clases).

Each information object in the glossary will become a concept. In the design phase, this concept will become a class if the software system is going to manipulate that information. From the specification of an information object in the process diagram, we will obtain the definition of the concept, that is, its attributes, relationships with other classes, and constraints. For example, from the specification of *Order* shown in Fig. 7, we could obtain i) the attributes *code, submissionDate, maximumDeliveryDate, totalPrice, state*, ii) the associations *Customer-Order* and *Order-Product*, and iii) the constraints that could be expressed textually or by means of OCL (*Object Constraint Language*) as {*maximumDeliveryDate > submissionDate*}.

Furthermore, it should be noticed that when a conceptual model evolves to a class diagram, responsibilities can be obtained from certain constraints already specified in the glossary. For example, the Order class could have responsibilities such as *getProducts, calculateTotalPrice, calculateMaximumDeliveryDate*, or *changeState*.

In the same way as we connected the activities with the use cases in the glossary, we will link each information object to the domain class that represents it in the system. The class diagram depicting the first conceptual model for our running example is shown in Fig. 10.

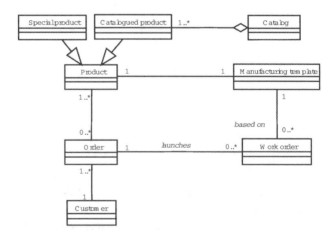

Fig. 10. Initial conceptual model from the *Register Order* business process

At this stage of the development, it is worth spending time on identifying the concepts rather than the relationships between them. We should concentrate on the *has to know* relationships. For example, from the glossary, we can state that an order *has to know* the related customer and the products being ordered (see Fig. 7).

Thus, some of the roles detected in the business model, and therefore specified in the role model, could be included as a class in the conceptual model. This is the case of the class *Customer* in our example.

Starting from the business model, it is also possible to identify some classes whose behavior depends on a rich set of reachable states. In this case, it could be of interest to define a state machine for them, represented by means of a UML *statechart diagram*. These classes are easily detected in the process diagrams, since they correspond to information objects labeled with several states. In our running example, *Order* would be a candidate for building a state machine that shows the states of an order (*proposed, under_review, reviewed, accepted* and *rejected*) and the events to change from one state to another.

6 Conclusions

In this paper, we have presented an approach for business modeling and requirements analysis, in which the use cases and the conceptual model are obtained in a straightforward using the business model as a starting point. Business modeling is centered on the use of UML activity diagrams.

With this guide, the modeler has available a systematic way to identify and organize use cases, and to identify and define the classes of the conceptual model. The business processes of the organization are identified from the goals proposed by Cockburn [3], and they are described by means of a flow of activities represented by a UML activity diagram. In this manner, system use cases are obtained from the activities of the business processes and they are organized into a hierarchy of levels, as proposed by Korson [7].

The classes of the conceptual model are obtained from the information objects flowing among the activities. We would like to highlight as an important feature of our approach that use cases are modeled at the same time that the conceptual modeling is done, in agreement with Korson [8], who states that this is crucial to get correct use cases, since understanding the domain is necessary to write useful use cases.

During the business and requirements modeling, the activities specification and the associated use cases, as well as the information objects and the corresponding domain classes, are gathered in a single glossary, which allows us to keep traceability between the different modeling artifacts.

In the Rational Unified Process (RUP) [6], defined by Rational for UML, business modeling is also included as a step within the iterations making up the process model. Jacobson et al. [6] present some steps that are similar to ours, but the hierarchical decomposition of the highest-level use cases is not considered, nor is a clear guide to discover the system use cases provided. Our approach to business modeling is a complete guide as opposed to the general sketch presented there.

References

1. Booch, G., Rumbaugh, J., Jacobson, I.: The Unified Modeling Language User Guide. Addison-Wesley (1999)
2. Ceri, S., Fraternalli, P.: Designing Database Applications With Objects and Rules. The IDEA Methodology. Addison-Wesley (1997)
3. Cockburn, A.: Using Goal-Based Use Cases" JOOP vol. 10 no.7. (Nov/Dec 1997) 56-62
4. Coleman, D.: A Use Case Template: Draft for discussion (1998)
   http://www.bredemeyer.com/use_case.pdf
5. Integration Definition for Function Modeling. Computer Systems Laboratory, National Institute of Standards and Technology, FIPS Pub. 183. (December 1993)
6. Jacobson, I., Booch, G. Rumbaugh, J.: The Unified Software Development Process. Addison-Wesley Longman, Inc. (1999)
7. Korson, T.: Misuse of Use Cases. (1998)
   http://software-architects.com/publications/korson/korson9803om.htm.
8. Korson, T.: Constructing Useful Use Cases (1999)
   http://software-architects.com/publications/korson/usecase3
9. Martin, J. Odell, J.J.: Object-Oriented Methods: A Foundation. Prentice Hall. (1997)
10. Ortín, M.J., García Molina, J., Martínez, A., Pellicer, A.: Combining OOram and IDEA for Information Systems Modeling. Technical Report TR-01-00. (December 1998)
11. Ortín, M.J., García Molina, J.: Role-Based Modeling with UML. IV Jornadas de Ingeniería del Software y Bases de Datos. Cáceres, Spain (1999)
12. Reenskaug, T.: Working with Objects: the OOram Software Engineering Method. Addison-Wesley / Manning Publications (1996)
13. Reenskaug, T.: Working with Objects: a Three-Model Architecture for the Analysis of Information Systems. JOOP vol. 10 no. 2 (May 1997) 22-30
14. Whitenack, B.: RAPPeL: A Requirements Analysis Process Pattern Language for Object-Oriented Development. In: Coplien, J.O., Schmidt, D.C. (eds.): Pattern Languages of Program Design. Addison-Wesley (1995) 259-291

A Conceptual Modeling Framework for Multi-agent Information Systems[1]

Ricardo M. Bastos [1] and José Palazzo M. de Oliveira [2]

[1] Faculdade de Informática, Pontifícia Universidade Católica do Rio Grande do Sul
Porto Alegre, RS, Brazil, e-mail: bastos@inf.pucrs.br

[2] Instituto de Informática, Universidade Federal do Rio Grande do Sul, Porto Alegre, RS,
Brasil, e-mail: palazzo@inf.ufrgs.br

Abstract. This paper presents a multi-agent conceptual model for resource allocation in a manufacturing environment. To attain this purpose a framework called M-DRAP – Multi-agent Dynamic Resources Allocation Planning – was developed. Multi-agent systems have been employed as a solution for problems that require decentralization and distribution in both decision-making and execution process. This is a premise in many information systems where (i) the domain involves intrinsic distribution of data, problem-solving capabilities and responsibilities; (ii) it is necessary to maintain the autonomy of the subparts, without lost of organizational structure; and (iii) the problem solution cannot be completely described *a priori* due to the possibility of real-time perturbations in the environment (equipment failures, for example) and also as a consequence of the natural dynamics of the business process. The main contribution of this work is the proposition of a set of activities and models defining a framework to represent multi-agent systems for business process under an enterprise model perspective.

1 Introduction

At the last two decades, the multi-agent approach has been developed in the artificial intelligence field. As a result of this development, it is possible to apply this paradigm to the information systems class that requires decentralization and distribution in its decision-making and execution process. According to Jennings, the application of agent technologies in information systems is justified by the following characteristics [Jennings 96]:

- The domain involves intrinsic distribution of data, problem-solving capabilities and responsibilities;
- It is necessary to maintain the autonomy of the subparts, without lost of organizational structure;
- The interactions are complex, including negotiation, information sharing and co-ordination;

---

[1] This work was partially supported by CNPq 523.883/96-0, 572021/97-6 and FAPERGS 99/0360.9 grants.

A.H.F. Laender, S.W. Liddle, V.C. Storey (Eds.): ER2000 Conference, LNCS 1920, pp. 295-308, 2000.

- The problem solution cannot be completely described *a priori*, due to the possibility of real-time perturbations in the environment (equipment failures, for example), and the natural dynamics of the business process.

Much work has being done applying multi-agent systems for information systems, especially at the production planning domain [Maturana 99] [Oliveira 97] [Cantamessa 97] [Baker 96] [Rabelo 95]. In these enterprise modeling works it was utilized informal methods to model the system. The object-oriented paradigm employed as reference results inadequate under a software-engineering point of view especially in regard of the conceptual model [Iglessias 98] [Jennings 00] [Taveter 99] [Wooldridge 99]. The object-oriented approach is not able to specify the autonomous solving-problem behavior, the structure of the agent's society organizational model and the aspects involved in the agent's interaction. In fact, the efforts in the direction of formal and rigorous approaches to model multi-agent systems are recent and consequently there is not a consensus or even consolidated methods and methodologies to model this kind of systems.

This paper examines the multi-agent paradigm under a *process conceptual modeling* point of view, considering the possibility to develop this kind of system integrated with an enterprise information system. For this purpose, we present a set of activities and models defining a framework to model multi-agent systems for business process. In section 2 the multi-agent paradigm is described and the main premises are addressed. At section 3 the activities and models that compose this framework are introduced in order to describe the modeling process and results. In section 4 we present some final considerations and conclusions.

2 The Multi-agent Paradigm

According Jennings an agent may be seen as being an entity with problem resolution capability encapsulated and with the following properties [Jennings 96]:
- Autonomy: the agents execute the majority of their activities to solve the problems without direct intervention, neither human nor from another agent. They have total control over their actions and internal status;
- Social ability: agents interact (when they find it appropriate) with other agents (artificial or human) so as to complete their problem resolutions or even to help others to execute their activities. This characteristic demands that agents have a minimum capability to communicate their requests to others and an internal mechanism to decide when and which interactions are appropriate;
- Reaction capability: agents perceive their environment and react in an opportune manner to the changes happening;
- Pro-active capability: agents do not simply act reacting to their environment but they demonstrate opportunism and behavior aimed at objectives as well as they take initiatives when it is adequate.

The agents must have autonomy and capacity to evaluate a requirement from another agent considering its objectives and local plans in order to define the convenience to attend that one. In the so-called multi-agent systems, the system's designer provides the agents with knowledge, decision procedures and interaction patterns in such a way that, when faced with a task, they are able to collectively organize themselves in order to perform that task efficiently. When regarded as a paradigm for software development, the main feature of multi-agent systems is the strong emphasis on autonomy and flexibility – important characteristics for environments subject to change.

An approach to model an industrial information system, independent of the paradigm (process-oriented, information-oriented or object-oriented), requires models that represent the static, functional and dynamic views. For the multi-agent paradigm it is also necessary to consider two levels: macro (societal) and micro (agent) [Wooldridge 99]. The macro level represents the society of agents where is defined the relationship among agents. The agent architecture and its behavior in the society are expressed in the micro level.

Several works review and develop methods and methodologies to multi-agent systems [Iglesias 97] [Taveter 99] [Wooldridge 99]. However, the conceptual modeling approaches available don't consider the integration of the agents with the information system database. This is a very important characteristic since the multi-agent systems must exploit the available information to achieve its objectives. In this sense, the macro level (multi-agent society) must be defined from the enterprise conceptual model. For each agent the behavior must be defined in order to implement the solution integrated with the information system as a whole model.

A natural way to model an information system is to apply an enterprise object-oriented model like CIMOSA– Computer Integrated Manufacturing Open Systems Architecture [Zelm 95] [Vernadat 94] and GERAM – Generalized Enterprise Reference Architecture and Methods [IFAC-IFIP 97]. This kind of model represents the reality of the domain problem under a structural and functional point of view. Each object of the model specifies a real component including its particular characteristics (attributes and operations) and its responsibilities at the enterprise information system. Therefore, the agents that compose the multi-agent model are related with these objects in order to maintain the consistence with the reality, although it would be necessary to define agents with attribution specifics for the multi-agent system. Some of the object classes from the enterprise object model will be used to specify the micro level (agents).

An agent class is defined as an entity that specify the agent behavior, attributions and architecture in order to create new agent instances at the society. The follow aspects must be considered in order to evaluate the potentiality of an object class to origin an agent class:

- Function: the application of the multi-agent paradigm in a information system establish decision decentralization and distributed planning of the activities that are executed in order to accomplish the objectives of the system as a whole. Thus, an object class could origin an agent class when its responsibilities, under a multi-agent perspective, will require decision-making capacity and autonomy to perform its attributions.
- Information: the attributes maintained by the objects define the knowledge of the system in terms of information. Consequently, this knowledge could be totally or partially relevant to the agents that forms the multi-agent system in order to

accomplish its objectives. Thus, all the object classes that maintain attributes required by the multi-agent solution are natural candidates to origin agent classes.

- Behavior: in the same way of the information knowledge, the behavior of the objects defined by its operations could be total or partially relevant for the solution based on the multi-agent paradigm. Thus, the operations could define an object class as a natural candidate to origin an agent class.

Based on this approach, some classes of objects that compose the enterprise model will be associated with agent classes in the multi-agent society. It means that for each new object instantiated of this class in the information system database must have a correspondent agent instance in the multi-agent society.

Although it is possible to establish a relation between the objects that belong to the enterprise model and the agents that compose the multi-agent society, in fact, the agents itself are entities which architecture and properties completely different. Moreover, the multi-agent society requires a different modeling approach in comparison with an object model due to the differences between both paradigms.

While an object is defined by its properties, attributes and operations, and its state is determined by its attribute values, an agent has architecture composed by a set of modules and its state [Shoham 93]. This architecture is determined by concepts like beliefs, desires and intentions. Each module implements a specialized function forming an integrated architecture that enables the agent to perform its attributions in the society.

When we specify a dynamic model of an object-oriented system, for each event received from an object there is precisely one possible reaction (defined by the operation that must be executed) to the receptor object. It means that the sender invokes a specific operation at the receptor. An agent has decision-making capacity and autonomy to decide perform or not a requirement issued by another agent according its local plans and goals [Jennings 00]. An agent does not know the internal operations of the another agent. Thus, a multi-agent society model only represent the communication channels required to exchange messages in order to permit the interaction among the agents.

3 A Conceptual Modeling Framework

Considering the limitations of the object-oriented methods to develop a conceptual modeling of multi-agent systems, we propose an approach with the next four activities: (i) Requirements Specification, (ii) Multi-Agent Solution Definition, (iii) Multi-Agent Society Model and (iv) Agent Model.

This approach was elaborated based on the modeling process to develop a multi-agent system called **M-DRAP** – Multi-agent Dynamic Resource Allocation Planning [Bastos 98, 98a, 99]. The M-DRAP defines an agent-based approach to dynamic resource allocation planning in production environments. The proposed architecture has a strong conceptual correspondence with the CIMOSA framework, representing the organizational structure of typical production systems in a consistent way. Thus, it allows organized derivation of models for specific environments.

At the sequence will be examined the activities that compose the conceptual modeling framework and illustrated its application through the M-DRAP system.

3.1 Requirements Specification

This is the first activity in any kind of software engineering methodology which objective is identifying and specifying the requirements of the system. For this purposes any kind of method like interview, observation, documentation exam may be employed. In fact, the main issue of this activity must be the knowledge of the domain problem considering its characteristics, constraints, procedures and expected results.

3.2 Multi-agent Solution Definition

Based on the requirements defined at the precedent activity, the next step comprehends the definition of a multi-agent solution for the domain. It involves the definition of a co-ordination mechanism for the multi-agent society and the specification of the solution. According Jennings, an adequate co-ordination mechanism (i) permits dependence between the agents actions, (ii) propitiates that the agents consider the global constraints in its decision-making process and planning and (iii) defines protocols for co-operation among the agents [Jennings 96].

Defined the multi-agent solution, it is possible to identify the agents required implementing the functionality of the system. At the M-DRAP system the resources allocation decisions are decentralized providing autonomy to the involved entities. In this sense, the system implements a distributed planning process in which every involved production entity is able to allocate itself to support the production requirements, considering the local and global constraints.

The present solution to the resources allocation problem through the multi-agents paradigm is builds on a *temporal planning strategy* based on PERT[2] networks. The temporal planning strategy proposed considers each production event as an independent project. For each project, a PERT network is build. Through the PERT Network, we can identify the relationships between the production activities that compose the production plan. These activities need production resources to be executed. Each activity must be planned considering its constraints and the availability of the necessary production resources. This whole planning must be done in such a way as to respect the production event conclusion's deadline.

Depending on the instant at which production event must be satisfied (deadline) it is possible to create a slack time for the critical path activities. To create this slack time it is necessary to anticipate the response to the production event. The slack time assures that the time between the beginning of the project and its end is greater than the sum of the execution times of the activities that compose the critical path. The agents at the production resources allocation process could use the slack time. The coordination among the agents is implemented through a market-oriented behavior [Welmann 95]. The market-oriented behavior defines the principles applied to the dynamic negotiation process among the agents. The main issue from the negotiation process concerns the production resources allocation to attend the production demand in a balanced way to the benefit of the global production planning.

A negotiation process is defined, which permit the agents to organize coalitions to participate in a bidding process. Both announcement and bid format are formally defined, and the bidding protocol is specified. The main objective of the agents is to

---

[2] PERT (Program Evaluation and Review Technique) [Hirschfeld 80].

attend the production demand in an economic, equilibrated and interactive way. Economic, as the bids must be defined using real production costs. Equilibrated, as the agent always tries to assign the production resource under its responsibility for activities that optimize its utility to the production system. Interactive, as to the allocation process is executed in real time and in a distributed way involving all the agents qualified to attend a production event.

In order to specify the structural and functional aspects involved in a production system in a consistent way we decided to adopt the CIMOSA as a reference model. This framework provides guidelines, architecture and a modeling language for enterprise representation involving the function, information, resource and organizational aspects.

Figure 1 exhibits a partial CIMOSA UML class diagram [Booch 98] representing the main classes of the enterprise model. Considering the CIMOSA, from a conceptual point of view, a Domain Process is functionally decomposed into Business Processes and Enterprise Activities. A Business Process represents an aggregation of Enterprise Activities. Both of them could be used in different Domain Processes. The Enterprise Activities involve one or more functional operations that are executed by the functional entities. Functional entities represent the production resources.

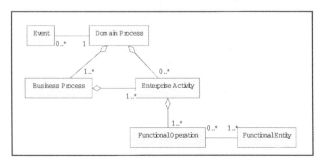

Fig. 1. CIMOSA Partial Class Diagram

The Domain Processes are directly triggered by the production events. Business Processes and Enterprise Activities compose a Domain Process defining a production route. This logical sequence of Business Processes and Enterprise Activities within a Domain Process is defined by a set of Procedural Rules.

3.3 Multi-agent Society Modeling

The objective of this activity is identifying the agents that compose the society and the relationships among them in order to permit the implementation of the solution defined at the precedent activity. The main result from this activity is a Multi-Agent Society Model defining the macro level. The Multi-Agent Society Model involves the Society Diagram and the Functions Model.

The Society Diagram

The Society Diagram is inspired at the UML Class Diagram [Booch 98]. This diagram represents the agents' classes and its communication channels define the hierarchical relationships. The hierarchical relationships characterize a possible authority relationship between the agents. The agents' classes are represented by rectangles where is defined the name and the maximal cardinality of instances. Arrows connecting the sender to the receiver represent messages between the agents.

Considering the aspects discussed at the section 2, the agents that compose the multi-agent society at the M-DRAP system corresponds to the classes of objects - processes and activities – defined at the CIMOSA model. The M-DRAP society diagram defines five agent classes (Figure 2): the Human Agents (HA), which represent the users involved in the interaction and four internal agents types. These agents are the Domain Process Agent (DPA), the Business Process Agents (BPA), the Enterprise Activity Agents (EAA) and the Functional Entity Agents (FEA). They are engaged in the planning process, and their ultimate goal is to satisfy the requests (represented by events) for those processes/activities. Each agent in the model plays a specific role; its autonomy is related to its position in the hierarchy. Thus, the architecture of each agent type reflects its role, according with the principle of dynamic decentralization.

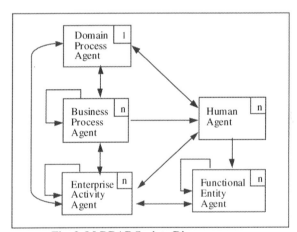

Fig. 2. M-DRAP Society Diagram

The internal agents are interrelated by a hierarchical structure, following the CIMOSA reference architecture. The hierarchy does not impose authority relations; instead, it defines restrictions for the agents' local plans. Each agent, when defining its local production plan, must consider (i) the constraints informed by the agent(s) immediately above it in the hierarchy, and (ii) the capability limitations and commitments of the agents immediately below it.

The resource allocation process is dynamically performed from the local resource allocation made by the agents. This allocation is accomplished based on the restrictions received from the upper levels. The agent's interaction is obtained using an extension of the contract net algorithm [Davis 88] and market-oriented behavior [Welmann 95]. There are two stages in the resource allocation process the temporal constraint definition and the bidding and contracting process. The first stage defines a pruning mechanism that selects the agents able to participate of the bidding process. The second stage specifies a selection process that identifies which agents will be contracted to execute the production plan.

The announcement elaboration process starts when the Domain Process Agent (DPA) receives a production event. The DPA identifies the corresponding domain process and selects the production plans adequate to this one. A production plan represents one of the possible production routes that can be applied to accomplish the domain process. The DPA sends announcements to BPAs and EAAs that compose each production plan requiring bids. These agents send announcements to leaves agents - a BPA send an announcement to the EAAs and the EAAs to the FEAs. The BPA and EAA bids are elaborated based on the FEAs bids.

The announcements are elaborated based in general and temporal constraints. The human agent defines the general constraints according to the production process characteristics, including activities sequence, event priority, quality level. The temporal constraint is defined according with the *temporal planning strategy* commented before.

From the market model view, the EAA responsible for the critical enterprise activity that makes use of the most constrained functional entity must be the first one to present a bid. This behavior is justified since the FEAs candidates to execute the functional operations required by the enterprise activity should use total or partially the slack time to elaborate its bid, having more flexibility to define its schedule. It means that just after the agent presents its bid, the EAA responsible for the next-to-last critical enterprise activity can elaborate and sent its bid, as it is necessary to know the slack time leftover.

Through this strategy, the dynamic re-planning of the FEAs activities as a response to eventual disturbances is feasible. This adaptable behavior is possible as the allocation process of the functional entity, executed by its associated FEA, is done considering the whole demands and the constraints from the announcement and not considering a particular production event. Consequently, the functional entity must organize itself to satisfy this demand, even if this implies the modification of a commitment already programmed considering that the constraints defined into the previous contracts will still be observed.

The Functions Model

The Functions Model defines for each agent its Role and Knowledge Base. The Agent Role establishes the agent Attributions at the society and the Constraints that must be observed in order to perform these attributions. The Attributions represent all the functions that the agent must be able to perform at the society in the sense to contribute for the achievement of its objectives and results. Each agent attribution is performed through the execution of the agent operations. In the same way, the object operations are implemented by the agents behavior, and can be described by any specification language.

The Constraints define aspects that must be observed by the agent in the execution of its attributions and in the access of the inherited agent knowledge. The constraints about the attributions specify the aspects that must be considered for the operations implementation, as for example, aspects related to the time (deadlines, execution time), rules, conventions, etc. The constraints about the inherited agent knowledge specify what kind of procedure the agent is qualified to perform at the related information system database (update the dates or just read that ones).

Two parts constitute the Agent Knowledge Base: inherited and specific. The first part involves both attributes and operations from the object that are relevant for the agent The second part comprehend the specific knowledge required by the agent in order to qualify it to perform its role at the society. In this activity, normally just the inherit knowledge is defined. The specific knowledge and the formation of the whole knowledge base just could be defined in the next activity where will be specified the agent architecture and its dynamic interaction model.

Considering that the attributes inherited from the objects are stored in its database, it is not necessary for the agent maintains this information in a particular database. We consider that the agent needs to access the information of the database information system just when the information is necessary to perform its attributions. However, the information that compose the specific agent knowledge base will only be accessed by the own agent, and consequently, it is stored in a particular database. Table 1 shows a simplified example of the Functions Model for the FEA.

Table 1. FEA Partial Functions Model

FUNCTIONS MODEL	
Agent: Functional Entity Agent	
Attributions Role	**Constraints**
• To elaborate and present a bid – individual or jointly with the agent coalition – in order to attend an announcement involving a functional operation which it is qualified; • To assume a commitment based on the contracts established with EAAs to execute an functional operation; • To treat disturbance events that affects its commitments.	• it is necessary to consider the announcement deadline; • the beginning and ending time of the commitment do not could coincide total or partially with the time interval of another commitment assumed before.
Inherited Knowledge Base	
Attributes	**Constraints**
• Functional Entity Identification • Functional Operations Qualified • Identification • Execution Time • Operation Cost per Time Unit • Set-up Time	• read • read • read, update • read • read, update
Operations	
• to calculate the operation execution time • to calculate the operation cost	

3.4 The Agent Model

The Agent Model defines the micro-level of the system representing the behavior of the agent through the definition of its interaction with another agents and its internal architecture. Three models compose the agent model: Agent Architecture, Agent Interaction and Agent Event-Response.

The Agent Architecture Model
The Agent Architecture Model defines the modules that constitute the agent structure with its relationships, responsibilities and properties. We consider that the agent model itself could be specified by an object-oriented approach where a class represents each module. Figure 3 presents the architecture of the FEA through a UML partial class diagram [Booch 98].

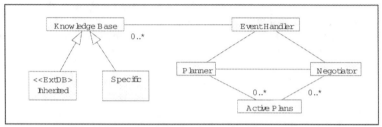

Fig. 3. Functional Entity Agent Architecture

In this activity, the specific knowledge base of the agent must be defined completing the Functions Model. The specific knowledge is defined by the set of the attributes and operations determined to each agent architecture class.

The Agent Interaction Model
The Agent Interaction Model is an extension of the UML Interaction Model [Booch 98], which permits to represent the message exchanged between the agents. For each agent class must be developed an Agent Interaction Model composed by one or more Agent Interaction Diagram representing the behavior of the agents according its attributions at the multi-agent society. Figures 4 and 5 present Agent Interactions Diagrams for the FEA. The horizontal dimension represent the agents and the vertical dimension the time.

The agents are represented by a rectangle divided into four slots: agent name, quantity of the agents involved at the interaction (1 or n), coalition name and agent role. In a multi-agent system, the agents could represent different roles according the circumstances involved at the interaction with another agents. In order to represent the situation the agent role at the Agent Interaction Diagram must be identified. Sometimes, the agents could form coalitions to perform the same activity jointly. This coalition is represented by a collection of agents that are responsible for co-operatively execute an activity in the same time interval. Eventually it could be necessary to identify an agent that executes a specific role among the agents' coalition (as for example, the coalition coordinator) in order to allow message exchange between them and to make feasible the activity execution. At the Figure 4 we have *n* FEA agents and one EAA involved at the interaction. The role of the EAA in this interaction is defined as a *Bidder*. At the Figure 5 an Agent Interactions Diagram for the FEA is represented involving the agents that form a coalition called *TaskForce*. There are two roles implicated: the *Coordinator* and the *Member* of the coalition.

The arrows represent the message exchanged between the agents. A bold arrow (Figure 4) represents a broadcast message, a message that the sender agent transmits for all the agents that belongs to an agent class. In this example, the EAA sends an announcement to all the FEA of the society. In answer, the EAA receive an individual message *Communication of habilitation* from the FEAs qualified to attend this announcement.

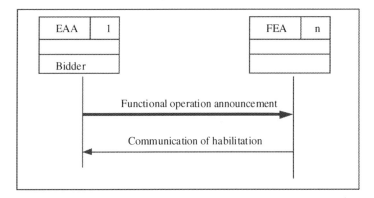

Fig. 4. Agent Interaction Diagram – Announcement Reception

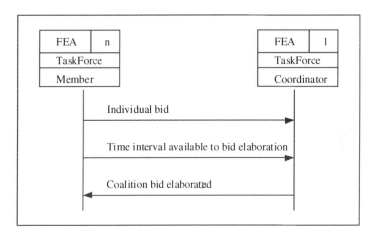

Fig. 5. Agent Interaction Diagram – Coalition Bid Elaboration

The Agent Event-Response Model

The Agent Event-Response Model is an extension of the Yourdon event-response model [Yourdon 96]. The Agent Event-Response Model identifies the possible responses of each event that has meaning for the agent according the coordination structure defined by the multi-agent system. In fact, in a multi-agent system an event could be considered as a result of an interaction between one or more agents. According Moulin, the interaction between agents could occur through explicit linguistic actions (communication) or through nonlinguistic actions (actions that modify the world which the agent is acting) [Moulin 96]. It is important to note that the agent reaction (response) depends of both its state and commitments at the moment of the event occurrence. At the Table 2 is presented an Event-Response

Model for the FEA involving the Agent Interaction Diagram presented at the Figure 5.

The Agent Event-Response Model and the Agent Interaction Model are complementary permitting the refinement of the agent model. In fact, the Agent Event-Response Model specifies in details the behavior of the agent during its interaction with another agents. Thus, an agent event-response model always is related with one or more Agent Interaction Model.

Table 2. Agent Event-Response Model

EVENT-RESPONSE DIAGRAM	
Agent: Functional Entity Agent	
Event	Announcement for Functional Operation
Responses	Communicate the intention to participate at the bidding process
Agent Interaction Diagram	Announcement Reception
Procedures Summary	• verify the capability to attend the required functional operation • elaborate the communication of candidature to participate at the bidding process and send to the announcer

Based on the Agent Event-Response Model the Agent Model could be refined, identifying new attributes and operations for the classes that compose the agent architecture. The attributes and operations of each class are completely defined to specify the final project of the agent. To model and refine the dynamic model of the agent architecture classes it may be employed the UML State Diagram [Booch 98].

4 Conclusion

In CIM systems, much work has being done in an empirical way from the CS point of view. A strong effort have being developed in Enterprise Modeling and Integration (EMI) standardization at the IFAC IFIP Task Force GERAM (Generalized Enterprise Reference Architecture and Methods), ISO (ISO IS 14258 Rules and Guidelines for Enterprise Models) as examples. The integration of these efforts with conceptual modeling is an immediate necessity. The central point of this paper is the definition of a conceptual model able to represents the complex industrial behavior. In this direction, the current work has presented a multi-agent conceptual model for resource allocation in a manufacturing environment which considers both temporal and synchronism aspects involved in the allocation process. In this model, the architecture supports the services encapsulation through the hierarchy of agents. One central aspect is the specification of a formal strategy for resource allocation planning supported through a multi-agent system that is able to reduce the complexity of the model considering real-time events that potentially affect the allocation. The architecture proposed for the system was developed considering the concepts defined in CIMOSA model.

The main contribution of this work is the proposition of a set of activities and models defining a framework to model multi-agent systems for business process under an enterprise model perspective. The approach considers the possibility to integrate the multi-agent system with the information system of the organization. For this purpose, we consider as a premise that the agents who compose the multi-agent model are related with the objects that compose the enterprise model in order to maintain the consistence with the reality.

References

[Baker 96] Baker, A.D. Metaphor or reality: a case study where agents bid with actual costs to schedule a factory, in Clearwater, S.C. (Ed.) *Market-Based Control: A Paradigm for Distributed Resource Allocation.* World Scientific, p.185-223, 1996.

[Bastos 98] O Planejamento de Alocação de Recursos Baseado em Sistemas Multiagentes, Porto Alegre. Doctoral dissertation in Portuguese, CPGCC/UFRGS, 1998.

[Bastos 98a] Bastos, R.M., Oliveira, J. Palazzo M. & Oliveira, F.M., Decentralised Resource Allocation Planning through Negotiation, in Camariha-Matos, L.M., Afsarmanesh, H. & Marik, V. (Eds.) *Intelligent Systems for Manufacturing: Multi-Agent Systems and Virtual Organization.* Kluwer Academic Publishers, p. 67-76,1998.

[Bastos 99] Bastos, R.M.; Oliveira, J.P.M.; Oliveira, F.M. Real-Time Resources Allocation Planning under a Market-Oriented Behaviour Perspective, in *Proceedings of the 15th ISPE/IEE International Conference on CAD/CAM, Robotics & Factories of the Future – CAR&FOF'99,* p. MW6-6–MW6-11, Águas de Lindóia-São Paulo, Brasil, 1999.

[Booch 98] Booch, G. Rumbaugh, J. & Jacobson, I. *The Unified Modeling Language User Guide.* Addison-Wesley Pub Co, 1998.

[Cantamessa 97] Cantamessa, M. Agent-based Modeling and Management of Manufacturing Systems. *Computers in Industry,* Amsterdam, v. 34, p. 173-186, 1997.

[Davis 88] Davis, R. & Smith, R.G. Negotiation as a Metaphor for Distributed Problem Solving, in Bond, A.H., Gasser, L. (Eds.), *Readings in Distributed Artificial Intelligence* Morgan Kaufmann, San Mateo, USA, 1988.

[Hirschfeld   80] Hirschfeld, Henrique *Planejamento com PERT-CPM e Análise do Desempenho,* Atlas, São Paulo, 1980.

[IFAC-IFIP 97] IFAC-IFIP Task Force. GERAM: Generalized Enterprise Reference Architecture and Methodology, Version 1.5, IFAC-IFIP Task Force on Architecture for Enterprise Integration, 1997.

[Iglesias 97] Analysis and Design of Multiagent Systems Using MAS-CommonKADS, in Muller, J.P, Singh, M.P. & Rao, A.S. (Eds.) *Intelligent Agents V,* Springer, p.317-330, 1998.

[Iglesias 98] A Survey of Agent-Oriented Methodologies, in Singh, M.P., Rao, A.S. & Wooldridge, M.J. (Eds.) *Intelligent Agents IV,* Springer, p.313-327, 1997.

[Jennings 96] Jennings, N.R. et ali. Using intelligent agents to manage business processes,. in *Proceedings of Practical Applications of Intelligent Agents and Multi-Agent Technology – PAAM'96,* London, UK, 1996.

[Jennings 00] Jennings, N. R. & Wooldridge, M Agent-Oriented Software Engineering, in Bradshaw, J. (Ed.) *Handbook of Agent Technology,* AAAI/MIT Press, 2000.

[Maturana 99] Maturana, F., SHEN, W. & NORRIE, D.H., MetaMorph: An Adaptive Agent-Based Architecture for Intelligent Manufacturing. *International Journal of Production Research,* 37(10), p.2159-2174, 1999.

[Moulin 96] Moulin, B.; Chaib-Draa, B. An overview of distributed artificial intelligence. In: O'hare, G.M.P.; Jennings, N.R. (Eds.). *Foundations of distributed artificial intelligence.* New York: John Wiley & Sons, p.3-55, 1996.

[Oliveira 97] Oliveira, E., Fonseca, J.M. & Steiger-Garção, A. MACIV: A DAI Based Resource Management System. *Applied Artificial Intelligence,* 11:525-550, 1997.

[Rabelo 95] Rabelo, R.; Camarinha-Matos, L.M. A Holistic Control Architecture Infrastructure for Dynamic Scheduling, in Kerr, R.; Szelke, E. (Eds.) *Artificial Intelligence in Reactive Scheduling*. Chapman & Hall, p. 78-94, 1995.

[Shoham 93] Shoham, Y. Agent-oriented programming. *Artificial Intelligence*, Amsterdam, v.60, p.51- 92, 1993.

[Taveter 99] Taveter, K. Business Rules' Approach to the Modelling, Design and Implementation of Agent-Oriented Information System, in *Proceedings of International Bi-Conference Workshop on Agent-Oriented Information Systems - AOIS'99*, Seattle, USA, 1999.

[Vernadat 94] Vernadat, F. Manufacturing systems modeling, specification and analysis, in *Proceedings of IFIP WG5.7 - Working Conference on Evaluation of Production Management Methods*, Gramado, Brasil, 1994.

[Wellman 96] Wellman,M.P. Market-oriented programming: some early lessons, in Clearwater, S.H. (Ed.) *Market-Based Control: A Paradigm for Distributed Resource Allocation,* World Scientific, 1996.

[Wooldridge 99] Wooldridge, M., Jennings, N. R., and & Kinny, D., A Methodology for Agent-Oriented Analysis and Design, in *Proceedings of. 3rd Int Conference on Autonomous Agents - Agents-99*, Seattle, USA, p.69-76, 1999.

[Yourdon 96]   Yourdon, E.; Argila, C. *Case Studies in Object-Oriented Analysis & Design*. USA:Yourdon Press, 1996.

[Zelm 95] Zelm, M. et ali. The CIMOSA business modelling process, *Computers in Industry*, **27**, p.123-142, 1995.

Object Role Modelling and XML-Schema

Linda Bird (nee Campbell)[1], Andrew Goodchild[1], and Terry Halpin[2]

[1]Distributed System Technology Center (DSTC),
Level 7, GP South, The University of Queensland, QLD, 4072, AUSTRALIA
{bird, andrewg}@dstc.edu.au
http://www.dstc.edu.au
[2]Microsoft Corporation, Seattle WA, USA
TerryHa@microsoft.com
http://www.orm.net

Abstract. XML is increasingly becoming the preferred method of encoding structured data for exchange over the Internet. XML-Schema, which is an emerging text-based schema definition language, promises to become the most popular method for describing these XML-documents. While text-based languages, such as XML-Schema, offer great advantages for data interchange on the Internet, graphical modelling languages are widely accepted as a more visually effective means of specifying and communicating data requirements for a human audience. With this in mind, this paper investigates the use of Object Role Modelling (ORM), a graphical, conceptual modelling technique, as a means for designing XML-Schemas. The primary benefit of using ORM is that it is much easier to get the model 'correct' by designing it in ORM first, rather than in XML. To facilitate this process we describe an algorithm that enables an XML-Schema file to be automatically generated from an ORM conceptual data model. Our approach aims to reduce data redundancy and increase the connectivity of the resulting XML instances.

1 Introduction

XML (eXtensible Markup Language) [1] is rapidly emerging as the premier encoding method for exchanging data in a portable fashion over the Internet. To date, the primary method for defining valid XML documents has been the 'Document Type Definition' (DTD). However, because DTDs were originally designed to describe semi-structured text-based documents, they have a number of limitations when it comes to describing the highly structured data commonly found in data-oriented applications. As a result, the World Wide Web Consortium (W3C) will soon be releasing a schema definition language – namely XML-Schema - designed specifically for the purpose of describing structured data. XML-Schema provides a richer set of data types, data constraints and data concepts than DTDs.

An important feature of XML-Schema is that it uses XML as the syntax for describing schemas. XML is text based and, while it is recognized as being "human-readable", a moderately sized XML-Schema can become difficult to understand. Graphical modelling languages are widely accepted as being a more visually effective means of specifying and communicating data requirements for a human audience. For

A.H.F. Laender, S.W. Liddle, V.C. Storey (Eds.): ER2000 Conference, LNCS 1920, pp. 309-322, 2000.

this reason, a number of companies have developed XML-Schema editors that will graphically present the main constructs of a schema to the user, including:

- XmlAuthority (from Extensibility);
- BizTalk Editor (from Microsoft); and
- Near and Far Designer (from OpenText).

All three of these editors rely on the tree-based nature of XML-Schema and present a graphical tree-like interface for editing XML schemas. Currently, the graphical languages used by these tools tend to lack an underlying methodology for constructing schemas, and make it difficult to represent some constraints that are often enforced on databases. Furthermore, the tree-like user interface forces schema designers to make decisions about the hierarchical structure of the schema too early in the modelling process.

With this in mind, this paper presents a new, conceptual approach to designing schemas for XML. In particular, we investigate the use of Object Role Modeling (ORM), a conceptual modeling method, as a means for designing XML-Schemas. By using ORM, we are able to model a rich variety of data constraints, and delay decisions about the tree-structure of the XML-Schema until after the conceptual analysis phase. Encoding an ORM schema in XML-Schema has benefits beyond facilitating the exchange of schemas between different CASE tools and repositories (in the way XMI[1] enables UML schemas to be exchanged). More importantly, the XML-Schema definition generated can be used to automatically validate the associated XML instance documents against the schema definition.

Section 2 of this paper gives a brief overview of the new XML-Schema language that is currently being developed by the W3C. Sections 3 and 4 describe ORM with an ongoing example and indicate how 'major object types' can be identified in an ORM model. These 'major object types' are then used in Section 5 to describe an algorithm for generating XML-Schema files from an ORM diagram. In Section 6 we enumerate some of the limitations of the mapping process, before concluding in Section 7.

2 XML-Schema

XML-Schema [2][3][4] is a new language being designed by the W3C to describe the content and structure of document types in XML. It serves the same purpose as the DTD language, but provides a more powerful method of describing and constraining the content of XML-documents. Although DTDs will continue to exist, XML-Schema should better meet the requirements of a wide-range of data-oriented applications that will use XML. In particular, XML-Schema provides the following features:

- **XML Syntax:** XML-Schema, unlike DTDs, uses an XML syntax, which means that existing XML-parsers can be used to build XML-Schema parsers.
- **Richer Data Typing:** DTDs provide only a primitive type system based on textual elements. XML-Schema extends this typing mechanism with an extensive range of

---

[1] http://www.oasis-open.org/cover/xmi.html

primitive types from SQL and Java, such as numeric, date/time, binary, boolean, URIs. Complex types can also be built from the composition of other types.

- **Support for Name Spaces:** XML-Schema is namespace-aware, enabling elements with the same name to be used in different contexts. Additionally, schema types and elements can be included (or imported) from a separate XML-schema using the same (or different) namespace.
- **Constraints:** XML-Schema provides an assortment of constraint types not supported by DTDs, including format-based 'pattern' constraints (e.g. "\d{3}[A-Z]{2}" represents three digits followed by two uppercase letters), key and uniqueness constraints, key references (foreign keys), enumerated types (value constraints), cardinality (or frequency) constraints and 'nullability'.
- **Other Features:** A number of other features, including anonymous type definitions, element content types, 'Any' elements and attributes, annotations, groupings and the use of derived types in instance documents, are also provided by XML-Schema.

3 ORM and XML-Schema

Object-Role Modelling (ORM)[2] [5][6] is a conceptual modelling approach that views the world in terms of objects, and the roles they play. Every elementary type of fact that occurs between object types in the Universe of Discourse (UoD) is verbalized and displayed on a conceptual schema diagram. ORM allows a wide variety of data constraints to be specified, including mandatory role, uniqueness, subset, exclusion, frequency and ring constraints.

Figure 1 shows an example ORM diagram that models a 'Conference Paper' UoD. In this diagram, object types are represented as named ellipses and relationship-types as named sequences of adjacent role boxes. Individual role names can also be used, but are omitted from this diagram for clarity. An arrowed-bar over a role or role sequence indicates an internal uniqueness constraint, and a circled 'U' or 'P' denotes an external uniqueness (or primary uniqueness) constraint. Value constraints are represented as a braced list of values, and frequency constraints as a numeric range attached to one or more roles. Subset constraints are shown as a dotted arrow, exclusion constraints as a circled 'x' between the relevant role-sequences, and subtype links as solid arrows between object types.

ORM was chosen for designing XML schemas for three main reasons. Firstly, its linguistic basis and role-based notation allows models to be easily validated with domain experts by natural verbalization and sample populations [7]. Secondly, its data modeling support is richer than other popular notations (Entity-Relationship (ER) or Unified Modeling Language (UML)), allowing more business rules to be captured [8]. Thirdly, its attribute-free models and queries are more stable than those of attribute-based approaches [9].

---

[2] http://www.orm.net

312   L. Bird, A. Goodchild, and T. Halpin

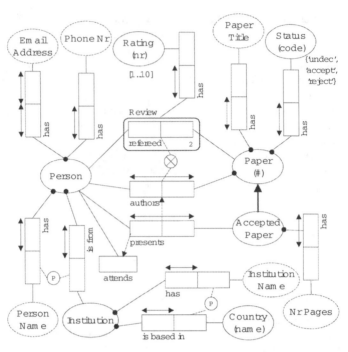

An 'Accepted Paper' is a 'Paper' that has a 'Status' of "accept".

Fig. 1. 'Conference Paper' ORM Schema

As XML schemas are hierachical, generating an XML-Schema definition from an ORM schema requires one or more object types to start the tree-hierarchy. One approach to this mapping problem could be to select a *single object type* as the XML root-node, and progressively define each ORM fact-type as sub-elements, producing an XML-instance such as:

```
<ConferencePaper>
   <Person name="Winnie the Pooh"> …
     <EmailAddress>pooh@hundwood.edu</EmailAddress>….
     <AuthoredPaper nr="27">
        <PaperTitle> A Macro-Economic Theory for Honey
              Distribution </PaperTitle> …
        <Status>undec</Status></AuthoredPaper></Person>
   <Person name="Eeyore"> …
     <EmailAddress> eeyore@hundwood.edu</EmailAddress>…
     <AuthoredPaper nr="27">
        <PaperTitle> A Macro-Economic Theory for Honey
              Distribution </PaperTitle> …
        <Status> undec </Status> </AuthoredPaper>
   </Person> </ConferencePaper>
```

However, as this example illustrates, this method leads to redundant data at the instance level. Here, the title and status of a paper are repeated for each Author of the paper.

Another approach would be to map *every object type* in the schema to a tag beneath the root node, and include its associated fact types as subelements, producing an XML-instance such as:

```
<ConferencePaper>
    <EmailAddress> eeyore@shadygrove.edu
        <Person name="Eeyore"> … </Person>
    </EmailAddress>
    <EmailAddress> pooh@shadygrove.edu
        <Person name="Winnie the Pooh"> … </Person>
    </EmailAddress>
    <Phone nr="+1-555-12348">
        <Person name="Eeyore"> … </Person>
    </Phone>
    <Phone nr= "+1-555-12345">
        <Person name="Winnie the Pooh"> … </Person>
    </Phone> …
</ConferencePaper>
```

However, as this example demonstrates, this approach leads to a difficult-to-read and disconnected XML instance (connected by an extensive list of 'key references').

In contrast to these two approaches, the approach presented in this paper minimizes redundancy in the XML-instance document, while retaining the connectivity of the XML data structures as much as possible. To achieve this, we use each of the 'most important' (or 'major') object types in the ORM model as a starting point for an XML hierarchy and associate each fact-type with exactly one of these hierarchies. To this end, we must first define the concept of a 'major object type'.

4 Major Object Types

The notion of a *major object type* is based on our previous work in [10] and [11], in which 'major object types' (or 'key concepts') were identified for abstraction purposes. It is also similar in idea to the process of mapping an ORM model into an Object-Oriented framework.

Intuitively, the 'major object types' are the 'most important' object types in a conceptual model. They are identified by selecting those object-types considered to be the 'most important participant' in some fact-type[3]. The 'importance' of a participant in a fact-type is determined by 'weighting' roles, based on the 'strength' with which they are 'anchored' to their player. The role with the highest weighting in a fact type is referred to as the *anchor* for that fact-type. This algorithm for weighting and anchoring fact types is summarized in the following twelve rules:

---

[3] A 'fact-type' is a relationship-type that is not part of the primary identification-scheme of any unnested object-type.

1. Any fact type role involved in a *non-implied mandatory role* constraint is weighted in inverse proportion to the number of roles participating in the constraint.
2. The player of the role in a *unary predicate* is 'the most important participant' in that predicate, and is weighted accordingly.
3. If only one role in a fact type is played by a 'non-leaf' object type, then this role is 'conceptually important' enough to be given a strong weighting.
4. If exactly one role within a fact type has the smallest maximum frequency[4] of that fact type, this role should be anchored.
5. If exactly one role in a fact type is played by a non-value type, then the fact type should be anchored on this role.
6. If exactly one role in a given fact type is played by an object type that became an anchor point via rules 1 to 5, the fact type is anchored on this role.
7. If a fact type is involved in exactly one single-role set constraint (ie. subset, equality or exclusion), and the role at the other end of the set constraint is anchored, then the constrained role in the given fact type should also be anchored.
8. If a fact type is involved in exactly one (possibly multi-role) set constraint and exactly one of the roles in the fact type is in the corresponding position within the set constraint as an anchored role, then this role is itself anchored.
9. If there exists a non-implied set constraint in which one of the roles involved in the constraint is the only involved role in its fact type to be played by an anchor point and the corresponding role's fact type in the other role sequence is not anchored, then this role becomes an anchor.
10. Those unanchored fact types, in which only one role is the 'join role' for some set constraint role sequence, should be anchored on this 'join role'.
11. The first role of each multi-role, non-implied set constraint becomes an anchor, if its fact type is not already anchored.
12. Any fact type not already anchored should be anchored on the first role involved in an internal uniqueness constraint.

After these 'anchoring' rules are applied, the major object-types are identified as those object types to which some fact type is anchored. Figure 2 shows the result of applying this anchoring algorithm to the ORM model in figure 1. The major object types are shaded, and the anchors are marked with thick arrow-tips.

Once this automatic anchoring procedure has been applied, it is suggested that the user be given the option to adjust the anchors as required. This allows additional human understanding of the UoD to impact on the final choice of 'major object types'. For more information and algorithm formalisms for automatically determining anchors and major object types, please refer to [10] and [11].

---

[4] 'Smallest maximum frequency' is calculated based on both uniqueness and frequency constraints.

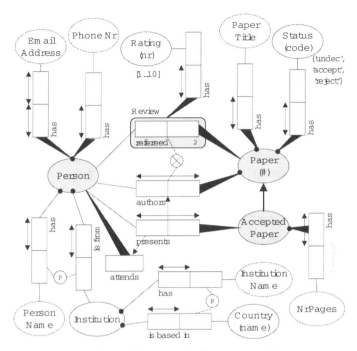

An 'Accepted Paper' is a 'Paper' that has a 'Status' of "accept".

Fig. 2. Anchored ORM Schema

5 ORM to XML-Schema Mapping

With the major object types of the conceptual schema identified, we can now describe our algorithm for generating an XML schema from an ORM diagram. The algorithm has three major steps:

5.1 **Step 1**: **Generate a Type Definition for Each ORM Object Type**

ORM value types (including those implied through reference modes) are represented in XML-Schema as simple types, and may include value or range constraints. For example, the value type "Email Address"[5] is mapped to:

```
<simpleType name="EmailAddress" base="string"/>
```

while the reference scheme "Rating (nr)" is mapped to:

```
<simpleType name="RatingNr" base="integer">
    <minInclusive value="1"/>
```

---

[5] ORM allows spaces in entity type names, but XML does not. To address this, we replace the spaces in entity type names with capitalisation.

```
<maxInclusive value="10"/> </simpleType>
```
and "StatusCode" is mapped to:
```
<simpleType name="StatusCode" base="string"/>
  <enumeration value="undec"/>
  <enumeration value="accept"/>
  <enumeration value="reject"/> </simpleType>
```

Entity types are mapped to complex types in XML-Schema, with the value types that form part of their primary identification scheme being represented as attributes[6], and the entity types that form part of their primary identification scheme being represented as sub-elements[7]. For example the entity type "Status" is mapped to:
```
<complexType name="Status">
  <attribute name="code" type="conf:StatusCode"
             minOccurs="1"/> </complexType>
```

while the entity type "Institution" is mapped to:
```
<complexType name="Institution">
  <attribute name="name" type="conf:InstitutionName"
             minOccurs="1"/>
  <element name="Country" type="conf:Country"/>
</complexType>
```

5.2 Step 2: Build a Complex Type Definition for Each Major Fact Type Grouping

As a general rule, each ORM fact type is mapped to a sub-element of the major object type to which it is anchored. For example, the fact-types anchored to "Person" map to the definition:
```
<complexType name = "PersonFacts" base= "conf:Person"
             derivedBy="extension">
  <element name="EmailAddress" type="conf:EmailAddress"
           minOccurs="0"/>
  <element name="Phone" type="conf:Phone"
           minOccurs="0"/>
  <attribute name="attends" type="boolean"
             minOccurs="1"/> </complexType>
```

In this example, the XML element names are based on the names of the associated entity types. The unary predicate "attends" is mapped to a boolean attribute, rather than to an element. Optional roles played by major object types have the constraint 'minOccurs="0"' and multi-role uniqueness keys have 'maxOccurs="unbounded"'.

While this approach (of using each major object type as the root of further subelements) produces a reasonable XML-Schema, in some cases the connectivity of the resulting XML-Schema can be improved by combining the fact types anchored

---

[6] In XML-Schema, attributes are, by default, assumed to be optional (minOccurs = 0) and functional (maxOccurs = 1). Only non-default occurrence constraints need be specified.

[7] In XML-Schema elements are, by default, assumed to be mandatory and functional.

Around two (or more) major object types—in particular, by nesting one major object type's fact types inside another's.

The method used to determine when (and in which direction) major object type groups may be nested, is based on the existence of a functional, mandatory role in the fact type connecting the major object types. While fact-type anchors are an effective method of identifying major object types, they should not be used to determine the direction of major object type nestings.

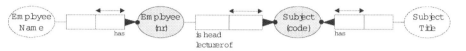

Fig. 3. 'Lecturer-Subject' ORM schema

To illustrate this, consider the 'Lecturer-Subject' schema shown in figure 3. If we were to combine the 'Employee' and 'Subject' fact types by nesting 'EmployeeFacts' inside 'SubjectFacts' (as indicated by the direction of the anchor), we would end up generating the following XML-Schema definition:

```
<complexType name="SubjectFacts" base="s:Subject"
            derivedBy="extension">
  <element name="Title" type="s:SubjectTitle"/>
  <element name="HeadLecturer">
   <complexType base="cp:Employee"
            derivedBy="extension">
     <element name="Name" type="cp:EmployeeName"/>
   </complexType> </element> </complexType>
```

An XML element ("Subject") based on this definition could have the following instances:

```
<Subject code="CS100">
   <Title> Intro to Programming </Title>
   <HeadLecturer empNr="5687">
      <Name>Helen March</Name></HeadLecturer></Subject>
<Subject code="CS210">
   <Title> Database Design</Title>
   <HeadLecturer empNr="5687">
      <Name>Helen March</Name></HeadLecturer></Subject>
```

There are two main problems with this mapping approach. Firstly, because each 'Employee' may be the head lecturer of more than one 'Subject' (as per the constraints in figure 3), the same set of 'Employee' facts may be associated with several different 'Subjects' (if they have the same head lecturer). This introduces redundancy, as evident in the example instance, in which both 'Subjects' include the fact that the Employee "5687" has the Name "Helen March".

The second main problem is that, based on the constraints in figure 3, not all 'Employees' are necessarily the head lecturer of a 'Subject' – and those Employees who are not a head lecturer can not be represented using the above XML-Schema.

318   L. Bird, A. Goodchild, and T. Halpin

Instead, there would need to be a separate list of 'Employees' who are not the head of any 'Subject', thus reducing the connectivity of the schema.

Instead of nesting major object types towards the anchors, as just shown, our approach is to nest the major object types away from mandatory, functional roles. Using our algorithm, the example in figure 3 maps to the XML-Schema definition:

```
<complexType name="EmployeeFacts" base= "s:Employee"
              derivedBy="extension">
  <element name="Name" type="s:EmployeeName"/>
  <element name="SubjectHeaded">
   <complexType base="s:Subject" derivedBy="extension">
     <element name="Title" type="s:SubjectTitle"/>
   </complexType> </element> </complexType>
```

It is possible to nest the major object type elements in this way because (a) the mandatory constraint requires each 'Subject' to be headed by *at least one* 'Employee', and (b) the uniqueness key requires each 'Subject' to be headed by *at most one* 'Employee'. The combination of these two constraints means that each 'Subject' must be headed by *exactly one* 'Employee'. Hence the 'Subject' fact types may be grouped with the 'Employee' fact types. To understand why these two constraints are so important in making this grouping possible, we consider each one in turn:

- *Uniqueness key*: If a 'Subject' could be headed by more than one 'Employee' (ie there was no functional uniqueness key), then nesting the 'Subject' facts inside 'Employee' would introduce redundancy into the schema. This is because the 'Subject Title' of a 'Subject' would be repeated every time that 'Subject' was headed by a different 'Employee'.
- *Mandatory constraint*: If a 'Subject' did not need to be headed by an 'Employee' (ie. no mandatory constraint), then nesting the 'Subject' facts inside 'Employee' would make it impossible to represent any 'Subject' not headed by an 'Employee'.

Therefore, when a single, mandatory, functional relationship type exists between two major object types, the fact types anchored to the object type on the functional, mandatory side can be nested inside the other.

A special case of this fact type grouping approach, is the nested fact type. In the ORM diagram in figure 2, each 'Review' object type has exactly one[8] 'Paper' being refereed and exactly one 'Person' refereeing it. This mandatory, functional relationship between 'Review' and both of its primary identifiers, makes it a candidate for the combining of major fact type groups. Since there are two mandatory functional roles involved (one on each 'implied' reference type[9]), we choose to combine the fact types towards the anchor of the nested fact type (ie. towards 'Paper'). Figure 4 shows the final fact type groupings.

---

[8] 'exactly one' means 'at least one' and 'at most one'.
[9] A reference type is an association that is part of the primary identification scheme of an object type.

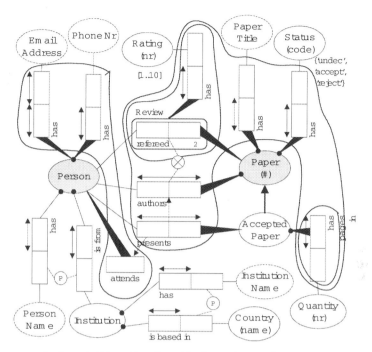

An 'Accepted Paper' is a 'Paper' that has a 'Status' of "accept".

Fig. 4. XML-Schema fact type groupings

The result of this combined fact type grouping on 'Paper' is the following XML-Schema definition:

```
<complexType name = "PaperFacts" base= "conf:Paper"
            derivedBy="extension">
  <element name="Title"  type="conf:PaperTitle"/>
  <element name="Status" type="conf:Status"/>
  <element name="Author" type="conf:Person"
          maxOccurs="unbounded" />
  <element name="Review" minOccurs="0" maxOccurs="2">
    <complexType>
       <element name="Referee" type="conf:Person" />
       <element name="Rating" type="conf:Rating"/>
    </complexType> </element>
  <element name="Presenter" type="conf: Person"
          minOccurs="0" maxOccurs="unbounded" />
</complexType>
```

As shown in the above example, 'Person' plays more than one role in the same fact type grouping —namely, the roles of 'referee', 'author' and 'presenter'. Therefore the associated XML elements are named using the role names, rather than the entity type names, to disambiguate the elements. Even when ambiguity does not arise, however, it is often preferable to name an element after the role rather than the entity type. For example, where appropriate, "BirthDate" would usually be a better element name than "Date".

Finally, subtypes are mapped to complex types that extend their supertype. For example:

```
<complexType name = "AcceptedPaperFacts" base=
        "conf:PaperFacts" derivedBy="extension">
    <element name="NrPages" type="conf:NrPages"/>
</complexType>
```

5.3 Step 3: Create a Root Element and Add Keys and Key References

Since each XML document must have a root node, we create a root node element to represent the whole conceptual model (in this case, called "Conference"). A subelement is then created, beneath the root node, for each major fact type grouping that was created in Step 2. Based on the example from figure 4, the resulting element definition generated is:

```
<element name = "Conference">
  <complexType>
    <element name="Person" type="PersonFacts"
            minOccurs="0" maxOccurs="unbounded"/>
    <element name="Paper" type="PaperFacts"
            minOccurs="0" maxOccurs="unbounded"/>
  </complexType> </element>
```

Finally, the primary identification scheme of each major object type is mapped to an XML-Schema "key". For example:

```
<key name="PaperKey">
  <selector>Conference/Paper</selector>
  <field>@paperNr</field> </key>
```

Multi-role uniqueness constraints and uniqueness constraints on non-anchored roles are mapped to XML-Schema "unique" constraints. For example:

```
<unique name="EmailUnique">
  <selector>Conference/Person</selector>
  <field>EmailAddress</field> </unique>
```

The fact types connecting major object type groupings are mapped to key references. For example:

```
<keyref name="AuthorPersonRef" refer="PersonKey">
<selector>Conference/Paper/Author</selector>
<field>@name</field> <field>Institution/@name</field>
<field>institution/Country/@name</field> </keyref>
```

6 Options and Limitations

When mapping ORM schemas to XML-Schemas, there are several different options available that have not been discussed so far. For example, all fact types anchored by a functional role, could be modelled as attributes rather than as sub-elements. Similarly, all direct primary identification schemes (simple or complex) could be represented as attributes. Alternatively, it may also be decided that introducing controlled redundancy into the schema is appropriate, or that decreasing the connectivity of the schema has some advantages. In the future, we would like to develop an approach to generating XML-Schemas from ORM that is configurable, so that modellers have greater control over the schemas they develop.

6.1 Limitations

While ORM and XML-Schema have many similar features, there are some features available in ORM that are not currently available in XML-Schema. For example:

- XML-Schema does not support exclusion constraints or subset constraints that target non-key elements;
- XML-Schema supports only a single inheritance model while ORM supports multiple inheritance;
- XML-Schema does not support disjunctive mandatory constraints;
- ORM subtype definitions cannot be fully represented in XML-Schema; and
- XML-Schema cannot represent some other ORM constraints (e.g. frequency or ring). For example, an optional role with a frequency constraint of exactly 2 (as in our Conference Paper example) cannot be fully represented in XML-Schema—the closest match to this is 'minOccurs="0" maxOccurs="2"'.

These issues could be addressed in a number of ways. One option would be to map constraint verbalisations to comments within the XML-Schema. While these comments would preserve the information in the original schema, they cannot be processed and used by an XML parser. Another approach would be to develop some non-standard extensions to XML-Schema to support the additional constraints. However, this would require that non-standard modifications be made to an XML-schema validator to support these extensions. XML-Schema also has a few features not supported by standard ORM. For example, XML-Schema supports format models (using 'patterns' such as "a field consists of two letters followed by three digits"). XML-Schema also allows mixed content models, which allow natural language text to be marked up with XML.

7 Conclusions and Future Work

This paper presented a method of mapping Object Role Models to XML-Schema. We believe that an ORM-based approach to designing XML-Schemas has advantages over current tree based XML-Schema editors for a number of reasons. Firstly, tree-based editors force designers to make decisions about the tree structure of a schema very early in the modelling process. Secondly, tree-based editors cannot graphically model many of the rich constraints available in ORM. Thirdly, ORM makes it easier to visualise, verbalise, populate and validate the model with the domain expert, thus making it easier to design a correct schema.

In developing the mapping algorithm, we discovered many ways to map an ORM schema to an XML-Schema. With this in mind, we distinguished our approach by aiming to minimize the data redundancy in the resulting XML-schema, while maximizing the connectivity of elements.

Future research plans include exploring alternative options for modelling n-ary and nested fact types, and additional configuration alternatives to the mapping process. We also plan to investigate the reverse procedure of generating an ORM schema from an XML-Schema. In particular, we wish to explore the notion of preserving the structure of an XML-Schema, thereby developing an ORM-to-XML-Schema editor that can map in both directions and produce the original schema. This will enable existing XML-Schemas to be edited with an ORM tool, while preserving the style and structure of the original schema.

References

1.   W3C.   Extensible   Markup   Language   (XML)   1.0.   Available   at:
     http://www.w3.org/TR/1998/REC-xml-19980210
2.   W3C. XML-Schema Part 0: Primer; Part 1: Structures; Part2: Datatypes. Available at:
     http://www.w3.org/TR/xmlschema-0/; - http://www.w3.org/TR/xmlschema-2/
3.   Halpin, T. 1998, 'Object-Role Modeling (ORM/NIAM)', *Handbook on Architectures of Information Systems*, Springer, Heidelberg, Ch. 4.
4.   Halpin, T. *Conceptual Schema & Relational Database Design*. 2nd edn, WytLytPub, 1999.
5.   Halpin, T. 1999, 'Fact-orientation before object-orientation: the case for data use cases', DataToKnowledge Newsletter, vol. 27, no. 6.
6.   Halpin, T. & Bloesch, A. 1999, 'Data modeling in UML and ORM: a comparison', Journal of Database Management, Idea group, Hershey.
7.   Bloesch, A. & Halpin, T. 1997, 'Conceptual queries using ConQuer-II', Proc. ER'97, Springer LNCS, no. 1331, pp. 113-26.
8.   Bird, L. Data Reverse Engineering: from a Relational Database System to a 3-Dimensional Conceptual Schema. Ph.D. Thesis, Department of Computer Science and Electrical Engineering, The University of Queensland. 1997.
9.   Campbell,L., Halpin,T., Proper,H., 'Conceptual Schemas with Abstractions: Making flat conceptual schemas more comprehensible', Data & Knowledge Eng. 20(1996), pp.39-85.

Constraints-Preserving Transformation from XML Document Type Definition to Relational Schema[*]

Dongwon Lee and Wesley W. Chu

Department of Computer Science
University of California, Los Angeles
Los Angeles, CA 90095, USA
{dongwon,wwc}@cs.ucla.edu

Abstract. As Extensible Markup Language (XML) [5] is emerging as *the* data format of the internet era, there are increasing needs to efficiently store and query XML data. One way towards this goal is using relational database by transforming XML data into relational format. In this paper, we argue that existing transformation algorithms are not complete in the sense that they focus only on *structural* aspects and ignoring *semantic* aspects. We present the semantic knowledge that needs to be captured during the transformation to ensure a correct relational schema. Further, we show a simple algorithm that can 1) derive such semantic knowledge from the given XML Document Type Definition (DTD) and 2) preserve the knowledge by representing them in terms of *semantic constraints* in relational database terms. By combining the existing transformation algorithms and our *constraints-preserving* algorithm, one can transform XML DTD to relational schema where correct semantics and behaviors are guaranteed by the preserved constraints. Experimental results are also presented.

1 Introduction

As the World-Wide Web becomes a major means of disseminating and sharing information, Extensible Markup Language (XML) [5] is emerging as a possible candidate data format because it is simpler than SGML and more powerful than HTML. To query XML data, one way is to reuse the established relational database techniques by converting and storing XML data in relational storage. Since the hierarchical XML and the flat relational data models are not fully compliant, the transformation is not a straightforward task.

To this end, several XML-to-relational transformation algorithms have been studied [8,9,16]. Although they work well for the given applications, to a greater or lesser extent, they miss one important point. That is, the transformation algorithms only capture the *structure* of the Document Type Definition (DTD) and

---

[*] This research is supported in part by DARPA contract No. N66001-97-C-8601.

A.H.F. Laender, S.W. Liddle, V.C. Storey (Eds.): ER2000 Conference, LNCS 1920, pp. 323–338, 2000.

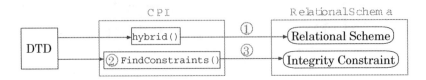

Fig. 1. Overview of our approach. Numbers 1) to 3) specify: 1) transforming schema, 2) discovering constraints via `FindConstraints()`, and 3) preserving constraints via `RewriteConstraints()`.

ignore the *semantic* constraints hidden in it. In this paper, via our *constraints-preserving inlining (CPI)* algorithm, we show the kinds of semantic constraints that can be derived from DTD during transformation, and how to preserve them by re-writing them in resulting schema notation. Since our algorithm to capture and preserve semantic constraints from DTD is orthogonal to transformation algorithms, ours can be applied to various transformation algorithms in [8,9, 16] with little change. Figure 1 presents an overview of our approach. First, given a DTD, we transform it to a corresponding relational scheme using an existing algorithm. Second, during the transformation, we discover various semantic constraints in XML notation. Third, we re-write the discovered constraints to conform to relational notation.

This paper is organized as follows. Section 2 gives background information and related work. In Section 3, one transformation algorithm is discussed in detail. Section 4 presents various semantic constraints that are hidden in DTD. Section 5 proposes our algorithm to preserve such constraints during transformation. Section 6 reports some experimental results that we have conducted and Section 7 summarizes with concluding remarks.

2   Background and Related Work

Relational Schema: We define a relational schema $\mathcal{R}$ to be composed of a *relational scheme (S)* and *semantic constraints (Δ)*. That is, $\mathcal{R} = (\mathcal{S}, \Delta)$. In turn, the relational scheme $\mathcal{S}$ is a collection of table schemes such as $r(a_1, ..., a_k)$, where a_i is the i-th attribute in the table r and the semantic constraints Δ is a collection of various semantic knowledge such as domain constraints, inclusion dependency, equality-generating dependency, tuple-generating dependency, etc.

XML and DTD: XML is a textual representation of the hierarchical data that is being defined by the World-Wide Web Consortium [5]. The meaningful piece of the XML document is bounded by matching starting and ending *tags* such as `<name>` and `</name>`. In XML, tags are defined by users while in HTML, permitted tags are pre-defined. Thus, XML is a meta-language that can be used for defining other customized languages. Using DTD, users can define the structure of the XML document of particular interest. A DTD in XML is very similar to a schema in a relational database. The main building blocks of DTD are *elements*

Table 1. A DTD for Conference.

```
<!DOCTYPE Conference [
  <!ELEMENT conf     (title,date,editor?,paper*)>
  <!ATTLIST conf     id      ID              #REQUIRED>
  <!ELEMENT title    (#PCDATA)>
  <!ELEMENT date     EMPTY>
  <!ATTLIST date     year    CDATA           #REQUIRED
                     mon     CDATA           #REQUIRED
                     day     CDATA           #IMPLIED>
  <!ELEMENT editor   (person*)>
  <!ATTLIST editor   eids    IDREFS          #IMPLIED>
  <!ELEMENT paper    (title,contact?,author,cite?)>
  <!ATTLIST paper    id      ID              #REQUIRED>
  <!ELEMENT contact  EMPTY>
  <!ATTLIST contact  aid     IDREF           #REQUIRED>
  <!ELEMENT author   (person+)>
  <!ATTLIST author   id      ID              #REQUIRED>
  <!ELEMENT person   (name,(email|phone)?)>
  <!ATTLIST person   id      ID              #REQUIRED>
  <!ELEMENT name     EMPTY>
  <!ATTLIST name     fn      CDATA           #IMPLIED
                     ln      CDATA           #REQUIRED>
  <!ELEMENT email    (#PCDATA)>
  <!ELEMENT phone    (#PCDATA)>
  <!ELEMENT cite     (paper*)>
  <!ATTLIST cite     id      ID              #REQUIRED
                     format  (ACM|IEEE)      #IMPLIED>
]>
```

and *attributes*, which are defined by the keywords <!ELEMENT> and <!ATTLIST>, respectively. In general, components in DTD are specified by the following BNF syntax:

<!ELEMENT> <element-name> <element-type>
<!ATTLIST> <attr-name> <attr-type> <attr-option>

For a detailed description of DTD model, refer to [12]. Table 1 shows a DTD for Conference which states that a conf element can have four sub-elements: title, date, editor and paper in that order. As common in regular expression, 0 or 1 occurrence (i.e., *optional*) is represented by the symbol "?", 0 or more occurrences is represented by the symbol "*", and 1 or more occurrences is represented by the symbol "+". A sub-element without any such symbols (e.g., title) represents a *mandatory* one.

Keywords #PCDATA and CDATA are used as *string* types for elements and attributes, respectively. For instance, the type of title element is defined as #PCDATA so that title element can be arbitrary character data. <attr-option> can be #REQUIRED or #IMPLIED among others. An attribute with a #REQUIRED option is a *mandatory* one while an attribute with a #IMPLIED option is an *optional* one. <attr-type> keywords ID and IDREF are used for the pointed and pointing attributes, respectively. IDREFS is a plural form of IDREF. For instance, the author element must have a mandatory id attribute and this attribute is used when other attributes point to this attribute. On the other hand, the contact element has a mandatory aid attribute that must point to the id attribute of

Table 2. A valid XML document conforming to the DTD for `Conference` in Table 1.

```
<conf id="er99">
  <title>Int'l Conference on Conceptual Modeling (ER)</title>
  <date> <year>1999</year> <mon>May</mon> <day>20</day> </date>
  <editor eids="sheth bossy">
    <person id="klavans">
      <name fn="Judith" ln="Klavans"/><email>klavans@columbia.edu</email>
    </person>  </editor>
  <paper id="p1">
    <title>Indexing Model for Structured...</title><contact aid="dao"/>
    <author><person id="dao"><name fn="Tuong" ln="Dao"/></person></author>
  </paper>
  <paper id="p2">
    <title>Logical Information Modeling of Heterogeneous...</title>
    <contact aid="shah"/>
    <author>
      <person id="shah"><name fn="Kshitij" ln="Shah"/></person>
      <person id="sheth">
        <name fn="Amit" ln="Sheth"/><email>amit@cs.uga.edu</email>
      </person>
    </author>
    <cite id="c100" format="ACM">
      <paper id="p3">
        <title>Making Sense of Scientific Information...</title>
        <author>
          <person id="bossy">
            <name fn="Marcia" ln="Bossy"/><phone>391.4337</phone>
          </person>
        </author> </paper> </cite> </paper>
</conf>
<paper id="p7">
  <title>Constraints-preserving Transformation from XML...</title>
  <contact aid="lee"/>
  <author>
    <person id="lee">
      <name fn="Dongwon" ln="Lee"/><email>dongwon@cs.ucla.edu</email>
    </person> </author>
  <cite id="c200" format="IEEE"/>
</paper>
```

the contacting `author` of the current paper. One interesting definition in Table 1 is the `cite` element; it can have zero or more `paper` elements as sub-elements, thus creating a cyclic definition. Table 2 shows a valid XML document conforming to the DTD for `Conference`. The document represents a portion of the fictional ER conference held in 1999. The first two `paper` elements are described with `id="p1"` and `id="p2"`, respectively. The `paper` element with `id="p2"` further has a `cite` element that describes the references in the paper. The `paper` element with `id="p7"` shows an example of the valid XML document that is *not* rooted at `conf` element. Note that a valid XML document can be rooted at any level of the DTD hierarchy as long as their sub-elements and attributes follow the DTD syntax.

Assumptions: Without loss of generality, to simplify our presentation, we assume that: 1) the input DTD has been already simplified using a technique in [16], 2) the input XML documents are all *valid*, and 3) the XML features such as *entities* or *notations* are not covered.

Related Work: [16] presents three transformation algorithms that focus on the table level of the schema while [9] studies different performance issues among eight algorithms that focus on the attribute and value level of the schema. Since our CPI algorithm provides a systematic way of finding and preserving constraints from a DTD, ours is an improvement to the existing transformation algorithms. Work done in STORED [8] deals with *non-valid* XML documents. When input XML documents do not conform to the given DTD, STORED uses a data mining technique to find a representative DTD whose support exceeds the pre-defined threshold. Since our algorithm to find and preserve constraints is not directly tied to a single transformation algorithm, ours can be applied to this algorithm as well. [13] also presents a DTD inference algorithm when it is not known. [4] discusses template language-based transformation from XML DTD to relational schema which requires human experts to write an XML-based transformation rule.

Some work has been done in [17] dealing with the transformation from relational tables to XML documents. There has been some transformation work in the OODB area as well [6]. Since OODB is a richer environment than RDB, their work is not readily applicable to our application. The logical database design methods and their associated transformation techniques to other data models have been extensively studied in ER research. For instance, [3] presents an overview of such techniques. However, due to the differences between ER and XML models, those transformation techniques need to be modified substantially.

3   Transforming DTD to Relational Schema

Transforming a hierarchical XML model to a flat relational model is not a trivial task. There are several difficulties including non 1-to-1 mapping, set values, recursion, and fragmentation issues [16]. For a better presentation, we chose one particular transformation algorithm, called the *hybrid inlining algorithm* [16] among many algorithms [4,8,9,16]. It is chosen since it exhibits the pros of the other two competing algorithms in [16] without severe side effects and it is a more generic algorithm than those in [4,8]. Since issues of discovering and preserving semantic constraints in this paper is orthogonal to that of transformation algorithms, our technique can be applied to other transformation algorithms easily.

3.1   Hybrid Inlining Algorithm

The *hybrid* algorithm [16] essentially does the following[1]:

1. Create a *DTD graph* that represents the structure of a given DTD. A DTD graph can be constructed when parsing the given DTD. Its nodes are elements, attributes, or operators in DTD. Each element appears exactly once in the graph, while attributes and operators appear as many times as they appear in the DTD.

---

[1] We have made a few changes to the hybrid algorithm for a better presentation (e.g., renaming, supporting "|" operator), but the crux of the algorithm remains intact.

2. Sub-elements in the choice model using the operator "|" are treated as if they are in the ordered sequence model with the following changes: 1) "+" operator is converted to "*" operator, 2) sub-elements without any occurrence operators are appended by "?" operator. For instance, `<!ELEMENT A ((a|b)+|c)>` is converted to `<!ELEMENT A (a*,b*,c?)>`. Further, an attribute with `#IMPLIED` or `IDREFS` type is converted to an operator node "?" or "+" in a DTD graph.

3. Identify *top nodes* in a DTD graph. A top node satisfies any of the following conditions: 1) not reachable from any nodes (e.g., source node), 2) direct child of "*" or "+" operator node, 3) recursive node with indegree > 1, or 4) one node between two mutually recursive nodes with indegree $= 1$. Then, starting from a top node T, *inline* all the elements and attributes at *leaf nodes* reachable from T unless they are other top nodes.

4. Attribute names are composed by the concatenated path from the top node to the leaf node using "_" as a delimiter. Use an attribute with ID type as a key if provided. Otherwise, add a system-generated integer key[2].

5. If a table corresponds to the shared element with indegree > 1 in DTD, then add a field `parent_elm` to denote the parent element to which the current tuple belongs. Further, for each shared element, a new field `fk_$X$` is added as a *foreign key* to record the key values of parent element X. If X is inlined into another element Y, then record the Y's key value in the `fk_$Y$` field.

6. Inlining an element Y into a table r corresponding to another element X (i.e., top node) creates a problem when an XML document is rooted at the element Y. To facilitate queries on such elements, a new field `root_elm` is added to a table r.

7. If an *ordered* DTD model is used, a field `ordinal` is added to record position information of sub-elements in the element. (For simplification, the `ordinal` field is not shown in this paper.)

Table 3 shows the output of the transformation by the hybrid algorithm.

Among eleven elements in the DTD in Table 1, four elements – `conf`, `paper`, `person`, and `eids` – are top nodes and thus chosen to be mapped to the different tables. For the top node `conf`, the elements `date`, `title`, and `editor` are reachable and thus inlined. Then, the `id` attribute is used as a key and the `root_elm` field is added. For the top node `paper`, the elements `title`, `contact_aid`, `author`, `cite_format` and `cite_id` are reachable and inlined. Since the `paper` element is shared by the `conf` and `cite` elements (two incoming edges in a DTD graph), new fields `parent_elm`, `fk_conf` and `fk_cite` are added to record who and where the parent node was. Note that in the `paper` table (Table 3), a tuple with `id="p7"` has the value `"paper"` for the `root_elm` field. This is because the element `<paper id="p7">` is rooted in the DTD (Table 2) without being embedded in other elements. Consequently, its `parent_elm`, `fk_conf` and `fk_cite` fields are null. For the top node `person`, the elements `name_fn`, `name_ln` and `email` are reachable

---

[2] In practice, even if there is an attribute with ID type, one may decide to have a system-generated key for better performance.

Table 3. A relational scheme ($\mathcal{S}$) along with the associated data that are converted from the DTD in Table 1 and XML document in Table 2 by the hybrid algorithm. Note that the hybrid algorithm does not generate *semantic constraints (Δ)*.

conf					
id	root_elm	title	date_year	date_mon	date_day
er99	conf	ER	1999	May	20

conf_editor_eids			
id	root_elm	fk_conf	eids
100001	conf	er99	sheth
100002	conf	er99	bossy

paper								
id	root_elm	parent_elm	fk_conf	fk_cite	title	contact_aid	cite_id	cite_format
p1	conf	conf	er99	–	Indexing ...	dao	–	–
p2	conf	conf	er99	–	Logical ...	shah	c100	ACM
p3	conf	cite	–	c100	Making ...	–	c200	IEEE
p7	paper	–	–	–	Constraints ...	lee	c200	IEEE

person								
id	root_elm	parent_elm	fk_conf	fk_paper	name_fn	name_ln	email	phone
klavans	conf	editor	er99	–	Judith	Klavans	klavans@cs...	–
dao	conf	paper	–	p1	Tuong	Dao	–	–
shah	conf	paper	–	p2	Kshitij	Shah	–	–
sheth	conf	paper	–	p2	Amit	Sheth	amit@cs...	–
bossy	conf	paper	–	p3	Marcia	Bossy	–	391.4337
lee	paper	paper	–	p7	Dongwon	Lee	dongwon@cs...	–

and inlined. Since the person is shared by the author and editor elements, again, the parent_elm is added. Note that in the person table (Table 3), a tuple with id="klavans" has the value "editor", not "paper", for the parent_elm field. This implies that "klavans" is in fact an editor, not an author of the paper.

4 Semantic Constraints in DTD

Domain Constraints. When the domain of the attributes is restricted to a certain specified set of values, it is called *Domain Constraints*. For instance, in the following DTD, the domain of the attributes gender and married are restricted.

```
<!ATTLIST author gender  (male|female) #REQUIRED
                 married (yes|no)      #IMPLIED>
```

In transforming such DTD into relational schema, we can enforce the domain constraints using SQL CHECK clause as follows:

```
CREATE DOMAIN gender VARCHAR(10) CHECK (VALUE IN("male","female"))
CREATE DOMAIN married VARCHAR(10) CHECK (VALUE IN("yes","no"))
```

When the mandatory attribute is defined by the `#REQUIRED` keyword in DTD, it needs to be forced in the transformed relational schema as well. That is, the attribute `ln` cannot be omitted below.

```
<!ELEMENT person  EMPTY>
<!ATTLIST person  fn CDATA  #IMPLIED   ln CDATA  #REQUIRED>
```

We use the notation "$X \not\rightarrow \emptyset$" to denote that an attribute X cannot be null. This kind of domain constraint can be best expressed by using the `NOT NULL` clause in SQL as follows:

```
CREATE TABLE person (fn VARCHAR(20), ln VARCHAR(20) NOT NULL)
```

Cardinality Constraints. In DTD declaration, there are only 4 possible cardinality relationships between an element and its sub-elements as illustrated below:

```
<!ELEMENT article (title, author+, reference*, price?)>
```

A. 1-to-$\{0,1\}$ mapping ("at most" semantics): An element can have either zero or one sub-element. (e.g., sub-element `price`)
B. 1-to-$\{1\}$ mapping ("only" semantics): An element must have one and only one sub-element. (e.g., sub-element `title`)
C. 1-to-$\{0, ...\}$ mapping ("any" semantics): An element can have zero or more sub-elements. (e.g., sub-element `reference`)
D. 1-to-$\{1, ...\}$ mapping ("at least" semantics): An element can have one or more sub-elements. (e.g., sub-element `author`)

For convenience, let us call each cardinality relationship as type A, B, C, and D, respectively. From these cardinality relationships, mainly three constraints can be inferred. First, whether or not the sub-element can be null. Similar to the attribute case, we use the notation "$X \not\rightarrow \emptyset$" to denote that an element X cannot be null. This constraint is easily enforced by the `NULL` or `NOT NULL` clause. Second, whether or not more than one sub-elements can occur. This is also known as *singleton constraint* in [18] and is one kind of equality-generating dependencies. Third, given an element, whether or not its sub-element should occur. This is one kind of tuple-generating dependencies. The second and third types will be further discussed below.

Inclusion Dependencies (IDs). An *Inclusion Dependency* assures that values in the columns of one fragment must also appear as values in the columns of other fragments and is a generalization of the notion of *referential integrity*.

Trivial form of IDs found in DTD is that "given an element X and its sub-element Y, Y must be included in X (i.e., $Y \subseteq X$)". For instance, from the `conf` element and its four sub-elements in DTD, the following IDs can be found as long as `conf` is not null: {`conf.title` $\subseteq$ `conf`, `conf.date` $\subseteq$ `conf`, `conf.editor`

$\subseteq$ conf, conf.paper $\subseteq$ conf}. Another form of IDs can be found in the attribute definition part of DTD with the use of the IDREF(S) keyword. For instance, consider the contact and editor elements in the DTD in Table 1 shown below:

```
<!ELEMENT person   (name,(email|phone)?>
<!ATTLIST person   id   ID       #REQUIRED>
<!ELEMENT contact EMPTY>
<!ATTLIST contact aid  IDREF   #REQUIRED>
<!ELEMENT editor  (person*)>
<!ATTLIST editor   eids IDREFS #IMPLIED>
```

The DTD restricts the aid attribute of the contact element such that it can only point to the id attribute of the person element[3]. Further, the eids attribute can only point to multiple id attributes of the person element. As a result, the following IDs can be derived: {editor.eids $\subseteq$ person.id, contact.aid $\subseteq$ person.id }. IDs can be best enforced by the "foreign key" concept if the attribute being referenced is a primary key. Otherwise, it needs to use the CHECK, ASSERTION, or TRIGGERS facility in SQL.

Equality-Generating Dependencies (EGDs). The *Singleton Constraint* [18] restricts an element to have "at most" one sub-element. When an element type X satisfies the singleton constraint towards its sub-element type Y, if an element instance x of type X has *two* sub-elements instances y_1 and y_2 of type Y, then y_1 and y_2 must be the same. This property is known as *Equality-Generating Dependencies (EGDs)* and denoted by "$X \to Y$" in database theory. For instance, two EGDs: {conf $\to$ conf.title, conf $\to$ conf.date} can be derived from the conf element in Table 1. This kind of EGDs can be enforced by SQL UNIQUE construct. In general, EGDs occur in the case of the 1-to-{0,1} and 1-to-{1} mappings in the cardinality constraints.

Tuple-Generating Dependencies (TGDs). *Tuple-Generating Dependencies (TGDs)* in relational model require that some tuples of a certain form be present in the table and use the "$\twoheadrightarrow$" symbol. Two useful forms of TGDs from DTD are the *child* and *parent constraints* [18].

1. **Child constraint:** "Parent $\twoheadrightarrow$ Child" states that every element of type *Parent* must have at least one child element of type *Child*. This is the case of the 1-to-{1} and 1-to-{1,...} mappings in the cardinality constraints. For instance, from the DTD in Table 1, since the conf element must contain the title and date sub-elements, the child constraint conf $\twoheadrightarrow$ {title, date} holds.

---

[3] Precisely, an attribute with IDREF type does not specify which element it should point to. This information is available only by human experts. However, new XML schema languages such as XML Schama and DSD can express where the reference actually points to [12].

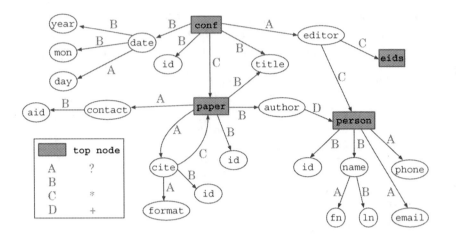

Fig. 2. An *Annotated DTD graph* for the Conference DTD in Table 1. The associated values of the nodes (i.e., indegree, type, tag, and status) are not shown.

2. **Parent constraint:** "Child →↠ Parent" states that every element of type *Child* must have a parent element of type *Parent*. According to XML specification, there is no notion of *root* in DTD. That is, XML documents can start from any level of elements without necessarily specifying its parent element. Therefore, parent constraints cannot be assured simply by looking at the DTD structure. Rather, it requires some semantic knowledge. In the DTD in Table 1, for instance, the editor and date elements can have the conf element as their parent. Further, if we know that all XML documents were started at the conf element level rather than the editor or date level, then the parent constraint {editor, date} →↠ conf holds. Note that the title →↠ conf does not hold since the title element can be a sub-element of either the conf or paper element.

5   Discovering and Preserving Semantic Constraints

To help find semantic constraints, we use the following data structure:

Definition 1. *An* **annotated DTD graph (ADG)** $\mathcal{G}$ *is a pair* $(\mathcal{V}, \mathcal{E})$, *where* $\mathcal{V}$ *is a finite set and* $\mathcal{E}$ *is a binary relation on* $\mathcal{V}$. *The set* $\mathcal{V}$ *consists of element and attributes in a DTD. Each edge* $e \in \mathcal{E}$ *is labeled with the cardinality relationship types (A to D) as defined in Section 4. In addition, each vertex* $v \in \mathcal{V}$ *carries the following information:*

1. indegree *stores the number of incoming edges.*
2. type *contains the element type name in the content model of the DTD (e.g., conf or* paper*).*
3. tag *stores a flag value whether the node is an element or attribute (if attribute, it contains the attribute keyword like* ID *or* IDREF*, etc.).*

Table 4. Cardinality relationships and their corresponding semantic constraints.

Relationship	Type	Symbol	Semantics	not null	EGDs	TGDs
1-to-{0,1}	A	?	at most	no	yes	no
1-to-{1}	B		only	yes	yes	yes
1-to-{0,...}	C	*	any	no	no	no
1-to-{1,...}	D	+	at least	yes	no	yes

4. status *contains "visited" flag if the node was visited in a depth-first search or "not-visited".*

Note that the cardinality relationship types in ADG considers not only element vs. sub-element relationships but also element vs. attribute relationships. For instance, from the DTD <!ATTLIST X Y #IMPLIED Z #REQUIRED>, two types of cardinality relationships (i.e., type A between element X and attribute Y, and type B between element X and attribute Z) can be derived. Figure 2 illustrates an example of ADG for the Conference DTD in Table 1. Then, the cardinality relationships can be used to find semantic constraints in a *systematic* fashion. Table 4 summarizes 3 main semantic constraints that can be derived from. The FindConstraints() algorithm can be immediately derived from the properties in Table 4. For detailed description, refer to [11].

Semantic constraints discovered by FindConstraints() have additional usage as we have shown in [11]. However, to enforce correct semantics in the newly generated relational schema, the semantic constraints in XML terms need to be re-written in relational terms. This is done by the algorithm Rewrite Constraints().

5.1   CPI: Constraints-Preserving Inlining Algorithm

We shall now describe our complete DTD-to-relational schema transformation algorithm: *CPI (Constraints-preserving Inlining) algorithm* is a combination of the hybrid inlining, FindConstraints() and RewriteConstraints() algorithms. The CPI algorithm is illustrated in CPI() and hybrid().

The algorithm first identifies all the top nodes from the ADG. This can be done using algorithms to find sinks or strongly-connected components in a graph [16]. Then, for each top node, the algorithm generates a corresponding table scheme using hybrid(). The associated constraints are found and re-written in relational terms using FindConstraints() and RewriteConstraints(), respectively. The hybrid() algorithm scans an ADG in a depth-first search manner while finding constraints and inlines a new field in the leaf node. The final output schema is the union of all the table schemes and semantic constraints.

Table 5 contains the semantic constraints that are re-written from XML terms to relational terms. As an example, the CPI algorithm will eventually spit out the following SQL CREATE statement for the paper table. Note that not only is the relational scheme provided, but the semantic constraints are also ensured by use of the NOT NULL, KEY, UNIQUE or CHECK constructs.

Algorithm 1: RewriteConstraints

> **Input** : Constraints Δ' in XML notation
> **Output**: Constraints Δ in relational notation
>
> **switch** Δ' **do**
>> **case** $X \not\rightarrow \emptyset$
>>> If X is mapped to attribute X' in table scheme A, then $A[X']$ cannot be null. (i.e., "CREATE TABLE A (...X' NOT NULL...)") ;
>>
>> **case** $X \subseteq Y$
>>> If X and Y are mapped to attributes X' and Y' in table scheme A and B, respectively, then re-write it as $A[X'] \subseteq B[Y']$. (i.e., If Y' is a primary key of B, then "CREATE TABLE A (...FOREIGN KEY (X') REFERENCES $B(Y')$...)". Else "CREATE TABLE A (...(X') CHECK (X' IN (SELECT Y' FROM B))...") ;
>>
>> **case** $X \rightarrow X.Y$
>>> If element X and Y are mapped to the same table scheme A (i.e., since Y is not a top node, Y becomes an attribute of table A) and Z is the key attribute of A, then re-write it as $A[Z] \rightarrow A[Y]$. (i.e., "CREATE TABLE A (...UNIQUE (Y), PRIMARY KEY (Z)...)");
>>
>> **case** $X \twoheadrightarrow X.Y$
>>> **if** (element X and Y are mapped to the same table) **then**
>>>> Let A be the table and Z be the key attribute of A. Then re-write it as $A[Z] \twoheadrightarrow A[Y]$. (i.e., "CREATE TABLE A (...Y NOT NULL, PRIMARY KEY (Z)...)") ;
>>>
>>> **else**
>>>> Let the tables be A and B, respectively and Z be the key attribute of A. Then re-write it as $B[fk\_A] \subseteq A[Z]$. (i.e., "CREATE TABLE B (...FOREIGN KEY (fk_A) REFERENCES $A(Z)$...)")
>
> **return** Δ;

```
CREATE TABLE paper (
   id          NUMBER      NOT NULL,
   title       VARCHAR(50) NOT NULL,
   contact_aid VARCHAR(20),
   cite_id     VARCHAR(20),
   cite_format VARCHAR(50) CHECK (VALUE IN ("ACM", "IEEE")),
   root_elm    VARCHAR(20) NOT NULL,
   parent_elm  VARCHAR(20),
   fk_cite     VARCHAR(20) CHECK (fk_cite IN (SELECT cite_id FROM paper)),
   fk_conf     VARCHAR(20),
   PRIMARY KEY (id),
   UNIQUE (cite_id),
   FOREIGN KEY (fk_conf)     REFERENCES conf(id),
   FOREIGN KEY (contact_aid) REFERENCES person(id)
);
```

Algorithm 2: CPI

Input   : Annotated DTD Graph $\mathcal{G} = (\mathcal{V}, \mathcal{E})$
Output: Relational Schema $\mathcal{R}$

$\mathcal{V} \leftarrow \texttt{topnode}(\mathcal{G})$;
for *each* $v \in \mathcal{V}$ **do**
   $table\_def \leftarrow \{\}$;
   if $v.tag = \,'element'$ **then**
      $\lfloor$ add('root_elm', $table\_def$); /* start where? */

   if $v.indegree > 1$ **then**
      add('parent_elm', $table\_def$);   /* shared elements case */
      add(concat('fk_', parent(v)), $table\_def$);

   $\mathcal{W} \leftarrow Adj[v]$; w $\in \mathcal{W}$;
   if *any* $w.tag = \,'$ID$'$ **then** add($w.type$, $table\_def$);
   else add('id', $table\_def$);   /* system-generated primary key */
   $\lfloor$ $\mathcal{R} \leftarrow \mathcal{R} + \texttt{hybrid}(v, table\_def, \emptyset)$;

return $\mathcal{R}$;

Algorithm 3: hybrid

Input   : Vertex v, TableDef $table\_def$, string $attr\_name$
Output: Relational Schema $\mathcal{R}$

$v.status \leftarrow \,'visited'$;
for *each* $w \in Adj[v]$ **do**
   if $w.status = \,'not\text{-}visited'$ **then**
      $\Delta' \leftarrow \texttt{FindConstraints}(v, w)$;
      $\Delta \leftarrow \texttt{RewriteConstraints}(\Delta')$;
      $\lfloor$ hybrid(w, $table\_def$, concat($attr\_name$, '_', $w.type$));

add($attr\_name$, $table\_def$); $\mathcal{R} \leftarrow table\_def + \Delta$;
return $\mathcal{R}$;

---

6   Experimental Results

We have implemented the CPI algorithm in Java using the IBM XML4J package. Table 6 shows a summary of our experimentation. We gathered test DTDs from "http://www.oasis-open.org/cover/xml.html" and [15]. Since some DTDs had syntactic errors caught by the XML4J, we had to modify them manually. Note that people seldom used the ID and IDREF(S) constructs in their DTDs except the XMI and BSML cases. The number of tables generated in the relational schema was usually smaller than that of elements/attributes in DTD due to the inlining effect. The only exception to this phenomenon was the XMI case, where extensive use of types C and D cardinality relationships resulted in many top nodes in the ADG.

Table 5. The semantic constraints in relational notation for the `Conference` DTD in Table 1.

Type	Semantic constraints in relational notation
ID	conf_editor_eids[eids] $\subseteq$ person[id], paper[contact_aid] $\subseteq$ person[id]
EGD	conf[id] $\rightarrow$ conf[title,date_year,date_mon,date_day] paper[id] $\rightarrow$ conf[title,contact_aid,cite_id,cite_format] person[id] $\rightarrow$ conf[name_fn,name_ln,email]
TGD	conf[id] $\twoheadrightarrow$ conf[title,date_year,date_mon,date_day] paper[id] $\twoheadrightarrow$ conf[title,contact_aid,cite_id,cite_format] person[id] $\twoheadrightarrow$ conf[name_fn,name_ln,email], conf_editor_eids[fk_conf] $\subseteq$ conf[id] paper[fk_conf] $\subseteq$ conf[id], paper[fk_cite] $\subseteq$ paper[cite_id] person[fk_conf] $\subseteq$ conf[id], person[fk_paper] $\subseteq$ paper[id]
not null	conf[id,title,date_year,date_mon,root_elm] $\not\rightarrow \emptyset$ conf_editor_eids[id,root_elm] $\not\rightarrow \emptyset$ paper[id,title,root_elm] $\not\rightarrow \emptyset$, person[id,name_ln,root_elm] $\not\rightarrow \emptyset$

Table 6. Experimental results of the CPI algorithm.

DTD		DTD Schema				Relational Schema				
Name	Domain	Elm	Attr	ID	IDREF(S)	Table	Attr	$\rightarrow$	$\twoheadrightarrow$	$\not\rightarrow \emptyset$
novel	literature	10	1	1	0	5	13	6	9	9
play	Shakespeare	21	0	0	0	14	46	17	30	30
tstmt	religious text	28	0	0	0	17	52	17	22	22
vCard	business card	23	1	0	0	8	19	18	13	13
ICE	content syndication	47	157	0	0	27	283	43	60	60
MusicML	music description	12	17	0	0	8	34	9	12	12
OSD	s/w description	16	15	0	0	15	37	2	2	2
PML	web portal	46	293	0	0	41	355	29	36	36
Xbel	bookmark	9	13	3	1	9	36	9	1	1
XMI	metadata	94	633	31	102	129	3013	10	7	7
BSML	DNA sequencing	112	2495	84	97	104	2685	99	33	33

The number of semantic constraints had a close relationship with the design of DTD hierarchy and the type of cardinality relationship used in the DTD. For instance, the XMI DTD had many type C cardinality relationships, which do not contribute to the semantic constraints. As a result, the number of semantic constraints at the end was small compared to that of elements/attributes in DTD. This was also true for the OSD case. On the other hand, in the ICE case, since it used many type B cardinality relationships, it resulted in many semantic constraints. For detailed discussions on the experimentation and the implementation of the CPI algorithm, please refer to [10].

7   Conclusion

This paper presents a method to transform XML DTD to relatonal schema both in *structural* and *semantic* aspects. After discussing the semantic constraints hidden in DTD, two algorithms are presented for: 1) discovering the semantic constraints using the hybrid inlining algorithm, and 2) re-writing the semantic constraints in relational notation. Our experimental results reveal that constraints can be systematically preserved during the conversion from XML to relational schema. Such constraints can also be used for semantic query optimization or semantic caching [11].

References

[1] Abiteboul, S., Buneman, P., Suciu, D. "Data on the Web: From Relations to Semistructured Data and XML", *Morgan Kaufmann Publishers*, 2000.
[2] Böhm, K., Aberer, K., Öszu, M. T., Gayer, K. "Query Optimization for Structured Documents Based on Knowledge on the Document Type Definition", *Proc. IEEE Advances in Digital Libraries (ADL)*, Los Alamitos, California, April, 1998.
[3] Batini, C., Ceri, S., Navathe, S. B. "Conceptual Database Design: An Entity-Relationship Approach", *The Benjamin/Cummings Pub. Inc.*, 1992.
[4] Bourret, R. "XML and Databases", *Internet Document*, September, 1999. http://www.informatik.tu-darmstadt.de/DVS1/staff/bourret/xml/XMLAnd Databases.htm
[5] Bray, T., Paoli, J., Sperberg-McQueen, C. M. (ed.), "Extensible Markup Language (XML) 1.0", *W3C Recommendation*, Feburary, 1998.
[6] Christophides, V., Abiteboul, S., Cluet, S., Scholl, M. "From Structured Document to Novel Query Facilities", *Proc. ACM SIGMOD*, Minneapolis, Minnesota, 1994.
[7] Deutsch, A., Fernandez, M. F., Florescu, D., Levy, A., Suciu, D. "XML-QL: A Query Language for XML", *Proc. The Query Language Workshop (QL)*, 1998. http://www.w3.org/TR/NOTE-xml-ql
[8] Deutsch, A., Fernandez, M. F., Suciu, D. "Storing Semistructured Data with STORED", *Proc. ACM SIGMOD*, Philadephia, Pennsylvania, June, 1998.
[9] Florescu, D., Kossmann, D. "Storing and Querying XML Data Using an RDBMS", *IEEE Data Engineering Bulletin*, 22(3), September, 1999.
[10] "XPRESS Home Page", 2000. http://www.cobase.cs.ucla.edu/projects/xpress/
[11] Lee, D., Chu, W. W. "Constraints-preserving Transformation from XML Document Type Definition to Relational Schema (Extended Version)", *UCLA-CS-TR 200001*, 2000. http://www.cs.ucla.edu/~dongwon/paper/
[12] Lee, D., Chu, W. W. "Comparative Analysis of Six XML Schema Languages", *UCLA-CS-TR 200008*, 2000. http://www.cs.ucla.edu/~dongwon/paper/
[13] Ludäescher, B., Papakonstantinou, Y., Velikhov, P., Vianu, V. "View Definition and DTD Inference for XML", *Proc. Post-ICDT Workshop on Query Processing for Semistructured Data and Non-Standard Data Formats*, 1999.
[14] Robie, J., Lapp, J., Schach, D. "XML Query Language (XQL)", *WWW The Query Language Workshop (QL)*, December, 1998.
[15] Sahuguet, A. "Everything You Ever Wanted to Know About DTDs, But Were Afraid to Ask", *Proc. 3rd Int'l Workshop on the Web and Databases (WebDB)*, Dallas, TX, 2000.

[16] Shanmugasundaram, J., Tufte, K., He, G., Zhang, C., DeWitt, D., Naughton, J. "Relational Databases for Querying XML Documents: Limitations and Opportunities", *Proc. VLDB*, Edinburgh, Scotland, 1999.

[17] Turau, V. "Making Legacy Data Accessible for XML Applications", *Internet Document*, 1999. http://www.informatik.fh-wiesbaden.de/~turau/veroeff.html

[18] Wood, P. T. "Optimizing Web Queries Using Document Type Definitions", *Proc. 2nd Int'l Workshop on Web Information and Data Management (WIDM)*, 1999.

X-Ray - Towards Integrating
XML and Relational Database Systems

Gerti Kappel, Elisabeth Kapsammer, Stefan Rausch-Schott, and
Werner Retschitzegger

Institute of Applied Computer Science, Department of Information Systems (IFS)
University of Linz, Altenbergerstraße 69, A-4040 Linz, Austria
{gk, ek, srs, wr}@ifs.uni-linz.ac.at

Abstract. Relational databases get more and more employed in order to store the content of a web site. At the same time, XML is fast emerging as the dominant standard at the hypertext level of web site management describing pages and links between them. Thus, the integration of XML with relational database systems to enable the storage, retrieval and update of XML documents is of major importance. This paper presents X-Ray, a generic approach for integrating XML with relational database systems. The key idea is that mappings may be defined between XML DTDs and relational schemata while preserving their autonomy. This is made possible by introducing a meta schema and meta knowledge for resolving data model heterogeneity and schema heterogeneity. Since the mapping knowledge is not hard-coded but rather reified within the meta schema, maintainability and changeability is enhanced. The meta schema provides the basis for X-Ray to automatically compose XML documents out of the relational database when requested and decompose them when they have to be stored.

1 Introduction

Web-based information systems no longer aim at purely providing read-only access to their content, which is simply represented in terms of web pages stored in the web server's directory. Nowadays, not least due to new requirements emerging from several application areas such as electronic commerce, the employment of databases to store the content of a web site turns out to be worthwhile 11, 20. This allows to easily handle both retrieval and update of large amounts of data in a consistent way on a large distributed scale 9. Besides using databases at the *content level*, the Extensible Markup Language (XML) 28 is fast emerging as the dominant standard for representing the *hypertext level* of a web site, i.e., the logical composition of web pages and the navigation structure 1, 6, 27. XML is a subset of SGML. As such, an XML document consists of possibly nested *elements* rooted in a single element. Elements, whose boundaries are delimited by *start-tags* and *end-tags,* may comprise *attributes*, whereby both are able to contain *values*. An XML document can be associated with a type specification called *document type definition (DTD)*, containing user-defined *element types* and *attribute specifications* which allow to describe the meaning of the content. Note, that there are already several efforts to replace DTDs by means of richer XML schema definition languages 28. However, since there is no standard up to now, the rest of the paper builds on DTDs.

A.H.F. Laender, S.W. Liddle, V.C. Storey (Eds.): ER2000 Conference, LNCS 1920, pp. 339-353, 2000.

Because of the increasing importance of XML and database systems (DBS), the integration of them with respect to storage, retrieval, and update is a major need 7, 27. Regarding the kind of DBS used for the integration, one can distinguish four different approaches 2, 12. First, *special-purpose DBS* are particularly tailored to store, retrieve, and update XML documents. Examples thereof are research prototypes such as Rufus 25, Lore 14, Strudel 11 and Natix 15 as well as commercial systems such as eXcelon 19 and Tamino 23. Second, because of the rich data modeling capabilities of *object-oriented DBS*, they are well-suited for storing hypertext documents 4, 29. Object-oriented DBS and special-purpose DBS, however, are neither in wide-spread use nor mature enough to handle large scale data in an efficient way. *Object-relational DBS* would be also appropriate for mapping to and from XML documents since the nested structure of the object-relational model blends well with XML's nested document model. Similar arguments as above, however, hold against their short-term usage. Thus, the more promising alternative to store XML documents are *relational database systems (RDBS)*. Such an integration would provide several advantages such as reusing a mature technology, seamlessly querying data represented in XML documents and relations, and the possibility to make legacy data already stored within an RDBS available for the web.

Concerning the kind of storage within an RDBS, there exist three basic alternatives. The most straightforward approach would be to *store XML documents as a whole* within a *single database attribute*. Another possibility would be *to interpret XML documents as graph structures* and provide a relational schema allowing to store arbitrary graph structures 12, 13, 22, 26. The third approach is that the *structure of XML documents* in terms of, e.g., a DTD *is mapped to a corresponding relational schema* wherein XML documents are stored according to the mapping 3, 5, 8, 10, 18, 24. Only the last of these alternatives allows to really exploit the features of RDBS such as querying mechanisms, optimization, concurrency control and the like. Thus, this approach is further investigated in the paper.

Despite of the benefits of the mapping approach, the problem is that when defining the mapping between an XML DTD and a relational schema, one has to cope with *data model heterogeneity* and *schema heterogeneity*. Data model heterogeneity refers to the fact that there are fundamental differences between concepts provided by XML and those provided by RDBS, which have to be considered when defining a certain mapping. These differences concern, e.g., structuring, typing and identification issues, relationships, default declarations, and the order of stored instances 17. Schema heterogeneity in our context means that, even if the DTD and the relational schema to which the DTD should be mapped represent the same part of the universe of discourse, the design of both is likely to be different. This could be because of different goals pursued during design like redundant representation of information versus normalization 17.

Existing approaches deal with these heterogeneity problems in various rather restricted ways. First, to reduce the heterogeneity a priori it is assumed that at least one of the schemata to be mapped to can be adapted to the other one 8. This, however, contradicts the requirement of autonomy. Second, there is only a certain pre-defined way of mapping provided 24, thereby preventing user-defined mappings which might eventually better resolve a certain heterogeneity with respect to, e.g., space or performance issues. Third, the mapping knowledge is often hard-coded within applications thus making maintenance in case of changes very difficult. With respect to these drawbacks, this paper proposes *X-Ray*, a *generic approach* for integrating XML with RDBS. The key idea is that mappings can be dynamically defined between *DTDs and relational schemata* thus coping with data model heterogeneity and schema heterogeneity. The integration *fully preserves the autonomy of both the DTD and the relational schema*, which in turn ensures the continuity of already existing applications working with the XML documents or the RDBS. This is made possible by introducing a *meta schema* storing information about the DTD, the relational schema and the mapping knowledge itself. The meta schema is responsible for mediation with respect to data model heterogeneity and schema heterogeneity and thus represents the core component of X-Ray. Since the mapping knowledge is not hard-coded but rather reified within the meta schema, maintainability and changeability is enhanced. This meta schema provides the basis for X-Ray to automatically compose XML documents out of the relational database when requested and decompose them when they have to be stored.

XML-DBMS introduced in 3 is closely related to X-Ray. Whereas in X-Ray the mapping knowledge may be specified in terms of tuples of the predefined meta schema, XML-DBMS provides a mapping language DTD. A specific user-defined XML document obeying this mapping language DTD, represents the mapping knowledge for yet another DTD and a relational schema. Based on our previous experience, however, using also a meta schema approach for mapping between objects and relations 16, working with a meta schema is quite intuitive and, thus, also suggested for X-Ray.

The remainder of the paper is organized as follows. Section 2 introduces different mapping possibilities between XML and RDBS. Based on these mapping possibilities, Section 3 defines a set of reasonable mappings to mediate between the different structuring mechanisms supported by XML and RDBS. The design of the meta schema is discussed in Section 4. Finally, Section 5 concludes the paper with a short summary and gives an outlook to future work.

2 Basic Kinds of Mappings between XML and RDBS

There are several possibilities for mapping a DTD to a relational schema. A straightforward way would be to map each element type to a relation and each XML attribute to an attribute of the respective relation. Due to data model heterogeneity and schema heterogeneity, however, such a one to one mapping is neither always possible nor desirable. For example, in the presence of deep element nesting directly mapping elements to tuples of different relations would lead to excessive fragmentation of the document over various relations, thus decreasing performance. This section proposes some basic mapping possibilities representing a prerequisite both for determining which kind of mapping is reasonable in a certain situation (cf. Section 3) and for designing the meta schema (cf. Section 4).

When considering the structuring mechanisms of XML and RDBS, three basic kinds of mappings may be distinguished, which are denoted in Fig. 1 together with an example. Note, that XML elements and attributes are represented in terms of UML objects 21.

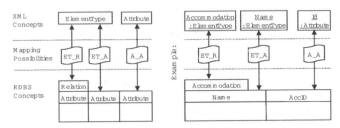

Fig. 1. Basic Kinds of Mappings

ET_R.   An element type (ET) is mapped to a relation (R), furtheron called *base relation*. Note, that several element types can be mapped to one base relation.

ET_A.   An element type is mapped to a relational attribute (A), whereby the relation of the attribute represents the base relation of the element type. Note, that several element types can be mapped to the attributes of one base relation.

A_A.   An XML attribute is mapped to a relational attribute whose relation represents the base relation of the XML attribute. Again, several XML attributes can be mapped to the attributes of one base relation.

It has to be emphasized that both element types and attributes can be mapped to a single base relation and a single attribute, only. Another point is that ET_A and A_A mappings determine that values of database attributes are mapped to values of XML elements or attributes. Thus, it makes sense that ET_R mappings occur together with ET_A and A_A mappings. Furthermore, it is not mandatory that all element types and attributes of a DTD as well as all relations and attributes of a relational schema have a mapping. An example at the relational side could be a foreign key that serves for establishing a relationship but might not be relevant within the XML document and therefore requires no mapping. The omission of mappings is imaginable not only in case that both DTD and relational schema have been developed independently from each other, but also if one has been derived from the other one. However, there are cases where a mapping is mandatory, e.g., if a certain constraint requires the existence of a value within the XML document (cf. Section 3.2).

The three basic kinds of mappings introduced above can be further refined with respect to the determination of an element type's base relation. For this, one has to look at the nesting hierarchy built by element types containing other element types. The former are furtheron called *composite element types*, the latter *component element types*. First, if an element type should be mapped, one has to consider the first of its direct or indirect composite element types that is mapped to a relation or an attribute, thus having a base relation. This base relation constitutes the *parent base relation* of the XML element type which should be mapped and is a candidate for being its base relation, too. If none of its composite element types is mapped, an arbitrary relation can be chosen as base relation. Concerning the example in Fig. 2 (cf. also the more comprehensive example given in Fig. 3), the element types address, street, and country all have the same parent base relation, namely Accommodation, which represents the base relation of the composite element type accommodation. Note, that aiming at an intuitive presentation, Fig. 2 depicts mappings between XML element types and relations in terms of a UML class diagram. To be able to distinguish between element types and relations, they are depicted as instances of the corresponding 'meta class' ElementType and Relation, respectively.

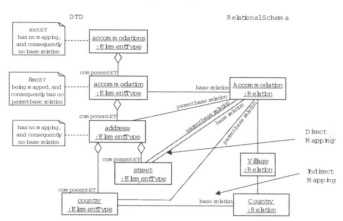

Fig. 2. Exemplary Mappings

Second, if an XML attribute should be mapped, its element type has to be considered first. If the attribute's element type is not mapped, its direct and indirect composite element types have to be considered as done for element types discussed above. Again, the relation which the first of these element types is mapped to represents the parent base relation of the XML attribute, thus being a candidate for being its base relation, too.

The parent base relation constitutes also the base relation, if the XML element type or the attribute, respectively, can be mapped to the relation or one of its attributes, which is furtheron called *direct mapping*. For an example, confer to the element type street in Fig. 2, which is directly mapped to an attribute of its parent base relation Accommodation. Otherwise, a proper base relation may be one of those relations, reachable by the parent base relation via foreign key relationships, which is furtheron called *indirect mapping*. For an example, consider the element type country, which is indirectly mapped to an attribute of relation Country reachable by its parent base relation Accommodation. Indirect mapping is reasonable in case that the relational attribute, which should be the mapping target, is factored out from the parent base relation, e.g., due to normalization reasons or because of vertical partitioning. Note, that element type address is used to group address data and thus has no relational counterpart and no base relation at all.

Both direct and indirect mapping is applicable to the three basic mapping possibilities introduced above thus resulting in $ET\_R_{direct/indirect}$, $ET\_A_{direct/indirect}$, and $A\_A_{direct/indirect}$ mappings. Furthermore, the possibility of a direct mapping always implies the possibility of an indirect mapping due to vertical partitioning.

3 Determining Reasonable Mappings between XML and RDBS

After introducing the basic kinds of mappings, this section discusses reasonable mappings. The determining factors can be categorized into *characteristics of the XML element type* (cf. Section 3.1) and *characteristics of the XML attribute* (cf. Section 3.2). In order to illustrate the subsequent investigations we provide a comprehensive running example building on the ones given in the previous section. The example is intended to show as many mapping possibilities as possible. Fig. 3 depicts the running example in terms of a DTD and in terms of a relational schema.

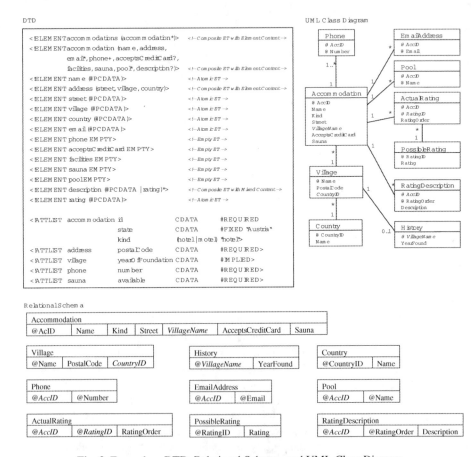

Fig. 3. Exemplary DTD, Relational Schema, and UML Class Diagram

The latter is depicted with a table structure and as UML class diagram better visualizing relationships. Concerning the relational schema, primary keys are prefixed with '@' and foreign keys are depicted using italic type. Even this small example shows that data model heterogeneity and schema heterogeneity prevent a simple one to one mapping. The description of this example is given from a mapping viewpoint throughout the forthcoming subsections.

3.1 Element Type Characteristics

As already mentioned, choosing a certain mapping is based on characteristics of the element type to be mapped. As illustrated in Fig. 4, these decisive characteristics can be categorized into three orthogonal dimensions comprising the *kind of element type*, if it *contains attributes*, and its *cardinality*.

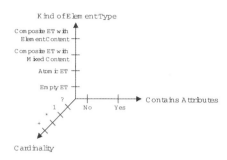

Fig. 4. Orthogonal Dimensions Characterizing XML Element Types

The most simple kind of element type contains an atomic domain (#PCDATA), only, and is furtheron called *atomic element type*. Composite element types (cf. Section 2) may have an atomic domain in addition to component element types, and thus are further distinguished into *composite element types with mixed content* and *composite element types with element content*. Concerning the latter, it has to be specified whether component element types occur in a *sequence* (","), or in a *choice* ("|") meaning that they are mutual exclusive. This is not applicable to the former since in this case component element types are allowed to occur in a choice with cardinality '*', only.

Element types that neither contain component element types nor have an atomic domain are called *empty element types*. Each element type no matter if it is an atomic, composite, or empty element type may contain *attributes*. Finally, *cardinality constraints* specify how often elements of a certain element type occur as component elements of its composite element. Since element types may be components of more than one composite element type, each of its occurrences as component element type can exhibit another cardinality. The cardinality symbols are '?' (null or 1), '*' (null or more), '+' (1 or more) and no symbol (exactly 1). Depending on the combination of these characteristics, certain reasonable mappings can be determined as shown in Table 1. In the following, these mappings are discussed by means of the running example.

First, we consider *composite element types with element content*. Mapping this kind of element type is neither influenced by cardinality nor whether it contains any attributes. Since there are no values associated with elements of this type, the only reasonable mapping possibility is ET_R. Depending on whether the element type can be mapped to its parent base relation or not, ET_R$_{direct}$ or ET_R$_{indirect}$ mapping can be used. In fact, the lack of any mapping would not result in a loss of information, since elements of this type contain no values which could be stored in the database.

Concerning our running example, whereas the root element type `accommodations` does not require any mapping, the element type `accommodation` is mapped to the relation `Accommodation` (ET_R mapping). Since `accommodation` does not have a parent base relation, we do not distinguish between a direct and an indirect mapping in this case.

Table 1. Reasonable Mappings of XML Element Types

Kind of Element Type	Contains Attributes	Cardinality	Reasonable Mapping
Composite ET with element content	No influence	No influence	$ET\_R_{direct/indirect}$; No mapping
Atomic ET	No influence	?, 1	$ET\_A_{direct/indirect}$
Atomic ET	No influence	+, *	$ET\_A_{indirect}$
Empty ET	No	1	No mapping
Empty ET	Yes	1	$ET\_R_{direct/indirect}$; No mapping
Empty ET	No influence	?	ET_A_{direct}
Empty ET	No influence	*, +	$ET\_A_{indirect}$
Composite ET with mixed content	No influence	No influence	$ET\_A_{indirect}$

Next, let us consider the mapping of an *atomic element type*. The reasonable mappings of such element types depend on the cardinality, only, and are not influenced by the existence of XML attributes. Since atomic element types contain values they always require a mapping to relational attributes, i.e., an ET_A mapping. In case of cardinality '?' and '1', an ET_A_{direct} mapping is possible, since no more than one element may occur. However, also an $ET\_A_{indirect}$ mapping may be necessary, when the relational attribute which the atomic element type should be mapped to is not part of the parent base relation. In case of cardinality '*' and '+', $ET\_A_{indirect}$ mapping is required due to normalization.

Concerning our running example, the most simple case is represented by element type `name` which has cardinality '1' and is mapped to attribute `Name` of base relation `Accommodation` representing an ET_A_{direct} mapping. `Accommodation` is mapped to element type `accommodation`, the direct composite element type of element type `name`, i.e., the base relation and the parent base relation are the same. This kind of mapping also applies to element type `street`. In this case the direct composite element type `address` has no mapping (cf. Section 2) and the indirect composite element type `accommodation` is mapped to the relation that contains the relational counterpart `Street`. The element types `village` and `country` require $ET\_A_{indirect}$ mappings, since their relational counterparts are stored in base relations different to the parent base relation `Accommodation` due to normalization reasons. The relational counterparts are attribute `Name` of base relation `Village` and attribute `Name` of base relation `Country`, respectively. This kind of mapping is possible, since `Accommodation` and `Village`, as well as `Village` and `Country` are directly connected via foreign key relationships. Element type `email` has cardinality '*' requiring an $ET\_A_{indirect}$ mapping and therefore is mapped to attribute `Email` of relation `EmailAddress`. The same holds true for element type `rating` with the difference that the parent base relation `Accommodation` and the base relation `PossibleRating` containing an attribute `Rating` are indirectly connected via the relation `ActualRating`.

Regarding *empty element types* with a cardinality '1', no matter if there are attributes or not, no mapping is required since a corresponding element occurs exactly once without carrying any value. However, if there were attributes, it would make sense to employ a direct or indirect ET_R mapping since the base relation could serve as the base relation for the attributes. In case of any other cardinality, the existence of attributes does not influence the reasonable mappings. An ET_A mapping is required in any case. It depends on the particular cardinality whether a direct or indirect mapping is reasonable.

Referring to our example, the empty element types `facilities` without attributes and `sauna` including a single attribute represent the most simple case both having a cardinality of one thus requiring no mapping. The attribute `available` of element type `sauna` is mapped to the relational attribute `Sauna` of the parent base relation of the element type `sauna`, namely `Accommodation`. The optional empty element type `acceptsCreditCard` contains no attributes and is mapped directly to the relational attribute `AcceptsCreditCard` of its parent base relation `Accommodation`. Finally, the empty element types `phone` and `pool` having a cardinality of '+' and '*', respectively, are mapped via ET_ A$_{indirect}$ to the relational attribute `Number` of the relation `Phone` and the relational attribute `Name` of the relation `Pool`, respectively.

Considering *composite element types with mixed content*, neither the existence of attributes nor the cardinality have any influence on the reasonable mappings. Since at the instance level, several values may occur within a single element, an ET_A$_{indirect}$ mapping is required. Our example contains one composite element type with mixed content, namely `description`, which is mapped to the attribute `Description` of the relation `RatingDescription`. The attributes `RatingOrder` of the two relations `ActualRating` and `RatingDescription`, which are not mapped to any XML concept, since they express an absolute order over both rating descriptions and actual ratings with respect to a certain accommodation.

3.2 XML Attribute Characteristics

The mapping of XML attributes depends on two orthogonal dimensions comprising the *type of the XML attribute* and its *default declaration*.

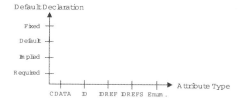

Fig. 5. Orthogonal Dimensions Characterizing XML Attributes

The type of the XML attribute may be a *string type* (CDATA), an *enumeration type*, or some special type including, e.g., ID and IDREF(S) responsible for unique identification of elements within an XML document and for referencing an element (IDREF) or several elements (IDREFS) having an attribute of type ID, respectively. For the sake of readability and space restrictions, we do not consider all possible types of XML attributes but rather the more important ones. The default declaration expresses whether a value is required (#REQUIRED), optional (#IMPLIED), fixed (#FIXED <ConstValue>) or default (<DefaultValue>).

For XML attributes with default declaration being #FIXED, no mapping is necessary independent of the type of the XML attribute. In our example, the XML attribute state of the element type accommodation has the constant value Austria. Regarding XML attributes which are not specified to be #FIXED, it has to be distinguished whether they are single-valued like CDATA or multi-valued defined by IDREFS. Single-valued attributes can be directly mapped to relational attributes (A_A_{direct}) or may require indirect mapping due to normalization reasons ($A\_A_{indirect}$), whereas multi-valued attributes may be mapped indirectly ($A\_A_{indirect}$), only. Considering ID and IDREF(S), it seems conceivable to map them to primary key attributes and foreign key attributes, respectively, of the relational schema. Due to data model heterogeneity, however, this is not always feasible, since there are differences concerning scope and composite keys 17.

Table 2. Reasonable Mappings of XML Attributes

Attribute Type	Default Declaration	Reasonable Mapping
No influence	#FIXED	No mapping
CDATA, ID, IDREF, enumeration	#REQUIRED, #IMPLIED, DefaultValue	$A\_A_{direct/indirect}$
IDREFS	#REQUIRED, #IMPLIED, DefaultValue	$A\_A_{indirect}$

In our example, directly mapped single-valued attributes comprise id and kind of element type accommodation, number of element type phone, and available of element type sauna. Single-valued attributes which have to be mapped indirectly are postalCode of element type address, and yearOfFoundation of element type village. Multi-valued attributes are not part of our example. It has to be emphasized that with one exception the reasonable mappings of an attribute are independent of the kind of mapping of its element type. In case that the element type of the attribute is not mapped and any of its composite element types that is not mapped depicts a cardinality of '*', the attribute can be mapped via $A\_A_{indirect}$, only.

4 The X-Ray Meta Schema

The different kinds of mappings proposed in the previous sections provide the basis for the design of the meta schema of X-Ray. The meta schema is the key mechanism for the genericity of X-Ray allowing to map DTDs and relational schemata. It mediates between heterogeneous concepts and provides the basis for X-Ray to automatically compose XML documents out of the relational database when requested and decompose them when they have to be stored. The mapping knowledge is not hard-coded within an application but rather reified and centrally stored within the meta schema, thus enhancing maintainability and changeability.

The meta schema consists of three components describing the relevant meta knowledge, namely DBSchema, XMLDTD and XMLDBSchemaMapping (cf. top part of Fig. 9). The DBSchema component is responsible for storing information about relational schemata that shall be mapped to DTDs to make their data available to XML documents or that shall be used to store XML documents. Analogously, the XMLDTD component stores schema information about XML documents as specified by means of DTDs. Finally, the XMLDBSchemaMapping component stores the mapping knowledge between DBSchema and XMLDTD. The goal of XMLDBSchemaMapping is to bridge both data model heterogeneity and schema heterogeneity in order to support a lossless mapping. This means that if an XML document is stored within the database, it should be possible to reconstruct it by retrieving the corresponding data out of the database and vice versa. It has to be emphasized that although the meta schema is designed on the basis of the concepts provided by DTDs, X-Ray does not require the existence of an explicit DTD. However, there must be at least a common implicit structure of the XML documents, which can be used by an administrator as input for XMLDTD and XMLDBSchemaMapping.

Concerning the storage of the meta knowledge itself, X-Ray comprises both a relational representation of the meta schema stored within the relational database and an object-oriented representation for main memory mapping. The latter is being initialized with the content of the relational meta schema at the beginning of an X-Ray session, herewith allowing an efficient composition and decomposition of XML documents at runtime. The object-oriented representation in terms of UML class diagrams is also used throughout this section to concisely and precisely depict the various meta schema components.

4.1 Database Schema Component

Concerning the database schema component, it has to be emphasized that it is not necessary to store meta knowledge about the complete relational schema, but only about those relations and attributes being relevant for the mapping to the DTD. However, not only base relations and their attributes are relevant, but also non-base relations which are the connecting relations between two base relations. As illustrated in Fig. 6 DBSchema contains at least one DBRelation, which consists of at least one DBAttribute. DBAttribute stores among others its *atomic domain* and whether it represents a *primary key attribute*. DBRelation and DBAttribute are generalized to DBConcept. Relationships (DBRelationship) connect two relations and specify one or more *join segments* (DBJoinSegment) comprising the join attributes, i.e., primary key and foreign key attributes of two relations that realize the relationship. The relationship comprises more than one join segment in case that the primary key is composed of two or more attributes. In case that parts of an XML document are stored within different relations, information about the proper join paths (DBJoinPath) is necessary.

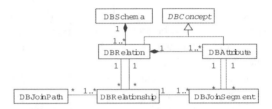

Fig. 6. Meta Schema of the Relational Schema

A DBJoinPath consists of one or more relationships. It comprises more than one relationship if more than two relations have to be joined for composing or decomposing a particular part of an XML document.

4.2 XML DTD Component

Similar to the database schema component, it is not necessary to store meta knowledge about the complete DTD, but only about those parts being relevant for the mapping to the relational schema. The meta knowledge specifies that a DTD (XMLDTD, cf. Fig. 7) has a certain element type (XMLElemType) that serves as root. For element types with attributes, XMLAttribute stores information about their *atomic domains* and their *default declaration*.

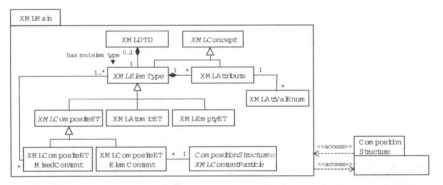

Fig. 7. Meta Schema of the DTD

Similar to the database schema component, XMLElemType and XMLAttribute are generalized to XMLConcept. For enumeration attributes the possible values are stored within XMLAttValEnum. According to the kinds of element types described in Section 2 and 3, XMLElemType is specialized into XMLAtomicET, XMLEmptyET, and XMLCompositeET. The latter is further specialized into XMLCompositeETMixedContent and XMLCompositeETElemContent.

The nesting structure of an XMLCompositeETElemContent is described by the package CompositionStructure (cf. Fig. 8). For an XMLCompositeETMixedContent the nesting structure needs not to be represented in the meta schema, since, as already mentioned, component element types are allowed to occur in a choice with cardinality '*', only.

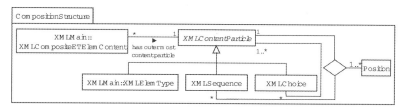

Fig. 8. Meta Schema of the XML Composition Structure

For component element types occurring in an XMLSequence or in an XMLChoice, the *cardinality* of the element type and in case of a sequence its *position* have to be stored. Furthermore, arbitrary combinations of sequences and choices can be described.

4.3 Mapping Knowledge

The mapping knowledge is expressed by various associations between the object classes of the XML DTD component and the database schema component. Fig. 9 illustrates these mapping relationships denoting them with bold lines.

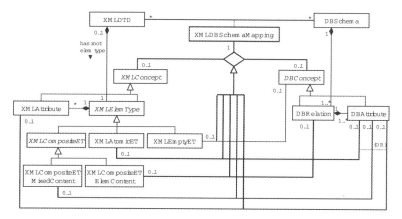

Fig. 9. Meta Schema Describing the Mapping Knowledge

For representation convenience, only those object classes are shown which are part of a mapping relationship. In order to meet the requirement that the meta schema is able to store mappings between different DTDs (XMLDTD) and different database schemata (DBSchema), the mapping between the class XMLConcept and the class DBConcept takes part in a ternary relationship with the association class XMLDBSchemaMapping. As discussed in Section 3, deciding on the exact kind of element type is a prerequisite for deciding a reasonable mapping to a database concept. Consequently, the leaf classes of the XMLElemType hierarchy are mapped to DBAttribute with two exceptions. The class XMLCompositeETElemContent is mapped to DBRelation, and the mapping of class XMLEmptyET is not further refined, since it inherits the (ternary) association to DBConcept. Besides the mapping relationships depicted in Fig. 9 there are also relationships to class DBJoinPath (cf. Fig. 6) which are not illustrated for representation convenience. Due to space restrictions, the attributes of the various object classes are also not shown. An example mapping in terms of the filled-in meta schema is given in 17.

5 Conclusion and Future Work

The main contribution of this paper is to describe X-Ray, an approach for mapping between XML DTDs and relational schemata. The mapping knowledge is not hard-coded but rather reified in terms of instances of a meta schema thus supporting autonomy of the participating DTDs and relational schemata as well as a generic integration thereof. On the basis of the meta schema, XML documents may be automatically composed out of data stored within an RDBS and vice versa decomposed into relational data without any loss of information. The X-Ray prototype builds on former experience in the area of data model heterogeneity and schema heterogeneity 16, and is currently used for case studies to investigate the validity of the developed meta schema.

Future work comprises short-term tasks such as supporting the whole set of XML concepts like implicit ordering and entity definitions, as well as long-term tasks such as integrating the XML Linking Language (XLink) and the XML Pointer Language (XPointer) 28. The latter will support the mapping of several XML documents and links between them to relational structures and vice versa. Another important aspect will be the investigation for simplifying the mapping between heterogeneous DTDs and relational schemata by, e.g., simplifying the given DTDs before mapping them 24. In this respect it will be also analyzed, how far the definition of the mapping knowledge may be automated on the basis of the reasonable mapping patterns described above. Leaving optimization issues aside, an automatically generated default mapping should be possible. If both legacy DTDs and legacy relational schemata are involved, however, schema heterogeneity will impede an automatic mapping.

References

1. Abiteboul, S., Buneman, P., Suciu, D.: Data on the Web: From Relations to Semistructured Data and XML. Morgan Kaufmann Publishers, 2000
2. Bourret, R.: XML and Databases. Technical University of Darmstadt, http://www.informatik.tu-darmstadt.de/DVS1/staff/bourret/xml/XMLAndDatabases.htm, June, 2000
3. Bourret, R., Bornhövd, C., Buchmann, A.P.: A Generic Load/Extract Utility for Data Transfer Between XML Documents and Relational Databases. 2nd Int. Workshop on Advanced Issues of EC and Web-based Information Systems (WECWIS), San Jose, California, June, 2000
4. Böhm, K., Aberer, K.: HyperStorM - Administering Structured Documents Using Object-Oriented Database Technology. Proc. of the ACM SIGMOD Int. Conf. on Management of Data, Montreal, Canada, June 1996
5. Carey, M., Florescu, D., Ives, Z., Lu, Y., Shanmugasundaram, J., Shekita, E., Subramanian, S.: XPERANTO: Publishing Object-Relational Data as XML. Int. Workshop on the Web and Databases (WebDB), Dallas, May, 2000
6. Ceri, S., Fraternali, P., Paraboschi, S.: Design Principles for Data-Intensive Web Sites. ACM SIGMOD Record, Vol. 24, No. 1, March 1999

X-Ray - Towards Integrating XML and Relational Database Systems   353

7. Ceri, S., Fraternali, P., Paraboschi, S.: XML: Current Developments and Future Challenges for the Database Community. Proc. of the 7th Int. Conf. on Extending Database Technology (EDBT), Springer, LNCS 1777, Konstanz, March, 2000
8. Deutsch, A., Fernandez, M., Suciu, D.: Storing Semistructured Data in Relations. Workshop on Query Processing for Semistructured Data and Non-Standard Data Formats, Jerusalem, Jan., 1999
9. Ehmayer, G., Kappel, G., Reich, S.: Connecting Databases to the Web - A Taxonomy of Gateways. Proc. of the 8th Int. Conf. on Database and Expert Systems Applications (DEXA), Springer LNCS 1308, Toulouse, September, 1997
10. Fernandez, M., Tan, W-C., Suciu, D.: SilkRoute: Trading between Relations and XML. 9th Int. World Wide Web Conf. (WWW), Amsterdam, May, 2000
11. Florescu, D., Levy, A., Mendelzon, A.: Database Techniques for the World Wide Web: A Survey. ACM SIGMOD Record, Vol. 27, No. 3, September, 1998
12. Florescu, D., Kossmann, D.: Storing and Querying XML Data Using an RDBMS. IEEE Data Engineering Bulletin, Special Issue on XML, Vol. 22, No. 3, September, 1999
13. Gardarin, G., Sha, F., Dang-Ngoc, T.-T.: XML-based Components for Federating Multiple Heterogeneous Data Sources. Proc. of the 18th Int. Conf. on Conceptual Modeling (ER), Paris, Nov., 1999
14. Goldman, R., McHugh, J., Widom, J.: From Semistructured Data to XML: Migrating the Lore Data Model and Query Language. Proc. of the 2nd Int. Workshop on the Web and Databases (WebDB), Philadelphia, June, 1999
15. Kanne, C.-C., Moerkotte, G.: Efficient Storage of XML Data. Proc. Of the 16th Int. Conf. On Data Engineering (ICDE), San Diego, March, 2000
16. Kappel, G., Preishuber, S., Pröll, E., Rausch-Schott, S., Retschitzegger, W., Wagner, R.R., Gierlinger, Ch.: COMan - Coexistence of Object-Oriented and Relational Technology. Proc. of the 13th Int. Conf. on the Entity-Relationship Approach (ER), Manchester, December, 1994
17. Kappel, G., Kapsammer, E., Retschitzegger, W.: X-Ray – Towards Integrating XML and Relational Database Systems. Technical Report, Department of Information Systems (IFS), JKU Linz, http://www.ifs.uni-linz.ac.at/ifs/research/publications/papers00.html, July, 2000
18. Klettke, M., Meyer, H.: XML and Object-Relational Database Systems - Enhancing StructuralMappings Based on Statistics. Int. Workshop on the Web and Databases (WebDB), Dallas, May, 2000
19. Object Design, Inc.: An XML Data Server for Building Enterprise Web Applications. http://www.odi.com/excelon/XMLResource/build_ent_web_apps.pdf, 1999
20. Pröll, B., Sighart, H., Retschitzegger, W., Starck, H.: Ready for Prime Time - Pre-Generation of Web Pages in TIScover. Proc. of the 8th Int. ACM Conference on Information and Knowledge Management (CIKM), Kansas City, Missouri, November, 1999
21. Raumbaugh, J., Jacobson, I., Booch, G.: The Unified Modeling Language Reference Manual. Addison-Wesley, 1999
22. Schmidt, A. R., Kersten, M. L., Windhouwer, M. A., Waas, F.: Efficient Relational Storage and Retrieval of XML Documents. Workshop on the Web and Databases (WebDB), Dallas, May, 2000
23. Schöning, H., Wäsch, J.: Tamino – An Internet Database System. Proc. of the 7th Int. Conf. on Extending Database Technology (EDBT), Springer, LNCS 1777, Konstanz, March, 2000
24. Shanmugasundaram, J., et al.: Relational Databases for Querying XML Documents: Limitations and Opportunities. Proc. of the 25th Int. Conf. On Very Large Data Bases (VLDB), Edinburgh, 1999
25. Shoens, K., et al.: The Rufus system: Information organization for semi-structured data. Proc. of the Int. Conf. On Very Large Data Bases (VLDB), Dublin, Ireland, 1993
26. Surjanto, B., Ritter, N., Loeser, H.: XML Content Management based on Object-Relational Database Technology. Proc. Of the 1st Int. Conf. On Web Information Systems Engineering (WISE), Hongkong, June 2000
27. Widom, J.: Data Management for XML - Research Directions. IEEE Data Engineering Bulletin, Special Issue on XML, Vol. 22, No. 3, September, 1999
28. W3C - World-Wide-Web Consortium. http://www.w3.org, 2000
29. VanZwol, R., Apers, P., Wilschutz, A.: Implementing Semi Structured Data with Moa. Workshop on Query Processing for Semistructured Data and Non-Standard Data Formats, Jerusalem, Jan., 1999

A Conceptual Model
for Remote Data Acquisition Systems

Txomin Nieva*and Alain Wegmann

Institute for computer Communications and Applications (ICA), Communication Systems
Department (DSC), Swiss Federal Institute of Technology (EPFL),
CH-1015 Lausanne, Switzerland
{Txomin.Nieva, Alain.Wegmann}@epfl.ch

Abstract. Data acquisition systems (DAS) are the basis for building monitoring
tools that enable supervision of local and remote systems. Unfortunately, DASs
are commonly based on proprietary technologies. The data format usually
depends on the industrial process, the fieldbus characteristics or the
development platform. Currently, there are many standards of DASs, but none
of them offer a well-accepted Application Programming Interface (API).
However, all of them comply with the same conceptual model. Understanding
this model allows for the significant improvement of the design of a specific
DAS. In this paper, we propose a conceptual model of a generic DAS. This
model gives researchers an abstraction of DASs and a quasi-formal
specification of a generic DAS. It also enables developers to compare the
existing standards and/or to propose a new open standard.

1 Introduction

In the last few years, companies from different business areas (such as power
engineering and transportation) have become increasingly interested in maintenance
management. Maintenance improves the reliability/availability of equipment and
therefore the quality of service, which managers have found provides substantial
benefits. However, maintenance management makes up anywhere from 15 to 40% of
total product cost [1]. Consequently, improving maintenance management can also
represent a substantial benefit to companies. Traditionally, there are two major
maintenance approaches: Corrective Maintenance and Preventive/Predictive
Maintenance (PPM). Corrective maintenance focuses on efficiently repairing or
replacing equipment after the occurrence of a failure. Corrective maintenance aims to
increase the maintainability of equipment by improving the speed of repair, or return
to service, after a failure. PPM focuses on keeping equipment in good condition in
order to minimize failures; repairing components before they fail. PPM aims to
increase the reliability of equipment by reducing the frequency of failures. A
successful management technique that can be applied for improving both techniques

---

* Corresponding author.

A.H.F. Laender, S.W. Liddle, V.C. Storey (Eds.): ER2000 Conference, LNCS 1920, pp. 354-368, 2000.

is the on-line supervision of the health of the equipment, which is usually known as condition monitoring. Condition monitoring is defined by *Davies* in [2] as:

"Condition monitoring is a management technique that uses the regular evaluation of the actual operating condition of plant equipment, production systems and plant management functions, to optimize total plant operation."

Condition monitoring, applied to maintenance tasks, provides necessary data in order to schedule preventive maintenance and to predict failures before they happen. Condition monitoring is based on direct monitoring of the state of equipment to estimate its Mean Time To Failure (MTTF). DASs and monitoring systems provide condition monitoring systems with the necessary information about the state of equipment. Remote access to this information provides significant benefits by allowing for the collection of condition-related data from wherever equipment is located. Remote monitoring systems exist in different business areas such as building [3], power engineering [4] and transportation systems [5]. Some remote monitoring systems make use of the Internet network and Internet technologies.

Substantial benefits can also be obtained by the intensive use of Asset Management Systems (AMS). Asset Management is defined by the *Government of Victoria* [6] as:

"The process of guiding the acquisition, use and disposal of assets to make the most of their service delivery potential (i.e., future economic benefit) and manage the related risks and costs over their entire life."

Asset management is a complementary task to maintenance. It provides support for the planning and operation phases. Similar to maintenance tasks, in AMSs access to utility data source is essential. *Draber et al.* proposed in [7] a data warehouse to homogenize data access from distributed and heterogeneous data sources.

In summary, DASs and remote monitoring systems build the infrastructure (see Fig. 1) needed to provide condition monitoring and AMSs with information about the state of equipment. Condition monitoring systems will analyze this data to estimate the MTTF. AMSs will propose or update preventive and predictive maintenance plans based on the information provided by the condition monitoring systems.

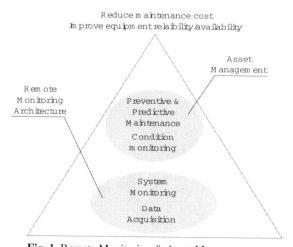

Fig. 1. Remote Monitoring & Asset Management

In this paper, we focus on the remote monitoring architecture and particularly on DASs. Unfortunately, DASs and monitoring systems are commonly based on proprietary technologies. The data format usually depends on the industrial process, the fieldbus characteristics or the development platform. Currently, there are many standards on DASs (see "OLE for Process and Control" (OPC) [8], "Interchangeable Virtual Instrument" (IVI) [9] and "Open Data Acquisition Standard" (ODAS) [10] among others). However, none of them provide a well-accepted API, which would enable for the communication of data between heterogeneous DASs. The "Object Management Group" (OMG) is currently addressing the issue of the lack of a universal API in [11]. Although there is not a well-accepted API for DASs, all DASs comply with the same information model. In [12] *Schenck* and *Wilson* propose the following definition of an information model:

"An information model is a formal description of types of ideas, facts and processes which together form a model of a portion of interest of the real world and which provides an explicit set of interpretation rules."

Understanding the information model will make DASs easier to understand, because it only focuses on the main aspects of the DAS by hiding low-level details that renders it difficult to understand. *Bubenko et al.* noted in [13] that:

"An effective approach to analyzing and understanding a complex phenomenon is to create a model of it. By a model is meant a simple and familiar structure or mechanism that can be used to interpret some part of reality. A model is always easier to study than the phenomenon it models, because it captures just a few of the aspects of the phenomenon".

In this paper, we present a conceptual model of a generic DAS. A conceptual model is an information model of a system from the object perspective that shows the relevant concepts of the system. Our conceptual model is the main result of a systematic review of DASs and software patterns, and from our practical experience with the development of a web-based monitoring tool applied to railway equipment [5, 14]. This model gives researchers a high level of abstraction of generic DASs. It also gives them a formal way to discuss about the main aspects of a DAS, and it enables them to compare existing standards and/or to propose a new open standard. We use Unified Modeling Language (UML) [15] as the modeling language to present our model.

Our conceptual model is inspired by several software patterns. A definition of software patterns is found in [16]:

"Patterns for software development are a literary form of software engineering problem-solving discipline that has its roots in a design movement of the same name in contemporary architecture, literate programming, and the documentation of best practices and lessons learned in all vocations".

This paper is organized as follows: First, we give a definition of DAS. Secondly, we present a web-based monitoring system applied to railway equipment that we developed. Then, we propose a generic conceptual model for DASs, which is based on this practical experience and some software patterns. Finally, we draw conclusions from the actual work.

2 Data Acquisition System

The following definition of a DAS is found in [17]:

"A DAS is a set of hardware and software resources designed to compute the internal representation and then, to deliver to the user the external representation".

Although this definition is appropriate, it does not reflect certain important aspects of a DAS. We postulate that a DAS is a system that gives:

- Means to *discover* and *access* system data.
- Means to *interpret* and *process* system data, in order to generate system information.
- Means to *publish* system information.

In order to clarify this definition we adopted the following definitions according to [18]:

- Discovering: "to obtain sight or knowledge of for the first time".
- Access: "to get at".
- Interpret: "to explain or tell the meaning of".
- Data processing: "the converting of raw data to machine-readable form and its subsequent processing".
- Publish: "to produce or release for distribution".

Therefore, our definition of a DAS is:

"A DAS is a set of hardware and software resources that provides the means to obtain knowledge of a system, provides the means to access to system data, converts system data to more useful system information and distributes this information to the user".

3 The RoMain System

We developed, in collaboration with *ABB Corporate Research* and in the frame of the *Railway Open System Interconnection Network (ROSIN)* European project, a web-based monitoring tool for trains that supports maintenance work. This monitoring tool was called Railway Open Maintenance tool (RoMain). The kernel of this tool is a DAS (developed on Java) installed on-board a train. The objective of this tool is not to replace the existing control network, but rather to enhance it with a parallel low-cost on-line data network for railways in order to support maintenance work. This data network will allow maintenance staff to supervise railway equipment from anywhere at anytime. It will also enable experts at different locations to collaborate and to get ahead of maintenance tasks. The user requirements for such a tool were: ubiquitous access, low cost, user friendly interface with textual and graphical views of the information and easy update of equipment documentation. Taking into account all these requirements, we decided to take an approach based on the Internet. The Internet has had a revolutionary impact on office automation, and now there is a clear trend towards using Internet technologies for industrial automation. The introduction of Internet technologies for accessing embedded systems is mostly cost driven, thus bringing significant benefits:

- Reduction of the development cost of an application, by enabling the use of Common Off-The-Self (COTS) software components.
- Elimination of the cost of a proprietary communication network, by using the common Internet network.
- Reduction of the cost of development of a client application for each different platform, by using a standard web-browser as a single client interface for heterogeneous platforms.
- Elimination of the cost of installing proprietary client applications, as the client interface is a standard web browser usually pre-installed on the client machine.
- Reduction of the cost of maintaining up-to-date equipment documentation, by offering a simple way (hyperlinks) to publish documents accessible immediately from anywhere in the world.
- Reduction of maintenance personal traveling costs, by the possibility of ubiquitous access to the information.
- Reduction of maintenance scheduling costs, by the possibility of ubiquitous access to the information at any time.

The architecture of the RoMain system, shown in Fig. 2, is composed of:

- *Train Gateways* - connected to the train network gather actual train data.
- *Ground Stations* - automatically establish connections to train gateways over wireless networks.
- *Name and Directory Servers* - provide information about the train component models and train directory.
- *Manufacturer Servers* - provide on-line information about train components, for example fact sheets, user manuals, or installation instructions.
- *Maintenance Stations* - run a standard web browser to access train data.

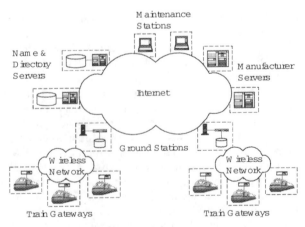

Fig. 2. The RoMain System

All the systems are interconnected by means of a secure TCP/IP network, usually the *Internet*, or eventually an *Intranet* or *Virtual Private Network*.

In this paper we expose a conceptual model for a generic DAS. This model has been developed as part of an iterative process consisting of the analysis, specification, implementation and deployment of a DAS for the RoMain system.

4   A Data Acquisition Conceptual Model

For a better understanding of the model, we group the concepts in four main packages, as shown in Fig. 3.

Fig. 3. Data Acquisition Main Packages

- *Device Models*: this package groups all the concepts regarding device models. A device model represents a model that characterizes a set of devices.
- *Device Items*: this package groups all the concepts regarding device items. A device item represents a real device that satisfies a device model.
- *Report Definitions*: this package groups all the concepts that allow the definition of criteria for generating monitoring reports.
- *Observations & Reports*: this package groups all the concepts regarding observations and reports taken on a system. Observations are classified as quantitative (measurements) or qualitative (category observations) according to the measurements and observations analysis pattern described by *Fowler* in [19]. Monitoring reports are classified as reports that record a snapshot of the system at a specific time (status reports) and reports that indicate a certain state of the system (event reports).

In the following sections, we describe in detail each of these packages. Finally, we conclude with a complete conceptual model diagram[1] representing the relevant concepts that make up any DAS, and their relationships.

---

1 In the models, we distinguish between *"operational"* and *"knowledge level"* concepts. We adopt this idea from *Fowler* according to [19]. At the operational level the model records the day-to-day events of the domain, while at the knowledge level the model records the general rules that govern this structure. We represent knowledge level concepts by using a box with a thick border, and a box with a thin border represents operational level concepts.

4.1 Device Models

An instance of Device Model is the representation of a model, created in the design process, that characterizes a set of devices. In this section, we describe the concepts and relationships related to device models and how these models are organized. The conceptual model corresponding to device models is shown in Fig. 4.

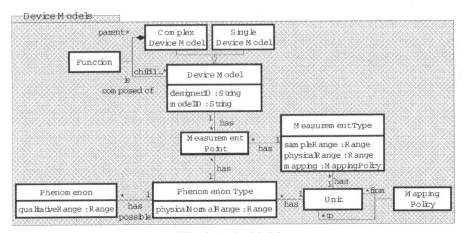

Fig. 4. Device Models

A device model defines many measurement points. An instance of "Measurement Point" defines a measurement point associated with a phenomenon type and a measurement type. An instance of "Phenomenon Type" represents something that can be quantitatively (e.g. *"temperature"*), or qualitatively (e.g. *"door_status"*), observed. A phenomenon type defines the units on which observations of this phenomenon type are expressed. We adopt the convention of using standard SI (metric) units, as proposed in [3], for phenomenon types. Eventually, a phenomenon type records the set of potential qualitative values that a measurement of such phenomenon type can take (e.g. *"temperature_low"*, *"temperature_medium"*, and *"temperature_high"*). Each of these values is an instance of "Phenomenon". Phenomenon records the range of quantitative observations of a phenomenon type that corresponds to a qualitative observation. This enables the automatic recording of an occurrence of this phenomenon upon a quantitative observation, of the corresponding phenomenon type, with a value within the range of the phenomenon. An instance of "Measurement Type" is associated with a measurement point to give some semantic information about the measurements taken at this measurement point. A measurement type defines the permissible ranges of sampled and physical values of a phenomenon type in a measurement point. In industrial DASs, it is very common that the value actually measured (we refer to this value as "sampled value") does not correspond to the physical value. Therefore, a measurement type also defines a mapping policy that makes it possible to calculate the physical value from the sampled value. A measurement type also defines the units in which a real measurement is taken. These units are not necessarily the same as the units of the corresponding phenomenon type. In this case, a mapping policy between units defines the conversion from measurement type units to phenomenon type units. In the following sub-sections we describe the device model composition, how to define and assign a global unique identifier (GUID) to a device model, and we give more details about mapping policies.

Device Model Composition. In DASs, tree structures allow us to efficiently define an industrial system because an industrial system is usually composed of many parts, which can also be composed of many other parts in a part-whole hierarchy. For example, an HVAC (heating, ventilation and air condition) system is composed of subsystems such as heating coil, cooling coil, supply fan, etc., that can be composed of other subsystems such as temperature sensors, ventilation sensors and so on. Par-Whole relationships are commonly used to model the construction of composite objects out of individual parts. Part-Whole relationship categories and their application in object-oriented analysis are further discussed by *Motchnig-Pitrik et al.* in [20]. We used the "Composite" pattern, detailed in [21], to compose objects into tree structures to represent part-whole hierarchies. "Device Model" implements default behavior for a device model. "Complex Device Model" defines behavior for a device model that is composed of other device models. A device model that is part of a complex device model implements a function on this complex device model. The association class "Function" allows us to record this information. A function must be unique in the naming space of the complex device model.

Device Model Identifier. A GUID must be assigned to a device model. Device designers are responsible for assigning a designer specific model identifier to their device models. This identifier, which we named "modelID", allows us to distinguish between two different models belonging to the same designer. A designer identifier, which we named "designerID", allows us to distinguish between two different manufacturers. As a result, a device model GUID is obtained from the combination of "designerID" and "modelID".

Mapping Policy. A mapping policy defines the conversion between two numerical values. "Function Mapping Policy" represents a mapping policy with a complex function $(y=f(x))$ while "Linear Mapping Policy" represents a mapping policy with a linear function $(y=Ax+B)$; where "x" corresponds to the original value and "y" to the calculated value. The mapping policy model is shown in Fig. 5.

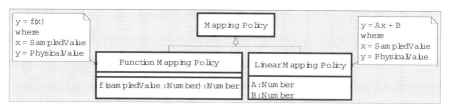

Fig. 5. Mapping Policy

4.2 Device Items

An instance of Device Item is a real item, created in the manufacturing process, that represents a real device. An instance of Device Item is described by an instance of Device Model. In this section we describe the concepts and relationships related to device items and how these items are organized. The conceptual model corresponding to device items is shown in Fig. 6.

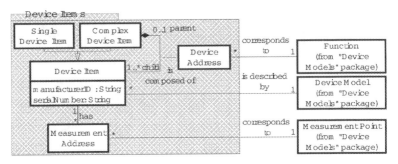

Fig. 6. Device Items

A device item is described by a device model. A device model can be associated with many device items. Each device item defines many measurement addresses. An instance of "Measurement Address" defines the actual location in a device item associated with a measurement point, where observations of a phenomenon type are taken. In the following sub-sections, we give more details about the device item composition and how to assign and define a GUID to a device item.

Device Item Composition. Similar to device models, device items are organized using the Composite pattern. An instance of Device Item is an occurrence of an instance of a Device Model. As a consequence, there is an analogous relationship between pairs of device items and pairs of the corresponding device models. A device item that is part of a complex device item is installed on a device address of the complex device item. "Device Address" allows us to record this information. A device address is unique within the naming space of a complex device item. A device address is associated with the corresponding function that a device item is implementing on a complex device item.

Device Item Identifier. A GUID must be assigned to a device item. Device manufacturers are responsible for assigning a unique identifier, which is named "serialNumber", to each device item. "serialNumber" uniquely identifies device items of the same device model. In order to be able to globally identify a device item, it is necessary to include the device model GUID. As a result, a device item GUID is obtained from the combination of its corresponding device model GUID and a "serialNumber".

4.3 Report Definitions

The ability to define reports, with a consistent status, of a set of data (the OMG DAIS RFP [11] refers to this as "DataSet", and OPC [22, 23] as "OPC Group") is one of the requirements of any DAS. Our approach for defining reports is inspired by *Mansouri-Samani* and *Sloman* [24]. The report model is shown in Fig. 7.

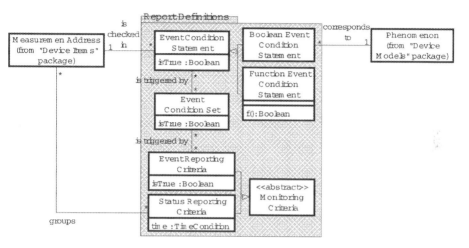

Fig. 7. Report Definitions

"Status Reporting Criteria" allows us to define a dataset with a list of measurement addresses where to observe periodically, or by schedule, phenomenon types. A report of this type is triggered by a time condition. "Event Reporting Criteria" allows us to define conditions in order to automatically generate reports when the system is under a certain state. In order to explain an event reporting criteria, we make use of the algebraic notation[2]. Then, we state that "Event Condition Statement" allows us to record $X=A$ and $X=A'$ condition statements; where A means that a certain phenomenon has been observed in a measurement address, as recorded by "Boolean Event Condition Statement", or that the result of a certain function "f()" checked in a measurement address returns true, as recorded by "Function Event Condition Statement". "Event Condition Set" allows us to record conditions such as $X=A.B$ and $X=(A.B)'$; where A, B are "Event Condition Statements". "Event Reporting Criteria" allows us to record reporting criteria such as $X=A+B$ and $X=(A+B)'$; where A, B are "Event Condition Sets". This allows recording any even criteria, because any criteria can be expressed by means of an algebraic combination of "Event Condition Statements" with the "AND" logical operator and an algebraic combination of "Event Condition Sets" with the "OR" logical operator. A transformation of any algebraic expression into these terms is possible by applying one of *De Morgan*'s law[3].

---

2 " . " corresponds to the "AND" logical operator; " + " corresponds to the "OR" logical operator; and " ' " corresponds to the "NOT" logical operator

3 The two laws, known as *De Morgan*'s, are: $(A+B)'=A'.B'$; and $(A.B)'=A'+B'$

Time Condition. "Time Condition" allows us to define periodical or by schedule time conditions to record status reports. The time condition model is show in Fig. 8. "Period" allows us to define a time condition as a period of time in milliseconds to enable taking periodical reports, while "Schedule" allows us to define a schedule when a status report will be generated. A schedule is defined by a recurrence time, a recurrence pattern and a recurrence range. "Recurrence Time" allows us to define a 24-hour period (with precision in milliseconds) when a status report will be generated. "Recurrence Pattern" allows us to define several ways to record a time pattern for occurrences of a status report; "Daily", "Weekly", "Monthly" and "Yearly" allow us to define a status report to be generated every certain number of days, every certain number of weeks on some specific days of a week, every certain number of months on a specific day of the month, and every certain number of years on a specific day of a certain month, respectively. Finally, "Recurrence Range" allows us to define a beginning time (just a "Begin Date") to start recording status reports and an end time to stop recording status reports. "End Time" allows us to define the end time by a number of occurrences, by a specific end date or with no end (meaning that the status report will be generated "forever").

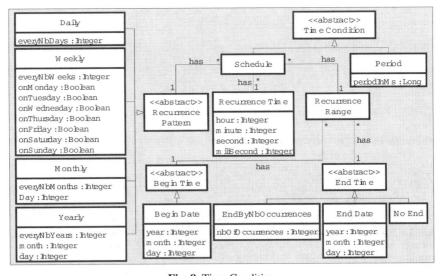

Fig. 8. Time Condition

4.4 Observations

The observation model, shown in Fig. 9, defines concepts that allow us to record observations taken on a device item. Our observation model is inspired by the "*observations and measurements*" analysis pattern described by *Fowler* in [19].

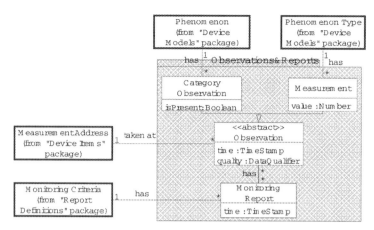

Fig. 9. Observations

In a device address we can record many observations. "Observation" is an abstract concept that represents both quantitative and qualitative observations. An observation records a timestamp to record the time an observation was taken. "Measurement" represents quantitative observations. A measurement records the physical value corresponding to the measurement, which is represented by the "value" attribute. A measurement is associated with a phenomenon type, while a phenomenon type can have many measurements. A "Category Observation" represents a qualitative observation. A category observation is associated with a phenomenon, while a phenomenon can have many category observations. Sometimes recording that a phenomenon is absent is as important as recording its presence. The "isPresent" Boolean attribute of category observation is added to enable recording the absence or presence of a phenomenon. In the following sub-sections we give more details about timestamps and data qualifiers associated with observations.

Timestamps. Recording the time an observation was taken is a key issue for enabling a subsequent analysis of observations. In order to avoid anomalies due to inconsistent time formats (e.g because of different time zones), we adopted the convention of storing all timestamps using UTC (Universal Coordinated Time) format. This, further discussed in [3], is a common practice in DASs.

Data Qualifiers. In DASs it is a common practice to include a data qualifier (see Fig. 10) with an observation. According to [11] a "Data Qualifier" includes information about the "Validity" ("valid", "held" from a previous value, "suspect", "not valid" or "substituted" manually), the "Current Source" ("metered", "calculated", "entered", or "estimated") and the "Normal Value" ("normal" or "abnormal") of an observation.

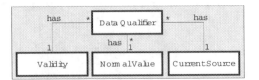

Fig. 10. Data Qualifier

4.5 Final Data Acquisition Model

The final conceptual model for DASs is shown in Fig. 11.

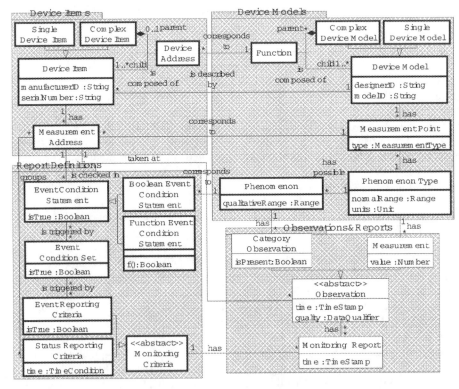

Fig. 11. Data Acquisition Conceptual Model[4]

---

4 To simplify the model we show only the main attributes of a concept. For the same reason, some concepts have been intentionally designed as attributes of higher-level concepts.

5 Conclusions and Future Work

In this paper, we propose a conceptual model of a generic DAS. This model gives researchers an abstraction of DASs and a quasi-formal specification of a generic DAS. It also enables developers to compare existing standards and/or to propose a new open standard. Understanding this model allows for the significant improvement of the design and development of a specific DAS. This model is generic enough to be applied for the design and development of similar DASs in many application domains such as building, power engineering and transportation systems.

This model has been developed as part of an iterative process consisting of the analysis, specification, implementation and deployment of a DAS for railway equipment. We use our own variation of the Catalysis [25] development process based on UML. The conceptual model specifies only the static aspects of a generic DAS. Currently, we are complementing this model with use case models that describe the dynamic aspects of a generic DAS. Conceptual and use case models are the foundation for specifying and designing specific DASs based on specific requirements of quality of service (QoS). We aim to provide some guidelines that, based on our generic conceptual and use case models, will help developer to make the right architectural choices depending on the specific requirements of a DAS for a specific system.

Acknowledgments

Thanks to all the members of the *ROSIN WP4* who brought a real framework to the discussions and the implementation of our hypotheses. To *Andreas Fabri* and *Hubert Kirrmann*, from *ABB Corporate Research*, for their valuable contributions to the research work of this paper. To *Guy Genilloud*, from *ICA*, for his reviews of the UML diagrams. And last but not least, to *Holly Cogliati*, from *ICA*, for proofreading this paper.

References

[1]   T.Wireman, *"Computerized Maintenance Management Systems"*, Industrial Press, Inc, 1994.

[2]   A.Davies, *"Handbook of Condition Monitoring - Techniques and Methodology"*, Kluwer Academic Publishers, 1997.

[3]   F.Olken, H.A.Jacobsen, C.McParland, M.A.Piette, and M.F.Anderson, *"Objects lessons learned from a distributed system for remote building monitoring and operation"* presented at Conference on Object-oriented Programming, Systems, Languages and Applications, Vancouver, Canada, October 18-22, 1998, http://www.lbl.gov/~olken/rbo/rbo.html.

[4]   R.Itschner, C.Pommerell, and M.Rutishauser, *"GLASS: Remote Monitoring of Embedded Systems in Power Engineering"* in IEEE Internet Computing, vol 2, 1998.

[5]   A.Fabri, T.Nieva, and P.Umiliacchi, *"Use of the Internet for Remote Train Monitoring and Control: the ROSIN Project"* presented at Rail Technology '99, London, UK, September 7-8, 1999, http://icawww.epfl.ch/nieva/thesis/Conferences/RailTech99/article/RailTech99.PDF.

[6]   Victorian Government, *"Asset Management Series: Principles, Policies and Practices"*, November, 1995, http://home.vicnet.net.au/~assetman/welcome.htm.

[7]   S.Draber, E.Gelle, T.Kostic, O.Preiss, and U.Schluchter, *"How Operation Data Helps Manage Lifecycle Costs"* presented at International Conference on Large High Voltage Electric Systems - CIGRE'2000, Paris , France, August 27 - September 2, 2000.

[8]   OPC Foundation, *"OLE for Process and Control Standard"*, 1997, http://www.opcfoundation.org.

[9]   IVI Foundation, *"Interchangeable Virtual Instruments Standard"*, 1997, http://www.ivifoundation.org/.

[10]   Open Data Acquisition Association, *"Open Data Acquisition Standard"*, 1998, http://www.opendaq.org/.

[11]   OMG, *"Data Acquisition from Industrial Systems (DAIS)"*, Request for Proposal (RFP), OMG Document: dtc/99-01-02, January 15, 1999, http://www.omg.org/techprocess/meetings/schedule/Data_Acquisition_RFP.html.

[12]   D.A.Schenck and P.R.Wilson, *"Information Modeling: The EXPRESS Way"*, Oxford University Press, 1994.

[13]   M.Boman, J.A.Bubenko Jr., P.Johannesson, and B.Wangler, *"Conceptual Modelling"*, Prentice Hall, 1997.

[14]   T.Nieva, *"Automatic Configuration for Remote Diagnosis and Monitoring of Railway Equipment"* presented at IASTED International Conference - Applied Informatics, Innsbruck, Austria, February 15-18, 1999, http://icawww.epfl.ch/nieva/thesis/Conferences/ai99/article/ai99.pdf.

[15]   J.Rumbaugh, I.Jacobson, and G.Booch, *"The Unified Modelling Language Reference Manual"*, Addison Wesley, 1999, http://www.rational.com, http://www.omg.org.

[16]   B.Appleton, *"Patterns and Software: Essential Concepts and Terminology"*, November 20, 1997, http://www.enteract.com/~bradapp/docs/patterns-intro.html.

[17]   J.Ehrlich, A.Zerrouki, and N.Demassieux, *"Distributed Architecture for Data Acquisition: a Generic Model"* presented at IEEE Instrumentation and Measurement Technology Conference - IMTC'97, Ottawa, Canada, May 19-21, 1997.

[18]   Merriam-Webster, *"WWWebster Dictionary"*, 2000, http://www.m-w.com/.

[19]   M.Fowler, *"Analysis Patterns: Reusable Object Models"*, Addison-Wesley, 1997, http://www2.awl.com/cseng/titles/0-201-89542-0/apsupp/index.htm.

[20]   R.Motschnig-Pitrik and J.Kaasboll, *"Part-Whole Relationship Categories and Their Application in Object-Oriented Analysis"* in IEEE Transactions on Knowledge and Data Engineering vol. 11, pp. 779-797, 1999.

[21]   Erich Gamma, Richard Helm, Ralph Johnson, and John Vlissides, *"Design Patterns - Elements of Reusable Object-Oriented Software"*, Addison-Wesley, 1995.

[22]   Utility Communications Specification Working Group, *"IEC 60870-6-503, Telecontrol Equipment and Systems - Part 6: Telecontrol Protocols Compatible with ISO Standards and ITU-T Recommendations - Section 503: TASE.2 Services and Protocols"*, August, 1996, ftp://ftp.sisconet.com/epri/iccp/iccp503.doc.

[23]   Utility Communications Specification Working Group, *"IEC 60870-6-802, Telecontrol Equipment and Systems - Part 6: Telecontrol Protocols Compatible with ISO Standards and ITU-T Recommendations - Section 802: TASE.2 Object Models"*, August, 1996, ftp://ftp.sisconet.com/epri/iccp/iccp802.doc.

[24]   M. Mansouri-Samani and M.Sloman, *"Monitoring Distributed Systems"* in Network and Distributed Systems Management, Addisson-Wesley, 1994.

[25]   D.F.D'Souza and A.C.Wills, *"Objects, Components, and Frameworks with UML - The Catalysis Approach"*, Addison-Wesley, 1999.

A Modeling Language for Design Processes in Chemical Engineering

Markus Eggersmann, Claudia Krobb, and Wolfgang Marquardt

Lehrstuhl für Prozesstechnik, RWTH Aachen
Turmstraße 46, 52056 Aachen, Germany
{Eggersmann, Krobb, Marquardt}@lfpt.rwth-aachen.de

Abstract. In the chemical industry a major task is to design manufacturing processes, which is a creative, ill–defined, complex, and incompletely understood problem. Currently, the design knowledge and experience are mostly located in the mind of the individual engineer. It is desirable to move this knowledge at least in part into a computer supported environment. The key element is a proper model of the design process itself, which is essential for understanding and supporting it. Examples of design processes in chemical engineering are analyzed in order to learn what is necessary to describe them completely. This leads to requirements a modeling language for design processes in chemical engineering has to fulfill. Because the requirements are not completely met by existing work process modeling languages, an existing language is modified and enhanced to allow representation of the processes under consideration. The new language is used to formally represent sample work processes from the application domain. It serves as a basis for the development of a support functionality which guides the designing engineer through his work within a prototypical environment for mathematical modeling of chemical processes, a major part of the design process.

1 Introduction

Almost all design processes share the common characteristics of being creative, ill–defined, complex and incompletely understood tasks. Therefore, there have been efforts to move the knowledge of design processes from the experienced design engineers into computer–based support environments in a number of domains in recent years. While these efforts mainly concentrate on software process and enterprise modeling, capturing knowledge about chemical engineering processes has gained increasing attention, since process design not only determines plant but also manufacturing cost. Therefore, there is a great need to render design knowledge of individual engineers explicit and to implement it in a support environment. Ideally, such an environment automates well understood parts of the design process, and guides the engineer through the rest of the design process without restricting his freedom to try new approaches. The expected result of such an environment is faster and better process design, because routine tasks are transferred from the engineer to the computer. At the same time this speeds things up and avoids mistakes, allows the engineer to devote more of his time to the creative parts of process design, and provides inexperienced users with the necessary assistance. In

A.H.F. Laender, S.W. Liddle, V.C. Storey (Eds.): ER2000 Conference, LNCS 1920, pp. 369–382, 2000.
© Springer-Verlag Berlin Heidelberg 2000

addition, incremental documentation can be integrated into the design process and even partly automated, thus simplifying later design revisions and plant debottlenecking by making the design process more transparent.

The key element in developing any design support environment is a model of the design process. To develop such a model it is necessary to identify the tasks that are performed while the chemical process is evolved from the initial design specifications to the final engineering design.

We therefore examine chemical process design processes from a work process oriented viewpoint, but keep in mind the product oriented perspective as well. We analyze examples of design processes with regard to the information they contain and use, to learn what is necessary to describe them completely. This leads to a set of requirements a modeling language for chemical process design processes has to fulfill. These requirements are detailed in the following section. Because the requirements are not met by existing work process modeling languages, a new process modeling language is proposed in Section 3. The language is used to formally represent example work processes. The formal models serve as a basis for the development of a support functionality which guides the design engineer during his work within the prototypical process modeling environment ModKit, as discussed in Section 4. An overview on related work is given in Section 5. Section 6 concludes the paper with a discussion of perspectives on future research.

2   Characteristics of Design Processes

As an example design process we consider the design of a polyamide6 production process. Polyamide6 is a widely used polymer material and its production is therefore important in the chemical industries.

2.1   Case Study of a Process Design

The case study of the polyamide6 process allowed us to identify the necessary work processes and the occurring product data, by which we mean all information which is created or used during the design of a chemical process. The general structure of the example design process is shown in Fig. 1 and discussed in the following paragraph.

Process design always starts with a *literature study* to gather information (e.g. about properties of the occuring substances, reaction data, etc.). Information gathering activities occur repeatedly throughout the design process because it is not possible to predict at the beginning exactly which information is necessary. Then a *decision* between batch and continuous operation has to be made. This decision influences all the following design activities. Further, it is based on inexact information because it has to be done at a very early stage when not all necessary data are present. These properties result in a complex decision process which triggers many sub–processes such as data collection and model–based analysis. The subsequent activities mainly depend upon the specific chemical process under consideration. In our case, the *flowsheet synthesis* reveals that there are three main parts of the process which have to be designed: *reaction, separation,* and *extrusion*. These parts may be designed separately but it has to be taken into account that they depend strongly upon each other. The design of the different parts is mainly based on mathematical models which describe their physical behavior. Models allow the prediction of the performance of the plant or parts of it and therefore lead to a faster design with less laboratory experiments required. A major task is to develop these models and to use them for the analysis of the process units in simulation studies. In a last step, the final *decision about the plant concept* is made based on the simulation studies. Details about the case study can be found in [8].

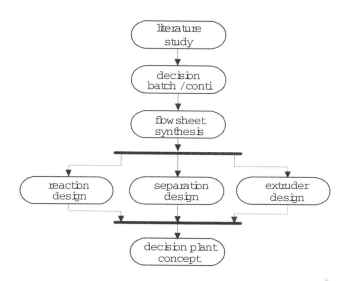

Fig. 1. Coarse structure of the scenario

The scenario may appear clear and well structured because the description given above is of a very coarse granularity and it is just an *example* design process. On a more detailed level or when describing design processes in general instead of an example many characteristics which are difficult to deal with will appear, as described in the following paragraphs.

2.2 Characteristics of Activities during the Design Process

Types of Activities. We identified three different types of activities, which are *synthesis*, *analysis*, and *decision*. During *synthesis* e.g. a flowsheet structure or a model of a process unit is generated. Generally speaking, an artifact is created on the basis of a specification during a *synthesis* task. The outcome is hardly predictable because the activity is highly creative and inventive. The result of the *synthesis* task is evaluated according to different criteria in an *analysis* task. *Analysis* is mostly based on a mathematical model of the chemical process. In this case the building of a model is a *synthesis* task and the investigation of the process is an *analysis task*. The *analysis* provides the information whether and to what extend the specific criteria are met and supplies arguments for a *decision* task. The *decision* serves to rate the criteria, to evaluate different arguments and to decide for one alternative. It may also require further *synthesis* or *analysis* activities. Each of these types has distinct properties that suggest a classification of the activities during process design according to this scheme.

Phases of Activities. We also identified different phases during the engineering process, which are *goal setting, planning, execution* and *documentation*. During *goal setting* the overall goal is fixed without considering how it can be achieved. Based on this goal, the activities which have to be performed to reach the goal are determined during the *planning* phase. It answers the question which activities are required and in what order. Planning might involve scheduling of unrelated activities, which have to be done within a given time frame. Once the activities have been fixed they have to be *executed*. During execution the goal description gradually looses its significance because it is refined by and by to become a description of the activities themselves before their execution. In the end, it is identical with the description of the activity itself. Only during goal setting and planning the goal has to be modeled explicitly. The *execution* has effects on the other phases, it might e.g. be realized that the goal is not attainable so that either the goal itself or the plans to reach it have to be modified. These phases occur on different and nested levels of detail. The first goal is to design a process with certain characteristics. The successive analysis steps supply results which allow to generate more detailed goals. This is only possible until a certain level of detail is reached, while several activities are necessary to reach a goal. The final phase is the *documentation* phase to trace important results and decisions.

The different types of activities occur in all the phases during process design. The phases as well as the types of activities have to be incorporated into the design process representation to explicitly capture the structure of the design process and to identify what data should be contained in an activity description.

2.3 Workflow in Design Processes

In the above mentioned scenario the simulation studies during the reactor section design (see Fig. 1) were investigated in detail, to reveal characteristics of the workflow. It is not possible to predict the order in which activities are executed in general. In some cases the order is irrelevant, whereas in others an activity depends very much on the results of the previous ones.

During the specification of the simulation it does not matter whether the flowrate of a stream or the corresponding temperature is specified first. Later, when different parameters are varied and sensitivity studies are performed, the result of a simulation run is analyzed by the designer. During this analysis the designer decides what to do next. This decision for the next activity may in some cases be automated but in many /most cases it requires the creativity and knowledge of the designer. Like the simulation studies many other work processes in chemical engineering do not follow a clearly defined algorithm. Decisions are made on a hunch, experience is implicitly employed, ingoing and outgoing information is not made explicit and might not be documented, new ideas are introduced into the process. This is caused by both the creative character of the work processes, where new ideas are continuously introduced and tested, and the fact that the engineers often simply cannot specify how exactly they will solve a given problem in advance. This situation makes it necessary to allow incomplete and imprecise models of work processes to be defined in a work process modeling language, which can be refined and made more precise when additional information has been gathered. Even if the activities related to one task are known, their order might be unknown in advance. An example for this case is the initialization of mathematical models for simulation. This is a trial–and–error procedure, where various heuristic strategies exist, but it can not be predicted which one will work, so the order in which the activities are executed depends on the intuition and experience of the engineer.

Besides such creative processes there are also activities which are almost of an algorithmic nature. When a mathematical model of a chemical process or a part thereof has been developed it might be necessary due to numerical problems or a lack of information to reduce the dimensionality of this model and thereby simplify it. These activities are fairly well understood and always performed in the same order [10].

There is a wide spectrum of design activities from algorithmic ones with a predefined order to activities which mainly depend on results of other activities so that their order is not fixed.

2.4   Dependencies between Work Processes and Products

We pointed out that activities may depend strongly on the results of other activities. These results are part of the *product data* [3,21], as we call all information which is generated or used during process design. The way how an activity is executed may vary considerably with the information available. In addition, a change in product data often requires the re–execution of an activity and checking the product data allows to determine whether an activity can be performed. This is possible because each activity is associated to certain product data that is required to execute the activity. Therefore, a detailed product model has to be available to the work process model. It is not sufficient to regard the products as some abstract objects or unstructured documents, because they directly influence the execution of the work process activities on a fine level of granularity. In order to allow computer–based support, it is necessary that executability of activities can be determined by checking the current state of product data. Most activities depend on the availability of input data, which is contained in the product on which the activity operates. The quality of the available information can also be of relevance to the activity. Another factor which influences the selection of an activity to be executed is the *goal* pursued. Goals are usually specified as a description of the desired product, which again results in a close tie between work process and product model.

3   A Representation of Design Processes

In the last section, we gave an overview over the characteristics of design processes, using chemical process design as an exemplary application domain. Since a lot of work has already been done concerning work process modeling in various domains, ranging from business process modeling to software engineering, there exist a lot of work process (meta–)models with varying focal points. The field ranges from purely graphical representations to formal logical formulas, with emphasis on any number of characteristics of processes in the application domain under consideration. An assessment of existing languages showed that none of them covers all the requirements identified above. This is illustrated for some of the formalisms in a very simplified way in Table 1. They include NATURE [11,23], which was developed for requirements engineering, Coordinates [19], an enterprise process modeling formalism with emphasis on the process industries, the Petri net variant Grafcet [6], often used to represent the function of supervisory control systems and adopted for work process representation [5,18] and APEL [9], which was developed for software engineering process modeling. A detailed evaluation of these and other representations can be found in [7].

Table 1. Characteristics of work process representations. + denotes satisfied, – denotes not satisfied, parentheses imply restrictions

	NATURE	Coordinates	Grafcet	APEL
types/phases of activities	–	–	–	–
predefined control flows	+	+	+	+
flexible control flows	–	–	–	+
product model	+	+	–	+
input information	+	+	(+)	+
qualitative information	–	–	–	–
goals	(+)	–	–	–
decisions	+	–	–	–
actors	–	–	–	+
resources	+	+	(+)	–

Obviously, NATURE as well as APEL meet most, but not all, of the requirements identified above. This at least partly results from the fact that both, requirements and software engineering processes, share a common characteristic with chemical process design: they are all highly creative processes.

Trying to reuse as many of the existing ideas as possible, but still accounting for all the peculiarities of the work processes under consideration, we decided to use the NATURE process metamodel as a basis and modify and enhance it to better meet the requirements of design process modeling in chemical engineering.

Activities in NATURE are represented by the *Context* in which they are executed. This includes the goal of the user as well as the current situation with regard to selected products. Depending on whether it consists of several subactivities, is an executable piece of code, or includes a decision about the next step to be executed, a further specification of the activity is given for each context. The order in which subactivities of a complex activity can be executed is defined using Petri nets, UML state charts or other formalisms. There is currently no option to omit specification of a control flow and nevertheless group related activities together. The outcome of an activity execution is not modeled in the NATURE framework, but a link between an activity and the product it modifies is maintained. Decisions are documented using the IBIS framework [24]. Goals are matched with menu items of the tools in which process support is to be offered. Situations are predefined product configurations, to which an activity is related if it is executable in that situation. Because of the level on which support is given, a single user system is assumed, and actors and qualifications are not modeled. Each activity contains references to the software which is involved in its execution.

In a first step during our work, the development of mathematical models of chemical processes was analyzed. This led to a number of changes to the original NATURE framework on which we based our efforts [18]. In a second step design processes were considered, which caused additional changes to the process metamodel. We give an account of the current consolidated version of our work process metamodel in the following. We emphasize the changes we made with regard to the original NATURE process metamodel. Our work process metamodel is part of the Conceptual Lifecycle Process Model CLiP [3], which was developed on the basis of the chemical engineering data model VeDa [20] and covers both product data and work processes related to chemical engineering.

Fig. 2 shows the process metamodel and an exemplary instantiation on the class level using UML notation.

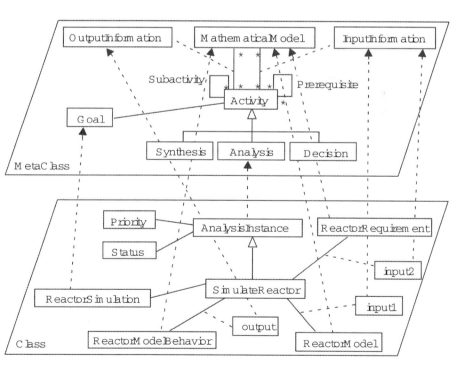

Fig. 2. Metamodel of the activity representation. Dotted arrows indicate instantiation

When the model is used by a process support tool like COPS (see Section 4), the activities defined on the class level are instantiated by the tool and refer to the specific instances of the product model on which they operate. Please note that not all attributes described in the following paragraphs are depicted in the diagram for reasons of readability.

The main classes of the metamodel are *Activity,* which describes a task, and the association classes *InputInformation* and *OutputInformation*, which provide a crosslink to the product metamodel. These are shown in more detail in Fig. 3. The topmost class in the product metametamodel is *System.* It's instance *MathematicalModel* is included in the figure to show the relationships between work process and product model on the metamodel level. The product metamodel is described in detail in [3]. To describe activities or work processes, detailed knowledge of the related product model is necessary to provide the *Input–* and *OutputInformation*.

An *Activity* relates to the *Context* in NATURE. We renamed it because we do not interpret it solely as a combination of a situation and an intention (*Goal*), but as a description of a task which can be executed. Execution can be achieved by either executing a set of *Subactivities*, by running a piece of code which is an *Implementation* of an algorithm to solve the task, or by prompting the user to do something, which is also done using executable code, i.e. an *Implementation*. One or more execution sequences for subtasks can be specified using *SchedulingConditions*, but it is also possible to omit any ordering relation between them, thus allowing flexible control flow. In this case the order of execution is determined on the basis of the available information at the time of execution. *Prerequisites* specify activities which have to be enacted before an activity, but which do not necessarily provide important information on which the second activity builds. This mechanism enables forced scheduling of activities which are not necessarily related via products.

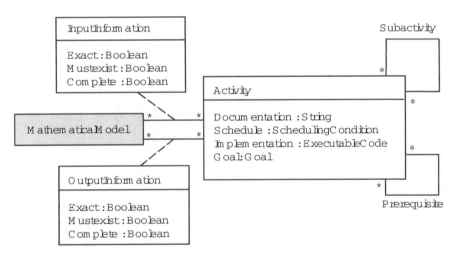

Fig. 3. Metaclasses Activity, InputInformation and OutputInformation with their attributes

The attributes *Status* and *Priority* were introduced on the class level for implementation purposes. The *Status* of an activity changes from non–enactable to enactable and then enacting, until the activity is finally terminated sucessfully, fails or is aborted. Activities may also become suspended during their execution. The priority attribute allows the support system to select the most fitting activity to be suggested to the user. The priority of an activity may change during the execution of a work process, e.g. because it is a prerequisite for another activity which the user wants to execute. The attributes of the metaclasses and classes are explained in more detail in [7].

Decisions are considered to be a special kind of *Activity,* therefore they are modeled as a subclass of *Activity* with additional attributes and relations to cover issues, positions, and arguments as defined in the IBIS model [24]. Observe that these are not displayed in Fig. 2 for reasons of readability. Other subclasses of *Activity* are *Synthesis* and *Analysis*, which reflect the specific characteristics and associations of these types of activities. The subclasses provide some structuring of the set of activities according to common characteristics of the activities. It is shown in [17] that it is not possible to define a hierarchy of activities so that a more specialized activity can always be substituted for the more general activity without influencing the executablity of a sequence of activities. Therefore, the class hierarchy of activities can not be used to draw conclusions about executability. The metamodel does not yet provide a mechanism to reflect phases of work processes as described in Section 2.2.

As pointed out in Section 2, an important characteristic of design processes is the lack of complete and reliable information before an activity is executed. We account for this by providing the *InputInformation* association class, which classifies the information it references concerning its exactness, completeness and existence. Thus it is possible to express that certain data have to be available before the activity can be executed, but that they do not have to be exact but might be estimates. Similarly, *OutputInformation*, which is not modeled in NATURE but which we regard as an important part of the description of an activity, also classifies information. It refers to the information generated or modified by an activity, which again does not necessarily have to be complete or accurate.

An important facet of design processes which is not yet covered by our metamodel are actors and their cooperation processes. Actors are characterized by their expertise, which restricts the set of actors who can possibly execute a given activity. Therefore, a qualification concept may link actors and activities in the future. Furthermore, availability of actors and their involvement in other projects may have to be taken into account.

The metamodel also lacks a structured goal representation which allows the creation of goal hierarchies and structuring of related goals. This is a complex issue which requires further research. For the time being, we decided to model goals as optional attributes of activities, since goals can not be specified on all granularity levels of work processes (see Section 2.2).

Our metamodel is formally specified in the form of O–Telos frames within the deductive object manager ConceptBase [16]. This representation ensures that the model is logically sound and consistent. The frame definitions of the *metamodel* and example models of parts of *chemical engineering design processes* are available online at http://www.lfpt.rwth–aachen.de/Research/Modeling/CBFrames/ .

4   Computer–Based Support

To allow verification and evaluation of the process modeling language introduced in the last section, we are currently implementing the prototypical Context Oriented Process Support tool COPS, which is based on this process representation. COPS aims at supporting users during work processes by offering guidance through predefined process fragments and suggestions for executable activities. In a first version, we concentrate on work processes occuring during the generation of mathematical models of chemical processes. These work processes are a part of the design process and create the foundation for simulation and evaluation of process alternatives. Mathematical model development was also used to evaluate the first version of VeDa and its implementation in ModKit as presented in [18]. Reasons for the selection of these processes as evaluation examples are their variation in complexity, the fact that they are better understood than chemical process design as a whole, and the low number of tools involved in their execution.

As depicted in Fig. 4, COPS interacts with the ModKit environment, which allows modeling and simulation of chemical processes [5]. ModKit consists of a number of modules which provide functionality for the specification of different model types, documentation, generation of simulator code etc. All product related models are stored in the Repository Of the ModKit Environment ROME [26]. The Central Coordination is responsible for starting and terminating tools on request by tools or users.

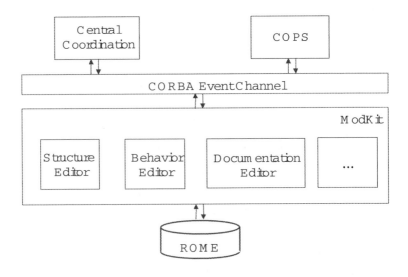

Fig. 4. Logical structure of the ModKit/COPS/ROME environment [21]

The communication between ModKit and COPS is realized using a CORBA event channel. COPS allows users to request suggestions concerning the next activity to be executed in a given context. The first prototypical implementation is restricted to an interaction between COPS and the *Behavior Editor*, in which mathematical equation systems specifying the behavior of a chemical process element can be modeled.

COPS receives support requests from ModKit components via a CORBA IDL interface. The call informs COPS about the tool issuing the request and the type of support requested, i.e. guidance through a predefined process fragment or suggestion of a single activity for the given situation. COPS can then query this tool as well as any other tool about their current situation (the selected products) and the available and executable actions provided by the tool. This information is used to map predefined activities or process fragments with the given situation. The activities are stored in different "modules" which are related to a given intention, such as modeling or simulation. Lookup is executed in the module(s) specified by the user, thus limiting the search space. The available product information is compared to the input information required by each activity. If too many executable activities are found, additional information may be requested from the user, e.g. about his goals. The result is a list of activities which are executable in the given situation and related to the goal the user currently pursues. Identified activities are suggested by highlighting the appropriate menu items in the respective tool if a corresponding menu item exists. In the case of automatically executable activities or activity sequences, an option to start the execution is offered to the user. Additional tools which are required for the execution of an activity or process fragment are automatically started by COPS through appropriate requests to the Central Coordination. If guidance was requested, the activities which are part of the identified process fragment are listed in a COPS agenda window in the correct order, and the respective required activity is highlighted both in the agenda window and the appropriate tool.

Due to the early stage of the implementation effort, COPS has not yet been applied to a complex real world problem, and empirical data concerning user acceptance is therefore not available at this time.

5   Related Work

The modeling of work processes has been investigated in a number of fields, ranging from business process modeling to software engineering.

We see two main differences between business processes and chemical process design processes. First, business processes are usually highly algorithmic and have only a small set of possible results per activity, while chemical design processes usually can produce a large number of unpredictable results. Second, the acquisition of information, which forms a major part of chemical process design, is usually not modeled in business process models, because information there is often just a link to a document.

The common characteristic of requirements engineering, software engineering and chemical process design is the creativity of their work processes. Nevertheless, supporting environments for these processes so far have been mainly prescriptive in the sense that they control the workflow and do not allow the user to deviate from the prescribed method. Environments of this type in the field of chemical engineering often do not even contain an explicit model of the design process itself [12]. Overviews over software process representation formalisms and process–centered environments can be found in [1,14].

Recent developments in support environments for chemical process design show a shift towards reducing the shortcomings listed above. The KBDS System [2], which was developed at the University of Edinburgh, provides a methodology and prototypical support environment for conceptual process design, but does not offer context–based suggestions of activities. Based on KBDS, the DRAMA system has been developed. DRAMA supports documentation of decisions during process design, but not the design process itself. On the other hand, ConceptDesigner aims at automating large segments of the design process, and reducing interaction with the human user [12]. The work of the *n*–dim group at Carnegie Mellon University is focused on the support of the collaborative and information management aspects of the design process [25,27].

Within the long–term, interdisciplinary research center IMPROVE at RWTH Aachen new methodologies towards a product and process integrated support of chemical process design are investigated [22]. The goal here is to integrate existing tools for chemical process engineering into a computer–based support environment and to provide support on both the management [13] and the fine–grained single user level [15]. The work presented here is part of these efforts. While the computer science oriented projects within IMPROVE mainly work on developing and implementing new support functionality, which is supposed to be valid for a wide range of engineering design activities, we concentrate on understanding chemical process design processes. Our ideas are implemented into prototypical software systems and evaluated by means of engineering case studies. These experiments provide feedback concerning the work process modeling activity on the one hand. On the other hand they form the basis for the development of a new generation of concepts for computer–based support by an integrated tool environment. In an iterative processes, our results are taken up by the computer science partners and integrated into their efforts within IMPROVE.

6   Conclusions

So far, neither our own system nor any of the other existing systems for design process support satisfy all needs of design engineers. While some approaches focus on the collaborative and distributed character of the design process, we currently do not cover this particular problem, but concentrate on the activities occurring during the design process and their representation, and especially on their relation to the product model.

Future work will have to integrate these ideas with concepts for representing actors and resources and an adequate representation of goals. Due to the evolutionary character of process design, version management for product models also plays an important role. However, before these problems can be tackled our process metamodel has to undergo another revision to allow representation of phases of activities and integrate them into the overall view of our process model.

Our long term goal is to provide an executable formalism for the representation of all aspects of design processes in chemical engineering which allows to capture design knowledge and enables context based support of engineers during process design.

Acknowledgement

The work presented in this contribution was sponsored by the Deutsche Forschungsgemeinschaft DFG within the work of *Sonderforschungsbereich 476 IMPROVE* and *Graduiertenkolleg Informatik und Technik* at RWTH Aachen.

References

1. Armenise, P., Bandinelli, S., Ghezzi, C., Morzenti, A.: A Survey and Assesment of Software Process Representation Formalisms. International Journal of Software Engineering and Knowledge Engineering, Vol. 3 No. 3, 401–426 (1993)
2. Bañares–Alcántara, R., Lababidi, H.M.S.: Design Support Systems for Process Engineering – II. KBDS: An Experimental Prototype. Computers chem. Engng, Vol. 19, No. 3, 279–301 (1995)
3. Bayer, B., Marquardt, W.: A Product Data Model for Design Data in Chemical Engineering. Technical Report LPT–2000–09, Lehrstuhl für Prozesstechnik, RWTH Aachen (2000)
4. Bayer, B., Schneider, R., Marquardt, W.:Integration of Data Models for Process Design – First Steps and Experiences. 7th International Symposium on Process Systems Engineering, Keystone, Colorado (2000)
5. Bogusch, R., Lohmann, B., Marquardt, W.: Computer–aided Process Modeling with ModKit. Submitted to Comp. Chem. Engng. (1999)
6. Davis, R., Alla, H.: Petri Nets for Modeling of Dynamic Systems – A Survey. Automatica, Vol. 30, No. 2, 175–202 (1994)
7. Eggersmann, M., Krobb, C., Marquardt, W.: A Language for Modeling Design Processes in Chemical Engineering. Technical Report LPT–2000–02, Lehrstuhl für Prozesstechnik, RWTH Aachen (2000)
8. Eggersmann, M., Schneider, R.: A Scenario for Design Processes in Chemical Engineering. Technical Report LPT–2000–06, Lehrstuhl für Prozesstechnik, RWTH Aachen (2000)
9. Estublier, J, Dami, S., Amiour, A.: APEL: A Graphical yet Executable Formalism for Process Modeling. Automated Software Engineering (ASE), March (1997)
10. Gerstlauer, A., Hierlemann, M., Marquardt, W.: On the Representation of Balance Equations in a Knowledge Based Process Modeling Tool. CHISA'93, Prag, September (1993)
11. Grosz, G., Rolland, C., Schwer, S., Souveyet, C., Plihon, V., Si–Said, S., Ben Achour, C., Gnaho, C.: Modelling and Engineering the Requirements Engineering Process: An Overview of the NATURE Approach, Requirements Engineering, Vol. 2, 115–131 (1997)
12. Han, C., Stephanopoulos, G., Douglas, J.M.: Automation in Design: The Conceptual Synthesis of Chemical Processing Schemes. In: Stephanopoulos, G., Han, C.: Intelligent Systems in Process Engineering, Part I: Paradigms from Product and Process Design, Academic Press, San Diego, pp. 93–146 (1995)
13. Jäger, D., Schleicher, A., Westfechtel, B.: AHEAD: A Graph–Based System for Modeling and Managing Development Processes. To appear in: M. Nagl, A. Schürr (Eds.): Proceedings AGTIVE (Applications of Graph Transformations with Industrial Relevance), Rolduc, LNCS, Springer–Verlag (1999)
14. Jarke, M., Marquardt, W.: Design and Evaluation of Computer–Aided Process Modeling Tools. In: Davis, J.F., Stephanopoulos, G., Venkatasubramanian, V.: Intelligent Systems in Process Engineering, AIChE Symp. Ser. 312, Vol. 92, 97–109 (1996)
15. Jarke, M., List, T., Weidenhaupt, K.: A Process–Integrated Conceptual Design Environment for Chemical Engineering, Proc. 18th Intl. Conf. on Conceptual Modeling (ER '99), Paris, France, 520–537 (1999)
16. Jeusfeld, M.A., Jarke, M., Nissen, H.W., Staudt, M.: Concept Base. Managing Conceptual Models about Information Systems. In: Bernus, P., Mertins, K., Schmidt, G. (Eds.): Handbook on Architectures of Information Systems, Springer, Berlin Heidelberg (1998)
17. Krobb, C.: Entwicklung einer Spezialisierungshierarchie für Modellierungsschritte im objekt–orientierten Datenmodell VeDa. Diploma project LPT–thes–1997–05, Lehrstuhl für Prozesstechnik, RWTH Aachen (1997).
18. Lohmann, B.: Ansätze zur Unterstützung des Arbeitsablaufes bei der rechnerbasierten Modellierung verfahrenstechnischer Prozesse. Dissertation, Fortschritt–Berichte VDI, Series 3: Verfahrenstechnik, No 531, Düsseldorf (1998)

19. Mannarino, G.S., Leone, H.P., Henning, G.P.: A Task–Resource Based Framework for Process Operations Modeling. In: Proceedings of FOCAPO '98, Snowbird, Utah (1998)
20. Marquardt, W. and the VeDa group: The Chemical Engineering Data Model VeDa. Part 1–6, Technical Reports LPT–1998–01 to LPT–1998–06, Lehrstuhl für Prozesstechnik, RWTH Aachen (1998)
21. Marquardt, W., von Wedel, L., Bayer, B.: Perspectives on Lifecycle Process Modeling. In: Proceedings of FOCAPD'99, Breckenridge, Colorado (1999)
22. Nagl, M., Westfechtel, B. (Eds.): Integration von Entwicklungssystemen in Ingenieuranwendungen – Substantielle Verbesserung der Entwicklungsprozesse. Springer, Heidelberg (1998)
23. Pohl, K.: Process–Centered Requirements Engineering. John Wiley & Sons, New York (1996)
24. Rittel, H., Kunz, W.: Issues as elements of information systems. Working Paper No. 131, Institute of Urban and Regional Development, Univ. of California, Berkeley, California (1970).
25. Subrahmanian, E., Konda, S.L., Dutoit, A., Reich, Y., Cunningham, D., Patrick, R., Thomas, M., Westerberg, A.: The n–dim Approach to Creating Design Support Systems. Proceedings of DETC '97, 1997 ASME Design Engineering Technical Conference, Sacramento, California (1997)
26 von Wedel, L., Marquardt, W.: ROME: A Repository to Support the Integration of Models over the Lifecycle of Model–based Engineering Processes, ESCAPE–10, Florence, Italy, May 7–10 (2000)
27. Westerberg, A., Subrahmanian, E., Reich, Y., Konda, S., and the n–dim group: Designing the Process Design Process. Computers chem. Egngn., Vol. 21, Suppl., S1–S9 (1997)

VideoGraph: A Graphical Object-Based Model for Representing and Querying Video Data

Duc A. Tran, Kien A. Hua, and Khanh Vu

School of Electrical Engineering and Computer Science
University of Central Florida
Orlando, FL 32816, USA.
{dtran, kienhua, khanh}@cs.ucf.edu

Abstract. Modeling video data poses a great challenge since they do not have as clear an underlying structure as traditional databases do. We propose a graphical object-based model, called VideoGraph, in this paper. This scheme has the following advantages: (1) In addition to semantics of video individual events, we capture their temporal relationships as well. (2) The inter-event relationships allow us to deduce implicit video information. (3) Uncertainty can also be handled by associating the video event with a temporal Boolean-like expression. This also allows us to exploit incomplete information. The above features make VideoGraph very flexible in representing various metadata types extracted from diverse information sources. To facilitate video retrieval, we also introduce a formalism for the query language based on path expressions. Query processing involves only simple traversal of the video graphs.

1 Introduction

We deal with the modeling aspect of *video database management systems - VDBMSs* [8] in this paper. Playing an important role in VDBMSs, this is the process of designing the high-level abstraction of raw video to facilitate various information retrieval and manipulation operations. It determines what features are to be used in the retrieval, and therefore in the indexing process. Other components, such as content analysis tools and query processing techniques, are also more or less dependent on it. A good model is essential to enabling a wider range of applications.

Much research has been done in the area of video modeling/retrieval based on audio-visual content, such as audio, color, texture and motion (e.g., [13,11, 3,4,18,7]). The advantage of this approach is that features can be extracted automatically. The low-level schemes, however, are very limited in expressing queries. For example, it would be difficult to ask for a video clip showing the sinking of the ship in the movie "Titanic" using only color, texture, and audio information. In contrast, video data models based on semantic content ([6,15,14, 10,1,9,5,17,12]) are capable of supporting more natural queries. They, however, must rely partially on manual annotation. A limitation of this approach is that semantic content can be ambiguous and context dependent. This problem can be

A.H.F. Laender, S.W. Liddle, V.C. Storey (Eds.): ER2000 Conference, LNCS 1920, pp. 383–396, 2000.

controlled by limiting the context and providing multiple semantic descriptions for different types of applications. We focus on the semantic level in this paper. In particular, we consider two types of video semantics:

- Event Description: This type of description indicates the video segments that show a particular event. Some examples of event description are ["Ship colliding with iceberg", 35th minute - 37th minute] and ["Captain dying", 48th minute - 49th minute]. The first description indicates that the 2-minute long segment, from time 35th minute to time 37th minute, shows the scene of a ship colliding with an iceberg. Similarly, the second description indicates that the scene of the captain dying begins at the 48th minute and ends at the 49th minute of the video.
- Inter-event Description: This type of description describes the temporal relationship between two events. Some examples of interevent description are ["Ship collides with iceberg before it sinks "] and ["Captain dies after the ship sinks"]. This type of semantic information is not associated with any video segment. Instead, it states the temporal relationship between two events. As an example, such information can be obtained from the script. Under this circumstance, we can report on the order of various events, but not the exact locations of their occurrences in the video stream.

We note that each event mentioned in an inter-event description may or may not have an explicit event description. In the first inter-event description given above, only the "colliding" event has an event description (i.e., from the 35th minute to the 37th minute). This situation arises in practice since information extractors are not perfect. As an example, an extractor based on explicit models may recognize a ship and the "colliding" scene, but lacks the knowledge to determine the "sinking" event.

Existing semantic-level video data models support only event, not the inter-event, descriptions. In this paper, we introduce a new model, called VideoGraph, which can accommodate both in one framework. This enables us to deduce implicit event descriptions, and therefore retrieve implicit video information as well. For instance, using the following metadata: ["Ship collides with iceberg before it sinks"], ["Captain dies after the ship sinks"], ["Ship colliding with iceberg", 35th minute - 37th minute], and ["Captain dying", 48th minute - 49th minute], we can imply the implicit temporal description ["Ship sinking", 37th minute - 47th minute] This implicit semantic allows us to answer queries such as "showing the scene of the sinking ship". Existing techniques based on only explicit descriptions would fail to process these queries. Our new capability is reminiscent of implicit information in a deductive database management system. To the best of our knowledge, video semantic implicity has not been exploited in the literature.

Another new feature considered in our model is the flexibility to associate an event with a temporal Boolean-like expression. This enhancement allows us to handle uncertainty. For instance, we can associate the event "Jack went aboard the ship" with the temporal description "[5th minute, 7th minute] or [9th minute, 10th minute]." This expression indicates that the event must be in one of the two video segments. Such conditions occur when we infer implicit video semantics from incomplete semantics. Thus VideoGraph can also deal with incomplete information, in addition to implicit information. To facilitate video retrieval, we present a formal query language for VideoGraph. The language is an extension to relational calculus with path expressions and has both a clear declarative and operational semantics. It is simple yet powerful enough to allow formulation of complex queries. Query processing involves only simple graph traversal.

The remainder of this paper is organized as follows. We formally present the details of VideoGraph in Section 2. The query language formalism is described in Section 3. The algorithm for computing the implicit video information is presented in Section 4. Finally, Section 5 summarizes our approach and indicates some directions for future research.

2   Video Data Representation

In this section, we introduce a new model called VideoGraph that has the desirable features discussed earlier. This is a follow-up of our previous work in [16]. Intuitively, a VideoGraph database is a set of edge labeled rooted graphs, each representing the semantics of a single video. Such a graph is a collection of nodes, each in turn representing a single event. Nodes are linked to each other based on their containment and temporal relationships [2]. The relationship between two nodes captures inter-event information involving the two corresponding events. A node may be associated with explicit temporal information or not. If not, its implicit information can still be obtained by reasoning on the graph. We discuss this in section 4.

Let us assume that **ATYPE** denotes the set of all integers, real numbers and strings. Let **TYPE** be a finite set containing special strings classified as data types in the video database, in other words, **TYPE** = $\{type_1, type_2, .., type_p,$ **TIME**$\}$ where $type_i$ is a string. Each type $type_i$ has a value domain, denoted as $\mathrm{dom}(type_i)$, such that $\mathrm{dom}(type_i) \subseteq$ **ATYPE** if $i \leq p$, and $\mathrm{dom}(\textbf{TIME}) = \Omega$ that is defined later. In what follows, unless we explicitly mention, concepts to be defined are considered within the same context of a single video.

Definition 1. *[Atomic object] Let l_1, l_2, .., l_n ($n \geq 1$) be strings in* **TYPE***, v_1, v_2, .., v_n values such that $v_i \in \mathrm{dom}(l_i)$. Let us consider a graph G containing a node O, called the* root *of G, and n nodes O_1, O_2, .., O_n with the following properties:*

- *Node O stores a unique integer value.*
- *For each $i \in \{1, 2, .., n\}$, node O_i stores value v_i.*
- *For each $i \in \{1, 2, .., n\}$, there is a directed link labeled l_i from O to O_i.*

Then node O is said to be the atomic object *represented by graph G and the value stored in the node is called the* identifier *of the atomic object.*

Definition 2. *[Object] Objects are recursively defined as follows.*

1. *Any atomic object is an object.*
2. *Let G_1, G_2, .., G_n represent n objects, O_1, O_2, .., O_m be m single nodes $(n+m>0)$ and l_1, l_2, .., l_{n+m} be $n+m$ strings in* **TYPE**. *Then the graph G built below represents an object.*
   - *G contains a node O, which is called the* root *of G, nodes O_i's and graphs G_j's for $i \in \{1, 2, .., m\}$ and $j \in \{1, 2, .., n\}$.*
   - *Node O stores a unique identifier for the object.*
   - *For each $i \in \{1, 2, .., n\}$, there is a directed link from O to the root of G_i labeled l_i.*
   - *For each $i \in \{1, 2, .., m\}$, there is a directed link from O to O_i labeled l_{i+n} and node O_i stores a value $v_i \in dom(l_{i+n})$*

We can say that node O is the object *represented by graph G.*

Here after, for flexibility, the terms object and internal node are interchangeably used. That is, we implicitly refer to an object as an internal node and vice versa. Links between any two nodes in the above definitions are called *c-links*. If a node O_1 has a c-link labeled l departing from it and going to another node O_2, then l is a *component* of O_1 and the *type* of node O_2 with respect to O_1. O_2 is the *value* of component l of O_1. Components of an object can be duplicated. Furthermore, if O_2 is an internal node, it is also called a *sub-object* of O_1 which in turn is said to be a *super-object* of O_2.

We divide the set of objects into two classes, key objects and non-key objects. Informally, a key object is an object that has temporal related information telling what parts of the video associate with the object. That can be complete or incomplete (that is, the exact video segment for an event is not known). Before giving the formal definition of a key object, we need to describe a new type for video temporal values.

Definition 3. *[I-expression] I-expressions (I stands for "interval") are defined as follows: (1) For any t_1 and t_2 integers ($t_1 \le t_2$), $[t_1, t_2]$ is an i-expression. It is also classified as an* interval. *(2) If p and q are two i-expressions, then so are (p&q) and ($p \mid q$). The meaning of operations "&" and "$\mid$" is that if an object associates with an i-expression p&q (or $p \mid q$), then it associates with both (or one) of p and q.*

Let Ω denote the set of all i-expressions. We recall that dom(**TIME**) $= \Omega$. I-expressions tell how to look up the video and can be used to express incomplete information. For instance, $[10, 20]$ corresponds to the video segment from time 10 to time 20; $[15, 18]$ & $[25, 30]$ corresponds to a set of two video segments, one represented by $[15, 18]$ and the other one by $[25, 30]$; $[25, 28] \mid [15, 20]$ corresponds to only one video segment, $[25, 28]$ or $[15, 20]$, but it is not known to be which.

Definition 4. *[Key object] If O is an object having* **TIME** *as a component and its corresponding* **TIME** *value is an i-expression, then O is categorized as a* key object.

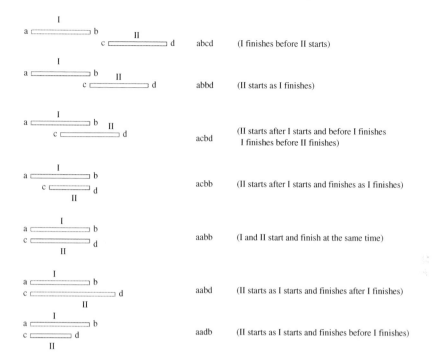

Fig. 1. Temporal relations between two intervals

A video graph is built on objects each corresponding to a single event in reality and their temporal relations reflects inter-event descriptions. An inter-event relation can be described as an element of the set **REL** = {ABBD, ABCD, ACBD, ACDB, AACD, AABB, ACBB} - the set of seven temporal interval relations illustrated in Figure 1 where I and II are two events in the video, and [a, b] and [c, d] are their video temporal segment information respectively.

Definition 5. *[Video graph] Given a video V, let graphs $G_1, G_2, .., G_n$ $(n \geq 1)$ represent its objects, that are not sub-objects of any others, the video graph of V is an edge labeled rooted graph G defined as follows: (1) The root A of the graph stores the identifier of V; (2) G contains every graph G_i for $i \in \{1, 2, .., n\}$; (3) For each $i \in \{1, 2, .., n\}$, there is a link from A to the root of G_i labeled **SEM**; (4) If two internal nodes O_1 and O_2 of G have a temporal relation $r \in$ **REL**, then there is a directed link from O_1 to O_2 labeled r and classified as an r-link.*
A video database *consists of a number of video graphs, each representing knowledge about an individual video in the database.*

The VideoGraph model encompasses both video data and the structure of them. Before going any further, let us give an example of a video database. We are interested in a single movie and the video graph for it is shown in Figure 2 where directed solid lines describe c-links and dotted curves describe r-links. In this video graph, the atomic objects are o_4, o_7, o_9, o_{10}, o_{11} and o_{12}. The key objects

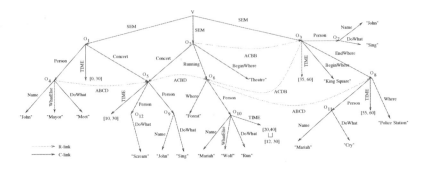

Fig. 2. An example of video graphs

are o_1, o_3, o_5, o_8 and o_{10}. The others are non-key objects. Note that o_{10} has incomplete temporal information ($[12, 30] \mid [20, 40]$) because it is not known that node o_{10} associates with which interval, $[12, 30]$ or $[20, 40]$. VideoGraph allows different kinds of temporal information, which are encapsulated in one type Ω (i-expressions). This distinct property was not possible in previous models. Hence, they have limited capabilities of utilizing semantics extracted from diverse and dynamic knowledge sources because all the events that do not have a temporal description or that have incomplete temporal information are not considered.

3   Query Language

We will now introduce a content-based mechanism to support users' querying the video database. We focus on answers of the following types: (1) Video segments: When the user searches for video segments that satisfy some semantic constraint. E.g., "I would like to watch those scenes where the ship was going to crash into an iceberg". (2) Semantic contents: When the user asks for information about a video clip. E.g., "Tell me what the last scenes of the movie are about?" We limit our queries to retrievals only, however update queries can be easily embedded. Our video query language is based on path expressions formulated according to the graphical structure of the database. The queries are also expressed on the inter-object temporal relationships. Prior to formally presenting what the syntax and semantics of queries are, we give some necessary definitions.

Without loss of generality, we consider only one video graph in the database which we are going to formulate queries on. Let o_1 and o_2 be two nodes and L labeled l be a link from o_1 to o_2. Then we can express $o_1 = \mathbf{from}(L)$ and $o_2 = \mathbf{to}(L)$.

Definition 6. *[Strict Path Sample] A strict path sample is of the form $l_1 \rightarrow l_2 \rightarrow ... \rightarrow l_n$ where l_1, l_2, .., $l_n \in \mathbf{TYPE}$ and $\forall\ i \in \{1, 2, .., n-1\}$, there exist two c-links L_i and L_{i+1} labeled l_i and l_{i+1} respectively such that $\mathbf{to}(L_i) = \mathbf{from}(L_{i+1})$.*

Definition 7. *[Path Sample] Let θ be either $\rightarrow$ or $\xrightarrow{*}$ and $l_1, l_2, .., l_n \in \textbf{TYPE}$.
A path sample is of the form $l_1 \ \theta \ l_2 \ \theta \ .. \ \theta \ l_n$ where there exist components l_{i1},
$l_{i2}, .., l_{iki}$ ($i \in \{1, 2, .., n-1\}$) such that $l_1 \rightarrow l_{11} \rightarrow l_{12} \rightarrow ..\rightarrow l_{1k1} \rightarrow l_2 \rightarrow$
$l_{21} \rightarrow l_{22} \rightarrow .. \rightarrow l_{2k2} \rightarrow .. \rightarrow l_n$ is a strict path sample.*

Our query language is based on the expression of paths. We allow the user
to formulate any path he or she is interested in. The path may or may not exist
in the graph, in other words, it either conforms to a path sample or does not
conform to any. In order to facilitate users' querying, we introduce the notion of
path expression.

Definition 8. *[Path Expression] Let O be an internal node, l be a label, θ be
either $\rightarrow$ or $\xrightarrow{*}$. A path expression is recursively defined as follows: (1) O. is a
trivial path expression, which contains only a node; (2) If α is a path expression,
then so are $\alpha\theta$ and $\alpha\theta(O)$.*

A query will be run successfully if it contains path expressions, each conforming to some path sample of the video graph. Otherwise, the query returns
nothing. For the semantics of the query, we define the validity property of a path
expression below.

Definition 9. *[Path Validity] Let θ be either $\rightarrow$ or $\xrightarrow{*}$. The validity of a path
expression is presented as follows:*

1. *Any trivial path expression is a valid path expression*
2. *A non-trivial path expression PE is valid if the conditions below hold:*
   - *If O. θl appears in PE, then there must exist a c-link L labeled l such
     that $O = \textbf{from}(L)$*
   - *If $l_1 \ \theta \ l_2$ appears in PE, $l_1\theta l_2$ must be a path sample*
   - *If $l_1(O) \ \theta \ l_2$ appears in PE, then $l_1\theta l_2$ must be a path sample and there
     must be two c-links, L_1 labeled l_1 and L_2 labeled l_2, such that $O = \textbf{to}(L_1)$
     $= \textbf{from}(L_2)$*
   - *If $l_1\theta l_2(O)$ appears in PE, $l_1\theta l_2$ must be a path sample and there must
     exist a c-link L labeled l_2 such that $O = \textbf{to}(L)$*

Having presented the above concepts, we are ready to give a description of
queries and their semantics. Our queries use three salient operators, $\triangledown, \oplus, \otimes$.
Given a node A, $\triangledown(A)$ gives the set of all nodes that are reachable from it by
traversing the graph (Node B is said to be reachable from node A if and only if
there exists a valid path expression starting with A and ending with B). $\oplus(A,
B)$ returns the temporal relation between two internal nodes A and B. $\otimes(A, ie)$
returns the temporal relation between the **TIME** value of internal node A and
an i-expression ie. The result of $\oplus$ and $\otimes$ operations must be an element of the
set **REL**.

3.1 Syntax of VideoGraph Queries

Definition 10. *[Atomic Condition] Let assume that $\odot$ is an operator in the set $\{<, >, =, \geq, \leq\}$, PE and PE' are path expressions, o_1 and o_2 are internal nodes, ie is an i-expression, $r \in$ **REL**, $v \in$ **ATYPE**. An atomic condition has one of the forms: (1) PE; (2) $PE \odot v$; (3) $PE \odot PE'$; (4) $\oplus(PE, PE') = r$; (5) $\otimes(PE, ie) = r$; (6) $o_1 \in \bigtriangledown(o_2)$.*

Definition 11. *[Condition] Let p and q be themselves conditions, $f(O)$ be a condition in which O appears, a condition is recursively defined to be one of the following: (1) any atomic condition; (2) $\neg p$, $p \wedge q$, $p \vee q$, or $p \Rightarrow q$; (3) $\exists O(f(O))$, where O is a variable representing an internal node; (4) $\forall O(f(O))$, where O is a variable representing an internal node.*

In this definition, $\exists$ and $\forall$ are two quantifiers in traditional logic and are said to *bind* to the variable O.

Definition 12. *[Free variable] A variable is said to be* free *in a condition or a sub-condition (a condition contained in a larger condition) if the (sub-)condition does not contain an occurrence of a quantifier that binds it.*

Now is time for the formal syntax of a VideoGraph query.

Definition 13. *[VideoGraph Query] A VideoGraph query is defined as an expression of the form:*

$$(TIME) \Leftarrow PE(O_1, O_2, .., O_n), SN, SC(O_1, O_2, .., O_n) \qquad (1)$$

or

$$(SEM) \Leftarrow PE(O_1, O_2, .., O_n), SN, SC(O_1, O_2, .., O_n) \qquad (2)$$

where SN is a predefined node, O_i's $(i = 1..n)$ are internal node variables and the only free variables in the formulas $SC(O_1, O_2, .., O_n)$ and $PE(O_1, O_2, .., O_n)$, $SC(O_1, O_2, .., O_n)$ is a condition containing one or more occurrence of each O_i, $PE(O_1, O_2, .., O_n)$ is a path expression containing one or more occurrence of each O_i, **TIME**, **SEM** *are special symbols describing what kinds of output are to be returned, an i-expression (temporal information) or a set of nodes (objects).*

3.2 Semantics of VideoGraph Queries

To complete the formalism of the query language, we must state which value assignments to free variables in a condition make the condition true. A query is evaluated in any given instance of the video database. Let each free variable O_i in a condition $SC(O_1, O_2, .., O_n)$ (we call it F for brevity) be bound to a value o_i (a node in the graph). With respect to the video database and for the assignments of values to variables, the condition F must be true if one of the following holds ($\odot \in \{<, >, =, \leq, \geq\}$):

- F is an atomic condition PE, and PE is a valid path expression.
- F is an atomic condition $PE \odot v$, and PE is a valid path expression which by traversing we can obtain a node value v' (an atomic value or an object identifier) that makes the comparison $v' \odot v$ true.
- F is an atomic condition $PE \odot PE'$, and PE, PE' are valid path expressions which by traversing we can obtain two nodes whose values, v_1 and v_2, make $v_1 \odot v_2$ true.
- F is an atomic condition $\oplus(PE, PE') = r$, and PE, PE' are valid path expressions which by traversing we can obtain two internal nodes such that their implicit or explicit **TIME** values are related to each other by relation r.
- F is an atomic condition $\otimes(PE, ie) = r$, and PE is a valid path expression which by traversing we can obtain an internal node whose implicit or explicit temporal relationship with the i-expression ie is equivalent to relation r.
- F is an atomic condition $o_1 \in \triangledown(o_2)$, and o_1 is reachable from o_2.
- F is of the form $\neg p$, and p is not true; or of the form $p \wedge q$, and both p and q are true; or of the form $p \vee q$, and one of them is true; or of the form $p \Rightarrow q$, and q is true whenever p is true.
- F is of the form $\exists O(f(O))$, and there is some assignment of values to the free variables in $f(O)$ and variable O, that makes it true.
- F is of the form $\forall O(f(O))$, and there is some assignment of values to the free variables in $f(O)$ that make it true no matter what value is assigned to variable O.

Now we need to be clear how the answer of a query is returned, that is, we need to formally define what the semantics of $PE(O_1, O_2, .., O_n)$ is, given an assignment of values to variables $O_1, O_2, .., O_n$. The query returns nothing if PE is not a valid path expression. Otherwise, PE represents a set of nodes that are obtained by traversing the video graph based on PE. We call those nodes instances of the path expression PE.

Definition 14. *[Instances] An object O is an* instance *of a path expression PE if one of the following holds:*

- *PE is O.*
- *PE is $PE' \to l$, and $\exists O'$ an instance of path expression PE' and a c-link L labeled l such that $O' = $ **from**(L) and $O = $ **to**(L).*
- *PE is $PE' \overset{*}{\to} l$, and $\exists O'$ an instance of path expression PE' and a c-link L labeled l such that $O = $ **to**(L) and **from**$(L) \in \triangledown(O')$.*
- *PE is $PE' \to l(O)$, and $\exists O'$ an instance of path expression PE' and a c-link L labeled l such that $O' = $ **from**(L) and $O = $ **to**(L).*
- *PE is $PE' \overset{*}{\to} l(O)$, and $\exists O'$ an instance of path expression PE' and a c-link L labeled l such that $O = $ **to**(L) and **from**$(L) \in \triangledown(O')$.*

Since the query outputs can be of two types, an i-expression or a set of node identifiers from which semantic contents are withdrawn, we consider the following cases: (1) **SEM filter**: The answer to a query is a set of objects (node identifiers), each being an instance of the path expression $PE(O_1, O_2, .., O_n)$ where O_i's $\in \bigtriangledown(SN)$ and O_i 's make the condition $SC(O_1, O_2, .., O_n)$ true. (2) **TIME** filter: The answer to a query is an i-expression which is computed by applying "&" operation on the **TIME** values of all the instances of the path expression $PE(O_1, O_2, .., O_n)$ where O_i's $\in \bigtriangledown(SN)$ and O_i's make the condition $SC(O_1, O_2, .., O_n)$ true.

3.3   Examples

Now we present some examples of queries. The video database is the video graph illustrated in Figure 2 where there is only one video. Suppose the identifier of that video is V.

Example 1. The scence going on in a police station where John does not appear.

$$(TIME) \Leftarrow o., V, o. \to Where = \text{``PoliceStation''} \wedge (\forall o'(o' \in \bigtriangledown(o)$$
$$\wedge o'. \to Person \to Name = \text{``John''}$$
$$\Rightarrow \oplus(o', o) = ABCD \vee \oplus(o, o') = ABCD) \quad (3)$$

Example 2. The information about the scene from time 15 to time 20 and from time 30 to time 32.
$$(SEM) \Leftarrow o., V, \otimes(o, [15, 20]\&[30, 32]) = AABB \quad (4)$$

Example 3. The information about the scene in which Mariah was trying to run away from a wolf.

$$(SEM) \Leftarrow o., V, \exists o'(o'. \to Person \to Name = \text{``Mariah''}$$
$$\wedge o'. \to Person \to DoWhat = \text{``Run''}$$
$$\wedge o'. \to Person \to WhatElse = \text{``Wolf''} \wedge \oplus(o', o) = ACDB) \quad (5)$$

Example 4. How the scene containing John's singing and Mariah's running begins.

$$(SEM) \Leftarrow o. \to BeginWhere, V, \exists o_1(\exists o_2(o_1 \in \bigtriangledown(o) \wedge o_2 \in \bigtriangledown(o)$$
$$\wedge o_1. \to Person \to Name = \text{``John''} \wedge o_1. \to Person \to$$
$$DoWhat = \text{``Sing''} \wedge o_2. \to Person \to Name = \text{``Mariah''}$$
$$\wedge o_2. \to Person \to DoWhat = \text{``Run''})) \quad (6)$$

The path expression $PE(O_1, O_2, .., O_n)$ in the query can contain more than one node variable. This capability makes our language more expressive and more powerful. The SN (start node) component is not necessarily the root, which helps limit the graph traversal scope for the results. An example of such queries is below.

Example 5. The following query returns the video segments that are included in a "running" scene and where Mariah appears.

$$(TIME) \Leftarrow o. \xrightarrow{*} o', o_2, \exists o''(o''. \rightarrow Running = o)$$
$$\wedge o'. \rightarrow Person \rightarrow Name = \text{``Mariah''} \qquad (7)$$

In this query, the PE expression contains two node variables o and o'. Note that, the start node is not the root, but node o_2 of the graph.

4 Implicit Information Inference

Having presented the video model and its related formal query language, one issue left is how to compute the implicit information from a VideoGraph database. This can be considered a preprocessing refinement phase. In our VideoGraph model, an internal node may or may not have a **TIME** link from it. If not, its temporal information can still be obtained by traversing the graph taking into account r-links to other nodes that have a temporal value. In other words, we are able to obtain implicit complete semantics from event descriptions, inter-event descriptions and incomplete information. In what follows, we introduce a simple way of how to do it. The algorithm converts an instance of the video data model to another called refined graph. The refined graph has direct temporal features associated with the internal nodes (that is, each node has a **TIME** link and a **TIME** value). Now we define what a refined graph is and shortly introduce a simple algorithm to compute the refined graph of a given video graph. Given a single video, its video graph can be represented as a tuple $G = (V, e, f, g, h)$ where V is the set of nodes; $e: V \rightarrow \{\text{INTERNAL, LEAF}\}$, a unary function returning the type of each node; $f: V \rightarrow \mathbf{ATYPE} \cup \Omega$ a unary function returning the value stored in each node; $g: V \times V \rightarrow \mathbf{TYPE} \cup \text{VOID}$, returning the label of the link from one node to another, VOID if there is no c-link between them; $h: V \times V \rightarrow \mathbf{REL} \cup \text{VOID}$, returning the temporal relationship between one node to another, VOID if there is no r-link between them.

Definition 15. *[Refined Graph] Given a video graph $G = (V, e, f, g, h)$ of a video. Its* refined graph *is also a VideoGraph $G_1 = (V_1, e_1, f_1, g_1, h_1)$ with the following properties.*

- $V \subseteq V_1$
- $card(V_1) = card(V) + card(V_2)$ where $V_2 = \{v \in V \mid e(v) = INTERNAL \wedge \forall v_1 \in V: g(v, v_1) \neq \mathbf{TIME}\}$
- *For each $v \in V_2$, there exists only a node $v_1 \in V_1 - V$ such that $g_1(v, v_1) = TIME$. Conversely, for each $v_1 \in V_1 - V$, there is only a node v of V_2 such that the above condition holds.*
- $e_1(v) = e(v)$ if $v \in V$, LEAF otherwise
- *If $v, v_1 \in V$, then $f_1(v) = f(v)$, $g_1(v, v_1) = g(v, v_1)$*
- *For $v, v_1 \in V_1$ such that g_1 is not yet defined for, $g_1(v, v_1) = VOID$*

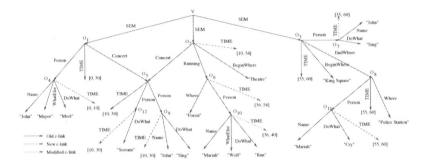

Fig. 3. A refined graph example

- $h_1 (v, v_1) = VOID \ \forall \ v, \ v_1 \in V_1$
- *If* $v_1, v_2 \in V$ *such that* $e(v_1) = e(v_2) = INTERNAL$, *and* $v_1', v_2' \in V_1$ *such that* $g_1(v_1, v_1') = g_1(v_2, v_2') = \textbf{TIME}$, *then the relationship between* $f_1(v_1)$ *and* $f_1(v_2)$ *must not conflict with* $h(v_1, v_2)$.

The corresponding refined graph of the video graph in Figure 2 is in Figure 3. We note that node o_{10} now has a more meaningful **TIME** value, which tells that it associates with i-expression [36, 40], not like in the source video graph where o_{10}'s certain temporal information was unknown. Now comes an algorithm to determine the refined graph of a video graph.

Algorithm 1 *[Refinement Algorithm] Given a video graph* $G = (V, e, f, g, h)$ *for a video. The corresponding refined graph is built as the following steps.*

1. V_2 *is initialized to the set of all non-key objects in the source video graph.*
2. *For each node* $v \in V_2$, *add a new node storing* $[0, \infty]$ *and add a link labeled* **TIME** *from* v *to the new node.*
3. *Initialize* $UpdateCounter$ *and* $UpdateCounter_1$ *to 0.*
4. *For any two internal nodes* v *and* v' *that are connected by an r-link, adjust the value of their* **TIME** *node so as to satisfy* $h(v, v')$. *If a node is a key object with a complete* **TIME** *value, do not change the value. If one of the values is changed, increase* $UpdateCounter_1$ *and* $UpdateCounter$ *both by 1.*
5. *If* $UpdateCounter_1 \neq 0$, *go back to step 3.*
6. *For any two internal nodes* v *and* v' *such that the former is a sub-object of the latter, adjust the value of their* **TIME** *node so that the* **TIME** *values of* v *and* v' *are related by the relation ACDB (containment relationship). If a node is a key object with a complete* **TIME** *value, do not change the value. If one of the values is changed, increase* $UpdateCounter_1$ *and* $UpdateCounter$ *by 1.*
7. *If* $UpdateCounter_1 \neq 0$, *assign it to 0 and go back to step 6.*
8. *If* $UpdateCounter \neq 0$, *go back to step 3.*
9. *Remove all the r-links resulting in a new graph, which is the refined graph of* G.

In a refined graph, all the r-links have been removed. It has another property that every object (internal node) has a temporal descriptor which is captured by applying the refinement algorithm above on the source video graph. Thus whenever a node is visited for its temporal information, no further graph traversal is needed. The merit of the algorithm is that it helps reduce the overhead of query processing. For example, if we only have the source video graph (i.e., without applying the refinement algorithm) as in Figure 3 and if the user very often wants to watch the scene associated with object o_{10}, then the query system will have to compute the implicit temporal description of o_{10} many times back and forth.

In environments where most objects in the video database are accessed with high frequency, it is a good idea to build the refined graph just once, ahead of time, and store it in the database. Whenever the video graph is updated, its corresponding refined graph is recalculated. Subsequent query processing steps will be taken on the refined graph, resulting in more processing overhead being reduced. However, it is not always necessary to save it permanently, especially if we take into account the cost of storing an additional graph. Depending upon the context where the video database is, we have to consider the tradeoff between the efficiency of using the refined graph and the storage cost charged. From that standpoint, we can decide to compute it, whether or not on the fly, as the user questions the video.

5   Conclusions

We have introduced in this paper a new video data model called VideoGraph. This model can capture not only descriptions of individual events, but also their interevent relationships. To fully benefit from these two types of semantics, we have provided an algorithm to compute the implicit information from the initial metadata. Another contribution is the support for uncertainty. This arises because information extraction tools usually fail to produce the complete metadata set and reasoning on incomplete information often results in uncertain information. To address this, VideoGraph associates each event with an i-expression. For instance, a disjunctive ("|") form can be used to indicate that the video event must occur in one of the listed video segments. Our scheme hence is more intelligent and expressible than existing techniques. To facilitate video retrieval, we have also presented in this paper a declarative query language based on path expressions. To the best of our knowledge, it is the first query language of this kind. Path expressions are easy to use yet powerful enough to allow the expression of fairly complex video queries.

This paper focuses on the formalism of the VideoGraph approach. For the time being, we are working toward an implementation of a networked video database system based on the model. In this effort, we are investigating various query processing techniques and different storage organizations suitable for storing video data. Another important direction is to extend VideoGraph framework to incorporate abstraction (such as classification, generalization and aggregation) and presentation mechanisms in which we believe our graphical approach will offer better flexibility and possibilities.

References

1. S. Adali, K. S. Candan, S.-S. Chen, K. Erol, and V. S. Subrahmanian. Advanced video information system: Data structures and query processing. *ACM-Springer Multimedia Systems Journal*, 1996.
2. J. F. Allen. Maintaining knowledge about temporal intervals. *Communications of the ACM, Springer-Verlag*, 26(11), 1983.
3. E. Ardizzone and M. L. Cascia. Automatic video database indexing and retrieval. *Multimedia Tools and Applications*, 4(1), 1997.
4. S.-F. Chang, W. Chen, H. Meng, H. Sundaram, and D. Zhong. Videoq: An automated content based video search system using visual cues. In *ACM Multimedia*, November 1997.
5. T.-S. Chua and L.-Q. Ruan. A video retrieval and sequencing system. *ACM Transactions on Information Systems*, 13(4), 1995.
6. C. Decleir and M.-S. Hacid. A database approach for modeling and querying video data. In *IEEE Data Engineering*, Australia, 1999.
7. N. Dimitrova, T. McGee, and H. Elenbaas. Video keyframe extraction and filtering: A keyframe is not a keyframe to everyone. In *Int'l Conf. on Information and Knowledge Management*, 1997.
8. A. K. Elmagarmid, H. Jiang, A. A. Helal, A. Joshi, and M. Ahmed. *Video Database Systems: Issues, Products, and Applications*. Kluwer Academic Publishers, 1997.
9. A. Hampapur, R. Jain, and T. Weymouth. Production model based digital video segmentation. *Journal of Multimedia Tools and Applications*, 1(1), 1995.
10. H. Jiang, D. Montesi, and K. Elmagarmid. Videotext database systems. In *IEEE Int'l Conf. on Multimedia Computing and Systems*, Ontario, Canada, June 1997.
11. J. Oh and K. A. Hua. An efficient and cost-effective technique for browsing, querying and indexing large video databases. In *ACM SIGMOD*, Dallas, TX, May 2000.
12. E. Oomoto and K. Tanaka. Ovid: Design and implementation of a video-object database system. *IEEE Trans. on Knowledge and Data Engineering*, 5, August 1993.
13. Y. Rui, S. Huang, and S. Mehrotra. Constructing table-of-content for videos. *ACM Springer-Verlag Multimedia Systems*, 7(5), 1999.
14. T. G. A. Smith and G. Davenport. The stratification system: A design environment for random access video. In *Proceedings of the 3rd Int'l Workshop on Network and Operating System Support for Digital Audio and Video*, La Jolla, CA, 1992.
15. D. Swanberg, C.-F. Shu, and R. Jain. Knowledge guided parsing in video databases. In *SPIE Conf. on Image and Video Processing*, volume 1908, San Jose, CA, February 1993.
16. D. A. Tran, K. A. Hua, and K. Vu. Semantic reasoning based video database systems. In *Proc. of 11th International Conference on Databases and Expert Systems Applications*, London, U.K, September 2000.
17. R. Weiss, A. Duda, and D. Gifford. Content-based access to algebraic video. In *IEEE Int'l Conf. on Multimedia Computing and Systems*, Boston, USA, 1994.
18. A. Yoshitaka, Y. Hosoda, M. Yoshimisu, M. Hirakawa, and T. Ichikawa. Violone: Video retrieval by motion example. *Visual Languages and Computing*, 7, 1996.

Object-Oriented Modelling in Practice: Class Model Perceptions in the ERM Context

Steve Hitchman

steve@eggconnect.net

Abstract. Whilst research has indicated that the practitioner use of entity-relationship modelling is problematic, proponents of object oriented modelling suggest that their paradigm offers both a new approach and also more effective modelling. This paper examines some practitioner perceptions of one object oriented modelling technique in the context of previous work on entity-relationship modelling. The findings show that there are similar practitioner issues arising from common underlying techniques and that object oriented modelling is problematic when used with project clients. However, the suggestion is that object oriented practitioners are not gaining insight from practitioner experience from more than twenty years of entity-relationship modelling practice.

1. Introduction

According to Silverston et al. [1] who promote a notation similar to the Oracle standard [2], entity-relationship modelling (ERM) has "... become the standard approach used towards designing databases. ... Currently, data modelling is a well-known and accepted method for designing effective databases". In a discussion of the principles of a relatively new method, Stapleton [3] states that "Certain models are fundamental to the process of development. ... the data model is crucial. ...In an object-oriented approach, it is difficult to imagine a system without classes defined in the core model set." However, some doubt has been raised about the practical advantages of entity-relationship modelling [4, 5]. For example, in a text on Data Warehousing, introduced by Inmon as "If you take time to read only one professional book, make it this one. ... There is wisdom contained within these pages [6], we can find these comments on the entity-relationship diagram:

"... but does it contribute to the understanding of the business ... unfortunately most people cannot hold a diagram like this in their minds and cannot understand how to navigate it usefully ... the truth is that relationships among data are best viewed dynamically on a screen, not by static road maps that the users try to hold in their minds ... (p.xxi - xxii) ... for queries that span many records or many tables, entity-relationship diagrams are too complex for users to understand and too complex for software to navigate ... entity-relation data models are a disaster for querying because they cannot be understood by users ..." (p.9) [7].

These comments are presumably based on a high level of practitioner experience, rather than on empirical evidence.

A.H.F. Laender, S.W. Liddle, V.C. Storey (Eds.): ER2000 Conference, LNCS 1920, pp. 397-408, 2000.
© Springer-Verlag Berlin Heidelberg 2000

Modelling within object oriented methodologies may be viewed as a new approach and is used because it is thought to be better or more appropriate. Class and object modelling (COM, sometimes called just class or more often object modelling) results in a similar diagram to the entity-relationship model. For example:

- classes are roughly analogous to entity-types;
- both classes and entity-types have attribute lists;
- there are similarly defined relationships between classes and entity-types;
- both classes and entity-types can be arranged in inheritance hierarchies (called sub-typing in entity-relationship modelling);
- and it is possible to transpose a diagram drawn using the Oracle notation to one drawn using the object modelling technique (OMT) notation [8], for example, and vice-versa.

However, there are certain differences in the notation available that means there may be some difficulty in transposition unless each diagram is restricted to using just the equivalent notation of the other. For example, there is no ternary relationship notation in the Oracle standard so the OMT modeller would need to refrain from its use and use a new entity-type (class) instead. Hitchman [9] discusses ternary relationships in detail in this context. Another example is the use of 'object responsibilities' in determining the functionality of particular classes in OMT, which would not directly occur in entity-relationship modelling and is related to the assignment of operations to classes.   OMT allows multiple inheritance networks as well as inheritance hierarchies. Therefore, the process of modelling is similar rather than the same. This similarity leads to the interesting question of whether class-modelling perceptions differ between OMT and ERM practitioners.   The literature on object oriented methods, which stresses the 'intuitive' nature of objects [8,10, for example] would suggest that using class modelling is better.

There is much to be learned from the practitioner literature on entity-relationship modelling. Assuming that class modelling is just new and different, perhaps because it is in some way 'intuitive', could lead to the wrong kind of training for class modellers. If class modellers are not exposed to the literature of ERM experience we would expect them to make the same mistakes as naïve entity-relationship modellers. Indeed, class modellers would simply be reinventing the wheel that entity model practitioners have taken so long to get into shape. The research questions asked here are firstly whether it is possible to examine practitioners using class modelling to see where the modelling process issues overlap with entity-relationship practice, secondly to see if the object oriented paradigm is perceived to be effective and finally whether practitioners are learning from experience.

2.   Methodology

Previous research [4] used a questionnaire to assess how practitioners used and perceived entity-relationship modelling. A similar research tool can be designed for an object oriented modelling technique and this means that the results of this research are additive, rather than comparative. This is an important point since the evidence gathered from the object oriented modellers adds to our knowledge of practitioners in general. It is not the intention here to *compare* findings with entity-relationship modellers, it would be difficult to find a comparative set of respondents in order to do this. The intention is to examine the complementary aspects of both modelling paradigms. In this context it is worth noting that DEKAF (Domain Expert Knowledge Approach Framework), which is used to help understand modelling in the practitioner domain [11], was used to help set the context for this study and proved effective in helping design the questionnaire, for example. Whilst the findings discussed here will be used to refine the DEKAF framework, that aspect of the research is beyond the scope of this paper.

There are many object oriented modelling analysis techniques in the literature, which have been available since at least 1991, for example [8,12]. Rather than examine the issue in general, one particular methodology was chosen since this offered the advantage of being able to be very specific with the research tool, and made analysis of responses easier. OMT was chosen as it was thought to be widely used and was also supported by at least one case tool that claimed a large user base - SELECT Enterprise [13]. OMT has been in evidence in, for example, the insurance industry [14]. It is also known that ideas from OMT were used in the development of UML [15].

The common elements of ERM and COM enable some re-use of the previous questionnaire, designed for entity-relationship practitioners, and many questions required little modification. There were specific new questions:

- questions were added to establish respondent's exposure to entity-relationship modelling;
- a multiple inheritance and ternary relationship question were added, representing some of the additional semantics available in OMT and not generally used by ERM practitioners.
- three questions were added to elicit information about the handling of association naming and class definitions;
- and the list of criteria to assess practitioner perceptions was refined by removing twelve of the original criteria and adding 5 others

3.   Analysis of Responses

This section has been written to enable the findings to be assessed in the context of the previous findings about ERM practitioners [4]. The underlying issues in interpretation of responses are the same as those discussed in the original paper and are not repeated here – these findings are best read in the context of the original research paper. It previously proved difficult to find an accurate list of ERM practitioners, and it similarly proved difficult to find an accurate list of active and experienced OMT modellers. A set of potential modellers resulted in 24 replies, representing a response rate of around 20%. As before, a telephone survey was used to boost response and to establish whether non-respondents were actually OMT modellers. The response rate is based on actual modellers, estimated to be 24% of the original survey. Although a small sample, this is a useful additive set of responses.

3.1   Survey Respondents

5 of the respondents considered their expertise to be mainly in real-time systems, 14 were business systems modellers, and 5 had experience of both. It is not known whether this is representative of OMT modellers as a whole, but these respondents are generally using COM in the same business context as ERM. The respondents had the following average experience:

	years of experience	formal days training	number of projects
Object Oriented Programming	2.8	5.1	4.2
Object Oriented Analysis	2.7	4.8	4.1
OMT	2	4.3	3.7
ERM	2.8	2.9	6

Figure 1. Respondents' experience.

Half of the respondents had no ERM experience, and the averages here are weighted by a few people with long ERM experience. Two respondents claimed 6 years OMT experience, and this reflects the relatively short availability of OMT. Only 5 respondents had no object oriented programming experience. All 24 respondents considered that they were 'still learning' about COM, and 13 thought they were reasonably expert.

3.2   Respondents' Perceptions of Data Model Use

Findings on the use of class models are similar to those for ERM. These are shown in Figure 2 and show the same perceived lack of confidence in understanding COM particularly by clients but also by other project analysts. This tends to contradict claims for COM as a 'natural' activity. However, there are a small number of modellers who seem to always succeed in using models which suggests that this is possible but difficult to achieve.

3.3   Respondents' Use of Model Semantic Constructs

To examine how respondents used the available semantic constructs, respondents were asked to model several short scenarios, defined completely by two way sentences. As before, these sentences could be directly translated to a diagram notation. It was decided to continue to use the original scenario descriptions as they were considered to be unambiguous 'English-like' definitions.

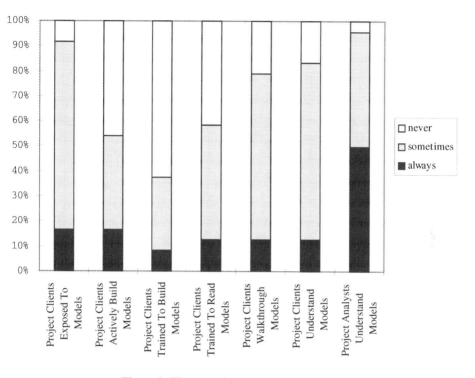

Figure 2 The use of data models (n=24)

This surprisingly caused difficulties for some respondents. For example, a scenario described by 'a project may include one or more employees' elicited the response 'what does 'may' mean" from one respondent. (In OMT terms, the normative language might be - there is some ambiguity in how relationships are defined - 'a project includes zero, one or many employees'.)

The first scenario examined four different one to many relationships in a simple task that required diagramming directly from unambiguous relationship sentences. Respondents were asked to deal with associations here as they normally did and the idea was to examine how the modellers dealt with simple associations. Formality with associations proved elusive with only 4 modellers being exactly correct in their cardinality and optionality, of which one respondent had 15 and another 10 years of ERM experience. The other two respondents had no ERM experience, but one was the most experienced OMT modeller. Seven modellers incorrectly modelled *every* relationship. This indicates that the notation does not facilitate accuracy, which seems to instead reflect practitioner experience. A large number of respondents (10) did not name associations at all.

It is worth stressing here that there is no evidence presented that the OO notation is better or worse in encouraging the precise specification of relationships, compared to

ERM notation. Even if the research from the questionnaires did show a statistically significant difference this does not necessarily help us understand the practitioner issue since:

- there is a problem that has to be addressed across notations anyway;
- the questionnaire process is not the same as modelling in the real world so we cannot be sure that the difference would be significant in practice;
- the difference in the experience of two groups may account for the difference in results;
- it is difficult to assess the effect of normative language on the results;
- OMT places no emphasis on using relationship naming to validate relationships in both directions, whereas some ERM techniques do – a technique could account for the difference rather than the notation.

Therefore, finding that both notations generate the same issue of relationship precision is more interesting than knowing whether a particular notation fares slightly better than another in a given artificial situation.

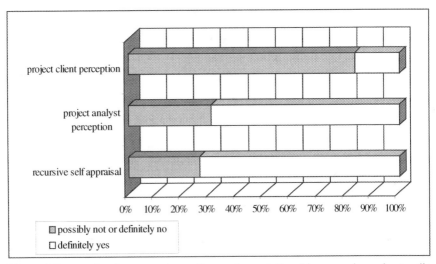

Figure 3 Respondent's perceptions of recursive relationship scenario understanding (n=24)

One to one recursive relationships were correctly modelled by 76% of respondents (slightly higher than the ERM modellers). Similarly many to many recursion was successfully modelled by 79% of respondents (exactly the same score for ER modellers) with 62% correctly modelling the cardinality. These figures include the 5 respondents who chose to (reasonably) model this with a whole part relationship – indicating that a wider range of notation will result in arbitrary decisions concerning how to model a particular scenario. No respondent decomposed the association with a new link class, although 2 answers used the OMT 'handbag' notation. Perceptions of how well the many to many recursive relationship is understood, shown in Figure 3, reflects the same doubt on the usability of the model notation as did the ERM practitioners. Just over 20% of respondents doubted their own understanding. The perceptions were that a slightly higher percentage of analysts would have problems and there was an overwhelming lack of confidence in clients (end users) being able to understand recursive relationships, which is consistent with the ERM modeller's perceptions.

A simple inheritance scenario was correctly modelled by 87% of respondents, which is noticeably more successful than the ER modellers. However, the same problems were encountered with more complex scenarios involving orthogonal sub-types, with only 37% of respondents registering correct diagrams. A new scenario with multiple inheritance produced an even lower figure of 29%, suggesting that the extra notation in OMT does not facilitate better diagramming. Indeed, there was some evidence here that OMT modellers tended to incorrectly promote an attribute to a new class when processing was confused with data. Inheritance notation perceptions reflect these findings and fare similarly to sub-typing in ERM. Figure 4 shows the respondent's perceptions. The results here are also similar to those of the ER modellers and again suggest that this notation is difficult to use with project clients and not well understood by analysts.

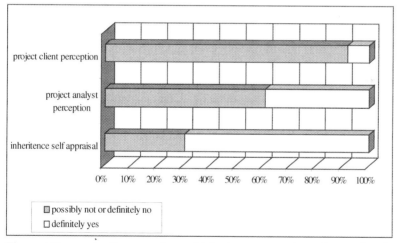

Figure 4 Respondent's perceptions of inheritance scenario understanding (n=24)

A new scenario that was meant to clearly imply the use of a ternary relationship notation only tempted 29% use, suggesting the notation is not popular. The same number of respondents choose to use a new class (the ERM answer) whilst there was a real variety of use of handbag notation in varying degrees (one to three handbags) and whole part. Given the simplicity of the scenario, the suggestion is that having a variety of notations that can be used even though a particular notation is provided promotes indiscriminate notation use.

The final scenario involved an exclusive relationship and was a little of a trick question as this notation is noticeably missing from OMT. 75% of respondents resorted to using sub-types (inheritance) which would be deemed incorrect in ERM.

In general, then, the findings are that respondents tend to use and perceive the notation in ways that are similar to those of ERM practitioners. One difference is that the greater variety of notation in OMT seems to result in a wide variety of notation used for essentially simple scenarios. This raises questions of how practitioners decide to use notations in complex situations.

3.4   Respondents' Perceptions of Modelling Usability for Analysis

This section examines how the modellers perceive the effectiveness of their notation and particularly whether the perception seems to be consistent with the ER modellers. Most of the criteria used are a subset of the ones previously used and the same scoring system based on [16, 17] is used here. OMT scores generally highly (Figures 5 and 6), and the results are surprisingly similar to the previous ERM scores. Confidence in scores in high, except for 'stability analysis' - a feature not directly used in OMT. Stability analysis also scored badly in ERM confidence results, this is part of some ERM methods.

> *KEY   Please **score** each criterion by ringing one of the following values:*
> 0   Makes things worse
> 1   No support / no effect
> 2   Poor support / not very beneficial
> 3   Good support / very beneficial
> 4   Excellent/ideal support

*Please indicate your level of **confidence** in answering each question by ringing one of the following values*:
1   Inadequate basis
2   Limited confidence
3   Confidence based on practical experience
4   Very confident, based on considerable experience in
5   a wide range of situations

Figure 5 Scoring system for respondent's perceptions of COM

Perception	Reference for Figures. 6&7	score	confidence
An aid to clear thinking	7	3.42	3.17
Elegant - generating a powerful sense of comprehension	5	3.13	3.0
Teachable and transferable to team members	16.8	3.09	2.96
Encourages good analysis practice	8	3.0	2.87
Communications for the analysis project	16	3.0	3.0
As a repository for future systems enhancement	23	2.96	2.74
As a basis for code generation	19	2.91	3.0
Quality improver	12	2.83	2.74
Consistently effective over many uses	10	2.74	2.73
As a basis for database generation	18	2.74	2.87
Elegant - ingeniously simple	6	2.7	2.96
Part of a defined IT process	20	2.7	2.83
To justify design and process	21	2.55	2.73
Easy to understand	16.1	2.43	2.91
Meaningful - results are clear, simple and unambiguous	16.3	2.41	2.96
To implement RAD philosophies	22	2.39	2.43
Formalised to enable validity checks	13	2.3	2.57
Facilitates Stability Analysis	14	2.29	1.81
Direct work saver	11	2.13	2.61
Ensures that the products of analysis are correct	9	2.09	2.91
Facilitate client understanding of the analysis area	17.1	2.04	2.87
Robust, resilient to misuse	15	2.0	2.78
An aid to client (end) user communication	16.2	2.0	2.96
To facilitate client (end) user project involvement	17	1.89	2.74
Facilitate client diagnosis of 'where we are now'	17.2	1.83	2.83
Facilitate client prognosis of 'where we want to be and why'	17.3	1.73	2.87

Figure 6 Respondent's perceptions of COM

Figure 7 shows the OMT perception results, ranked in score order, together with confidence scores. Noticeably the highest scores were for 'an aid to clear thinking' and 'elegant, generating a powerful sense of comprehension'. So OMT gets a definite thumbs up in this respect. On the other hand, it is very noticeable that any mention of 'client' in the criteria downgrades the score to the lowest, and in with these low scores is 'robust and resilient to misuse'. This clearly indicates that OMT helps thinking - but is not the formal client based solution. This tends to confirm the wider promotion of use cases (or prototyping) in several OO methods.

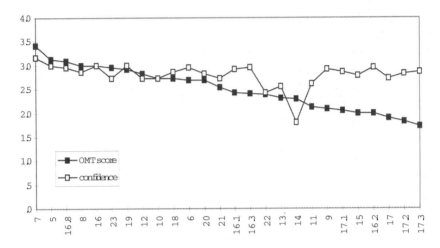

Figure 7 Perception and confidence scores for OMT respondents

Figure 8 shows the scores of ERM and OMT practitioners and whilst there can be no direct comparison in statistical terms it is possible to make the following observations. The scores are generally similar and similar trends are apparent - as would be expected from the assertion that these are complementary techniques. OMT scores markedly lower for 'database generation' which is consistent with ERM being used to specify relational databases. Similarly, client involvement scores markedly lower in OMT - implying that it will be less likely to be used with clients in requirements specification. However, OMT scores markedly higher as 'elegant, ingeniously simple' - despite the fact that ERM notation is less complex. This might imply that complex notation does not result in complex modelling. On the other hand OMT scores lower in terms of 'ensures that the products of analysis are correct' suggesting that the highest scoring 'an aid to clear thinking' may imply a corresponding lack of formality.

3.5   Using Previous ERM Knowledge

About half of the modellers had already experienced ERM, and half had only been exposed to OMT (possibly with other OO analysis notations). Respondents were asked how many textbooks, on entity-relationship modelling (ERM) they had either looked at and browsed through, or read thoroughly. As a check they were also asked to write down the title and author of what they thought the most useful book. The respondent with the most experience of ERM claimed to have browsed through 10 books, but not read any thoroughly. Only 7 respondents claimed to have thoroughly read *a* book on ERM. On average, respondents had browsed 2 books, but 9 respondents had not seen any ERM book. Only four respondents were able to name an ERM text. Four respondents named a text on object oriented modelling, confusing the issue, and the remaining 16 respondents could not name a text. This displays an overall lack of interest in what has been written about ERM. Combined with the few number of direct training days, this indicates that practitioners are unlikely to learn from either academics or (perhaps more importantly) practitioners who provide guidance and best practice advice. There is clear evidence here that OO practitioners are not likely to learn from the experience of ERM practitioners.

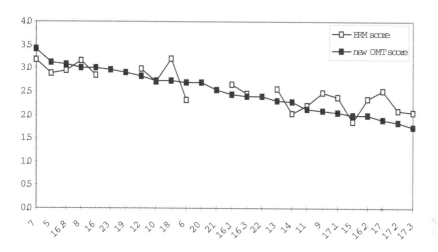

Figure 8 Perception scores for OMT respondents with scores from ERM practitioners

4.   Conclusions

It proved possible to use a very similar research tool to examine apparently diverse techniques. There is a noticeable similarity in the use of notation and perceptions of OMT and ERM practitioners, suggesting the activities are complementary rather than competitive. The normative language of the model, however, may have a strong impact on its use. The findings show particular differences, such as the arbitrary use of diverse notation available in OMT, whilst reinforcing the opinion that project clients are unlikely to benefit from the use of OMT notation - although a few modellers always use models with clients. This strongly contradicts the idea of classes and objects as being intuitive or natural thinking devices within diagram notations, and a perceived lack of understanding in use includes project analysts. The findings also suggest that OMT practitioners are not learning from the experience of ERM practitioners and the accumulated wisdom of practitioner texts.

References

1.   Silverston, L, Inmon, W.H. & Graziano, K (1997)   *The Data Model Resource Book* John Wiley
2.   Barker, R. (1989) *Case*Method: Entity Relationship Modelling* Addison Wesley
3.   Stapleton, J. (1997) Dynamic Systems Development Method Addison-Wesley
4.   Hitchman, S. (1995) Practitioner perceptions on the use of some semantic concepts in the entity-relationship model. *European Journal of Information Systems,* **4**, 31-40
5.   Goldstein, R., C. & Storey, V., C. (1990) *Some Findings On The Intuitiveness Of Entity-Relationship Constructs* in Lochovsky, F. H.(Editor) (1990) *Entity-Relationship Approach To Database Design and Querying* Elsevier Science
6.   Inmon, W. H. (1995) From the introduction in Kimball (1996)
7.   Kimball, R. (1996) *The Data warehouse Toolkit: Practical Techniques For Building Dimensional Data Warehouses* John Wiley
8.   Rumbaugh, J., Blaha, M., Permalani, W., Eddy, F. and Lorensen, W. (1991) *Object Oriented Modelling and Design* Prentice Hall
9.   Hitchman, S (1999) Ternary Relationships – to three or not to three, is there a question ? *European Journal Of Information Systems* Vol. 8 December pp. 224-231
10.   Coad, P. & Yourdon, E. (1991) *Object Oriented Analysis* Prentice Hall
11.   Hitchman, S. (1997) Using DEKAF To Understand Modelling In The Practitioner Domain *European Journal of Information Systems*, **6**(3) pp.181-189.
12.   Jacobson, I., Christerson, M., Jonsson, P., Overgaard, G. (1992) *Object-Oriented Software Engineering: A Use Case Driven Approach* Addison-Wesley
13.   SELECT, (1995) *Object Oriented Analysis & Design for Client-Server Development* Version 2.1, published by SELECT Software Tools Limited, England.
14.   Hitchman, S (1998) Is Class Modelling A Hidden Agenda For The Domain Expert ? - A Case Study in *Proceedings of the Seventh International Conference of Information Systems Development (ISD98)*, Bled, Slovenia, 21-23 September
15.   Rumbaugh, J., Jacobson, I., & Booch, G (1998) *The Unified Modeling Language: Language Reference Manual.* Addison-Wesley
16.   Law D, 1988 *Methods For Comparing Methods, Techniques in Software Development* NCC Manchester
17.   STARTS, 1987 *The STARTS Guide* 2nd Edition NCC Manchester 1987

ROVER: A Framework for the Evolution of Relationships *

Kajal T. Claypool, Elke A. Rundensteiner, and George T. Heineman

Worcester Polytechnic Institute, 100 Intitute Road,
Worcester, MA, USA
{kajal, rundenst, heineman}@cs.wpi.edu
http://davis.wpi.edu/dsrg

Abstract. Relationships have been repeatedly identified as an important object-oriented modeling construct. Today most emerging modeling standards such as the ODMG object model and UML have some support for relationships. However while dealing with schema evolution, OODB systems have largely ignored the existence of relationships. We are the first to propose comprehensive support for relationship evolution. A complete schema evolution facility for any OODB system must provide (1) primitives to manipulate all object model constructs; (2) and also maintenance strategies for the structural and referential integrity of the database under such evolution. We hence propose a set of basic evolution primitives for relationships as well as a compound set of changes that can be applied to the same. However, given the myriad of possible change semantics a user may desire in the future, any pre-defined set is not sufficient. Rather we present a flexible schema evolution framework which allows the user to define new relationship transformations as well as to extend the existing ones. Addressing the second problem, namely of updating the schema evolution primitives to conform to the new set of invariants, can be a very expensive re-engineering effort. In this paper we present an approach that de-couples the constraints from the schema evolution code, thereby enabling their update without any re-coding effort.
Keywords: Schema Evolution, Relationships, Object-Oriented Databases, Consistency Management.

1 Introduction

Object-oriented database (OODB) systems today are popular with communities that require the modeling of complex data types, such as CAD, multimedia applications and e-commerce applications [16]. The very nature of these applications has also brought forth the need for the OODB to provide mechanisms for (1) modeling associations between types and (2) for dynamically changing not

* This work was supported in part by the NSF NYI grant #IRI 94-57609. We would also like to thank our industrial sponsors, in particular, IBM for the IBM partnership award and Excelon Inc. for software contribution.

A.H.F. Laender, S.W. Liddle, V.C. Storey (Eds.): ER2000 Conference, LNCS 1920, pp. 409–422, 2000.
© Springer-Verlag Berlin Heidelberg 2000

only the data but also the structure of the types that contain them [18]. Research and industry alike have rushed forth to fulfill these requirements but have approached them as two independent problems with independent solutions. Relationship modeling has been a much studied topic such as in composite objects [10] or in aggregation relationships [2,5] and has been adopted to some degree in commercial OODB systems [14,15]. On the other hand, while most OODB systems to date provide some basic evolution support [19,14,1], the OODB community at large does not treat the relationship construct as a first-class citizen. Thus, while there is some relationship support at the data level, evolution support for relationships has not been investigated. Our work is the first to bridge this gap and to provide a complete framework for the support of schema evolution for relationships.

A complete schema evolution facility for any OODB system must provide the ability to (1) manipulate and change all of the object model constructs, i.e, a new set of schema evolution primitives to evolve the new construct(s); (2) and to maintain the structural and referential integrity of the database, i.e., it must modify the existing schema evolution primitives to assure that they obey the new constraints of the new object model augmented with the relationship construct. For the relationship construct [5] we now propose, a set of schema evolution primitives to handle their addition and deletion. One example of such a primitive is `form-relationship` that creates a bi-directional relationship between two classes that already have two independent uni-directional relationships.

Evolution primitives maintain the consistency of an OODB by preserving the invariants of the object model at all times. However, an update to the object model changes the set of invariants. Hence, the existing evolution primitives must be updated to conform to the invariants of the new object model. Consider for example, the existing evolution primitive `delete-class`. Today, while most OODB systems provide support for relationships by modeling them as reference attributes, the evolution primitives such as `delete-class` treat them as ordinary literal attributes. And while at the object level most OODB systems do referential integrity checks to ensure referential consistency, at the schema level, referential relationships are largely mis-treated during schema evolution, resulting in structurally inconsistent schemas. And while most systems may be able to trap and deal with the referential problems, the schema inconsistencies are largely ignored.

However, updating the schema evolution primitives to now conform to the new set of invariants is a huge re-engineering effort. To diminish the cost of updating an existing schema evolution facility while still providing a consistency preserving schema evolution facility, we propose Evolution Wrappers based on the concept of Software Contracts [12] to effectively maintain existing schema evolution primitives in the light of a changing object model. These contracts, declarative in nature, are easy to update and can be changed without the overhead of re-coding.

For relationships, a complex construct with numerous special properties such as cardinality, uni- vs bi-directional, etc., a pre-defined primitive set of operations is not sufficient. Many different primitives may be combined to compose more complex transformations. Moreover, this must all be executed *atomically* like any pre-defined system-provided primitive. We provide a framework that offers such composite evolution operations while maintaining the integrity and the atomicity of the operations themselves.

A big drawback of the compound primitives is their lack of flexibility. A user such as a DBA administrator, may wish to create a bi-directional relationship between two completely disjoint classes. Two options for the user are to (1) have the system provide a schema evolution primitive to handle this scenario or (2) write an ad-hoc program to accomplish the same. From a system's perspective to apriori assess the needs of the users and to plan for it in a hard-coded fashion is a hard if not an impossible problem. On the other hand, ad-hoc programs by the users do not have any guarantee of consistency or atomicity. Thus it is necessary to offer extensibility in our framework that would allow not just the system but also the users to express desired transformations while still providing the guarantee of database integrity as well as the atomicity of the transformations. The solution we present in this paper based on SERF [6] provides an extensible and flexible schema evolution framework for supporting complex relationship evolution.

Summary. In this paper we present a complete solution framework, *ROVER* for the evolution of relationships which provides for consistent evolution of relationships while supporting extensibility for user-defined evolution operations. Here, (1) We identify the problem that current OODBs while offering relationship support in their object models do not handle evolution of such relationships. We characterize the *consistency problems* that arise from the use of the existing evolution primitives in these systems; (2) We propose a minimal set of *new evolution primitives* needed for supporting the evolution of both uni-directional and bi-directional relationships; (3) We propose a set of *compound evolution primitives* for relationships; (4) We generalize our approach such that users as opposed to the system can now extend the set of schema evolution operations provided by the OODB with their own complex schema evolution operations for relationships; and (5) We present ROVER Wrappers to reduce the re-engineering costs that would otherwise be incurred in updating the old schema evolution primitives to conform to the new set of invariants.

2   Background

In this section, we briefly introduce relationship constructs as defined by ODMG [5], while full details can be found in [5]. Paralleling the concept of foreign keys in relational databases, object models almost always have support for the association between two classes. Most models support the notion of a reference attribute which defines a one-way association between two classes. The ODMG object model also defines the notion of a bi-directional association wherein if class A refers to class B then class B must refer to class A. The user can define the cardinality of these references as one-to-one, one-to-many or many-to-many. To capture this notion of association, we use the *referential relationship* ($\longrightarrow$) that specifies when one type refers to another type; and a *bi-directional relationship* ($\longleftrightarrow$) that specifies a referential relationship and its inverse.

Table 1 presents some of the notation and functions for the meta information as used in the paper. More details can be found in [7].

Table 1. Notation for Axiomatization of Schema Changes

Term	Description
types($\mathcal{C}$)	The set of all types in the system
s, t, T, $\perp$	Elements of **types**($\mathcal{C}$)
super(t)	The set of all direct supertypes of type t
sub(t)	The set of all direct subtypes of type t
in-paths(t)	The set of all paths $<c,r>$ referring to type t
in-degree(t)	The count of all paths referring to type t
out-paths(t)	The set of all paths $<t,r>$ going out of type t
obj-in-degree(o_i)	The number of objects referring to the object o_i

In Table 2 we list the basic schema evolution operations that are supported by most OODBs [14,19].

Table 2. Taxonomy of Basic Schema Evolution Primitives for Classes, Attributes and Inheritance Hierarchy.

Evolution Primitive	Description
add-class(c, $\mathcal{C}$)	Adds new class c to $\mathcal{C}$ in the schema **S**
delete-class(c)	Deletes class c from $\mathcal{C}$ in the schema **S**
add-ISA-edge(c_x, c_y)	Adds an inheritance edge from c_x to c_y
delete-ISA-edge(c_x, c_y)	Deletes the inheritance edge from c_x to c_y
add-attribute(c_x, a_x, t, d)	Add attribute a_x of type t and default value d to class c_x
delete-attribute(c_x, a_x)	Deletes the attribute a_x from the class c_x

3   Minimal Primitives for Relationship Evolution

In this section we present a set of evolution primitives needed for the evolution of uni-directional (unary) as well as for bi-directional (binary) relationships. The primitives presented here are *minimal* in that they cannot be decomposed into any other evolution primitives and *essential* in that they are all required for the evolution of relationships. They can be composed together with other evolution primitives to form more complex transformations, as we will discuss in Section 5.

3.1 Evolution of Unary Relationships

As per the ODMG object model and other object models [19], unary relationships are generally modeled via the use of a reference attribute. For example, Figure 2 shows an unary relationship between classes Teacher and Course via the reference attribute teaches. We propose two evolution primitives *add-reference-attribute* and *delete-reference-attribute* that allow us to add and delete an unary relationship.

Fig. 1. Graphical Schema Description of the Uni-Directional Relationship between Two Classes Teacher and Course.

```
class Teacher
{
        attribute Course teaches;
}
```

Fig. 2. ODMG Syntax for a Uni-Directional Relationship, teaches.

The add-reference-attribute primitive adds a uni-directional relationship between two types. For example, the primitive *add-reference-attribute (Teacher, teaches, Course, null)* adds a complex attribute teaches of the type Course to the class Teacher. Its default value is set to null. The domain type of the reference attribute signifies the cardinality of the relationship. Thus, a Set signifies a *many* relationship and a type signifies a *one* relationship. In this example, we thus have an *one-to-one* relationship as Course is an application type signifying a cardinality of one.

The delete-reference-attribute primitive allows the deletion of an existing uni-directional relationship between two types. For example, the primitive *delete-reference-attribute(Teacher, teaches)* deletes the complex attribute teaches from the class Teacher. This primitive is an inverse operation of the primitive *add-reference-attribute*.

3.2 Evolution of Bi-directional Relationships

The ODMG relationship syntax for Figure 3 is given in Figure 4. In the class Teacher, the attribute teaches is a *reference* attribute of type Course and the inverse of this relationship is given by the attribute is-taught-by in class Course. A binary relationship thus is modeled by the following characteristics: a *source-class* (Teacher), *inverse-class* (Course), the *source-relationship-name* (teaches), the *inverse-relationship-name* (is-taught-by), cardinality of the relationship in the *source* class referred to as *source-card* (many), cardinality of the relationship in the *inverse* class referred to as *inverse-card* (one), type of storage (class or collection) for the source relationship *source-type* (set), and type of storage for the inverse relationship *inverse-type* (Teacher).

```
class Teacher {
    relationship set<Course> teaches
        inverse Course::is-taught-by;
}
class Course {
    relationship Teacher is-taught-by
        inverse Teacher::teaches;
}
```

Fig. 3. Graphical Schema Description of the teaches - is-taught-by Relationship.

Fig. 4. ODMG Syntax for Specifying a Bi-Directional Relationship.

A bi-directional relationship can be broken down into a pair of uni-directional relationships between the two classes. Hence we propose that the only two additional primitives needed for the manipulation of bi-directional relationships are: *form-relationship* and *drop-relationship* (see Table 3).

The form-relationship primitive elevates the status of two already existing uni-directional relationships between two types. For example, in Figure 3 *form-relationship(Teacher, teaches, Course, is-taught-by)* indicates that from now onwards the two uni-directional relationships `Teacher.teaches` and `Course.is-taught-by` are to be modeled as a bi-directional relationship. Hence, any update of objects in one class `Teacher` will be automatically reflected in the other class `Course`.

The drop-relationship primitive transforms a bi-directional relationship to a pair of uni-directional relationships. For example, *drop-relationship(Teacher, teaches, Course, is-taught-by)* drops the binary relationship down to two separate uni-directional relationships `Teacher.teaches` and `Course.is-taught-by`. Henceforth, the two relationships are maintained independently of the other. Any update to one relationship will not effect the other relationship.

Other evolution operators needed for the support of relationships such as changing the cardinality or changing the domain type can be accomplished via a combination of the already available basic primitives and hence do not require any additional primitives (see Section 5 for more details). Table 3 summarizes the new evolution primitives for relationships.

4   Compound Evolution Operations for Relationships

While providing some basic capability to evolve relationships these evolution alone are not adequate to manipulate all the properties of a relationship. For instance, using only the primitives in Section 3 the users are unable to change the cardinality of the relationship, i.e., change from a `collection` to an application `type` or vice versa. This change could however be accomplished by a

Table 3. Taxonomy of Basic Evolution Primitives for Relationships.

Evolution Primitive	Description
$add\text{-}reference\text{-}attribute(c_x,\ r_x,\ c_y,\ d)$	Add unary relationship from class c_x to class c_y named r_x with default value d
$delete\text{-}reference\text{-}attribute(c_x,\ r_x)$	Delete unary relationship in class c_x named r_x
$form\text{-}relationship(c_x,\ r_x,\ c_y,\ r_y)$	Promotes the specified two unary relationships to a binary relationship
$drop\text{-}relationship(c_x,\ r_x,\ c_y,\ r_y)$	Demotes the specified binary relationship to two unary relationships

combination of other existing schema evolution primitives. Similar to Breche [4] and Lerner [11] we now support atomic system-defined evolution operations that are composed of several (basic) evolution primitives.

```
change-cardinality-m1(cₓ, rₓ, t_d, inverse-cₓ, inverse-rₓ)
{
        // Drop the relationship to two individual uni-directional relationships
        drop-relationship(cₓ, rₓ, inverse-cₓ, inverse-rₓ)

        // Create a new attribute in class cₓ
            add_atomic_attribute(cₓ, tmp-attr, t_d, null);

        // For each object in the extent, find the
        // most-common-value of the set rₓ and use
        // that as a representative value for the tmp-attr.
        while ((o = extent(cₓ).nextElement()) != null)
            o.tmp-attr = most-common-value(o.rₓ);

        // Remove the attribute rₓ from class cₓ
        delete_attribute(cₓ, rₓ);

        // Rename the attribute attr to rₓ
        rename_attribute(cₓ, attr, rₓ);

        // Form the bi-directional relationship from two uni-directional relationships
        form-relationship(cₓ, rₓ, inverse-cₓ, inverse-rₓ)
}
```

Fig. 5. A Compound Primitive to Change the Cardinality of a Relationship.

In our work we identify the set of complex operations to evolve the different properties of a relationship construct. As an example, Figure 5 depicts a system-defined operation to accomplish a cardinality change, i.e, it changes the cardinality of a relationship from `many` (set) to `one` (an application type). In this example, we first use the `drop-relationship` primitive to break the bi-directional relationship into two individual uni-directional relationships. The primitive `add_atomic_attribute` adds a temporary attribute `tmp-attr` whose domain is representative of the cardinality change. Thus, the domain of `tmp-attr` is t_d. In the next step, for all objects in the extent of the class c_x, we need to convert the set-value (r_x) to a single value. We use a system-defined method `most-common(r_x)` that finds the most commonly occurring object a_i in a given collection r_x. Thus, for all objects in the extent of c_x the single-value attribute `tmp-attr` is assigned a, the most common occurring value in the object's collection r_x. We then delete the attribute r_x and rename the temporary attribute a to r_x. As a last step we re-formulate the bi-directional relationship between the two classes. This evolved relationship is now a one-to-one relationship as opposed to a one-to-many relationship.

Table 4 summarizes the set of compound changes necessary to manipulate a bi-directional relationship.

Table 4. Taxonomy of Compound Evolution Operations for Relationships.

Compound Evolution Operation	Description
$change\text{-}cardinality\text{-}1m(c_x,\ r_x,\ m_d)$	Changes cardinality of relationship r_x in class c_x from its current type to collection m_d
$change\text{-}cardinality\text{-}m1(c_x,\ r_x,\ t_d)$	Changes cardinality of relationship r_x in class c_x from collection m_d to type t_d
$change\text{-}rel\text{-}name(c_x,\ r_x,\ y_d)$	Changes name of relationship r_x in class c_x to y_d
$change\text{-}type(c_x,\ r_x,\ t_d)$	Changes type of relationship r_x in class c_x to t_d

5   Flexible Evolution of Relationships

One of the biggest drawback of the compound evolution primitives presented in Section 4 is their *pre-determined* semantics for the compound changes. Consider for example the rather simple compound change, `cardinality-change` shown in Figure 5. We arbitrarily use the most commonly occurring object value as the single-value representative. However, other conversion at the object level such as taking the first, last or median value of the set values are possible. By hard-coding these compound changes, the user has no flexibility to alter their semantics.

This set of evolution primitives, simple or compound, is *pre-determined* and *fixed* not allowing the user[1] any flexibility. Consider that the user now wants to build a bi-directional relationship between two types where only one uni-directional relationship exists [2]. Their only alternative would be an ad-hoc program to combine existing primitives but without the atomicity and consistency guarantees that a system-defined primitive would have. To address these issues we present our framework which gives users the *flexibility* to alter semantics for compound changes as well as *extensibility* to define their own schema evolution operations within a framework guaranteeing the atomicity of the operation and the consistency of the database.

5.1   SERF: A Framework for Extensible Schema Evolution

Our solution [6] is based on the hypothesis that complex schema evolution transformations can be broken down into a sequence of basic evolution primitives, (an invariant-preserving atomic operation) glued together by a query language such as OQL [5].

The user could now on the fly introduce any new transformation when desired. For example the transformation in Figure 6 converts the schema of Figure 1 to the schema depicted in Figure 3. To accomplish this we first create the uni-directional relationship between the class `Course` and the class `Teacher` with the name `is-taught-by`. Once we have two uni-directional relationships we can use the *form-relationship* evolution primitive to convert it to a bi-directional relationship. This operation is defined by the user within the confines of our proposed framework that can guarantee the atomicity and consistency of the system.

SERF Template. An OQL transformation as in Figure 6 *flexibly* allows a user to define different schema transformation. However, these transformations are *not generic*, i.e., they cannot be applied to other existing classes or different schemas. For example, the transformation shown in Figure 6 is valid only for the classes `Teacher` and `Course`. To address this, we introduce the notion of templates [6]. A template uses the query language's ability to query over the meta data (as stated in the ODMG Standard) and is enhanced by a name and a set of parameters to make transformations *generic* and *re-usable*. Figure 7 shows a templated form of the transformation presented in Figure 6. Here we query over the meta-data to discover the **type** of the attribute `source-attrib-name` and then add the `inverse-attrib-name` to it (shown in Courier font). The object transformation here remains the same as in the transformation of Figure 6. This template shows how we can take advantage of the single referential relationship and perform the object transformations all within our template structure.

---

[1] By user we imply a database administrator who has knowledge and permission to alter the structure of the database.
[2] This is as opposed to two uni-directional relationships between the two types, as required by schema evolution primitive `form-relationship`.

// This transformation adds a relationship between two partially disjoint classes
// i.e., one class has a one-sided relationship to the other. This transformation
// also performs the object transformations.

add_reference_attribute (Course, is-taught-by, Teacher, null);

// Get the extent of the class

define extents (cName) as
    select c
    from cName c;

// Do the object transformations
// set: Course.is-taught-by = Teacher
for all obj in extents (Teacher):
    obj.teaches.set(obj.teaches.is-taught-by, obj);

// form a binary relationship
form-relationship(Teacher, teaches, Course, is-taught-by);

Fig. 6. Transformation From Uni-directional to Bi-directional Relationship

```
begin template add-inverse-relationship ( source-class, source-rel-name, inverse-name)
{
    // find the inverse class
        refClass = select c.attrType
                   from Attribute c
                   where c.name = $source-relName and
                   c.parent.name = $source-class;


    // add the reference attribute to the inverse class
    add-reference-attribute (refClass, $inverse-name, $source-class, null);


    // get the extent of the class
    define extents (cName) as
        select c
        from cName c;

    // object transformations
    for all obj in extents ($source-class):
        obj.$source-rel-name.set( obj.$source-rel-name.$inverse-name, obj);

    // create the bi-directional relationship
    form-relationship ($source-class, $source-rel-name, refClass, $inverse-name);
}
end template
```

Fig. 7. Template for Converting a Uni-directional Relationship to a Bi-directional Relationship

6   Contract-Based Solution for Consistent Relationship Evolution

Conventionally, schema evolution primitives contain hard-code constraints that parallel the invariants of the object model. These constraints must be satisfied in order to guarantee the consistency of the system. Changes in the object model result in changes to the invariants which may in turn be reflected as an update to schema evolution operations [7]. This is an expensive process requiring the re-engineering of the affected software. Our approach *ROVER* instead focuses on de-coupling the constraints from the actual implementation code of the schema evolution operations. To accomplish this we introduce the notion of *contracts*, a declarative mechanism for expressing the constraints for a template. A template with contracts is termed a *ROVER Wrapper*. Changes to the invariants of the object model now merely result in the update of the declarative contracts associated with the evolution operations rather than the update of the actual system code. Below we briefly introduce *contracts* and show how the de-coupling of constraints can be achieved while more details can be found in [7].

Contracts provide a declarative description of the behavior of a template (or primitive) as well as a mechanism for expressing the constraints that must be satisfied prior to the execution of the actual evolution primitive. Contracts are divided into two categories **pre-conditions** and **post-conditions**.

The constraints, termed *pre-conditions*, are placed prior to any body of template code (OQL statement including system-defined schema evolution primitive). The *pre-conditions* are separated from the actual OQL statements by means of the keyword **requires**. *Post-conditions*, a set of contracts that appear after the body of the actual schema evolution operation at the end of the SERF template, specify the behavior of the primitives. These post-conditions are preceded by the keyword **ensures:** and describe the exact changes that are made to the schema by the evolution operator and hence its behavior.

Example: As an example consider the addition of relationship constructs to the object model of an OODB system. An upgrade to the schema evolution facility in this case requires new schema evolution primitives to handle the creation, modification and deletion of uni-directional and/or bi-directional relationships. This upgrade cannot be circumvented and hence new schema evolution primitives must be added to the system. However, an update of all existing schema evolution operations to conform to the new set of invariants is also required. For example, to delete a class prior to the existence of relationships, the constraint that a class needed to be a *leaf* class was necessary to ensure that the resulting schema and database was consistent, i.e., the `delete-class` preserved the database consistency. With the addition of relationships, this constraint alone is not sufficient. We now also need to ensure that the `to-be-deleted` C_i class is not referred to by another class. Moreover, no objects in the database must refer to the objects of the C_i class. So while the conditions that need to be enforced prior to the execution the schema evolution operation have to be upgraded, the

actual actions of the operations do not change. Hence the evolution primitive
`delete-class` itself does not change.

Figure 8 shows the constraints after the addition of a relationship for the
delete-class primitive as pre-conditions[3]. Thus, in this model it is easy to extend
or modify the constraints without re-writing the code for the evolution primitive.

delete-class (C_i)
{

    requires:
        $C_i \in \mathcal{C} \wedge$
        $\sigma(C_i) \in \textbf{types}(\mathcal{C}) \wedge$
        $sub(C_i) = 0 \wedge$
        $in\text{-}degree(C_i) = 0 \wedge$
        $\forall\ o_i \in \textbf{extent}(t)$
            $obj\text{-}in\text{-}degree(o_i) = 0$

    template body here
}

Fig. 8. Pre-Conditions for Delete-Class Primitive in Contractual Form

delete-class (C_i)
{   template body here

    ensures:
        $C_i \notin \mathcal{C} \wedge$
        $\sigma(C_i) \notin \textbf{types}(\mathcal{C}) \wedge$
        $\forall <C_x, r_x> \in out\text{-}paths(C_i)$
            $(<C_i> \notin in\text{-}paths(C_x))$
$\wedge$
        $\forall\ C_x \in super(C_i)$
            $(C_i \notin sub(C_x)\ )$
}

Fig. 9. Post-Conditions for the Delete-Class Primitive Template

Advantages of Contracts. ROVER Wrappers provide faster updates to the
OODB system when the underlying object model is updated for example with relationships. Namely, we would simply now add additional declarative constraints
and behavior to existing schema evolution primitive templates rather than having to update the system code. Moreover these contracts can detect erroneous
conditions prior to the execution of the schema evolution primitives. This would
help avoid the cost of roll-backs in cases of failure. The *post-conditions* represent
the desired final state and can hence be used for testing the correctness of the
ROVER Wrapper.

7   Related Work

Semantic modeling research has looked into the modeling of relationships and
the different semantics that can be applied for these relationships [3]. In object
databases, Kim, Bertino and others [9] have examined the part-whole relationship (composite objects). Bertino et al. [2] have presented a formal composite
object model that now supports referential integrity constraints for the ODMG
object model. None of them have studied the issue of schema evolution on such
object models with relationships.

---

[3] The notation used here is a set-theoretic version of the contract language.

Most research and commercial systems provide schema evolution in the form of a fixed set of evolution primitives [1,14,19]. In recent years, research has begun to focus on the issues of supporting more complex schema evolution operations. Breche and Lerner [4,11] studied the design of a set of more complex operations. Lerner [11] has proposed compound type changes like *Inline, Encapsulate, Merge*, etc. from the aspect of discovering the transformation sequences to map between these given two schemas. Lastly, our previous work on SERF [6] has provided a framework that allows the user to define arbitrarily complex schema changes by composing them out of the basic set of evolution primitives and OQL. None of these previous approaches have looked at the problem of evolution (be it simple or complex) in the context of object models with relationships.

Consistency management is often done at runtime and is normally handled by transaction roll-backs. Work has been done towards providing behavioral consistency, i.e., the consistency of class methods under evolving environments [8,13]. While there are similarities in that their algorithms are also detecting broken references, their approach is also hard-coded. With our approach we push these constraints/invariants to a higher level thus making them easy to update and also allowing for pre-execution verification checks.

8   Conclusion

In this work, we present the first solution for the evolution of relationships. We provide a flexible mechanism to handle the evolution of relationships. We incorporate the notion of *contracts* for ROVER Wrappers as an alternative to hard-coding the constraints into the schema evolution primitive and thus helping to offset the re-engineering cost. We provide implementation details on ROVER in [7]. The core SERF system for flexible transformation was demonstrated at SIGMOD 2000 [17].

References

1. J. Banerjee, W. Kim, H. J. Kim, and H. F. Korth. Semantics and Implementation of Schema Evolution in Object-Oriented Databases. *SIGMOD*, pages 311–322, 1987.
2. E. Bertino and G. Guerrini. Extending the ODMG Object Model with Composite Objects. In *OOPSLA*, pages 259–270, 1998.
3. G. Booch. *Object-Oriented Analysis and Design*. Benjamin Cummings Publications, 1994.
4. P. Bréche. Advanced Primitives for Changing Schemas of Object Databases. In *Conference on Advanced Information Systems Engineering*, pages 476–495, 1996.
5. Cattell, R.G.G and et al. *The Object Database Standard: ODMG 2.0.* Morgan Kaufmann Publishers, Inc., 1997.
6. K.T. Claypool, J. Jin, and E.A. Rundensteiner. SERF: Schema Evolution through an Extensible, Re-usable and Flexible Framework. In *Int. Conf. on Information and Knowledge Management*, pages 314–321, November 1998.

7. K.T. Claypool, E.A. Rundensteiner, and G.T Heineman. Extending Schema Evolution to Handle Object Models with Relationships. Technical Report WPI-CS-TR-99-15, Worcester Polytechnic Institute, March 1999.
8. C. Delcourt and R. Zicari. The Design of an Integrity Consistency Checker (ICC) for an Object Oriented-Database System. In P. America, editor, *ECOOP*, pages 97–117, 1991.
9. W. Kim, E. Bertino, and J. F. Garza. Composite objects revisited. *SIGMOD*, pages 337–347, 1989.
10. W. Kim, J. F. Garza, N. Ballou, and D. Woelk. Architecture of the ORION Next-Generation Database System. *IEEE Transactions on Knowledge and Data Engineering*, 2(1), March 1990.
11. B.S. Lerner. A Model for Compound Type Changes Encountered in Schema Evolution. Technical Report UM-CS-96-044, University of Massachusetts, Amherst, Computer Science Department, 1996.
12. B. Meyer. Applying "Design By Contract". *IEEE Computer*, 25(10):20–32, 1992.
13. M.A Morsi, S. Navathe, and Shilling J. On Behavioral Schema Evolution in Object-Oriented-Database System. In *Int. Conference on Extending Database Technology (EDBT)*, pages 173–186, 1994.
14. Object Design Inc. *ObjectStore - User Guide: DML. ObjectStore Release 3.0 for UNIX Systems*. Object Design Inc., December 1993.
15. Objectivity Inc. White Paper, Schema Evolution in Objectivity, February 1994.
16. Joan Peckham, Bonnie MacKellar, and Michael Doherty. Data model for extensible support of explicit relationships in design databases. *VLDB Journal*, 4(2):157–191, 1995.
17. E.A. Rundensteiner, K.T. Claypool, and L. et. al Chen. SERFing the Web: A Comprehensive Approach for Web Site Management. In *Demo Session Proceedings of SIGMOD'00*, 2000.
18. D. Sjoberg. Quantifying Schema Evolution. *Information and Software Technology*, 35(1):35–54, January 1993.
19. O_2 Technology. *O_2 Reference Manual, Version 4.5, Release November 1994*. O_2 Technology, Versailles, France, November 1994.

Improving the Reuse Possibilities of the Behavioral Aspects of Object-Oriented Domain Models

Monique Snoeck and Geert Poels

MIS Group, Dept. Applied Economic Sciences, K.U.Leuven,
Naamsestraat 69, 3000 Leuven, Belgium
{monique.snoeck, geert.poels}@econ.kuleuven.ac.be

Abstract. Reuse of domain models is often limited to the reuse of the structural aspects of the domain (e.g. by means of generic data models). In object-oriented models, reuse of dynamic aspects is achieved by reusing the methods of domain classes. Because in the object-oriented approach any behavior is attached to a class, it is impossible to reuse behavior without at the same time reusing the class. In addition, because of the message passing paradigm, object interaction must be specified as a method attached to one class which is invoked by another class. In this way object interaction is hidden in the behavioral aspects of classes. This makes object interaction schemas difficult to reuse and customize. The focus of this paper is on improving the reuse of object-oriented domain models. This is achieved by centering the behavioral aspects around the concept of business events.

1. Introduction

Domain modeling is an essential requirement capturing activity, prior to information systems modeling. As such, the main objective of any domain model is to be a vehicle for communication between system developers and business people, facilitating the mutual perception and understanding of important aspects of the business reality [17]. Although domain models are the particular representation of one or more aspects of a specific type of business (e.g. manufacturing, transportation, ...), the reuse of models from one domain to another is feasible (and supposedly also beneficial) when domains share a common knowledge structure. This principle of analogical reuse [13] has been supported by research contributions from various fields, including analysis patterns [7],[20], generic data models [8],[16], generic components [2], enumerative and faceted classification schemas [12], and automated pattern retrieval and synthesis [19]. Most of this work aims at facilitating the reuse of structural aspects of a domain (e.g. data models). Sometimes, in particular with respect to object-oriented modeling, it also concerns the reuse of functionality (e.g. object operations). In general however, the proposals that have been made do not concern the reuse of behavioral aspects related to the interaction of domain objects [14]. In object-oriented analysis, reuse is often centered around the reuse of class definitions. This type of reuse however, focuses on the reuse of design and code. Reuse at earlier stages of software development should focus on the reuse of analysis models. In object-oriented analysis there are typically at least three types of models, one for each view on the Universe of Discourse: a static model, an interaction model and a behavioral model. The goal of this research is to facilitate the reuse of the latter two types of models.

A.H.F. Laender, S.W. Liddle, V.C. Storey (Eds.): ER2000 Conference, LNCS 1920, pp. 423-439, 2000.
© Springer-Verlag Berlin Heidelberg 2000

In this paper we present some research experiences with analogical reuse in the context of event-based domain modeling. In an event-based approach, the dynamic perspective of the domain is modeled by identifying the real-world events that are relevant to the universe of discourse. Domain objects are modeled in terms of their participation in real-world events (also called business events). In this way the dynamic perspective is modeled independently and at a high level of abstraction. This contrasts with the prevalent approach in OOA that models the dynamic perspective through the concept of class-method, which is at a lower level of abstraction and subordinated to the concept of class. An important issue regarding the reuse of event-based domain models concerns the reuse of the participation of domain objects in real-world events, both in terms of the effect events have on domain objects as in terms of interaction between domain objects. The main focus of this paper is the improved reuse of such interaction aspects when an event-based approach is taken to conceptual domain modeling.

Our research concerns both the abstracting and customization of event-based domain models. We use an example throughout the paper to illustrate the possibilities and particularities of analogical reuse of domain object interaction schemas. In section 2 we present domain models for a library and a hotel administration along with a generic model that is a domain abstraction for these two analogous domains. The models in this section only represent structural aspects of the domain and take the form of UML class diagrams. Some issues regarding generalization and customization are illustrated and discussed. In section 3 the focus shifts to the modeling of behavioral aspects. First, behavior is added to the generic model following the rules prescribed by a formal method for object-oriented enterprise modeling [22],[23]. Next, the reuse of this behavior is illustrated and discussed. Section 4 then investigates the effect of required customization on the reuse of object interaction schemas. It is shown that an event-based approach to conceptual domain modeling improves the reuse possibilities of the object interaction schemas. More in particular, the effect of customization is shown to be less pervasive in an event-based interaction schema compared to a message passing interaction schema. Conclusions are presented in section 5.

2. A Generic Domain Model for Product Usage

Consider the following (simplified) domain descriptions for a library and a hotel administration:

> "In the library we have a catalogue with titles and for each title the library has one or more copies. People can register to the library and become members. Members can borrow and return copies. Loans can be renewed. If a book is not on shelf, a reservation can be made for that title: the first copy that is returned to the library will then be put aside."

> " A hotel offers a set of rooms that are categorized into room types. Customers make reservations for a particular room type. When the reservation is confirmed, a specific room is assigned for the customer's later stay."

The structural aspects of these domain descriptions are shown in Fig. 1 and Fig. 2 respectively.

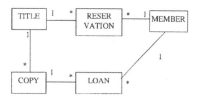

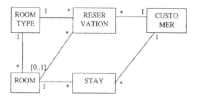

Fig. 1. A simple Library Domain Model

Fig. 2. A simple Hotel Administration Domain Model

As one can immediately notice, the class diagrams for the library and the hotel show a very similar structure. In both types of businesses products are categorized to product types. Customers can "use" a product during a certain period of time, after which the product must be returned. Prior to this usage there may or may not be an "order" or reservation for the product's type.

The generic domain model for Product Usage is shown in Fig.3[1]. In this model, the association between USAGE_INTENTION and PRODUCT represents the allocation of products to reservations or orders. The association between USAGE_INTENTION and USAGE allows tracking how many of the effective usages are the consequence of a prior usage intention.

The generic model can also be extended to support multiple branches of one business. In the model of Fig. 4 we assume that product types are company-wide. However, the characteristics of a product type can be different from branch to branch: a double room in New York will have another (higher) price than a double room in Las Vegas. This requires the introduction of the class PRODUCT_TYPE_IN_BRANCH. Individual products are the materialization of such a PRODUCT_TYPE_IN_BRANCH and are as such located in one branch.

Although this generic domain model can be reused in many types of 'renting business', each domain will have its own particularities that must be taken care of. Tailoring the generic structural model to the particularities of the own domain can be done by **adding** or **dropping** classes and/or associations, and by **considering additional business rules**. For example, in the library we will probably not be interested in keeping track of how many loans are the consequence of a reservation. As a result, the association between the RESERVATION class and the LOAN class has not been retained. In the case of the hotel administration, the decision whether or not to retain this association depends on the information needs of the specific company. For instance, the association must be retained if the hotel manager wishes to know for how many stays there was a prior reservation.

---

[1] In the classification framework of Lung and Urban [8] the domain abstraction for a library system and a hotel reservation system is called 'Object Allocation'. It is described as (p. 173) "an analogy for domains that allocate an object to another object (usually an agent). The allocated objects are returned after a period of time". Other example domains include car rental and airline reservation systems. Note that Lung and Urban do not propose generic models for their domain abstractions.

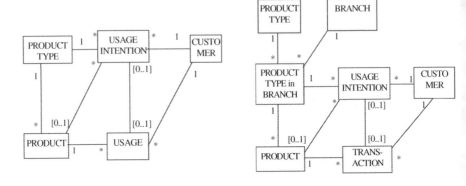

Fig. 3. A generic domain model for Product Usage

Fig. 4. An extended generic domain model for Product Usage

As another example, in a car rental domain model, which is another domain of the type Product Usage, it would also make sense to add an association between BRANCH and PRODUCT (i.e. a car) that records the current location of a car. This would allow customers to return the car to another branch than where it was rented. For example, it would allow customers to rent a car in the Brussels office and return it in the Paris office. In a library, the concept of PRODUCT_TYPE_IN_BRANCH makes less sense. It is sufficient to keep track of the location of each copy by directly linking COPY to LIBRARY (the branch) (Fig. 5).

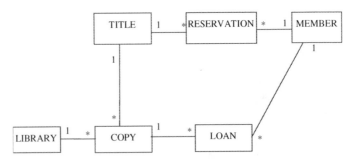

Fig. 5. Extended Library Domain Model

3. Adding Behavioral Aspects

3.1   Motivation for an Event-Based Approach

In the case of object-oriented conceptual modeling, domain requirements will be formulated in terms of business or enterprise object types, associations between these object types and the behavior of business object types. The definition of desired object behavior is an essential part in the specification process. On the one hand, we have to consider the behavior of individual objects. This type of behavior will be specified as methods and statecharts for object classes. On the other hand, objects have to collaborate and interact. Typical techniques for modeling object interaction aspects are interaction diagrams or sequence charts, and collaboration diagrams.

These techniques are based on the concept of message passing as interaction mechanism between objects. The main disadvantage of this concept is that in the context of domain modeling, message passing is too much implementation biased. We propose an alternative communication paradigm, namely, object interaction by means of joint involvement in business events. This type of interaction is modeled with an object-event table. Let us illustrate this with an example. In the context of a library, we can identify (among others) the two domain object types MEMBER and COPY. A relevant event type in this domain is the *borrowing* of a copy. This event affects both domain object types: it modifies the state of the copy and it modifies the state of the member. When using message passing as interaction mechanism, two scenarios are possible. Either the member sends a message to the copy, or the copy sends a message to the member (see Fig. 6). If in addition LOAN is recognized as a domain object type as well, then the *borrow*-event will create loan objects. In this case, three objects are simultaneously involved in one event and should be notified of the occurrence of the *borrow*-event. With message passing, this leads to 9 possible interaction scenarios as depicted in Fig. 7. With each additional object type, the number of possible message passing scenarios further explodes. For example, if four objects have to synchronize on the occurrence of one event, we already have 64 possible message passing scenarios. Of course, from a systems design perspective, some scenarios can be considered more adequate than others. Domain modeling should however never be concerned with design aspects and business domain modelers should not be burdened with design considerations.

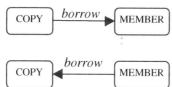

Fig. 6. Two possible scenarios for borrowing a copy

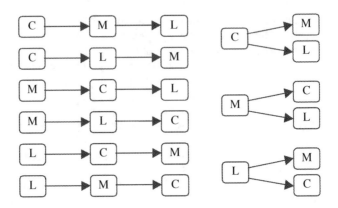

Fig. 7. Possible scenarios when three objects are involved in a single event
(C = COPY, M = MEMBER, L = LOAN)

Table 1. Object-event table for the library

	MEMB ER	CO PY	LO AN
cr_memb er	×		
acquire		×	
borrow	×	×	×
return	×	×	×
...			

The alternative that we propose in this paper is to model only the essence of the interaction: some objects are affected by a given event, others are not. To model which objects are involved in which event types, we can use a very simple technique: the object-event table. Table 1. shows a possible object-event table for the library example. The table clearly shows that a *cr_member* event affects only the member object, that the acquisition of a copy only affects a copy, but that the borrowing and return of a copy affect a member, a copy and a loan object.

The use of the object-event table to model object interaction implies that the notion of **event** plays a central role. Some object-oriented analysis methods agree that events are a fundamental part of the structure of experience [4][6][21]. Events are atomic units of action: they represent things that happen in the real world. Without events nothing would happen: they are the way information and objects come into existence (creating events), the way information and objects are modified (modifying events) and disappear from our Universe of Discourse (ending events). As we are concerned with domain modeling, we will only consider **business** events (i.e. real world events) and, for example, not consider information systems events like keyboard and mouse actions. The concept of the object-event table allows to model interaction at a much higher level of abstraction than is the case with message passing. Moreover, the interaction pattern is independent of the number of objects involved in an event. At domain modeling level, we should not burden ourselves with event notification schemas. How exactly objects are notified of the occurrence of an event is a matter of implementation. When using object-oriented technology this will be done with messages, but when using other technologies, both traditional and modern (e.g. distributed component technologies), (remote) procedure calls can do as well.

3.2. The Generic Behavioral Schema for Product Usage

The specification of the behavioral aspects of the domain model consists of one object-event table and a set of lifecycle models, one for each of the domain classes. The object event table identifies the relevant event types for the Universe of Discourse and specifies the involvement of objects in events. In the object-event table (OET), events are not attached to a single domain class. One event can affect more than one object. In the object-event table, there is one column for each domain class and one row for each type of event relevant to the Universe of Discourse. A row-column intersection is marked with a 'C' when the event creates an object of the class, with an 'M' when it modifies the state of an object of the class and with an 'E' when it ends the life of an object of the class. A marked entry in a column means that, in an object-oriented implementation of the domain model, the domain class has to be equipped with a method to implement the effect of the event on the object. In this way the object-event table identifies the methods that have to be included in the class definition of domain objects.

For the (extended) generic domain model for Product Usage (Fig. 4) we identify the following event types:

create_customer, modify_customer, end_customer, create_branch, modify_branch, end_branch, create_product_type, modify_product_type, end_product_type, allocate_product_type_to_branch, modify_product_type_in_branch, end_product_type_in_branch, create_product, modify_product, end_product, cr_usage_intention, allocate_product, confirm_availability, cancel_usage_intention, start_usage, normal_return, abnormal_return, modify_conditions, invoice_usage, receive_payment, end_usage

The OET is represented in Table 2. A detailed discussion of the rules governing the construction of this OET is beyond the scope of this paper, but can be found in [22], [23]. We merely note here that each marked entry identifies a possible place for information gathering. If for example, we wish to keep track of how many product types are offered in a branch, it makes sense to mark the entries BRANCH/*allocate_product_type_to_branch* and BRANCH/*end_product_type_in_branch*. Similarly, if within the class CUSTOMER we wish to keep track of the total amount of payments made by this customer (e.g. to identify "golden" customers, or to specify some discounting rules), we need to mark the entry CUSTOMER/ *receive_payment*. At implementation time, methods that are empty because no relevant business rule was identified, can be removed to increase efficiency.

Another behavioral aspect that is modeled concerns the specification of object lifecycle models. In the library for example, a copy should be returned before it can be borrowed again. With each class we will thus associate a lifecycle expression. The default lifecycle is that objects are first created (a choice between the C-entries), then modified an arbitrary number of times (an iteration of a choice between the M-entries) and finally come to an end (choice between the E-entries). In most object-oriented methods such lifecycles are represented using state charts. It is however also possible to represent such lifecycles as regular expressions, using a '+' to denote choice, a '.' to denote sequence and a '*' to denote iteration. From a mathematical and formal point of view, regular expressions are equivalent to state charts.

The lifecycle expression of a domain class should contain all events for which an entry has been marked in the corresponding column of the OET. In addition, the lifecycle expression should respect the type of the entries: events marked with a 'C' should appear as creating events, events marked with an 'M' should appear as modifying event types and events marked with an 'E' should terminate the life of the object[2]. For example the lifecycle expression for the class USAGE is represented in Fig. 8 as state chart and is specified as follows by means of a regular expression:

USAGE = *start_usage* . (*modify_conditions*)* . (*normal_return* + *abnormal_return*). *invoice_usage* . (*receive_payment* + *end_usage*)

Table 2. OET for the extended generic domain model for Product Usage.

	CUST O MER	BRAN CH	PRODU CT TYPE	PROD. TYPE IN BRANCH	PROD UCT	USAGE INTENTION	U SA GE
Create_customer	C						
Modify_customer	M						
End_customer	E						
Create_branch		C					
Modify_branch		M					
End_branch		E					
Create_product_type			C				
Modify_product_type			M				
End_product_type			E				
Allo-cate_prod_type_to_branch		M	M	C			
Modify_prod_type_in_branch		M	M	M			
End_product_type_in_branch		M	M	E			
Create_product		M	M	M	C		
Modify_product		M	M	M	M		
End_product		M	M	M	E		
Cr_usage_intention	M	M	M	M		C	
Allocate_product	M	M	M	M	M	M	
Confirm_availability	M	M	M	M		M	
Cancel_usage_intention	M	M	M	M		E	
Start_usage	M	M	M	M	M	E	C
Normal_return	M	M	M	M	M		M
Abnormal_return	M	M	M	M	M		M
Modify_conditions	M	M	M	M	M		M
Invoice_usage	M	M	M	M	M		M
Receive_payment	M	M	M	M	M		E
End_usage	M	M	M	M	M		E

---

[2] Additional rules that guarantuee consistency between the object-relationship schema, the object-event table and the lifecycle expressions can be found in [22], [23].

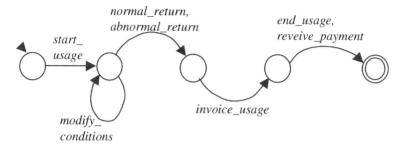

Fig. 8. State chart for USAGE

That is, after a usage has started, the conditions can be modified (e.g. postponing the return date) zero, once or more times. The product is then returned either in a normal state or in an abnormal state (e.g. crashed car). The usage is then invoiced and ends with the payment of the invoice or with the default *end_usage* event if the invoice gets never paid. The lifecycle for USAGE_INTENTION is:

USAGE_INTENTION = *create_usage_intenion* . *allocate_product* . *confirm* . (*cancel_usage_intention* + *start_usage*)

When classes show some parallel behavior the '||' symbol is used to denote parallel composition in regular expressions, such as in the lifecycle of product:

PRODUCT = *create_product* .
[(*modify_product* + *allocate_product* + *invoice* + *receive_ payment* + *end_usage*)*
|| (*start_usage* . (*modify_conditions*)* . (*normal_return* + *abnormal_return*))*].
end_product

That is, after a product has been created, its life is determined by two parallel threads. On the one hand there is the usage cycle and on the other hand there are a number of events that can occur randomly and independent from the usage cycle. The life of the product is terminated by the *end_product* event. Notice that constraints on event types such as *invoice* and *receive_payment* are already specified in the lifecycle of USAGE and need not be re-specified in the lifecycle of PRODUCT. The equivalent state chart is given in Fig. 9.

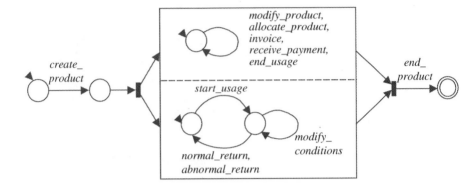

Fig. 9. State chart for PRODUCT.

The OET and object lifecycle models of the generic Product Usage model can be reused in the library, hotel administration and car rental domains. Again, some tailoring might be needed. For example, the way products are allocated to an intended transaction is similar in the hotel and car rental domains, but very different from the library domain. In a car rental and hotel business it is good practice to confirm the reservation to ensure that the requested product (i.e. a car or room) is available on the requested date. In a library however, such confirmation is not required: the member will simply receive the first copy that is returned and no firm assurance can be given on the data a copy will be available.

Reuse of the behavioral parts of the generic domain model is achieved in different ways. At the most abstract level behavior is reused by deciding which events to reuse and how. First, events can be reused as such by simply **renaming** them. For the car rental company, most event types can be reused by simply renaming them. For example, in the car-rental case *create_product_type* becomes *create_car_model*, *modify_product_ type* becomes *modify_car_model* and so on. Secondly, events can be **refined**. In the car-rental example, the *abnormal_return* can be split in two event types: *crash_car* and *total_loss*. Thirdly, events can be **added,** e.g. the event type *repair* can be added to allow putting a car back in circulation after a crash. Finally, events can be **dropped**. For example in a library there is no need to allocate free books to reservations. Hence, the event types *allocate_ product* and *confirm_availability* are dropped.

At a more detailed level of specification individual class behavior is reused by refining the life cycle expressions of object types according to the modified event type definitions. For example, the life cycle of CAR becomes more complex as we want to specify that after a total loss a car can never be rented again and that after a crash, the car needs repairing.

CAR = *buy_car* .
[(*modify_car_details* + *allocate_car* + *invoice* + *receive_ payment* + *end_rental*)*
||(*rent* . (*change_return_date*)* . (*normal _return* + *crash_car.repair*))*
.(1 + (*rent* . (*change_return_date*)* .*total_loss*)]
. *end_car*

In this lifecycle the '1' stands for the empty event. The lifecycle thus specifies that after an arbitrary number of rent-cycles either nothing special happens or we have one final rent cycle that ends with the total loss of the car. The equivalent state chart is given in Fig. 10.

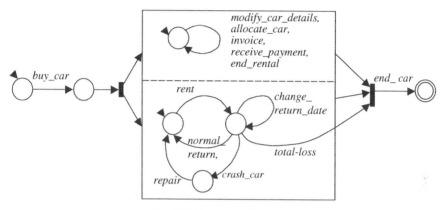

Fig. 10. State chart for CAR.

In the library example, the life cycle of COPY is refined to specify that after a copy has been lost it can never be borrowed again:

COPY = *classify_copy* .
    [(*modify_copy_details* + *fine* + *receive_ payment* + *end_loan*)*
  || (*borrow* . (*renew*)* . *return*)*.(1 + (*borrow* . (*renew*)* .*lose*))]
  . *end_copy*

4. Improved Reuse of the Object Interaction Schema

The most important implication of the use of the object-event table resides in the modeling of object interaction. In the approach proposed in the previous section, it is assumed that events are broadcasted to objects. This means that when an event occurs and is accepted, all corresponding methods in the involved objects will be executed simultaneously provided each involved object is in a state where this event is acceptable. This way of communication is similar to communication as defined in the process algebras CSP [9] and ACP [1] and has been formalized in [5], [23]. Message passing is more similar to the CCS process algebra [15]. There exist various mechanisms for the implementation of such synchronous execution of methods. For the purpose of analyzing the effects on reuse, we will assume that there is an event handling mechanism that filters the incoming events by checking all the constraints this event must satisfy. If all constraints are satisfied, the event is broadcasted to the participating objects; if not it is rejected. In either case the invoking class is notified accordingly of the rejection, acceptance, and successful or unsuccessful execution of the event. This concept is exemplified in Fig. 11. for part of the generic schema of Fig. 3. For each type of business events, the event handling layer contains one class that is responsible for handling events of that type. This class will first check the validity of the event and, if appropriate, broadcast the event to all involved objects by means of the method 'run'.

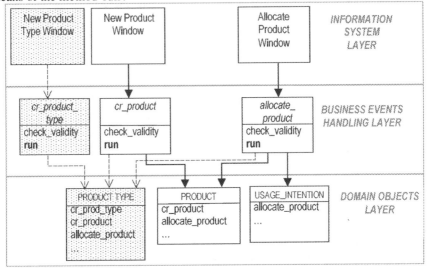

Fig. 11. Part of an event broadcasting schema for the generic schema

In a conventional object-oriented approach, object interaction is achieved by having objects send messages to each other. This is documented by means of collaboration diagrams. Because of the absence of the broadcasting paradigm, events must be routed through the system in such a way that all concerned objects are notified of the event. As there is no generally accepted schema, the routing schema must be designed for each type of event individually. An additional problem is the identification of the object where the routing will start. In most examples given in object-oriented analysis textbooks, the business events are initially triggered by some information system event. For example, in an ATM system, the *withdraw_amount* business event is triggered by the information system event *insert_card*. Such interactions can be represented by including information system objects such as user interface objects in the collaboration diagram. From a domain modeling perspective, we would prefer object interaction to be independent from information system services. For example, the business event *withdraw_amount* can also be triggered by other information system services such as the counter application. In order to represent interaction independent from information system services, in the collaboration diagram below, a dummy class is included that represents the business event invocation. The routing of the event starts in that class and is then routed through the domain model in such a way that all domain classes affected by this type of event are notified. Fig. 12 shows possible interaction schemas for the *cr_product* and the *allocate_product* event types. Notice that because the *allocate_product* event type affects four different domain classes for this event type there are 64 possible routing schemas that allow to notify all 4 objects of the occurrence of an *allocate_product* event (see discussion in section 3.1).

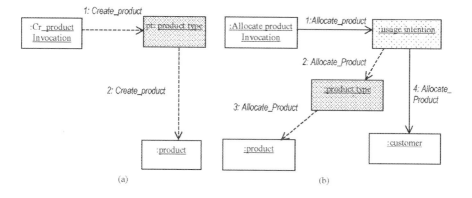

Fig. 12. Collaboration diagrams for the generic schema of Fig. 3.

When the generic schema is customized, object types and event types can be added to, refined or dropped from the generic schema. Such changes turn out to be less pervasive for the broadcasting paradigm than for the message passing paradigm. Let us assume for example that the generic schema is reused for a Small Car rental Company, were the object class PRODUCT_TYPE is not required: in this Small Car rental Company reservations are made directly for individual cars. For the broadcasting schema this means that except for the removal of the product type domain class, all modifications are localized in the event handling layer. The required modifications are shown as shaded areas in Fig. 11. The effect on the collaboration diagrams is more pervasive: the whole interaction schema must be redesigned (see shaded areas in Fig. 12). The modification of the interaction schema even requires modifications in other domain classes. For example, in Fig. 12 (b), the removal of the PRODUCT_ TYPE domain class, requires a modification of the USAGE_INTENTION domain class as this class must now propagate the *allocate_product* event directly to PRODUCT rather than to PRODUCT_TYPE (i.e. send a message to PRODUCT instead of PRODUCT_TYPE). Similarly, adding a domain class has a more limited effect on the broadcasting schema compared to the classical approach. Let us for example add the BRANCH and PRODUCT_TYPE_IN_BRANCH domain classes such as to obtain the generic schema of Fig. 4. In the broadcasting schema the effect for the existing classes is limited to the event handling layer as exemplified in Fig. 13. For the conventional interaction schema documented with collaboration diagrams, the effect is again more pervasive. Depending on the new routing schema for events, the modifications also propagate to one or more existing domain classes. Fig. 14 shows an example of a new collaboration diagram for the *allocate_product* event type, which requires a modification of the PRODUCT domain class.

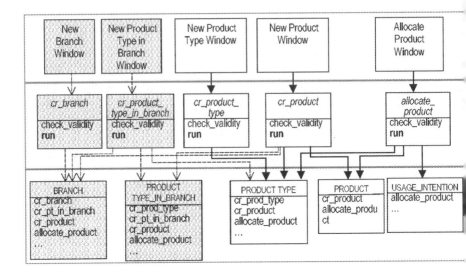

Fig. 13. Effect of adding the domain classes BRANCH and PRODUCT_TYPE_IN_
BRANCH to the event broadcasting schema for the generic schema

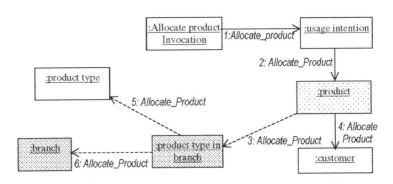

Fig. 14. Modified collaboration diagram for *allocate_product*

5. Conclusions

In this paper we considered various issues related to the generalization of 'analogous' domain models and the customization of the resulting generic domain models. The example indicates that the reuse of both structural and behavioral aspects of domain models is certainly possible. The most difficult part to reuse is the object interaction aspect [14]. It is a well-known fact that if no special effort is taken to minimize the number of collaborations, interaction diagrams quickly take an aspect of spaghetti

[25]. In addition, when interaction between domain object classes is not centered around the concept of business events, it is hidden in the methods of object classes. It then is very difficult to evaluate the impact of customization on the interaction schema. Depending on the chosen scenario for event propagation, one or more domain classes will require some adaptation to implement the modified scenario. The use of the broadcasting paradigm simplifies the reuse of object interaction aspects substantially. Moreover, the required modifications are more systematic and hence easier to trace, and the domain classes are better isolated from modifications such as the addition or removal of domain classes.

It is also important to notice that the broadcasting paradigm allows a system with a layered structure as shown in Fig. 11: information system objects and domain objects are kept in separate layers, with the event handling layer playing the role of "event broker" [24]. These layers also reflect an appropriate separation of concerns, namely the separation of domain knowledge and business rules from information system support. In addition, as the services of objects in a layer are only used by objects of an upper layer, modifications in the upper layer do not propagate to lower layers. This not only makes customization of generic models easier, but also facilitates system maintenance.

Because of the modeling of business events as first-class citizens in the domain model, the reuse of behavior can be considered by looking only at the event types. In a way, the choice of columns and rows in the OET to reuse can be done independently of each other. In a classical approach you would choose the classes in the structural model and hope that they contain the required behavior.

It must be noticed that the reuse of domain models cannot be considered on its own. Domain models must be seen as reusable software requirements. Defining a domain model is part of the requirements engineering step in the development of an information system: all business rules described in the domain model have to be supported by the information system. Methods such as JSD [10], OO-SSADM [21], Syntropy [4], Catalysis [6], and MERODE [23][22] even explicitly define domain modeling as a separate step in the development process. Jacobson [11] assumes the existence of a domain model that serves as a basis to identify entity objects. As such object interaction schemas, which capture a major part of the business rules governing a domain, can be considered as reusable specifications. The event-based approach to conceptual domain modeling assumed in this paper greatly enhances their reuse possibilities.

The generic domain model presented in this paper models domains of the type Product Usage, included as the domain abstraction Object Allocation in the classification framework of Lung and Urban [12]. We have worked on, and continue to work on, generic models for other domain abstractions. A related topic of research is the definition of distance measures for event-based, object-oriented domain models [18]. Such measures, similar in concept to the similarity measures for components of Castano et al. [3], allow to quantify and evaluate the conceptual distance between domains. This information can for instance be used to decide whether analogical reuse is feasible, i.e. whether it is worth reusing from a generic domain model.

Acknowledgements

Geert Poels is a Postdoctoral Fellow of the Fund for Scientific Research - Flanders (Belgium)(F.W.O) and wishes to acknowledge the financial support of the Fund for Scientific Research.

References

[1]   Baeten, J.C.M.: Procesalgebra: een formalisme voor parallelle, communicerende processen. Kluwer programmatuurkunde, Kluwer Deventer (1986)

[2]   Castano, S., De Antonellis, V.: The F³ Reuse Environment for Requirements Engineering. ACM SIGSOFT Software Eng. Notes 19 (1994) 62-65

[3]   Castano, S., De Antonellis, V., Pernici, B.: Building Reusable Components in the Public Administration Domain. In: Proc. ACM SIGSOFT Symposium Software Reusability (SSR'95). Seattle (1995) 81-87

[4]   Cook, S., Daniels, J.: Designing object systems: object-oriented modeling with Syntropy. Prentice Hall (1994)

[5]   Dedene, G., Snoeck, M.: Formal deadlock elimination in an object oriented conceptual schema. Data and Knowledge Eng. 15 (1995) 1-30

[6]   D'Souza, D.F., Wills, A.C.: Objects, components, and frameworks with UML: the catalysis approach. Addison-Wesley (1998)

[7]   Fowler, M.: Analysis Patterns: Reusable Object Models. Addison-Wesley (1997)

[8]   Hay, D.C.: Data Model Patterns: Conventions of Thought. Dorset House Publishers, New York (1996)

[9]   Hoare, C. A. R.: Communicating Sequential Processes. Prentice-Hall (1985)

[10] Jackson, M.A.: System Development. Prentice Hall (1983)

[11] Jacobson, I. et al.: Object-Oriented Software Engineering, A use Case Driven Approach. Addison-Wesley (1992)

[12] Lung, C.-H., Urban, J.E.: An Approach to the Classification of Domain Models in Support of Analogical Reuse. In: Proc. ACM SIGSOFT Symposium Software Reusability (SSR'95). Seattle (1995) 169-178

[13] Maiden, N.A., Sutcliffe, A.G.: Exploiting Reusable Specifications Through Analogy. Communications of the ACM 35 (1992) 55-64

[14] Mili, H., Mili, F., Mili, A.: Reusing Software: Issues and Research Directions. IEEE Trans. Software Eng. 21 (1995) 528-561

[15] Milner R.: A calculus of communicating systems. Springer Berlin, Lecture Notes in Computer Science (1980)

[16] Mineau, G.W., Godin, R.: Automatic structuring of knowledge bases by conceptual clustering. IEEE Trans. Data and Knowledge Eng. 7 (1995) 824-829

[17] Nellborn, C.: Business and Systems Development: Opportunities for an Integrated Way-of-Working. In: Nilsson, A.G., Tolis, C., Nellborn, C. (eds.): Perspectives on Domain modeling: understanding and Changing Organisations. Springer Verlag, Berlin (1999)

[18] Poels, G., Viaene, S., Dedene, G.: Distance Measures for Information System Reengineering. In: Proc. 12th Int'l Conf. Advanced Systems Eng. (CAiSE*00). Stockholm (2000) 387-400

[19] Purao, S., Storey, V.C.: Intelligent Support for Retrieval and Synthesis of Patterns for Object-Oriented Design. In: Proc. 16th Int'l Conf. Conceptual Modeling (ER'97). Los Angeles (1997) 30-42

[20] Robertson, S.: Mastering the Requirements Process. Addison-Wesley (1999)

[21] Robinson, K., Berrisford, G.: Object-oriented SSADM. Prentice Hall (1994)

[22] Snoeck, M., Dedene, G.: Existence Dependency: the key to semantic integrity between structural and behavioral aspects of object types. IEEE Trans. Software Eng. 24 (1998) 233-251

[23] Snoeck, M., Dedene, G., Verhelst, M., Depuydt, A.: Object-oriented Enterprise Modeling with MERODE. University Press, Leuven (1999)

[24] Snoeck, M., Poelmans, S., Dedene, G., A Layered Software Specification Architecture. In: Proc. 19th Int'l Conf. Conceptual Modeling (ER2000). Salt Lake City (2000)

[25] Wirfs-Brock, R., Johnson, R.E.: Surveying current research in OO design. Communications of the ACM 33 (1990) 105-124

Algebraic Database Migration to Object Technology

Andreas Behm, Andreas Geppert, Klaus R. Dittrich

Database Technology Research Group
Department of Information Technology, University of Zurich
Winterthurerstr. 190, CH-8057 Zurich, Switzerland
[abehm|geppert|dittrich]@ifi.unizh.ch

Abstract. Relational database systems represent the current standard technology for implementing database applications. Now that the object-oriented paradigm becomes more and more mature in all phases of the software engineering process, object-oriented DBMS are seriously considered for the seamless integration of object-oriented applications and data persistence. However, when reengineering existing applications or constructing new ones on top of relational databases, a large semantic gap between the new object model and the legacy database's model must be bridged. We propose database migration to resolve this mismatch: the relational schema is transformed into an object-oriented one and the relational data is migrated to an object-oriented database. Existing approaches for migration do not exploit the full potential of the object-oriented paradigm so that the resulting object-oriented schema still "looks rather relational" and retains the drawbacks and weaknesses of the relational schema. We propose a redesign environment which allows to transform relational schemas into adequate object-oriented ones. Schemas and transformation rules are expressed in terms of a new data model, called semi object types (SOT). We also propose a formal foundation for SOT and transformation rules. This formalization makes it possible to automatically generate the data migration process.

1 Introduction

The presence of new technologies like the World Wide Web, E-commerce or data warehousing is a key argument for many companies to reengineer legacy information systems. Now that the object-oriented method is prevailing in modern software development, almost all components of new information systems are developed within an object-oriented software engineering life cycle. The notable exception is in many cases the relational (or even hierarchical and CODASYL-) database component. Many organizations still principally refrain from using object-oriented database management systems (OODBMS), for reasons which are beyond the scope of this paper. Other organizations are willing to give OODBMS a try, but then require the existing relational data to be available in the object-oriented database system (OODBS). This requirement raises the research problem of how to convert schemas and databases stored in an existing (relational) database system into those of an OODBS. To that end, mainly hybrid approaches such as object-oriented views over relational schemas have been proposed, but these do not resolve the data model mismatch, and the required conversions at runtime lead to performance degradations. We therefore propose *database migration*, i.e. the transformation of a relational schema into an object-oriented one and the subsequent migration of the data to the object-oriented DBMS.

Several database vendors offer so-called connectivity tools to relational databases, some of them providing also load facilities. However, these and additional approaches

A.H.F. Laender, S.W. Liddle, V.C. Storey (Eds.): ER2000 Conference, LNCS 1920, pp. 440–453, 2000.
© Springer-Verlag Berlin Heidelberg 2000

proposed in research [1, 13, 6] are not flexible enough, especially in supporting extensive schema restructuring operations. They primarily map each relation to a class and replace inclusion dependencies by references or inheritance relationships. All approaches have in common that they provide a *one-step mapping*, i.e., every element of the target object schema is directly derived from an element of the relational source schema. These approaches are adequate at best for small and well-designed schemas which, e.g., are derived from an entity-relationship schema and/or are in third Normal Form (3NF). Furthermore, connectivity tools and other proposals exhibit two major inadequacies:

- **Weaknesses and drawbacks of relational schemas:** Relational legacy databases typically contain "obscure" optimizations or are denormalized in order to avoid expensive join operations in queries. For example, multiple tables conceptually related by aggregation or inheritance relationships are often collapsed into a single table. A direct mapping of such structures into object-oriented ones preserves inadequacies; consequently, schema transformation must be flexible (e.g., not in all cases semantics-preserving), as overcoming such drawbacks enhances the semantic expressiveness of the schema.

- **Different design strategies:** Relational and object-oriented database design are principally different and follow different design strategies. Besides additional semantic features in object-oriented schemas, such as aggregation or inheritance, object-oriented models comprise the concepts of methods and encapsulation, objects as an abstraction level, but usually do not provide a view mechanism. As a consequence, the straightforward transformation of a well-designed relational schema does not necessarily result in a well-designed and intuitively understandable object schema. These differences are discussed in [8]. Hence, flexible transformations need to be supported in order to obtain a well-designed object-oriented schema.

Approaches supporting both, flexible schema transformation and an automatically generated data migration process, do not yet exist. The main contributions of our approach therefore consist of:

- **Flexible schema transformation:** This allows the relational schema to be transformed into (any) "adequate" object-oriented schema as obtained by forward engineering, rigorously using an object-oriented design method like OOD [9]. We therefore introduce a *redesign environment*.

- **Automatically generated data mapping:** The *formal foundation* of our approach is given by an algebra. This makes it possible to successively define and rewrite data mappings during schema redesign. In consequence, the data migration process can be generated automatically as soon as the schema transformation process is completed.

Recently, object-relational DBMS (ORDBMS) have started to offer some object-oriented features defined in the SQL:1999 standard such as inheritance, user-defined types, etc. Other features are likely to be addressed in SQL4 (e.g., collection-valued attributes). ORDBMS will have the same problem as discussed here for OODBMS, namely to convert an existing relational schema into one exploiting object-oriented features, and to adapt existing databases. Thus, the results presented here for OODBS will also be important for (future) ORDBS.

The remainder of this paper is structured as follows. The basis of our approach is the *redesign environment* which is illustrated in section 2. We introduce a new data model, called semi object types (SOT), and an algebra, in which the activity of schema transformation is embedded. The migration process, embedded in this redesign environment, is presented in section 3. The essential element of the migration process is the concept of *transformation rules* which supports both, schema transformation and data mapping. Section 4 contains concluding remarks and open issues.

2   Redesign Environment

The redesign environment contains an intermediary model for the transformation of a relational schema into an object-oriented one. It consists of two major parts: a data model in which schema transformations are expressed, and an algebra which supports to formulate data mapping expressions.

The central modelling construct of the data model are *semi object types* (SOT), which are comparable to relations or classes. The purpose of the SOT model is to express all (static) properties of object-oriented schemas in a way that makes restructuring operations as easy as possible.

The SOT model is introduced because neither the relational nor the object-oriented data model fulfils the requirements for schema transformation and data migration. The relational data model lacks (unique) identifiers and the support of complex structures. Identity is simulated through key attributes. Since key attributes may also represent contents of the universe of discourse, transformation rules relaxing the uniqueness requirement of these attributes cause problems. As regards complex structures, aggregates and sets may be simulated through additional relations, however list or array structures cannot be expressed directly.

On the other hand, the object-oriented model is too restrictive with respect to object identity and inheritance. Moreover, schema restructuring cannot easily be propagated to the data level. The interface of an object or its class membership (usually) cannot be changed once it is created. No influence can be taken on object identifiers.

An SOT schema consists of a set of SOTs $S = \{s_1, ..., s_n\}$. Every SOT s_i consists of a set of attributes $s_i.A = \{a_1, ..., a_m\}$. SOTs and attributes are identified by unique (system-generated) identifiers. In this way we avoid the problem of name conflicts and the need for defining default names for attributes and SOTs being created during the transformation process. However, in subsequent examples we use names instead of identifiers for better readability. Attributes can be divided into basic attributes (of type integer, string, etc.), collection attributes (of set, list or array type) and reference attributes. Binary relationships between SOTs are expressed through reference attributes, which also define the cardinalities for both SOTs participating in the relationship. A reference attribute of an SOT s_i is a triple $r = (s_j, C_1, C_2)$, where s_j is the referenced SOT, C_1 denotes the cardinality of s_i-references to instances of s_j, and C_2 denotes the inverse cardinality. The cardinalities C_1 and C_2 are pairs $C_i = (c_{i_1}, c_{i_2})$ where $c_{i_1} \in \{0, 1\}$ and $c_{i_2} \in \{1, n\}$. In other words, the cardinalities denote whether the relationship is injective, total, surjective or functional. Inheritance is expressed as a special kind of one-to-one relationship between SOTs.

Data is expressed in the form of objects which are elements of extensions. An object o is an instance of an SOT and consists of an identifier and a tuple value: $o = (id, [a_1:v_1, ..., a_n:v_n])$. An extension is defined as a set of objects $\{o_1, ..., o_n\}$. The SOT algebra defines a number of operators for both metadata and data manipulation. Various algebras for object-oriented data models exist [3, 12, 18], most of them with the purpose of formalizing a query language. The SOT algebra was derived from NO_2 [14] and extended with restructuring and metadata operators. Operators are distinguished in schema operators and data operators. The application of schema operators produces side effects on the SOT schema. Data operators are free of side effects and provide various data restructuring operations. In the following, we briefly introduce a subset of those operators of the SOT algebra, which are used in subsequent examples. The complete and formal definition of the algebra is presented in [5].

- *Concatenation*: Objects can be constructed through concatenation of an identifier and a tuple value:

$$id \oplus [a_1:v_1, ..., a_n:v_n] = (id, [a_1:v_1, ..., a_n:v_n])$$

- *New identifier*: The operator *newid* creates a new identifier, derived from one (or more) existing identifier and one SOT identifier, e.g: $id_2 = newid(id_1, s_1)$. This way we can avoid side effects, i.e., applying the operator for a certain identifier id_1 and a certain SOT s_1 always results in the same identifier id_2. In contrast, traditional (object creating) algebras always yield a new identifier value when invoking operators creating a new object. Details and variations of this operator can be found in [5].

- *Image operator*: The image operator $\iota[\lambda x . f(x)]$, as known from [3], applies the function f to every object within an extension argument:

$$\iota[\lambda x . f(x)]\{o_1, ..., o_n\} = \{f(o_1), ..., f(o_n)\}.$$

- *Projection*: The projection operators has the same purpose as in the relational algebra. We distinguish *identifier projection* π_I and *tuple projection* π_T, which can be applied to objects, and *object projection* π_O, which can be applied to both, objects and extensions:

$$\pi_I(id, [a_1:v_1, ..., a_n:v_n]) = id$$

$$\pi_{T\{a_{\pi_1}, ..., a_{\pi_m}\}}(id, [a_1:v_1, ..., a_n:v_n]) = [a_{\pi_1}:v_{\pi_1}, ..., a_{\pi_m}:v_{\pi_m}] \text{ where}$$
$$\{\pi_1, ..., \pi_m\} \subseteq \{1, ..., n\}$$

$$\pi_{O\{a_{\pi_1}, ..., a_{\pi_m}\}}(id, [a_1:v_1, ..., a_n:v_n]) = (id, [a_{\pi_1}:v_{\pi_1}, ..., a_{\pi_m}:v_{\pi_m}]), \text{ and}$$

$$\pi_{O\{a_{\pi_1}, ..., a_{\pi_m}\}}\{o_1, ..., o_n\} = \iota[\lambda x . \pi_{O\{a_{\pi_1}, ..., a_{\pi_m}\}}x]\{o_1, ..., o_n\}$$

- Selection: This operator can be applied to all kinds of collection values, for example to extensions:

$$\sigma[\lambda x . f(x)]\{o_1, ..., o_n\} = \{o_i \mid o_i \in \{o_1, ..., o_n\}, f(o_i)\}$$

- *Map (extend)*: The *map extend* operator $\chi_{a : \lambda x . f(x)}$ extends an object with an attribute a whose value is computed by a function f:

$$\chi_{a:\lambda x.f(x)}(id, [a_1:v_1, ..., a_n:v_n]) = (id, [a_1:v_1, ..., a_n:v_n, a:f((id, [a_1:v_1, ..., a_n:v_n]))])$$

$$\chi_{a : \lambda x . f(x)}\{o_1, ..., o_n\} = \iota[\lambda y . \chi_{a : \lambda x . f(x)}y]\{o_1, ..., o_n\}$$

- *Map (rename)*: The *map rename* operator $\chi_{\langle a_{1_1}, ..., a_{1_m}\rangle : \langle a_{2_1}, ..., a_{2_m}\rangle}$ replaces attributes in an object without changing their values such that attribute a_{1_1} is replaced by a_{2_1} and so on:

$$\chi_{\langle a_{1_1}, ..., a_{1_m}\rangle : \langle a_{2_1}, ..., a_{2_m}\rangle}(id, [a_{1_1}:v_{1_1}, ..., a_{1_m}:v_{1_m}, a_2:v_2, ..., a_n:v_n]) =$$
$$(id, [a_{2_1}:v_{1_1}, ..., a_{2_m}:v_{1_m}, a_2:v_2, ..., a_n:v_n])$$
$$\chi_{\langle a_{1_1}, ..., a_{1_m}\rangle : \langle a_{2_1}, ..., a_{2_m}\rangle}\{o_1, ..., o_n\} = \iota[\lambda x . \chi_{\langle a_{1_1}, ..., a_{1_m}\rangle : \langle a_{2_1}, ..., a_{2_m}\rangle}x]\{o_1, ..., o_n\}$$

- *Union*: The union operator unifies two extensions:

$$\{o_{1_1}, ..., o_{1_n}\} \cup \{o_{2_1}, ..., o_{2_m}\} = \{o_{1_1}, ..., o_{1_n}, o_{2_1}, ..., o_{2_m}\}$$

An example of an SOT schema with two SOTs, Person and Employee, is shown on the left side of Fig. 1. The attributes name and firstname are basic attributes of type string. The type of the collection attribute titles is a list of strings. The reference attribute superv_of denotes a recursive relationship between employees. It is represented as an arrow with cardinalities at both ends. An employee can supervise zero or more other employees whereas every employee is supervised by at most one other employee. The reference attribute person denotes an inheritance relationship, represented as a bold arrow. Instances of both SOTs are presented on the right side.

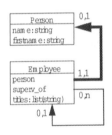

Person

id	name	firstname
id_1	Benn	Chris
id_2	Berge	Glenn
id_3	Felli	Paul
id_4	Kumar	Irina

Employee

id	person	superv_of	titles
id_5	id_1	$\{id_6, id_7\}$	<Prof,Dr>
id_6	id_2	{}	<Dr>
id_7	id_3	{}	<>

Figure 1   SOT Schema and Instances

3   The Migration Process

In this section, the complete migration process from relational schemas and data to object-oriented schemas and data is presented. An overview of the migration process within the SOT redesign environment is shown in Fig. 2. The schema transformation process is subdivided into three sequential tasks:

1. Transformation of the relational schema into an SOT schema
2. Redesign of the SOT schema
3. Transformation of the SOT schema into an object-oriented schema

The data migration process is generated automatically after completing these three steps of schema transformation.

The three steps of the schema transformation process are illustrated by means of a concrete example. The relational schema in Fig. 3 is composed of five relations and contains information about institutes, employees, students and hardware. For simplicity we omitted the type of each attribute. Most of the relations are self-explanatory. The relation Institute contains five description attributes representing a textual field of at most five lines. The relation Employee contains information about name, office and address of employees. Students are characterized by a name, a local address and a home address. The relation Hardware contains information about workstations and monitors, which are distinguished by the attribute type. An IP is a property of workstation entities and a screen size is a property of monitors. The relation InstEmp defines a many-to-many relationship between Institute and Employee.

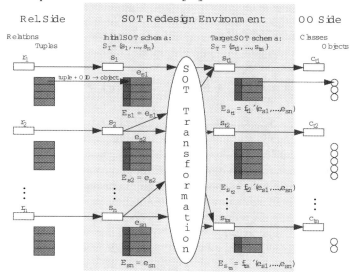

Figure 2   Overview of the Migration Process

Employee (emp_id, name, office_floor, office_nr, adr_street, adr_zip, adr_city)
Institute (name, street, zip, city, descr1, descr2, descr3, descr4, descr5)
Student (stud_id, name, adr_street, adr_zip, adr_city, home_street, home_zip, home_city)
Hardware (serial_nr, office_floor, office_nr, type, ip, screen_size)
InstEmp (emp_id, inst_name)

Foreign Keys: InstEmp (inst_name) -> Institute (name)
                      InstEmp (emp_id) -> Employee (emp_id)

Figure 3   Relational Schema

3.1   Transformation of the Relational Schema into an SOT Schema

The transformation of the relational schema into an SOT schema is a straightforward process. For each relation r_i in the relational schema $\{r_1, ..., r_n\}$, one SOT is created containing the same attributes. The initial SOT schema thus consists of a set of SOTs $S_I = \{s_1, ..., s_n\}$. Then, a set of *initial extensions* $\{e_{s_1}, ..., e_{s_n}\}$ is computed, which contains all the instances of the initial SOTs. These instances are composed by concatenating tuples of the corresponding relation and (unique) object identifiers. The initial extensions remain constant throughout the entire process.

During the schema transformation process, every SOT s_i within the schema has its own *current extension* E_{s_i} which is defined by an algebraic expression determining the current instances of s_i. The current extensions of the initial SOT schema are initialised as $E_{s_i} = e_{s_i}$ for $s_i \in S_I$.

The initial SOT schema of the institute example is presented in Fig. 4.

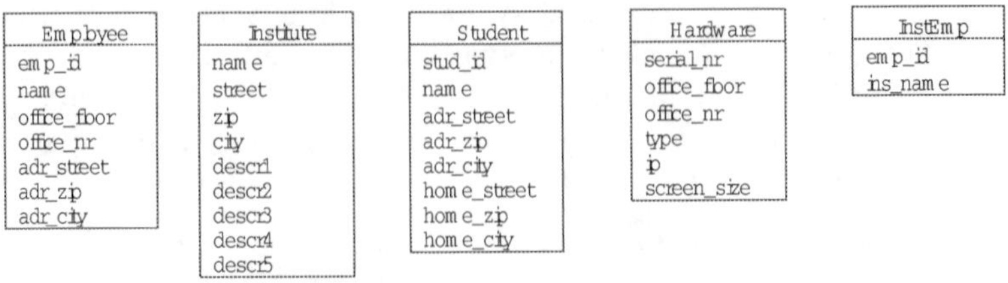

Employee	Institute	Student	Hardware	InstEmp
emp_id	name	stud_id	serial_nr	emp_id
name	street	name	office_floor	ins_name
office_floor	zip	adr_street	office_nr	
office_nr	city	adr_zip	type	
adr_street	descr1	adr_city	ip	
adr_zip	descr2	home_street	screen_size	
adr_city	descr3	home_zip		
	descr4	home_city		
	descr5			

Figure 4   Initial SOT Schema

3.2   Redesign of the SOT Schema

Redesign of the SOT schema means transforming the initial SOT schema into a schema having a more object-oriented flavour, i.e., an object-oriented schema as created by forward engineering using rigorously an object-oriented design method like OOD [9]. We now introduce the concept of *transformation rules*, which supports a consistent modification of schema and data.

Schema transformations are well-known in both, the relational [15, 16, 2] and the object-oriented [7] context. On the relational side, schema transformations have been used for reverse engineering [15] or quality improvement [2], in order to reduce deficiencies of the relational schema like denormalization or optimization. On the object-oriented side, schema transformations have been used for the detailed design phase [7]. Some of those transformation rules have been adopted. We currently propose 22 transformation rules, which are presented in detail in [5]; new transformation rules can be added easily. All transformation rules have a common structure which consists of five parts: a pattern, definitions, preconditions, schema operations and data operations:

- **Pattern:** The pattern defines to which elements of the SOT schema the rule can be applied. These elements also serve as input of the subsequent operations parts.

- **Definitions:** In this part, new SOTs or attributes are defined which result from applying the transformation rule. In addition, specific schema values required for the data operations can be computed.

- **Preconditions:** This part contains a number of conditions which must hold when the transformation rule is applied for a certain pattern. All predicates are first order logic expressions of the SOT algebra and contain pattern elements as free variables.

- **Schema Operations:** The effect on schema level of applying transformation rules is defined in the schema operations parts. This part contains a set of SOT schema operations which add, modify or remove SOTs or attributes.

- **Data Operations:** The last part of a transformation rule contains a set of data operations modifying the extension expressions E_{s_i} of modified or added SOTs s_i. Each data operation is expressed as $E_{s_i} = f(E_{s_{c_1}}, ..., E_{s_{c_k}})$, where $\{s_{c_1}, ..., s_{c_k}\}$ are variables referring to elements of the current SOT schema, and f is an algebraic expression in terms of the SOT algebra.

In the following we informally introduce those transformation rules which have been used in the institute example. In addition, the formal definitions of the first two transformation rules are given, since these two rules are used later for a data migration example.

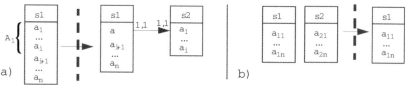

Figure 5 a) Vertical Split of an SOT, b) Merging two SOTs

- **Vertical Split of SOT**: The first transformation rule vertically splits an SOT, as shown in Fig. 5 a). Vertical split was applied for all SOTs except InstEmp, such that the address attributes street, zip and city, and the office attributes of ce_oor and of ce_nr have been moved to new SOTs. It can be distinguished whether duplicates lead to multiple objects or a single object. In case of the SOT Address, every address is mapped into one object, such that identical addresses lead to distinct objects. In case of the SOT Office, duplicates have been removed, such that every combination of of ce_oor and of ce_nr leads to one object. Applying the first variation results in a one-to-one relationship, whereas the second variation results in a one-to-many relationship.

Pattern	$s_1 : S,\ A_1 : set(A)$
Definitions	$s_2 = new_{SOT}(A_1)$ $a = new_{Ref}(s_2, (1,1), (1,1))$ $A_2 = s_1.A - A_1$
Preconditions	$A_1 \subset s_1.A$
Schema Operations	$s_1.set\_attributes(A_2 + set\{a\})$ $SOT = SOT + set\{s_2\}$
Data Operations	$E_{s_1} = \chi_{a\,:\,\lambda x\,.\,newid(\pi_l x,\, s_2)} \pi_{O_{A_2}} E_{s_1}$ $E_{s_2} = \iota[\lambda x\,.\,newid(\pi_l x,\, s_2) \oplus \pi_{T_{A_1}} x] E_{s_1}$

Figure 6 Transformation Rule: Vertical Split of SOT

The formal definition of the first variation of this transformation rule is presented in Fig. 6. Vertically splitting an SOT s_1 means creating a new SOT s_2 which contains a proper subset of the attributes of s_1. In turn, these attributes are removed from s_1. The pattern consists of the SOT s_1 which has to be split, and a set of attributes A_1 which are moved to the new SOT. The *definitions* part creates the new SOT s_2 which contains the attributes A_1 and a new attribute a, referencing s_2 in an exactly-one-to-exactly-one relationship. The precondition states that A_1 must be a proper subset of the attributes of s_1. The effects of schema operations are modifying the attributes of s_1 and including s_2 in the SOT schema, denoted by a global variable SOT. The data operators compute the extensions of s_1 and s_2. The extension E_{s_1} results from a projection on those attributes remaining in s_1 and a mapping of the reference attribute a. The map operator $\chi_{a\,:\,\lambda x\,.\,f(x)}$ adds the attribute a to each object in the argument extension. The value of the attribute is computed by applying the function $f(x)$ captured in the lambda expression to the object. The extension E_{s_2} is computed by creating one new object for every existing object within E_{s_1}. Each new instance of E_{s_2} is composed of a new object identifier and the attributes A_1.

- **Merge SOTs**: The second transformation rule merges two SOTs s_1 and s_2 into a single one, as shown in Fig. 5 b). Merging two SOTs into a single one was applied for those SOTs obtained from splitting addresses and offices, resulting in single SOTs Address and Office.

Pattern	$s_1 : S, s_2 : S, al_1 : list(A), al_2 : list(A)$
Definitions	$s_1 \neq s_2$ $members(al_1) = s_1.A$ $members(al_2) = s_2.A$ $\forall(i \in indexlist(al_1))(type(al_1[i]) = type(al_2[i]))$
Preconditions	$s_1 \neq s_2$ $members(al_1) = s_1.A$ $members(al_2) = s_2.A$ $\forall(i \in indexlist(al_1))(type(al_1[i]) = type(al_2[i]))$

Schema Operations	$SOT = SOT - set\{s_2\}$
	$for\ (s \in SOT)$
	   $for\ (a \in s.R)$
	     $if\ (a.S = s_2)$
	      $a.set\_S(s_1)$
	     $if\ (a.S = s_1)$
	      $a.set\_c21(0)$

Data Operations	$E_{s_1} = E_{s_1} \cup \chi_{al_2 : al_1} E_{s_2}$

Figure 7   Transformation Rule: Merge SOTs

The formalism of this transformation rule is presented in Fig. 7. The original rule is rather complex and thus has been simplified for the sake of exposition. Here, we assume that all attributes of s_1 are mapped into the attributes of s_2 and both SOTs do not contain reference attributes. The pattern consists of two SOTs s_1 and s_2, denoting s_2 is merged into s_1. Two attribute lists $al_1 = \langle a_{1_1}, ..., a_{1_n}\rangle$ and $al_2 = \langle a_{2_1}, ..., a_{2_n}\rangle$ denote the order in which the attributes are mapped, i.e., a_{2_1} is mapped onto a_{1_1} and so on. The definitions part is empty. The preconditions state that the participating SOTs must be different, all attributes of both SOTs s_1 and s_2 appear in the pattern's attribute lists, and the types of corresponding attributes must be equal, i.e., the type of a_{1_1} must be the same as the type of a_{2_1} and so on. Schema operators remove s_2 from the SOT schema and modify all reference attributes referencing either s_1 or s_2. The new extension of s_1 is computed by unifying E_{s_1} and E_{s_2} whereby the instances of E_{s_2} undergo the atttriute mapping.

- **Reference creation**: This transformation rule allows to create reference attributes. They are determined, e.g., through inclusion dependencies, which have been gathered from reverse engineering of the relational database. The cardinalities of a reference attribute can be determined through the presence of primary keys, candidate keys and null values in the relational schema.

- **Transitive References**: A transitive reference is defined as a path of two existing references between three SOT. For example, after creating two references in the initial SOT InstEmp referencing Employee and Institute, respectively, a reference from Employee to Institute representing a many-to-many-relationship can be created.

- **Inverse References**: This transformation rule allows to created inverse relationships as known from the ODMG standard [10]. In the SOT schema, inverse references are denoted as bidirected arrows like the relationship between Employee and Institute.

- **Horizontal split of SOT**: As an example of horizontal splitting, hardware has been specialized into workstations and monitors, depending on the value of the attribute type. In addition, the attributes ip and screen_size have been moved to Workstation and Monitor, respectively.

- **List construction**: Several rules support the creation of collection attributes. For example, the five description attributes of Institute have been transformed into a single list attribute.

- **Remove SOT/attributes**: SOTs and attributes which became obsolete can be removed, for example, the SOT InstEmp.

- **Inheritance**: References representing a one-to-one relationship can be declared as inheritance references. This way, the Person and Hardware hierarchies have been defined.

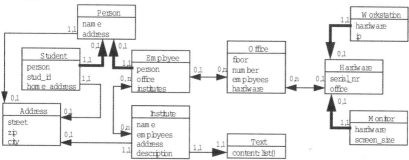

Figure 8   Target SOT Schema

The resulting target SOT schema of the institute example is shown in Fig. 8. Two aspects are worth mentioning in this example. First, traditional migration tools are only able to replace the relation InstEmp by a many-to-many relationship. The remaining relations will be mapped to classes having the same attributes. Second, the schema contains three typical examples of transformation patterns. The first one is the multiple occurrence of common attribute groups like the address attributes street, zip and city, or the office attributes of ce_oorand of ce_nr. The second example is the implementation of a multivalued attribute through multiple separated attributes as in the description of an institute. Finally, the last example consists of the variant attribute type of the relation Hardware, whose purpose is to distinguish workstation and monitor entities.

3.3   Transformation of the SOT Schema into an Object-Oriented Schema

In the last step of the schema transformation process, the target SOT schema $S_T = \{s_{t_1}, ..., s_{t_m}\}$ is transformed into an object-oriented schema, again in a straightforward process. The class structure of the object-oriented schema of the institute example is presented in Fig. 9, expressed by the interface notation of the ODMG object definition language (ODL) [10].

Afterwards, the data is migrated to an object-oriented database. First, the target extensions $\{E_{s_{t_1}}, ..., E_{s_{t_m}}\}$ have to be computed, either by successive computation of the in-

```
interface Person {
    attribute string name;
    attribute Address address;);

interface Student : Person {
    attribute string stud_id;
    attribute Address parent_address;);

interface Employee : Person {
    attribute string emp_id;
    relationship Office office
        inverse Office:employees;
    relationship set<Institute> institutes
        inverse Institute:employees;);

interface Institute {
    attribute string name;
    attribute Address address;
    attribute Textdescription;
    relationship set<Employee> employees
        inverse Employee::institutes;);

interface Text {
    attribute list<string> content;);
```

```
interface Address {
    attribute string street;
    attribute string zip;
    attribute string city;);

interface Hardware {
    attribute string serialnr;
    relationship Office office
        inverse Office:hardware;);

interface Workstation : Hardware {
    attribute string ip;);

interface Monitor : Hardware {
    attribute string screen_size;);

interface Office {
    attribute string floor;
    attribute string number;
    relationship set<Employee> employees
        inverse Employee::office;
    relationship set<Hardware> hardware
        inverse Hardware::office;);
```

Figure 9   Object-Oriented Schema

curred data operations (without the possibility of optimization), or by recursively rewriting the extension expressions such that all target extensions are expressed as functions defined over the initial extensions: $E_{S_{ti}} = f_{ti}'(e_{s_1}, ..., e_{s_n})$. These expressions can be optimized, for example, predefined algebraic rewriting rules can be applied to single terms, and common subexpressions of different extensions have to be computed only once. All resulting target extensions are independent of each other, therefore parallel computation is possible. Optimization for relational and object-oriented algebras has been extensively studied in [4, 11, 12, 17, 18, 19]. The resulting expressions are then compiled into queries against the relational database. The object-oriented database is populated by creating objects from the SOT target extensions. This can be performed by creating code in the ODMG object interchange format (OIF) [10], or by generating a migration program.

An example of OIF code for the instances presented in Fig. 1 is shown in Fig. 10. One employee object is composed of one SOT Employee object and one SOT Person object. This step concludes the migration process.

For the following subset of the institute example, we demonstrate the use and the effects of two transformation rules. The example in Fig. 11 presents an initial SOT schema comprising two SOTs Institute and Person. The initial extensions of both SOTs are shown on the right side. The first two transformation rules as introduced in section 4.2 have to be applied to obtain the target SOTs presented in Fig. 13.

The first transformation rule applied to the initial SOT schema is the vertical split of an SOT as shown in Fig. 6. The effects on schema and instance level of applying this transformation rule for both Institute and Person are illustrated in Fig. 12. The pattern is initialized with s_1 = Institute and A_1 = {Street, ZIP, City} as well as with s_1 = Person and A_1 = {AdrStreet, AdrZIP, AdrCity}. The resulting current extensions of the SOTs Institute, Person, Address and PersAddr are defined by the following algebraic expressions, based on the initial extensions $e_{Institute}$ and e_{Person}:

id4 Person {
    name "Kumar",
    firstname "Ina"}

id5 Employee {
    name "Benn",
    firstname "Chris",
    superv_of{id6,id7},
    titles {"Prof","Dr"}}

id6 Employee {
    name "Berge",
    firstname "Glenn",
    superv_of{},
    titles {"Dr"}}

id7 Employee {
    name "Felli",
    firstname "Paul",
    superv_of{},
    titles {}}

Figure 10   Example of OIF Code

Institute

id	Name	Street	ZIP	City
id_1	Ifl	Winterthurerstr. 190	8057	Zurich
id_2	GIUZ	Winterthurerstr. 190	8057	Zurich

Person

id	Name	AdrStreet	AdrZIP	AdrCity
id_3	Oeler	Gujerstr. 10	9200	Gossau
id_4	Berger	Rigistr. 23	3855	Brienz

Institute	Person
Name	Name
Street	AdrStreet
ZIP	AdrZip
City	AdrCity

Figure 11   Subset of SOT Schema with Extensions

$$E_{Institute} = \chi_{Address : \lambda x . newid(\pi_I x, Address)} \pi_{O_{\{Name\}}} e_{Institute}$$

$$E_{Address} = \iota[\lambda x . newid(\pi_I x, Address) \oplus \pi_{T_{\{Street, ZIP, City\}}} x] e_{Institute}$$

$$E_{Person} = \chi_{Address : \lambda x . newid(\pi_I x, PersAddr)} \pi_{O_{\{Name\}}} e_{Person}$$

$$E_{PersAddr} = \iota[\lambda x . newid(\pi_I x, PersAddr) \oplus \pi_{T_{\{AdrStreet, AdrZIP, AdrCity\}}} x] e_{Person}$$

The second transformation rule merges the SOT PersAddr into Address according to Fig. 7. The effects on schema and instance level of applying this transformation rule is illustrated in Fig. 13. The pattern is initialised with s_1 = Address, s_2 = PersAddr, al_1 = $\langle$Street, ZIP, City$\rangle$ and al_2 = $\langle$AdrStreet, AdrZIP, AdrCity$\rangle$. The current extension of the SOT Address is defined by the following algebraic expression, in the second case again based on the initial extensions $e_{Institute}$ and e_{Person}:

$$E_{Address} = E_{Address} \cup \chi_{\langle AdrStreet, AdrZIP, AdrCity \rangle : \langle Street, ZIP, City \rangle} E_{PersAddr}$$

$$E_{Address} = \iota[\lambda x . newid(\pi_I x, Address) \oplus \pi_{T_{\{Street, ZIP, City\}}} x] e_{Institute} \cup$$
$$\chi_{\langle AdrStreet, AdrZIP, AdrCity \rangle : \langle Street, ZIP, City \rangle} (\iota[\lambda x . newid(\pi_I x, PersAddr) \oplus$$
$$\pi_{T_{\{AdrStreet, AdrZIP, AdrCity\}}} x] e_{Person})$$

Institute

id	Name	Address
id_1	Ifl	id_5
id_2	GIUZ	id_6

Person

id	Name	Address
id_3	Oeler	id_7
id_4	Berger	id_8

Address

id	Street	ZIP	City
id_5	Winterthurerstr. 190	8057	Zurich
id_6	Winterthurerstr. 190	8057	Zurich

PersAddr

id	AdrStreet	AdrZIP	AdrCity
id_7	Gujerstr. 10	9200	Gossau
id_8	Rigistr. 23	3855	Brienz

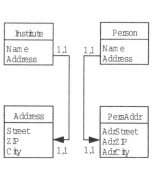

Figure 12   Effect of Applying Vertical SOT Split

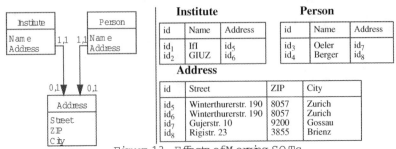

Figure 13   Effects of Merging SOTs

4 Conclusions and Future Work

In this paper we studied the problem of migrating relational databases to object-oriented databases. In relational and object-oriented database design different design strategies are followed. Moreover, legacy databases often contain specific drawbacks which should be overcome when reengineering them. The differences in design strategies motivate schema transformation techniques being more powerful and flexible than existing ones, which mostly generate structurally identical object-oriented schemas.

In the second part, we presented a framework which provides a formal foundation for the migration of legacy relational schemas and data into object-oriented databases. This is the first approach which supports automatic data migration for complex schema transformations. In particular we described:

- The complete migration process from relational schema and data to object-oriented schema and data.
- A data model and a redesign environment in which the migration process is embedded.
- The concept of transformation rules supporting complex schema transformations.

The contents of the SOT redesign environment have been implemented using the functional programming language Scheme. In particular, the implementation comprises the SOT algebra, transformation rules, and a program which allows to apply transformation rules to an SOT schema.

In addition, we currently evaluate and validate the flexibility of our approach by means of a relational schema managing bank accounts which consists of 100 relations with altogether 3.500 attributes and 180 views. First experiences indicate that the proposed transformation rules allow to construct well-designed object-oriented schemas. Furthermore, we are currently working on the following extensions:

- Extended tool support: how can a tool support or simplify the transformation of large schemas and how can it control the transformation process? To what extent can heuristics support choosing the best-suited transformation rules?
- Updates: for simplicity we assumed that there are no updates on the database during the migration process. How can updates on the source database be propagated through the migration process such that incremental migration is possible?
- Optimization: How can the cost of the migration process be reduced by using algebraic rewriting and parallelization?

Acknowledgements

We thank Ruxandra Domenig, Dirk Jonscher and Martin Schönhoff for their helpful comments on an earlier version of this paper.

References

[1]   S. Amer-Yahia, S. Cluet, and C. Delobel. Bulk loading techniques for object databases and an application to relational data. *Proc. Int'l Conf. on Very Large Databases (VLDB)*, New York, August 1998.

[2]   C. Batini, S. Ceri, and S. Navathe. *Conceptual database design: an entity-relationship approach*. Benjamin/Cummings Publishing Company, Inc., 1992.

[3]   C. Beeri. Query languages for models with object-oriented features. In *Advances in Object-Oriented Database Systems*, NATO ASI Series, chapter 3. Springer-Verlag, 1994.

[4]   C. Beeri and Y. Kornatzky. Algebraic optimization of object-oriented query languages. *Proc. 3rd Int'l Conf. on Database Theory*, Paris, France, December 1990.

[5]   A. Behm. *Migrating relational databases to object technology*. PhD thesis (forthcoming), University of Zurich, Switzerland, 2000.

[6]   A. Behm, A. Geppert, and K. R. Dittrich. On the migration of relational schemas and data to object-oriented database systems. *Proc. 5th Int'l Conf. on Re-Technologies for Information Systems*, Klagenfurt, Austria, December 1997.

[7]   M. Blaha and W. Premerlani. A catalog of object model transformations. *Proc. 3rd Working Conf. on Reverse Engineering*, Monterey, California, November 1996.

[8]   M. Blaha and W. Premerlani. Detailed design. In *Object-Oriented Modeling and Design for Database Applications*, chapter 10. Prentice-Hall, 1998.

[9]   G. Booch. *Object-Oriented Analysis and Design with Applications*. Benjamin/Cummings, 1994.

[10] R. G. G. Cattell. *The object data standard: ODMG 3.0*. Morgan Kaufmann, 2000.

[11] S. Cluet and C. Delobel. A general framework for the optimization of object-oriented queries. *Proc. SIGMOD Int'l Conf. on Management of Data*, New York, June 1992.

[12] S. Cluet and G. Moerkotte. Classification and optimization of nested queries in object bases. In *Bases de Donnees Avancees*, 1994.

[13] J. Fong. Converting relational to object-oriented databases. *SIGMOD Record*, 26(1), 1997.

[14] A. Geppert, K. R. Dittrich, V. Goebel, and S. Scherrer. The NO2 data model. Technical Report 93.09, Institut fuer Informatik der Universität Zürich, 1993.

[15] J-L. Hainaut, C. Tonneau, M. Joris, and M. Chandelon. Transformation-based Database Reverse Engineering. *Proc. 12th Int. Conf. on the ER Approach*, Dallas, December 1993.

[16] R. J. Miller, Y. E. Ioannidis, and R. Ramakrishnan. The use of information capacity in schema integration and translation. Proc. *19th VLDB Conf.*, Dublin, Ireland, 1993.

[17] G. Mitchell, S.B. Zdonik, and U. Dayal. Optimizations of object-oriented query languages: Problems and approaches. In *Advances in Object-Oriented Database Systems*, NATO ASI Series, chapter 6, Springer-Verlag, 1994.

[18] G. M. Shaw and S. B. Zdonik. A query algebra for object-oriented databases. *Proc. Int'l Conf. on Data Engineering*, Los Angeles, CA, February 1990.

[19] S. L. Vandenberg and D. J. DeWitt. Algebraic support for complex objects with arrays, identity, and inheritance. In *Proceedings of the ACM SIGMOD*, May 1991

A Layered Software Specification Architecture

M. Snoeck, S. Poelmans, and G. Dedene
Management Information Systems Group
Katholieke Universiteit Leuven,
Naamsestraat 69, 3000 Leuven
email: {monique.snoeck, stephan.poelmans,
guido.dedene}@econ.kuleuven.ac.be

Abstract. Separation of concerns is a determining factor of the quality of object-oriented software development. Done well, it can provide substantial benefits such as additive rather than invasive change and improved adaptability, customizability, and reuse. In this paper we propose a software architecture that integrates concepts from business process modeling with concepts of object-oriented systems development. The presented architecture is a layered one: the concepts are arranged in successive layers in such a way that each layer only uses concepts of its own layer or of layers below. The guiding principle in the design of this layered architecture is the separation of concerns. On the one hand workflow aspects are separated from functional support for tasks and on the other hand domain modeling concepts are separated from information system support. The concept of events (workflow events, information system events and business events) is used as bridging concept between the different layers.

1. Introduction

The proponents of object-oriented software development attribute a number of qualities to object-oriented software development such as improved adaptability, maintainability and reuse. In spite of more than a decade of experience with object-oriented technology, the improvements are not really overwhelming. Although separation of concerns is recognized as a determining factor for the adaptability and maintainability of software, there is not yet a generally accepted way to achieve this. Separation of concerns can be pursued at different levels of abstraction in the software development process. In this paper we present a layered software architecture that represents a separation of concerns at a high level of abstraction: it classifies specifications in different layers such that each layer only relies on concepts of the same layer or on concepts of the layers below. At the same time, this architecture outlines a basic implementation architecture. To obtain this layered architecture, we start from the assumption that a full-fledged information systems development method should take all aspects into account: aspects of business process modeling and aspects of the functional part should be linked together. Because most object-oriented analysis and design methods do not yet integrate business process modeling aspects in an adequate way we motivate this assumption and briefly present the approach in section 2. In section 3 we present the four basic layers of our architecture. In section 4 these layers are further refined. Finally section 5 presents some conclusions and topics for further research.

A.H.F. Laender, S.W. Liddle, V.C. Storey (Eds.): ER2000 Conference, LNCS 1920, pp. 454-469, 2000.

2. The Need for Integration of Workflow Aspects in Information System Modeling

The layered architecture proposed in this paper encompasses all the aspects of IT-support in an organization. On the one hand there is the required support for supervising, recording and controlling business activities. On the other hand there is the required functional support for these business activities. The first type of support constitutes the Workflow System, whereas the functional support constitutes the Business Information System. Current object-oriented analysis and design methods focus primarily on the analysis and design of business information systems. In this paragraph we first look at the requirements for adequate business process support. We then argue the statement that current object-oriented analysis and design methods lack a business process view. Finally, we briefly explain how the concept of "business event" can be used to link concepts from business process modeling with object-oriented analysis and design concepts.

2.1 Requirements for an Adequate Workflow System Support

In the literature on workflow modeling, several techniques are proposed to define and represent the structure of a business process (petri-nets, flow charts, etc.). In most cases, the method to be followed is imposed by the vendor of the workflow package [13]. Nevertheless, some general requirements can be put forward that are necessary to be able to model a process:

Requirement 1. Processes and activities need to be defined in a hierarchical manner. Process design typically requires a top down decomposition of high level processes into subprocesses down to atomic activities. It is the division of labor in the organization that determines the subdivision in sub-processes and activities. Activities and tasks might have a different meaning in different organizational theories. In the field of workflow modeling however, both terms are often used interchangeably and we will do the same in this paper.

Requirement 2. The modeling of dependencies between activities and between activities and agents is crucial. The main goal of a workflow system is in fact the automation or support of the co-ordination between activities and between activities and agents. Co-ordination can be defined as the management of dependencies [16]. In what way does one activity depend on the results of another activity? The modeling of dependencies constitutes the heart of workflow modeling. The existence of dependencies implies a certain order of execution. Some (sub)tasks cannot be performed before previous tasks have been completed; other tasks need to be executed in parallel, and so on.

Requirement 3. Agents are humans or computer applications that are assigned to roles. Also the interaction between activities and agents needs to be planned ahead. Agents can be human end-users or computer applications that perform activities. When human agents execute certain tasks, they might be assisted by computer applications to support them. Only when applications are directly coupled to the workflow system, they are considered as agents. When an application is invoked by the workflow system and when the application performs a certain activity without any intervention of the end-user, it is called an autonomous agent. An agent is called semi-autonomous when it is directly coupled to a workflow system, although an intervention of the end-user is still required. Agents are assigned to activities via the construction of roles. A role defines the responsibility for the performance of a (collection of) task(s)[14].

Requirement 4. The specification of the business process (or workflow) needs to be a persistent artefact able to control, supervise and record performed activities. The workflow process is specified as a model in a formal textual and/or visual language. This model specification is used whenever a new workflow instance needs to be created. Each time a workflow instance is created, the persistent workflow model is needed for controlling, supervising and recording the performed activities. Moreover, in order to monitor and improve performance, it is often also required to save the states of instantiated processes that have been enacted. Historical data regarding the actual course of processes can be useful and even necessary to improve the persistent process model.

The dependencies between activities and between activities and agents can be considered as the control logic of business processes. The functional part contains the necessary data and the applications that (partly) perform the activities (the non-human agents). The isolation of the control structure from the data and functional structure is a typical characteristic of workflow systems [25].

2.2   The Lack of a Business Process View in Object-Oriented Systems Development

Workflow systems and object-oriented technology have undoubtedly been some of the most important domains of interest of information technology over the past decade. Both domains however, have largely evolved independently, and not much research can be found in which workflow modeling principles and concepts have been applied to OO systems development. In object-oriented development, the primary emphasis is on the specification and development of the functional part of the information system, whereas the business process part is largely neglected or supposed to be given [e.g. 2, 3, 4, 6, 7, 11, 19, 23]. Although recently there is an increased support for the *software development* process and its workflows [9, 18], *business* process modeling is still treated in a fairly limited way.

In the first place, the top down decomposition of processes (requirement 1) is barely supported in object-oriented development. Functional decomposition, a vital concept from the structured programming world, is often considered as old-fashioned and ineffective by object-oriented developers [26]. One way to introduce some of the business process aspects in information systems analysis is the use of Use Cases [11, 2]. Use cases describe the functional requirements by identifying actors and scenarios of system usage by these actors. As such this technique is a valid candidate to model the interaction between a user and the system. The technique offers some possibilities for modularization by allowing use cases to "include" and "extend" other use cases, although this is not a functional decomposition as described in requirement 1 above. In the UML approach [11, 2], use cases are mainly a support for information system design: they are used for finding objects and determining the systems structure. More importantly, use cases are not intended to model the assignment of agents to activities and the co-ordination between activities (requirement 2) and can therefore not be considered as a workflow modeling technique. In addition the process logic is not designed to be implemented as an (persistent) application (requirement 4).

Because of their affinity with petri-nets, activity diagrams are much better suited for modeling activities and their dependencies. Although activity diagrams can be stored as persistent artefacts in a CASE-tool, they are not used to realize a workflow engine that controls, supervises and records the performed activities (requirement 4).

Finally, several other dynamic representations (like state transition diagrams and sequence diagrams) are created in the development phase. The process logic in this type of diagrams is however mainly relevant for the functional aspects of the application. In some cases, aspects of a business process can be found in this type of diagrams. However, such business process logic is not explicitly and separately implemented as described in requirement 4 above.

2.3 Advantages to Gain

A separation of concerns is a key element in keeping systems maintainable and adaptable. In current object-oriented system development practice, the organizational aspect of an information system is often not explicitly modeled. And when it is, it is not always taken as an important element in guiding design decisions. By integrating business modeling concepts into object-oriented modeling, the link between the services that an information system has to render and the organizational elements becomes more apparent. This can be an important help in designing more adaptable systems. In addition, when workflow elements are not modeled separately, they are often hidden in the procedural logic of class-methods. The explicit separation of workflow elements from process elements that are inherent to the domain or to the procedural logic of an implementation also allows for more adaptable systems. For example, sequence constraints on events that result from the business logic are part of the domain model (e.g. in a library, the return of a copy to the library must be preceded by a borrowing event). These type of sequence constraints are less likely to change over time than sequence constraints that are the result of workflow aspects (e.g. if a member of the library does not show up after five reminders, set all the books (s)he borrowed to the state "lost").

2.4 Using Business Events as Bridging Concept

Events play a central role in the set of layers proposed in the next paragraphs. In most object-oriented approaches, events are subordinate to objects: they only serve as triggers for the execution of an object's method. In the approach proposed below, events are raised to the same level of importance as objects. Indeed, events are a fundamental part of the structure of experience [4]. Events are atomic units of action that represent things that happen in the world. Without events nothing would happen: they are the way information and objects come into existence (creating events), the way information and objects are modified (modifying events) and disappear from our universe of discourse (ending events). In the context of the business information system that gives the functional support for the workflow system, we make a difference between business events (also called real-world events) and information-system events such as keystrokes and mouse actions. The separation between these two types of events allows a more user-oriented and task-oriented view of information system design. Business events are those events that occur in the real world, even if there is no information system around. Information-system events are directly related to the presence of a computerized system. They are designed to allow the external user to register the occurrence of or invoke a real-world event. For example, the use of an ATM-machine to withdraw money from one's account will invoke the business event "*withdraw*" by means of several information-system events such as "*insert card*", "*enter PIN code*", "*enter amount*", and so on. Using events as a fundamental concept integrates well with the object-oriented approach as demonstrated in methods such as Syntropy [4], Catalysis [6], OO-SSADM [19] and MERODE [21, 23].

Business process modeling takes an action and process-oriented view on the domain. As a result, task and activities are easier to formulate in terms of business events than in terms of business objects (which are better for modeling structural aspects). From a business modeling perspective, only business events are of particular interest. Information system events such as keyboard actions and mouse clicks are modeled as elements in the information system, but are not relevant elements in a business process model.

As business events appear both in the functional part and in the business process part, they can serve as the bridging concept between workflow activities and information system design. The figures below represent a meta-model for the concepts used in the proposed system development approach. In a first step, business processes are modeled at a conceptual level by decomposing them down to the activity level and by indicating which business event each activity invokes. The meta-model for business process modeling only shows the BUSINESS PROCESS and WORKFLOW ACTIVITY classes. We assume however that the complete set of workflow concepts are modeled in an object-oriented way, such as for example in the TriGS$_{flow}$ model [14]. For the functional aspects, the considered domain is modeled at the conceptual level by identifying domain object classes and by indicating by which business event they are affected. As a result, business domain object interaction can be modeled by joint involvement in business event types, rather than by message passing. This makes the domain object interaction scheme more implementation independent and more adaptable to changed requirements. The effect of an event on a domain object class is recorded in a domain object class method. Fig. 1 shows a (simplified) meta-model relating the modeling concepts at this stage of the specification process. At the highest conceptual level, business events directly link business process modeling concepts to domain modeling concepts .

In a refining step the business processes and the business domain are analyzed in search for information system support. So, next to the description of the domain of interest in the domain model, we need a specification of the services (also called user functions) that the information system has to render to the prospective users. This part of the specification is closely related to the specification of the workflow model: it is the description of the functional support for the activities of the workflow model. The activities that have to be performed by agents can be further classified as manual, interactive or fully automated. Interactive and automated activities are realized by means of an information system service. In this refined model, the information system services interface the workflow system with the domain model by giving computerized support for the invocation of business events. Fig. 2. represents the meta model for this more detailed level of specifications.

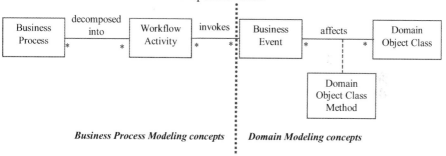

Business Process Modeling concepts : **Domain Modeling concepts**

Fig. 1. Meta-model for conceptual modeling

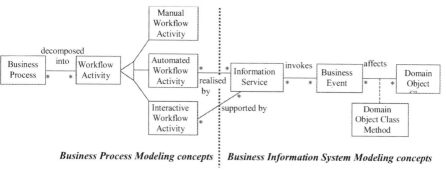

Business Process Modeling concepts : **Business Information System Modeling concepts**

Fig. 2. Information systems modeling meta-model.

3. First Set of Basic Layers

The first set of layers are dictated by the separation of the workflow aspects from the functional support of tasks by information system services. Hence, at the highest level of abstraction, we have one layer for the workflow aspects and another layer for the information system aspects. A further refinement of the layers is obtained by applying the principle of model-driven development. Model-driven development is based on the idea that requirements should be captured in different models, according to their origin, as described in the work of Zachman [27, 24] and Maes[15]. Here we retain the separation of domain modeling from information system support modeling. Some specifications stem from fundamental business requirements (including business objects, business events as well as business constraints). This type of requirements is also valid if there is no information system. Other specifications are typically related to the presence of an information system. They describe the required functional support such as input facilities, generation of reports, EDI formatting, and so on. The specifications of the first type constitute a **business domain model** that contains the relevant domain knowledge for running a business. On top of this business domain model, a **information service model** is built as a set of input and output services, offering the desired information functionality to the users of the information system. Output services allow users to extract information from the business domain model, and present it in the right format on paper, on a workstation or in an electronic format. Input services allow users to register new or modified information that is relevant for the business.

The model-driven approach can also be applied to the workflow layer. The workflow domain model describes the essential concepts of business process modeling such as the concepts of agent, workflow activity, business process, task dependencies, worklists, and so on. It is by populating the workflow domain model with instances of agents, tasks, dependencies, business processes, etc. that the business processes of a particular organization are defined. At the same time the populated domain model is a persistent model of the organization's business processes (requirement 4). The workflow service layer describes the information system support offered by the workflow system such as facilities to view a worklist, to add a work-item to a work list, to pass on work items, to register the accomplishment of a task in a work list, to create new tasks, and so on, but also services that allow to control supervise and record performed activities (requirement 4). As a result, we obtain an architecture with 4 different layers (Fig. 3).

The dynamic aspects of each layer are triggered by four different kinds of events. Workflow-system events are information-system events related to the workflow system. They can for example be keyboard actions and mouse clicks to be captured by the workflow system and aiming at the invocation of a business-process event. Business-process events are real-world events that trigger the dynamic aspects of the workflow domain, such as the creation of a new task, assigning a person to a task, finishing a task and so on. They are related to the organizational aspects of the business. Information-system events trigger the dynamics of the information system. They will be used to invoke the execution of information system services that give functional support for the tasks in the business process model. Finally, the business events are triggering the behavior of domain objects. They are the real-world events that constitute the dynamic part of a business.

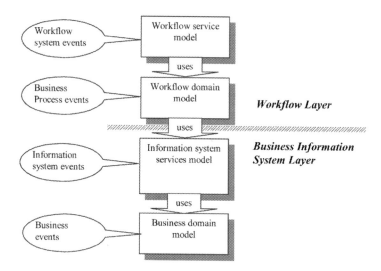

Fig. 3. Four basic layers.

The link between the workflow layer and the business information system layer is achieved by linking tasks in the workflow domain model with the supporting information system services. Execution of a task or work item means that either this service is automatically invoked by the workflow system (autonomous software agent) or that the user invokes the service him(her)self through the business information system user interface.

4. Refining the Layers

The four basic layers are each refined in two sub-layers. The business domain layer is subdivided in a business event layer and a business domain objects layer. As mentioned in section 2, it is assumed that business domain objects interact by being jointly involved in business event types rather than interacting through message passing. To realize this kind of interaction, it is assumed that events are broadcasted to objects. This means that when an event is invoked, each object that is involved in the event checks whether the constraints imposed by this object on events of that type are satisfied. If all the involved objects accept the event, all corresponding methods in the involved objects are executed simultaneously. This way of communication is similar to communication as defined in the process algebras CSP [10] and ACP [1] and has been formalized in [5, 23]. Message passing is more similar to the CCS process algebra [17]. There exist various mechanisms for the implementation of such synchronous execution of methods. In the layered architecture proposed in this paper, we assume that there is an event handling mechanism that filters the incoming events by checking all the constraints this event must satisfy. If all constraints are satisfied, the event is broadcasted to the participating objects; if not it is rejected. In either case the invoking class is notified accordingly of the rejection, acceptance, and successful or unsuccessful execution of the event. For each type of business events, the event handling layer contains one class that is responsible for handling events of that type. This class will first check the validity of the event and, if appropriate, broadcast the event to all involved objects by means of the method 'broadcast'. This approach to domain object interaction enhances the adaptability of the domain model compared to a conventional approach were domain objects dispatch the event by sending messages to each other.

The information services layer can be further subdivided by separating user interface aspects from the transaction aspects. This separation is similar to the classical three tier architecture were control logic is also separated from user interface aspects. In this approach however, the control logic is partly in the transaction layer, partly in the business event layer, and partly in the methods of the business domain objects. User interface objects are responsible for all presentation aspects and for syntactical user input validation. Input transactions invoke one or more business events using parameter values received from the user interface objects. Output transactions query the set of domain objects to retrieve the requested information. The transaction layer can be used to group event invocations according to task requirements. Commit and roll-back features can also be implemented in the transaction layer.

To better illustrate the responsibilities of the different layers, we will exemplify objects in the four business information system layers by considering four examples of information system services for an order handling system. Let us assume that the domain model contains the four object types CUSTOMER, ORDER, ORDER LINE and PRODUCT, ORDER being existence dependent on CUSTOMER, and ORDER LINE being existence dependent on both ORDER and PRODUCT. The corresponding ER-schema is given in Fig. 4.

Fig. 4. ER-schema for the order handling system

Business event types are *create_customer, modify_customer, end_customer, create_order, modify_order, end_order, create_orderline, modify_orderline, end_orderline, cr_product, modify_product, end_product.* The object-event table (see Table 1) shows which object types are affected by which types of events and also indicates the type of involvement: C for creation, M for modification and E for terminating an object's life. For example, the *cr_orderline* creates a new occurrence of the class

ORDERLINE, modifies an occurrence of the class PRODUCT because it requires adjustment of the stock-level of the ordered product, modifies the state of the order to which it belongs and modifies the state of the customer of the order. Notice that Table 1 shows a maximal number of object-event involvements. If we do not want to record a state change in the customer object when an order line is added to one of his/her orders, it suffices to simply remove the corresponding object-event participation in the object-event table. Full details of how to construct such an object-event table and validate it against the data model and the behavioral model is beyond the scope of this paper but can be found in [21, 23].

Table 1. Object-event table for the order handling system

	CUSTOMER	ORDER	ORDERLINE	PRODUCT
create_customer	C			
modify_customer	M			
end_customer	E			
create_order	M	C		
modify_order	M	M		
end_order	M	E		
create_orderline	M	M	C	M
modify_orderline	M	M	M	M
end_orderline	M	M	E	M
create_product				C
modify_product				M
end_product				E

We consider four possible information system services: viewing a list of customers, creating a new customer, creating a new order with one or more order lines and deleting an order. For each of the services we identify relevant objects in each of the business information system layers and explain the interaction between these objects. Fig. 5 represents this graphically.

- Viewing a list of existing customers
  This will require an output transaction that queries the set of domain objects and presents the result of the query in a window. The execution of this transaction is invoked by means of user interface objects. Possibly, the user interface can allow to enter some search criteria the syntax of which is validated by the user interface objects. The result of the transaction is passed to user interface objects, responsible for presenting the resulting list of customers on screen.

- Creating a new customer
  This service requires an input transaction that will invoke the *create_customer* business event. The service is requested via the user interface layer, which is responsible for accepting user input of parameter values (e.g. customer name, ad-

dress, phone number) and for syntactical validation of these values (e.g. format of phone number, name must be alphabetical, …). The user interface objects pass the values to the transaction object Create New Customer in the transaction layer. The transaction object is responsible for creating an occurrence of the *create_customer* event type and invoking the execution of this event. In the event handling layer, the CREATE_CUSTOMER object is responsible for the further validation of the event against business rules (e.g. checking a uniqueness constraints on customer name and address) and for broadcasting the event to the involved objects in the business domain objects layer. In this case, the create_customer event will create a new occurrence of the CUSTOMER class.

- *Creating a new order with one or more order lines*

A possible implementation would allow users to enter the required data for an order together with a number of order lines on a single screen. These data are then passed to a multiple-event transaction object in the transaction layer. This example illustrates that a single transaction New Order can be further subdivided in sub-transactions such as Create Order and Add line to Order. The New Order transaction invokes the Create Order sub-transaction and one or more times the Add line to Order transaction. These sub-transactions will in their turn invoke a create_order event and the create_orderline events respectively. The integration of an additional service allowing to view a list of products that can be ordered (which would be a separate output transaction) must be done in the user interface layer.

- *Deleting an order chosen from a list of orders.*

This service requires the combination of an output transaction that generates a list of orders with an input transaction that invokes the *end_order* and *end_orderline* events. Commit and roll-back features can also be implemented in the transaction layer. In this service the transaction should, for example, only be committed if all order lines and the order were deleted successfully. If something went wrong during the transaction e.g. one of the *end_orderline* events was not invoked and broadcasted successfully, the roll-back feature allows to put all objects back to the state before the transaction was invoked.

The workflow domain layer and the workflow services layer are subdivided in exactly the same way. As a result, we obtain the four layers of Fig. 7 for the workflow layer.

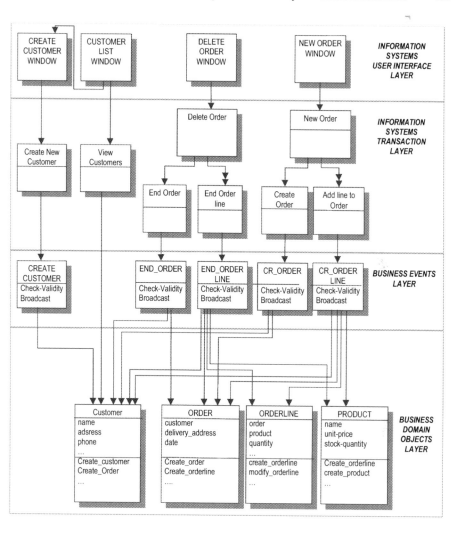

Fig. 5. Sublayers within the Business information system layer

5.   Conclusion

This paper proposes a layered software specification architecture, guided by the principle of separation of concerns. The first set of layers was obtained by separating workflow aspects from functional support. By explicitly incorporating a workflow layer, we ensure that business process aspects are modeled separately. In the absence of such a layer, control aspects of business processes are often hidden in the objects that constitute the functional support for tasks, which is against the principle of separation of concerns.

These two layers were refined by separating domain modeling from information system modeling. Defining a domain model is an important requirements engineering step in the development of an information system: all business rules described in the

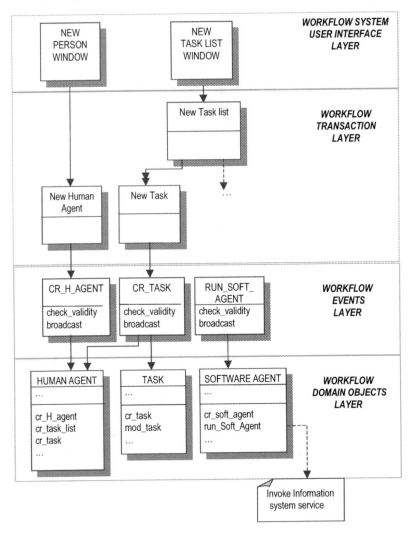

Fig. 6. Sublayers for the Workflow Layer

domain model must be supported by the information system. Methods such as JSD [12], OO-SSADM [19], Syntropy [4] and MERODE [23] even explicitly define domain modeling as a separate step in the development process. Interestingly, these method methods also recognize events as fundamental modeling concepts. The information system services are modeled as layer on top of the kernel layer constituted by the domain model. More importantly, information system services are independent units that can be plugged in and out the system without affecting the underlying domain layer. The services are glued together by means of the user interface layer. Again, this layer can be stripped off, without affecting the lower layers.

The layers for the business information system aspects are an integral part of the MERODE approach to software development which means that there is about 10 years of real-life experience with the business information systems layers. A survey amongst MERODE-users of this approach reveals that the separation of information system services aspects from domain modeling aspects indeed has a positive impact on modularity and hence on maintenance costs [22]. But according to the same survey, the separation of business knowledge from functional support has also other advantages: it results in a better understanding of the functioning of the business and it results in more transparent systems.   The business information system layers can easily be compared to classical three tier architectures. Such architecture have for example, an application tier which contains the user interface aspects and part of the application logic, a domain tier and a persistent tier [7]. Jacobson [11] identifies three types of objects, presumably located in three corresponding tiers: entity objects (which constitute the domain model), control objects and user interface objects. The main difference with such three tier architecture is the identification of different types of control logic. In our approach the control logic is spread across different layers, according to the aspects it refers to. Business rules are stored as methods of domain classes or as general event constraints in the event handling layer. Application control aspects are located in the transaction and user interface layer. Control logic related to the organization of business process is stored in the definition of workflow domain objects, and finally, the workflow service layer captures control logic related to the use of a workflow system.

The model-driven approach which is one of the cornerstones of this paper is derived from the Zachman Information System Architecture [27, 24]. This architecture also contains a scope layer and a technology layer. The proposal of Maes [15] only retains the three lower layers, arguing that scope is a matter of ICT strategy development. As presented in this paper, the architecture does not yet include the third layer, that is to say, the technology aspects. In the Zachman and Sowa architecture [24] data, functionality, network, and other aspects are considered as orthogonal dimensions to the four basic dimensions (see Fig. 7). We expect that technology aspects have to be considered as an orthogonal dimension to the architecture of this paper. Indeed, different technology choices can be made for the realization of the different layers. For example, the business domain model and the workflow domain model can be realized with different database management systems and/or a different programming language. Also network aspects can be very different from one layer to another. One approach to deal with technology aspects is to combine code generation with the reuse of patterns and frameworks as proposed in [8].

	Data	Function	Network	People	Timing	Motivation
	what	how	where	who	when	why
SCOPE						
ENTERPRISE						
SYSTEM						
TECHNOLOGY						

Fig. 7. The extended Zachman framework for information systems architecture

References

1. Baeten, J.C.M., *Procesalgebra,* Kluwer programmatuurkunde, 1986
2. Booch, G., Rumbaugh, J., Jacobson, I., *The unified modeling language user guide*, Addison Wesley, 1999
3. Coleman, D. et al, *Object-oriented development: The FUSION method*, Prentice Hall, 1994
4. Cook, S., Daniels, J., *Designing object systems: object-oriented modeling with Syntropy,* Prentice Hall, 1994
5. Dedene G. Snoeck M. Formal deadlock elimination in an object oriented conceptual schema, *Data and Knowledge Engineering*, Vol. 15 (1995) 1-30.
6. D'Souza, D.F., Wills, A. C. Wills, Objects, Components and Frameworks with UML, The Catalysis Approach, Addison-Wesley, 1999, 785 pp..
7. Fowler, M., Analysis Patterns, Reusable Object Models, Addison Wesley Longman, 1997, 357 pp.
8. Goebl, W., Improving productivity in building Dat-Oriented Information Systems - Why Object Frameworks are not enough, Proc. of the 1998 Int'l Conf. On Object-Oriented Information Systems, Paris, 9-11 September, Springer, 1998.
9. Graham I., Henderson-Sellers B., Younessi H., The Open Process Specification (Open Series), Addison Wesley, 1997, 336 pp.
10. Hoare C. A. R., *Communicating Sequential Processes*, Prentice-Hall International, Series in Computer Science, 1985.
11. Jacobson, I., Christerson, M., Jonsson P. et al., *Object-Oriented Software Engineering, A use Case Driven Approach*, Addison Wesley, Rev. 4th pr., 1997.
12. Jackson, M.A., *System Development*, Prentice Hall, Englewood Cliffs, N.J., 1983.
13. Joosten, S., Werkstromen : een overzicht, in Informatie, jaargang 37, nr. 9, pp. 519-528.
14. Kappel, G., P. Lang, S. Rausch-Schott, & W. Retschitzegger, Workflow management based on objects, rules and roles, In Bulletin of the Technical Committee on Data Engineering, March 1995,18(1), pp. 11-18.
15. Maes, R., Dedene, G., *Reframing the Zachman Information System Architecture Framework*, Tinbergen Institute, discussion paper TI 96-32/2, 1996.
16. Malone, T. W., Crowston, K. The Interdisciplinary Study of Co-ordination, In ACM Computing Surveys, Vol. 26, No. 1, March 94, pp.87-119.
17. Milner, R., *A calculus of communicating systems,* Springer Berlin, Lecture Notes in Computer Science, 1980.
18. Rational Software Corporation, *The rational Unified Process*, http://www.rational.com/
19. Robinson, K., Berrisford, G., Object-oriented SSADM, Prentice Hall, 1994

20. Snoeck, M., Poels, G., Improving the Reuse Possibilities of the Behavioral Aspects of Object-Oriented Domain Models, In: Proc. 19th Int'l Conf. Conceptual Modeling (ER2000). Salt Lake City (2000)
21. Snoeck M., Dedene G. Existence Dependency: The key to semantic integrity between structural and behavioral aspects of object types, IEEE Transactions on Software Engineering , Vol. 24, No. 24, April 1998, pp.233-251
22. Snoeck M., Dedene G., Experiences with Object-Oriented Model-driven development, Proceedings of the STEP'97 conference, London, July 1997
23. Snoeck M., Dedene G., Verhelst M; Depuydt A.M., Object-oriented Enterprise Modeling with MERODE, Leuven University Press, 1999
24. Sowa J.F., Zachman J.A., Extending and formalizing the framework for information systems architecture, *IBM Systems Journal*, 31(3), 1992, 590-616.
25. Vaishnavi, V., Joosten, S. & B. Kuechhler, Representing Workflow Management systems with Smart Objects, 1997, 7 pp.
26. Wolber D., Reviving Functional Decomposition in Object-oriented Design, JOOP, October 1997, pp. 31-38
27. Zachman J.A., A framework for information systems architecture, *IBM Systems Journal*, 26(3), 1987, 276-292.

A Reuse-Based Object-Oriented Framework Towards Easy Formulation of Complex Queries

Chabane Oussalah and Abdelhak Seriai

IRIN, Université de Nantes, 2 rue de la Houssinière,
BP 92208, 44322 Nantes cedex 3, France
{oussalah, seriai}@irin.univ-nantes.fr

Abstract. Corollary to the development of new kinds of application - like decision support ones - manipulating large quantities of data structured in new kinds of data supports -like data warehouses, queries formulated to access this data have been growing in their complexity. However, with the exception of approaches facilitating syntax problems, there is a noticeable absence of models taking charge of the user through from specification of their needs to formulation of their queries. In the case of object-oriented database systems, whose data models and concepts are more complex than those of relational databases, database users need assistance with several kind of formulation problems. In this article, we take a novel look at object-oriented queries - reifying the ones which users formulate as components then reusing them, by means of strategies for selecting, assembling and adapting them to help in the formulation of new complex queries.

Introduction

In database or knowledge-base systems, query formulation languages are the means offered to users for accessing data or knowledge. Queries allow selection of data under particular – either simple or complex - constraints, resulting from one or several structures. However, query formulation remains obstinately difficult for large classes of users [11, 22 , 23] in spite of a whole panoply of languages propositions - formal, declarative, graphic, e.g., [24, 9, 4, 18]. This complexity is related to several things:

- The size of the database schemas handled by users often exceeds their assimilation capacity. This difficulty of assimilation is now accentuated by the emergence of new technologies such as data warehouse [10].
- There is a structural and semantic gap between the vision that users have of their data and the database schema model representing this data. The evolution of user needs amplifies this gap even more [12].
- There is a need to formulate queries using semantics which are increasingly difficult to express. Decision support applications accessing data-warehouses are a good example [2, 15].

A.H.F. Laender, S.W. Liddle, V.C. Storey (Eds.): ER2000 Conference, LNCS 1920, pp. 470-483, 2000.

In face of the growth in query formulation complexity, we notice the absence of models taking charge of the user through from specification of needs to formulation of queries. In the case of object-oriented database systems which have more complex data models and concepts then relational databases, the database user need to be assisted in several formulation problems. For example, which are the classes in the schema which their instances best meet a request for information? Which are the properties of these classes- in terms of attributes and methods - to use? Which relations between these classes best correspond to the required information? Which are the conditions that the instances of these classes must check and how to combine them together? Is the joining of two or several classes necessary? How to structure the query predicates? And so on.

We think that users can be given effective help in query formulation by means of re-using former queries. Indeed, the objective of this work is to propose a method of assisting users to formulate queries by storing, accessing and re-using existing ones. It consists of making available to users component queries plus a strategy for re-using them.

The structure of the model we call QUERYAID is developed using object concepts [6] which have incontestable advantages in making reusable components. In particular, object classes are considered as strongly reusable elements [14].

The rest of this paper is organized as follows. In section 2 we present the concepts, the structure and the functions of QUERYAID. Our approach to formulating queries in order to be able to reuse them is presented in section 3. In section 4 we show how to formulate queries by re-using already existing ones. In section 5, we detail some similar approaches, and we discuss their similarities and differences relative to this one. The conclusion and some prospects are the subject of section 6.

Concepts, Structure, and Functions of QUERYAID

Concepts

QUERYAID is an object-oriented query model aiming to assist users with the formulation of their queries. It is based on the one hand on making explicit the distinction between query expressions and query answers, on the other hand on reification of these two facets as what we call query-definitions and query-results.

A *query expression* is a syntactic description of a query given in a specific query formulation language. *Query answers* are the data returned by the database manager system as the results of evaluation of a query.

We define a *query-definition* as an abstract composite-class resulting from the reification of a query expression. Each query expression will thus be represented as a composite class with its own attributes and its own component classes representing its various parts (projection attributes, selection conditions, join classes, join conditions, etc.). A *query-result* on the other hand is the class resulting from the reification of the answers to a query. It consists in representing the answers to a query as objects and in creating an object-class representing them. Attributes of a query-result are specified in the corresponding query expression. They will be a selection or a composition of those defined in the classes of the database schema.

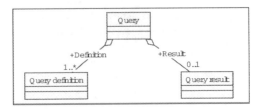

Fig 1. The query composite class using UML notation[1]

We define a *query composite-class* as an abstract composite-class encapsulating query-definition and query-result classes resulting from reification of expression and answers to a query.

In this way, classes representing former query-definitions and those representing former query-results can be considered as component classes which can be reused to assist users in the formulation of new queries. Query composite-classes can be considered as an interface for accessing these reusable component classes.

Structure and Functions of QUERYAID

QUERYAID is structured around three modules: a library storing and organizing query-definition components, a number of user-oriented views storing and organizing in an incremental fashion query-results specific each to one workgroup, and a query reuse manager (cf. figure 2).

This architecture assures two principal functions which consist of:

- Designing query-result and query-definition components on the basis of existing query formulations - *formulation of queries for reuse*. Query-results are used for the incremental construction of a local databases via the incremental construction of a user-oriented views. Query-definitions are reified and organized in a library.
- Using existing query-result and query-definition components to help users in the formulation of new queries - *formulation of queries by reuse*.

Query formulation for reuse. Formulating queries for reuse is the mechanism offered by QUERYAID for representing and organizing queries as components to be reused. It consists in the enrichment of the view and the library by incorporating into them new components , i.e., query-results and query-definitions.

The library constitutes the structural and organizational support for query-definition components. It allows representing them, organizing them and inferring links between them (sharing of parts of queries, query specialization and generalization hierarchy, query composition hierarchies). Assistance to the user is based on the reuse of query-definition components drawn from the library, easing the transition

---

[1] [7].

'rom a partial or informal formulation of his query, couched perhaps in a natural lan-
ʒuage, to a complete and formal one formulated using the specific database (or knowl-
ᵌdge base) query language. Assistance to the user is based on the assumption that
ɪsers of any business or group of similar businesses often formulate queries which are
ᵴemantically similar.

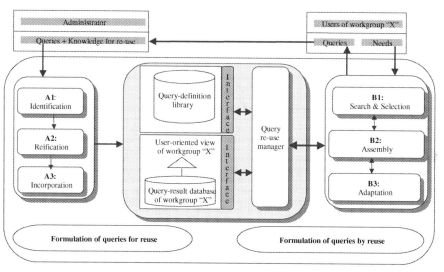

Fig 2. Structure of QUERYAID

The views are the support of the query-result classes. Each of them is composed of
 he classes representing all the data handled by users of any one workgroup. Indeed,
he user's views are constructed by incremental incorporation of the classes represent-
ɪng objects resulting from the reification of the data returned in response to users'
ɟueries. Each view will contain solely the entities conceived and chosen by its group
ᵊf users - through their queries. This method reduces both the size of the data model
ᵻandled by the users and the mismatch between the perception which users have of
ᵗheir data and the handled data model. The user-oriented views thus implicitly facili-
ate formulation via favoring reuse of their components - query-result classes.

The enrichment of the view and the library is assured by the for-reuse query for-
nulation process of QUERYAID which consists of:

- Identifying the structures of the query-definition, query-result and query composite-classes corresponding to query expressions formulated by users and judged by
  the administrator to be "good" for reuse - and thus for being incorporated into the
  various supports of QUERYAID (step A1).
- Reifying the query-definition, the query-result and the query descriptions following their specific structures identified at the precedent step. The classes resulting
  from reification are enriched by a body of knowledge directed at offering help
  through reuse of former queries (step A2).
- Integrating each component resulting from the reification step into its own support - the query-definition components into the library, the query-result compo-

nents in the specific workgroup view and the query composite-classes into the re-use manager hierarchy (step A3).

Query formulation by reuse. Formulating queries by means of reuse of former ones is the mechanism offered by QUERYAID to guide and assist the user during query formulation by re-using the various query-definition and query-result components available in the library and the view.

The by-reuse structure of QUERYAID is instantiated by the query reuse manager which is the support for representation and organization of query composite-classes (see § 4). The function of this one consists:

- First, of the implementation of the reuse mechanism of components exiting in view and library - i.e. query-results classes and query-definitions classes.
- Second, of the assurance of both the independence and the co-operation between the library and the view by managing access to their components and the links between them; it thus plays the role of QUERYAID interface.

The by-reuse query formulation process of QUERYAID can be called upon by a user who finds difficulty in formulating a given query. The help proposed consists of:

- Assistance to users in the selection from library and view of the component que-ries semantically closest to their needs (step B1).
- Assistance to users in the assembly of some selected component queries together (step B2).
- Assistance to users in the adaptation of some selected existing component queries (step B3).

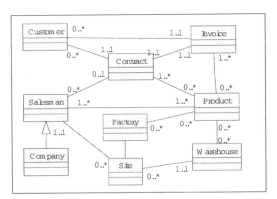

Fig 3. Example of a database schema using UML notation

In what follows, due to space limitation, detail of the incremental conception process of views is not presented. For more detail, readers can refer to [23]. Following query-expressions illustrating concepts of QUERYAID are formulated using the OQL lan-guage [6]. They use the part of the schema presented in figure 3.

QUERYAID: A Query Formulation Model For Reuse

QUERYAID is a model designed to implement process and containers structuring and organizing existing queries as reusable component classes. These ones are resulting from the reification of the existing queries as query-result, query-definition and query component classes.

The role of the library is to store query-definition component qualified by the administrator as useful for formulation of new queries. These query-definition components are incorporated in the library after being identified and reified.

Identification of Query-Definition Components

The identification of a query-definition component consists in both transforming the corresponding query expression into a standard or canonical format using defined transformation rules and in determining the elementary components composing the corresponding query composite-class.

Canonical Format of Query-Definitions. A query expression can be formulated according to various forms. The transformation of query expressions into canonical format aims at writing them under a common format - to facilitate the detection of the semantic links involved. To do that, we have defined various transformation rules having some similarities with those of [3]. Among them:
- Change each query predicate to ensure that the lower-valued attribute (using alphanumeric order) is on the left hand side of it, e.g., *(change (Y>X) to (X<Y))*.
- Delete any redundancies present, e.g., *((attribute-1 >= attribute-2 & attribute-1 > attribute-2) becomes (attribute-1 > attribute-2))*.
- Alter variables names to their instantiation class names, e.g., *(X in The_Customers, Y in X.Invoice becomes Customer in The_Customers, Invoice in Customer.Invoice)*.
- Transform query predicates to disjunctive normal format (disjunction of conjunction), e.g. *((attribute-1 | attribute-2) & attribute-3 becomes (attribute-1 & attribute-3) | (attribute-2 & attribute-3))*.

Example 1. The query expressions *Query 1.1* and *Query 1.2* presented below both select "*those customers who are French companies*". These two query expressions are syntactically different but semantically equivalent.

Query 1.1:	Query 1.2:
Select X	*Select Y*
From X in The_Customers	*From Y in The_Customers*
Where X.Country = "France"	*Where Y.Country ="France"*
and X.type = "Company".	*And Y in [Select Z*
	From Z in The_Customers
	Where Z.type="Company"]

The result of transforming these two queries into canonical format is given below.

> Canonical query:
>   *Select Customer*
>   *From Customer in The_Customers*
>   *Where Customer.Country ="France"*
> *Intersect*
>   *Select Customer*
>   *From Customer in The_Customers*
>   *Where Customer.type="Company"*

Decomposition of Canonical Expressions. Query-definition components identification also consists in decomposing the corresponding canonical expression into elementary sub-expression. Decomposition of this expression is carried out by rewriting it in algebraic format. This representation allows extraction of existing relations between query-definition components by allowing semantic comparisons between them. Indeed, the comparison principle outlined in § 3.3 is based on the resemblance of the elementary components of the comparative queries and on the relations between the parameters of these ones, e.g., the class used in a projection operator, the classes used in a join operator, etc..

Example 2. The algebraic decomposition of the canonical query expression presented in example 1 is given below, using a symbolic language. The term *composite* means the attribution of composite class status to the parameter in the expression. The term *component* means the attribution of component-class status to the parameter in the expression.

> Query 2.1: Algebraic object format.
>   *Composite(R) = Intersect [Component (R1), Component (R2)]*
>   *Composite(R1) = [*
>   *Component = Projection (Customer, All)*
>   *Component = Selection (Customer, "Customer in The_Customers")*
>   *Component = Selection (Customer, "Customer.Country="France " ")]*
>   *Composite (R2)=[*
>   *Component = Projection (Customer, All)*
>   *Component = Selection (Customer, "Customer in The_Customers ")*
>   *Component = Selection (Customer, "Customer.type="Company " ")]*

Reification of Query Definitions

A query-definition component class is the result of the reification of a query canonical expression, following the result of the decomposition step. A query-definition is then represented by a composite class. The different parts of this are represented by elementary-component classes. The parameters of the algebraic expressions forming part of the canonical expression are represented as attributes of the elementary-component classes. The example in figure 4(a) shows the reification model of an algebraic join expression. The join class attributes[2] represent the various parameters of the join expression i.e. the two classes used in the join, the two access paths corresponding to these classes and the join condition. The example in figure 4(b) shows the reification model of a *Select-From-Where* query-definition.

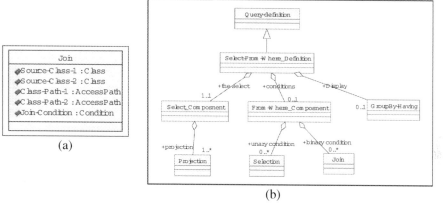

(a)

(b)

Fig 4. Query-definition reification models

Incorporation of Query-Definitions into the Library

Query-definition composite classes and their elementary-component classes are organized in the library as inheritance and composition hierarchies. Inheritance hierarchies translate the specialization/generalization relations between whole query expressions on the one hand and between their various parts on the other. Composition hierarchies translate the part-of relation between query expressions and their different parts.

The principle of query subsumption. Given two queries Q_i and Q_j, Q_i subsumes Q_j if A_j - the response to Q_j - exists in A_i - the response to Q_i.
The queries being reified, we transform the problem of query subsumption into a subsumption of query-definition composite-class problem. Indeed, we deal with the question of inclusion of query results by analyzing subsumption of query-definition classes. We note that this is possible because all query expressions are rewritten under the same canonical format. Thus, the syntactic structures of the query expressions don't influence the result of the subsumption algorithm.

---

[2] The terms *Class, AccessPath,* and *Condition* represent the object types of these attributes

The principle of subsumption of query-definition composite. A query-definition composite class R_1 subsumes another query-definition composite class R_2 if:
- The elementary-components set of R_2 is included in the elementary-components set of R_1. R_2 is then known as subsumed by addition of elementary components.

    Example 3. The composite class resulting from reification of query expression *Query 3.1* is a specialization by addition of the *selection* elementary-component resulting from reification of the selection expression *"which are products produced during the year 2000 ?"* to the composite class issuing from the reification of the query expression *Query 3.2*.

    > Query 3.1: "which are the luxury products produced during the year 2000?"
    > Query 3.2: "which are luxury products?"

- Each elementary component of R_1 specializes or is equal to an elementary component of R_2. R_2 is known as specialized by specialization of its elementary components.

    Example 4. *Query 4.1* query-definition class is a specialization of the *Query 4.2* query-definition class by the fact that the elementary-component forming part of *Query 4.2* and representing projection on *Company* class specializes the elementary-component forming part of *Query 4.1* and representing projection on *Salesman* class.

    > Query 4.1: *"which companies drew up contracts during the year 2000?"*
    > Query 4.2: *"which are the salesmen who drew up contracts during the year 2000?"*

QUERYAID: A Query Formulation Model by Reuse

QUERYAID is a model designed to assist users to reuse existing component queries. This is done by facilitating the refinement of informally or partially specified expression formulated by a user into a query formulation in the specific database query language. This task is carried out by the query-reuse manager. Indeed, the query reuse manager implements via query composite classes the mechanism intended to help the user in his query formulation by helping him to select, assemble and adapt components corresponding to preexistent queries.

Searching for and Selection of Component Queries

The tasks of searching and selecting consist in searching the query-definition library and the user-oriented view for components which match a requirement formulated by a user in natural language. These activities are based mainly on the items of knowledge enabling selection which are encapsulated by the query composite classes. This knowledge guides the user towards the component queries most likely to meet their needs. Indeed, each query composite class encapsulates a set of attributes, among them:

- A set of key words summarizing the role of the query. These key words are either introduced by the administrator or inferred automatically by the model from an analysis of the question which generating the query. All keywords corresponding to user's business are structured in a thesaurus.
- The set of schema classes used in the query expression. The model also infers them from the query expression.
- The set of class attributes used in the query expression.
- The domain name. The users in a given workgroup can in effect be structured in various fields, as also can their queries.

When one user uses QUERYAID in by-reuse mode, the model extracts from his query specification expression formulated in natural language information necessary for the selection of component queries. This consists of the set of values of attributes presented above. Classification is used to select, among the component queries those having a certain semantic proximity to the desired one by measuring proximity between their encapsulated attributes. We employ a thesaurus materializing the semantic relations between the keywords (specialization relations of various fields, equivalence between various classes names, equivalence between various attributes names, etc.).

In addition, query specification expressions expressed in natural language may include a partial formal query expression. In this case the model does its classifying into the library hierarchy using non-complete class corresponding to the partial query expression. The classification results end up a set of composite classes representing preexistent query-definitions complementing the partial expression formulated by the user.

The selected query are visualized to user on the two hierarchy graphs: query-definition and query-result hierarchy graphs. We propose navigation through their links as complement to the selection done by the model. Links between component queries are in effect directed towards the expression of semantic proximity between queries. These ones are lied by specialization/generalization links, composition links or elementary-component sharing links. The user can consult semantically close queries when passing from one query-definition or query-result graph node to another.

Assembly of Component Queries

When component queries selected by the model to meet the user's request prove unsatisfactory, the combination of several component queries can yield a one closer to the user's needs. The assembly of queries is carried out using a set of heuristic rules implemented as methods of the query reuse manager. The model breaks up the specification expression of the user's needs into several parts, each of them corresponding to an expression of an existing component query specification. By applying one or succession of heuristic rules, the model proposes possible connections between these components. Such a connection might be an intersection, a union, a join and so on.

Example 5. *Query 5.1* specification expression can be broken up into two sub-query specification expression, *Query 5.2 and Query 5.3.* If the two component queries representing *Query 5.2 and Query 5.3* already exist, the model can suggest making a join between them (joining them using *customer-identity and salesman-identity* attributes) for obtaining a query-definition corresponding to *Query 5.1* specification.

> Query 5.1: *"Which customers drew up contract concerning luxury products and in the same time are salesmen of type company and selling only economical products?"*.
> Query 5.2: *"Which customers drew up contract concerning luxury products?"*
> Query 5.3: *"Which salesmen of type company selling only economical products?"*.

To put this into practice, the by-reuse process executes the part of the heuristic rule listed below.

> *If Description (Query-1)   includes Description (projection (Class-1))*
> *If Description (Query-1)   includes Description (Query-2)*
> *If Description (Query-1)   includes Description (Query-3)*
> *If Description (Query-2)   is   Description (Selection (Class-1))*
> *If Description (Query-3)   is   Description (Selection (Class-2))*
> *If (attribute-1, attribute-2) is   TheMostCloseAttributes(Class-1, Class-2)*
>   *Then*
>   *Definition (Query-1)   is*
>   *Projection(Class-1)*
>   *Where [Selection (Definition (Query-1))*
>   *And Selection (Definition (Query-2))*
>   *And Classe1.attribute-1 (=, >, <, ...) Classe2.attribute-2]*

This allows the selection of the query-definition components *Query-2* and *Query-3* and their assembly by a join operation to give adequate formulation of a new query called *Query-1* which its specification is expressed in a natural language. The term *Description(Query-Name)* corresponds to the set of attributes presented in § 4.1 and which are extracted from the query specification expression specified in parameter. The term *Definition(Query-Name)* corresponds to the query definition expression formulated in the database query language. *Include* operator calculate the semantic proximity between its two operands (set of query-keywords) using the defined thesaurus. Thus, *Query-1* will be substituted for *Query 5.1*, *Query-2* for *Query 5.2* and *Query-3* for *Query 5.3*, *Class-1* for *Customer* class, Class-2 for *Salesman* class, *attribute-1* for *Customer.identity* attribut, *attribute-2* for *Salesmen.identity* attribute.

Adaptation of Component Queries

Adapting component queries is a matter of extracting from a set of candidate component queries their common properties. An example might be the fact that they all carry out a join between a set of classes or that they realize all their selections following certain conditions, etc. It is useful for a user to be able to extract these properties and to divide the set of selected component queries into a certain number of generic models. These models provide the user with indications of possible modifications. Indeed, generic query models can be regarded as query-formulation and -adaptation frameworks.

To do that, each query composite class encapsulates a set of methods which enables adaptation. These methods allow extraction from a set of component queries which can be candidates for meeting a user's need of one or more generic models.

Example 6. *Query 6.1* below is a generic query expression structure obtained by abstracting a set of existent component-queries. The *Specializing(parameter)* and the *Enumerate-Values* keywords indicate respectively, classes specializing the class cited in parameter, and an enumerate list - which can be given - of possible values.

```
Query 6.1:
    Select X: Specializing(customer). identity, Y: Specializing (salesman).site
    From X in The_Customers,
    Where X.identity in
        [Select Y.identity
        From Y in The_salesmen
        Where Y.Product.type= Enumerate-Values-1]
    And X.Product.type = Enumerate-Values-2
```

If this generic query model is obtained by abstraction of some candidate component-query expressions, selected to satisfy a user specification, then it gives a good indication of the general structure of the needed query. Particular queries close of the selected ones can be thus obtained by instantiation of the generic query expression parameters.

Thus, the query-definition expression of *Query 5.1* presented in the example 5 can be instantiated from *Query 6.1*. Indeed, it can be obtained by instantiation of *specializing (customer)* in *Customer* class, *specializing (salesman)* in *Company* class, and by chose of *"Economic"* value from the *Enumerate-value-1* list, and of *"luxury"* value from *Enumerate-value-2* list.

Discussion and Related Work

A large number of approaches facilitating the task of query formulation have been proposed. Recent tendencies of query formulation languages are one such. These languages, which are declaratory or graphical, try to relieve the user of the necessity of knowledge of a rigid syntax, e.g., [24, 11, 18, 4, 9]. Such a language though gives users assistance only over syntactic difficulties.

Another form of assistance comes with systems based on cooperative answering, e.g., [20, 21, 16]. These try to give users explanations of the answers to their queries - especially if these fail (e.g., an empty response to a selection query). Some co-operative approaches offer syntheses or generic model answers. The assistance given in this case arrives at least after one query formulation attempt. It is matter of using the explanations and the syntheses generated by the co-operative answering system to formulate a new query which is semantically more correct or which better meets the expressed need.

Approaches introducing the concept of view, e.g., [1], can also be regarded as implicit ways of helping with query formulation. Views have the advantage of allowing users to structure their database (or knowledge-bases) according to their own vision of the data. They thus allow easier handling of the data, i.e., query formulation. In addition, our technique of organizing queries into a query-definition library resembles query containment techniques and others approaches making use of it - for rewriting queries using views for instance. These techniques are all based on a logical approach, e.g., [8, 19]. Our technique of verifying query subsumption (query containment) goes by way of reification of query-definition and verification of their subsumption by use of object oriented (OO) mechanisms of abstraction, specialization, classification, component share, etc. The OO query subsumption approach adapted here takes advantage of database semantic - hierarchy graph and integrity constraints.

Conclusion

The approach developed here lies within the scope of reuse development environments [5, 17, 13]. The model developed incorporates knowledge defining query components and ways to select, adapt or combine them to meet particular requirements. The tasks of formulation for-reuse and formulation using reuse are interwoven. Indeed, the queries obtained by assembly and adaptation can serve as entries into the for-reuse formulation process. Like any reuse-based approach the power of this one depends on the quantity of reusable components available. We are currently developing a graphical framework which implements the concepts introduced here. This framework may in addition included some ideas for extending the by-reuse phase to optimize the evaluation of user queries using stored results of former ones (persistent or temporary ones).

References

1. S.Abiteboul, A. Bonner. *Objects and views*. Proc. ACM SIGMOD Management of Data, 1991.
2. S.Acharya, P.B.Gibbons, V.Poosala. *Aqua: A fast Decision-Support Systems Using Approximate Query Answers*. Proc. VLDB conference, pp.54-57, 1999.
3. M.Al-Qasem, S.M.Deen. *Query Subsumption*, Proc. FQAS, pp.29-42, 1998.
4. E. Andonoff, C.Mendiboure. *Help tools for database querying: the OHQL proposal*. Proc. 2nd BIWIT95. San Sebastian, Spain, 1995.
5. M.Arikawa. *A View Environment to Reuse Class Hierarchies in an Object-Oriented Database System*. Proc. DASFAA conference, pp.259-268, 1991.
6. T.Atwood, D.Barry, J.Duhl, J.Eastman, G.Ferran, D. Jordan, M.Loomis, D.Wade. *The object Database Standard: ODMG-93, Release 1.2*. Morgan Kaufmann, San Francisco, 1994.

7. G.Booch, J.Rumbaugh, I.Jacobson. *The Unified Modeling Language User Guide*, Addison-Wesley, 1998. ISBN 0-201-57168-4.
8. M.Buchheit, M.A.Jeusfeld, W.Nutt, M.Staudt. *Subsumption between Queries to Object-Oriented Databases*. Proc. EDBT conference, pp.15-22, 1994.
9. T.Catarci, S.K.Chang, G.Santucci. *Query Representation and Management in a Multi-paradigmatic Visual Query Environment*. Journal of Intelligent Information Systems 3(3), 1994.
10. S.Chaudhuri and U.Dayal. *An overview of Data Warehousing and OLAP technology*. ACM SIGMOD Record, 26(1), pp.65-74, 1997.
11. M. Chavda and P. T.Wood. *Towards an ODMG-Compliant Visual Query Language*. Proc. of 23rd Int. VLDB conference, Athens, Greece, pp.25-29, 1997.
12. J.Chen, D.McLeod. *Schema Evolution for Object-based Accounting Database Systems*. Proc. ISOOMS, pp. 40-52, 1994.
13. P.Constantopoulos, Matthias Jarke, John Mylopoulos, Yannis Vassiliou: *The Software Information Base: A Server for Reuse*. VLDB Journal 4(1): 1-43 (1995)
14. F.D'Souza., A.C.Wills, *Objects, Components, and Frameworks With UML, the catalysis Approach*. Addison-Wesley, 1998.
15. C.Faloutsos, H.V.Jagadish, N.Sidiropoulos. *Recovering Information from Summary Data*. Proc. of the 23rd VLDB, Athens, Greece, pp.36-45, 1997.
16. T.Gasterland, P.Godfrey, J.Minker. *An Overview of Cooperative Answering*. Journal of Intelligent Information Systems, Kluwer Academic Publishers, vol. 1, N° 2, pp. 123-157, 1992.
17. P.A.V. Hall, *Overview of Reverse Engineering and Reuse Research*. Information and Software Technology, Vol 34, N° 4, 1992.
18. E.Keramopoulos, P.Pouyioutas, C.Sadler. *GOQL, a Graphical Query Language for Object-Oriented Database Systems*. Proc. BIWIT 97, Biarritz (France), 1997.
19. Alon Y. Levy, Dan Suciu. *Deciding Containment for Queries with Complex Objects*. Proc. PODS conference, pp.20-31, 1997.
20. J.Minker. *An Overview of Cooperative Answering in Databases*. Proc. FQAS conference, pp.282-285, 1998.
21. A.Motro. *Cooperative Database Systems*. Proc. FQAS conference, p.1-16, 1994.
22. N.Murray, C.Goble, N.Paton. *A Framework for Describing Visual Interfaces to Databases*. Journal of Visual Languages and Computing, Vol. 9, N° 4, p. 429-456, 1998.
23. A.Seriai, C.Oussalah. *Query reification based approach for object-oriented query formulation aid*. Proc. of The 11th IEEE International Conference on Tools with Artificial Intelligence (ICTAI'99) Chicago IL, 1999.
24. K.Vadaparty, Y.A.Aslandogan, G.Ozsoyoglu. *Towards a unified Visual Database Access*. SIGMOD, Washington, USA, pp.357-366, 1993.

Evaluating the Quality of Reference Models

Vojislav B. Mišić[1] and J. Leon Zhao[2]

[1] The Hong Kong University of Science and Technology
Clear Water Bay, Kowloon, Hong Kong
vmisic@ust.hk
[2] University of Arizona, Tucson, Arizona
jlzhao@u.arizona.edu

Abstract. The process of system design often begins with the selection of an appropriate reference model. Model selection necessitates a good understanding of the system to be developed, as well as of the reference models available for that particular type of systems. In this paper, we propose a conceptual framework for comparing reference models, based on an elaboration of a linguistics-based classification approach. This framework is applied to the comparative analysis of two well known reference models for electronic commerce.

1 Introduction

The term 'system architecture' denotes the description of the structure (or structures) of a computer-based information system; the structures in question comprise software components, the externally visible properties of those components, and the relationships among them [1]. A reference model, on the other hand, describes a standard decomposition of a known problem domain into a collection of interrelated parts, or components, that cooperatively solve the problem. It also describes the manner in which the components interact in order to provide the required functions. To put it briefly, a reference model is a conceptual framework for describing system architecture [11], thus providing a high-level specification for a class of systems [8].

Once developed, a reference model for a particular problem domain can be utilized in different ways.

1. A reference model can provide the framework for the identification, development, and coordination of related standards. (The existence of widely accepted models and standards are the necessary foundation for top-down development of flexible, interoperable, and coherent applications and systems.) It certainly facilitates communication among the stakeholders.
2. A reference model can be used to develop more specialized models to support specific requirements and scenarios [10].
3. A reference model may be mapped onto a collection of software components and data flows between those components, to obtain a so-called reference architecture. The reference architecture may, then, be further refined—by adding sufficient implementation detail—to arrive at system architecture.

A.H.F. Laender, S.W. Liddle, V.C. Storey (Eds.): ER2000 Conference, LNCS 1920, pp. 484–498, 2000.
© Springer-Verlag Berlin Heidelberg 2000

4. Finally, a widely accepted reference model enables the use of an architecture-based development process. Architecture-based development is characterized by composing or assembling separately developed components into a functional entity, while respecting the constraints and organizational structure imposed by the system architecture (and, ultimately, the underlying reference model). Such development is generally considered to lead to increased quality and/or cost savings, both during system development and in normal operational life [1].

We stress that the reference model alone cannot guarantee any particular level of functionality or quality required of a system. An appropriate foundation in the form of a reference model is necessary, but not sufficient, to ensure quality. Systems designed without an underlying reference model, however, are almost certain to offer reduced quality in terms of coherence, extendibility, and interoperability with other systems.

Although a number of reference models for different types of systems have been (and are currently being) proposed by individual companies and organizations and industrial consortia, they offer vastly different viewpoints and feature sets. So far, little has been done to analyze and compare these models in order to describe, abstract, and compare their basic properties, and offer some guidelines for selecting the most appropriate reference models for a particular type of system to be developed.

To this end, we propose a set of criteria to analyze and evaluate a given reference model. Such analysis may serve as the basis for choosing the right model for a specific system. Such analysis could also be a convenient starting point for making improvements in some of the existing models, or for merging of models in order to combine their individual strengths. The abstraction and generalization of desirable model properties could provide a foundation for creating new and, hopefully, more useful models. Finally, existing system architectures, mostly proprietary and usually incompatible with one another, could also be analyzed in the context of such analysis framework.

Quality of reference models becomes an important issue in the context of modern system development. Namely, system development times are becoming progressively shorter, while at the same time the expected quality levels are steadily increasing. This poses stringent requirements on the quality of system architectures and the underlying reference models. Quality assessment requires an appropriate quality framework, which is the issue we are trying to deal with in this paper.

Our quality framework is an extension of the linguistics-based classification approach originally proposed by Lindland et al. [7]. This approach is based on the notion of a model as a communication vehicle between the problem domain, the language used to describe it, and the human stakeholders. Therefore, the quality properties are classified into syntactic, semantic, and pragmatic categories. In each of these categories, we outline the properties that constitute quality, and investigate their attributes and possible interaction. In this way, we arrive at a quality framework that can be applied to a range of reference models.

As an initial application of our quality framework, we have chosen the reference models for electronic commerce systems. With the foundation provided by modern information technologies for data processing and communications, implementation of electronic commerce systems is well under way in many companies: merchants, financial institutions, and technology providers alike. Yet there is little overall cohesion in all these efforts, and current electronic commerce systems leave much to be desired in terms of interoperability, reusability, extensibility, and scalability. One of the reasons certainly is the lack of a common, widely accepted architecture or reference model (although a number of such models have been proposed). We analyze three of the well-known models in the context of the quality framework and analyze their relative strengths and weaknesses.

The paper is organized as follows. In Section 2, we establish a framework for classification and selection of model properties relevant to quality of reference models. The three subsections deal with syntactic, semantic, and pragmatic properties, respectively. Section 3 presents two reference models for electronic commerce; a comparative evaluation of their properties is given in Section 4. Section 5 concludes the paper and outlines some directions for future research.

2   The Linguistics-Based Comparison Framework

All conceptual models, including the reference models we are about to analyze, possess a number of properties through which they can be analyzed and compared to one another. In particular, properties related to model quality could be used as criteria for assessment and comparison of reference models. However, there seems to be little consensus in the research community as to which properties constitute quality, and different authors have proposed different sets of properties and criteria by which models can be assessed and compared. In most cases, properties that constitute quality are vague or informally defined, and the lists of such properties show little structure, or no structure at all [7]. Moreover, some of the previously proposed properties that correspond to quality are clearly inapplicable because:

1. They strive towards goals that are unrealistic or even impossible to reach (e.g., total completeness is not a realistic goal in most systems).
2. They mix properties of the specification with those of the language and method used to arrive at the specification (e.g., expressiveness is a property of the language, not of the model built with it).
3. They presuppose the existence of a design or even an implementation (e.g., testability as the availability of procedures that verify the design and/or implementation of a component against its requirements). Note that we are discussing models at the conceptual level – they have yet to be utilized as conceptual foundation for implementing the actual systems.
4. Finally, some of the properties may be irrelevant or insignificant in the application context under consideration (e.g., executability is not a significant property for electronic commerce reference models).

Therefore, existing lists of properties cannot be readily used in our analysis, although they could provide a good starting point.

In order to define a set of properties specifically tailored to reference models, we adopt the simple linguistics-based classification framework [7] in which the properties are grouped into three major categories: syntactic, semantic, and pragmatic. These categories roughly correspond to the relationships of the three cornerstones of the modeling process:

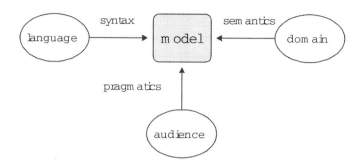

Fig. 1. Properties of a model are defined with respect to its language, domain, and audience (adapted from [7]).

language, domain, and audience, with the model itself, as shown in Fig. 1. Then, we will try to identify properties relevant for a particular application context, provide appropriate definitions, and structure them according to the classification framework.

Such a procedure is not entirely straightforward, nor without difficulties. First and foremost, each reference model has a number of properties in each of the aforementioned categories. Not all of them are relevant for our analysis, and a decision has to be made as to which properties should be included in our framework. In other words, we have to parameterize the set of properties according to the specific needs of the class of models under consideration. This parameterization is performed with the main objectives of a particular class of models in mind. Only properties that contribute to those objectives should be included in the analysis – those that don't, should be left out. Different classes of models may, and probably will, require different sets of properties. For example, extendibility is a relevant property for models analyzed in this paper—they are high-level specifications of a class of systems [8]—but formality is not, and can be discarded.

Second, properties are often interrelated, and appropriate definitions should help the modeler to clearly distinguish between related, or even similar properties. For example, scope and completeness can be considered as two facets of a single property, or two different properties. (We decided to analyze them separately, for reasons discussed below.)

Moreover, some properties may appear to belong to more than one category (for example, technology dependence could be either a syntactic or a pragmatic property, or maybe both at the same time). In most cases, the ambiguity could be resolved by distinguishing the property itself from its implications that belong to a different category.

We note that different quality properties usually have different sets of allowable values. Those sets may be finite (and, in some cases, even binary – the property may be present or not) or infinite. Even though quantitative evaluation may not always be possible in absolute terms, comparison in relative terms always is, as shown later. Note, also, that most quality attributes are too complex and amorphous to be evaluated on a simple scale. They do not exist in isolation but rather have meaning only within a context: for example, a model is extendible (or not) with respect to certain classes of changes.

All of the aforementioned limitations should be kept in mind before we describe key properties to be used in assessment and comparison of the reference models, in particular those relevant for electronic commerce reference models.

2.1 Syntactic Properties

Syntactic properties describe the model in terms of the modeling language constructs used to define it, without considering its meaning [7]. Syntactic quality, then, describes how well the model corresponds to the rules of the language used. For example, an UML object model [3] can comply (or not) with the rules of the language – we can analyze the degree of compliance, as well as the use of syntactic (i.e., language) constructs in that model. However, the models we are about to analyze are usually not expressed in a language with strict, formally defined rules, against which compliance or noncompliance can be easily evaluated. Rather, these models offer very high-level, sometimes informal descriptions of system functionality, partitioning, and interconnection. Therefore, we must restrict our analysis to syntactical properties that describe the structure and organization of the model itself. Those properties do not correspond to quality *per se*: for example, a more detailed model is not better by definition than a less detailed one, or vice versa. Syntactic properties do, however, define a context within which the quality of the model may be assessed – by examining other, semantic and pragmatic properties, and comparing it to other models.

In terms of content, different models provide different **levels of abstraction**: some of them define a few high-level components only, whereas others decompose these into more primitive components and services, and provide more details about the interconnection and interaction of these components and services.

The level of abstraction of a model is closely related to its **level of detail**, or granularity: less abstract models tend to provide finer granularity than the more abstract ones. (Still, two models with the same level of abstraction may differ somewhat—but not too much—by the level of detail.) More abstract models give system developers more freedom in the design process, but also more responsibility, since freedom of choice includes the freedom to make bad choices (for example, a nonstandard interface). We stress again that quality does not correspond directly to a particular level of abstraction and/or detail, but it can be assessed within the context defined by these properties.

In terms of **stratification**, reference models may be classified as flat or layered (stratified), i.e., their components may be described at a unified level of abstraction, or may be structured into several layers at different abstraction levels. In a layered structure, each of the lower layers delivers a well-defined functional support to the higher layer(s) whilst concealing unnecessary details, and each layer relies on the lower layer(s) for more primitive functions [11,17].

Layered decomposition has been used many times to analyze, design, and develop very complex systems. Layered models do possess subtler structural properties such as number of layers, partitioning of system functionality between them, the level of abstraction of each layer, the manner in which components interact (both within each layer and between different layers), and the like, do provide a basis for comparison. Again, any actual value does not equate with quality, or lack thereof; it is the totality of the properties and the context in which they exist that determines the quality. For example, one might ask what is the optimal number of layers in a reference model; the well-known reference model for database system architecture [6] defines three layers, whereas the ISO/OSI reference model[11] defines no less than seven. Although neither of them has been strictly followed in actual implementations, both are considered as significant milestones in the development of their respective application areas.

Some stratified models are hierarchical while others are not; in the latter case, any module can communicate with several modules in the layer immediately above, whereas in the latter case a module may communicate with a single higher-level module only. Most reference models for electronic commerce are stratified, but not hierarchical.

Finally, we note that abstraction and hierarchical decomposition are among the basic techniques for managing complexity in all phases of system development [5].

2.2 Semantic Properties

Semantic properties describe the model from the viewpoint of the domain being modeled, focusing on the meaning of the model. Semantic quality of a model depends on how well the model corresponds to its domain [7]. Some semantic properties are generally applicable to all reference models, and all conceptual models in general, whereas others are limited to electronic commerce ones.

The most important among general semantic properties are the following.

Consistency or unambiguity is an important property that all conceptual models must possess. Inconsistencies in the model may lead to difficulties in implementing applications, or sometimes even render implementation impossible.

Coherence describes how well the components of a given system contribute to a common purpose or objective pursued by the system [2]. This may well be the most difficult property to analyze, since that purpose or objective (which is, by definition, outside the system itself) may not be well, or even uniquely defined. Coherence has definite pragmatic implications, as incoherent models will lead to difficulties in implementing applications. Coherence is reduced if the model contains components without a clearly defined purpose or components that do not seem to integrate seamlessly with the rest of the model.

Completeness shows to what degree all relevant aspects of the system have been specified by the model, within the boundaries defined by model orientation, balance, and scope. Although completeness is a very important semantic characteristic for any conceptual model, the lack of it does not necessarily imply low quality for an electronic commerce model. Namely, electronic commerce is still rapidly evolving, and new features and new requirements, such as market models, customer interaction modes, and the like, are emerging quite often. Therefore, any given model is incomplete by definition – but it must be sufficiently open and extendible in order to accommodate those new features. (Openness and extendibility are important pragmatic properties, to be discussed in more detail in the next subsection.)

Note that models targeted in our analysis are not built using rigid mathematical formalisms. Therefore, general semantic properties—consistency, coherence, and completeness—are impossible to prove in a formal-mathematical sense, and we must rely on informal analysis instead.

Other semantic properties are electronic commerce-specific. Among them, we note the following:

Orientation determines whether a particular model is primarily concerned with business or technology aspects of electronic commerce systems. The distinction is not always clear nor easy to make, as electronic commerce encompasses a wide range of issues in a continuously changing spectrum from those with pure business orientation to the purely technology-oriented ones [13,17].

Although orientation is a semantic property, it has profound pragmatic implications. Namely, technology-oriented models, by definition, should be better suited for actual implementation.

Scope defines how much of the relevant issues are covered by the model – the 'breadth' of the model, informally speaking. Scope should be assessed together with some other properties of the model, such as completeness and orientation. These properties are, to a certain degree, interdependent: for example, orientation of the model determines its scope, whereas the scope defines the context within which model completeness can be evaluated. In particular, the scope of electronic commerce models may be characterized through some additional aspects or properties, such as:

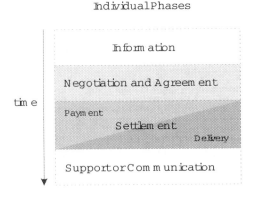

Fig. 2. Phases of an electronic commerce transaction (adapted from [15]).

1. **Perspective (viewpoint) support** – The existence of different perspectives or viewpoints was recognized in several areas. For example, the ISO Open EDI Standard describes 'two perspectives of business transactions' through business aspects and information technology aspects [10]; the description of a business system usually includes the following perspectives: functional, informational, behavioral, and organizational [4]; finally, the ISO/IEC Reference Model for Open Distributed Processing defines no less than five viewpoints: enterprise, information, computational, engineering, and technology [12]. (Note that electronic commerce is both a business

system and an open distributed processing application *par excellence*.) Therefore, a model of electronic commerce should specify the perspective or perspectives it takes into account – and it should account for as many different perspectives as possible.

2. **Transaction type support** – The participants in electronic commerce transactions are generally classified into business entities, individual customers, and public administration, and individual transactions may be classified according to the category of the participants involved. (Some of these classes—actually three, out of six possible, classes: business to business, business to individual, and business to public administration—are obviously more relevant to electronic commerce than the others.) Some models support one type of transactions only, while the others support two or all three of the aforementioned types. Intuitively, the quality of a model is higher if more transaction types are supported; however, completeness of the model must also be taken into account before a definitive judgment is made.

3. **Support for transaction phases** – It is generally agreed that the execution of an electronic commerce transaction consists of the following phases: information (or information collection), negotiation and agreement, and settlement [15], as shown in Fig. 2. (Other schemes have been proposed, but they can easily be mapped to this one.) As some models are confined to a single phase, while others attempt to cater for interactions in two or more phases, we may distinguish between single- and multi-phase models; obviously, the more phases a model supports, the more usable it is likely to be.

Note that all of these properties (and the last two in particular) are an interesting mix between semantic and pragmatic considerations related to electronic commerce. First, they are primarily determined by practical issues of transaction support; second, they do belong to the semantic properties category; and third, they have significant pragmatic implications. Such a mix should not come as a surprise, as we are discussing conceptual models with immediate applicability in practice.

Finally, **balance** refers to the ratio of weights given to different perspectives within the system description, and it is related to both orientation and scope of the model. It is well known that some models may concentrate on a particular perspective and pay little attention to others – for example, a data model describes the system from the informational perspective only. However, models offering a more balanced treatment are generally considered to be of better quality; again, this has to be assessed in the context defined by model orientation, scope, and completeness.

2.3 Pragmatic Properties

Finally, pragmatic properties describe the relationship of the model with its intended audience. (The audience here denotes the set of users who have to interpret it.) Pragmatic quality, then, describes how well the model corresponds to its audience interpretation [7]. In our case, the audience consists of system modelers and developers. Modelers will use a reference model to develop more specialized models, and developers will design and develop individual applications and/or complete systems. Pragmatic properties, then, determine whether the model is easy to comprehend and use. The most important pragmatic properties include the following.

Extendibility refers to the ability of the model to accommodate new or changed requirements, which may (and ultimately will) emerge over time. Changes must be absorbed and seamlessly integrated into the model, if it is to remain a coherent foundation for system development. Note that extendibility is an internal property of a model – it does not depend on any external models and/or standards.

Openness describes the degree to which a model is capable of interfacing with the outside world – systems that are not part of the model itself. All other properties being equal, a more open model will always be considered to be of better quality than a less open one. Note that interfacing is a two-way relationship and openness cannot depend solely on the model under consideration, let alone be easily described with a single numerical or textual attribute. Still, we should be able to compare any two systems in terms of openness, and form some kind of relative ordering according to this property.

At the implementation level, openness translates into interoperability: the ability of the system to interface with other systems and services. Again, this ability depends on those other systems and services as well. Given the abundance of commercially available systems and standards, interoperability is arguably among the most important properties (if not the most important property) of an electronic commerce model.

Technology dependence describes the relationship of a model with the existing technologies and/or standards. It is indeed closely related to openness, but in this case the importance of technology dependence (or independence) provides enough justification to consider it a property in its own right.

We note that technology dependence is related to the orientation of the model, as technology-oriented models are also more likely to be technology-dependent than the business-oriented ones. Technology dependence is related to the model scope as well (though to a lesser degree), as models with a narrow scope are more likely to depend on a particular technology than those with wider scope. For example, consider the choice of cryptographic algorithm (or algorithms) in the payment phase: a model that concentrates on the payment phase might prescribe a particular algorithm, whereas a model that covers all transaction phases may prescribe just the interface(s), but not the actual algorithm. Given the rate of change of computing and communication technologies, it does seem reasonable to relate quality to technology independence, and to favor the models that exhibit less dependence on specific technology, or technologies.

Pragmatic properties are difficult, or even impossible, to assess formally, as is the case with properties from other categories. But even an informal analysis can give us valuable insights, and help the developers make an informed choice of the model best suited to a particular problem.

3   Application of the Framework: Electronic Commerce Reference Models

As noted before, a number of reference models related to electronic commerce have been proposed so far; however, given the scope and space limitations, we will apply our quality assessment framework on two among the best known ones: the OMG Reference Model for electronic commerce [8] and Secure Electronic Market for Europe [14].

3.1   OMG Electronic Commerce Reference Model

The OMG (Object Management Group) Reference Model for electronic commerce [8,9] has been developed by the OMG Electronic Commerce Domain Task Force (EC-DTF). It provides a high-level object-oriented framework for specification of requirements for electronic commerce systems, designed in accordance with OMG's Object Management Architecture – a generic set of components, interfaces, and protocols for distributed object-oriented applications [16].

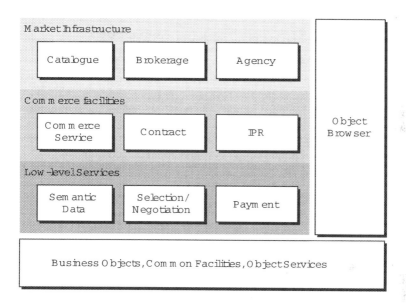

Fig. 3. OMG Electronic Commerce Reference Model – principal facilities (adapted from [8]).

The OMG architecture provides a multi-layer interoperability framework, wherein related functional requirements are grouped into a number of containers called facilities. Facilities are, in turn, categorized into market infrastructure services (catalogues, brokerage, and agencies), commerce facilities (such as contract, service management and related desktop facilities, and management and administration of intellectual property rights – IPR), and low level services (these include payment, semantic data facility, profile management, and selection/negotiation). A schematic view of the facilities and their respective categories is shown in Fig. 3. Additionally, the object browser and navigator facility, available to all facilities regardless of the layer they belong to, introduces an extendible framework for the inspection, presentation and execution of electronic commerce entities.

Since the model is based on an object-oriented systems architecture, each facility is handled as a real object offering interfaces to other objects. Detailed semantics for the facilities' interfaces are provided, including (in some cases) high-level protocols of their usage [9].

In this manner, the reference model serves as a high level framework, allowing individual components to be developed on a per-market basis, and 'plugged in' in accordance with the requirements of the market. The emphasis is on interoperability of different solutions, instead of imposing a single unified solution—be it a protocol, payment method, product dictionary, or anything else—to all market participants. Given the number of systems already present on the market, this would be an impossible task anyway.

Detailed CORBA IDL specifications have already been defined for interfaces of some of the facilities, and work is under way for defining others.

3.2   Secure Electronic Market Place for Europe (SEMPER)

Secure Electronic Marketplace for Europe (SEMPER) is a project sponsored by the European Commission, with the ultimate goal of developing and providing an open and comprehensive solution for secure commerce over the Internet and other open networks [14]. The first phase concentrated on the development of a framework and architecture for secure electronic commerce, and the provision of fundamental electronic commerce services (such as offering, ordering, payment and delivery for information services) within this framework. The subsequent phases concentrate on extending the architecture and developing more advanced services.

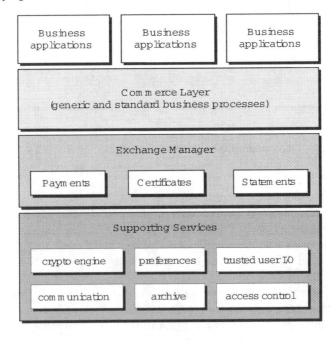

Fig. 4. The SEMPER architecture (adapted from [14]).

The SEMPER model assumes that any business scenario consists of a number of standard business processes, which may be further decomposed into a sequence of unidirectional and/or bidirectional exchanges of business items. (SEMPER documentation uses the terms 'transfers' and 'fair exchanges', respectively.) Business items include all kinds of items that can be exchanged in a business transaction: payments, credentials, and documents or statements. These assumptions lead to the architecture shown in Fig. 4. Front-end business applications are built on top of the Commerce Layer containing services that directly implement business scenarios such as 'mail-order retail' or 'on-line purchase'. Since the service set(s) cannot be fixed in advance, provisions are made for secure downloading of additional services to extend the functionality of the existing applications. In fact, entire applications can be downloaded, thus allowing customers to participate in business scenarios they never encountered before.

Exchanges of business items, such as payments, credentials, and/or documents, are managed by a separate service in the Exchange Management Layer. Thus multiple existing implementations can be integrated into a unified service framework. For example, different payment systems may be simultaneously installed and any one of them can be used in an actual transaction, while the appropriate negotiations that must take place before the transaction may be entirely transparent to user.

Exchange managers use a set of low-level basic services provided by the Supporting Services Layer: communication, cryptographic services, user profile management, access control, and others. Different services are specified in the SEMPER architecture, although the services dealing with security requirements are given a prominent role. Again, future extensions are deemed inevitable, hence the standardization activity has focused on the interfaces between individual service blocks, rather than on the details of the blocks themselves.

4   Comparison of Models

The OMG and SEMPER models presented in the previous section will now be evaluated according to the criteria outlined in Section 2.

4.1   Syntactic Properties

In terms of stratification and integration, both OMG and SEMPER models are layered, non-hierarchical models, as components from different layers can and should collaborate to provide the required system functionality. The OMG model is stratified in three layers, supported by the common functionality from the Common Facilities layer. The SEMPER framework is also stratified, but with a slightly different decomposition of components and services. Actually, the main differences between the two is that the SEMPER model uses a separate layer to manage exchanges of business items, whereas the OMG framework gives this activity a much less prominent role.

Integration of different components and layers in the OMG model is implemented through the use of CORBA IDL specifications, an essentially object-oriented technique. The SEMPER models does not address this issue, although adherence to the object technology seems almost inevitable.

OMG and SEMPER models appear equivalent in terms of granularity and level of abstraction. Both offer a rather detailed decomposition of components and services needed for electronic commerce applications, although they differ in the partitioning approach.

For example, the communication services, together with security, authentication, and identification services, are present as components within the supporting services layer of the SEMPER model. The OMG model does not identify them separately but delegates most of them to the so-called Common Services layer; maybe the authors of the OMG model have assumed that those services are part of the Common Services infrastructure and therefore have to be specified, designed, and implemented outside of electronic commerce applications. Which of these approaches is better, and under what circumstances, remains an interesting topic for further research, both in theory and practice.

In the OMG model, integration of components and services from different layers is achieved with object-oriented techniques, using CORBA IDL specifications. The manner of integration of different components and layers is not specifically prescribed by the SEMPER framework, although adherence to the object technology seems almost inevitable.

4.2   Semantic Properties

OMG and SEMPER models appear equivalent in terms of general semantic properties, such as consistency, coherence, and completeness.

In terms of orientation, generic treatment of business items and exchanges of these items (to the point of devoting a separate Exchange Management layer) implies a certain business orientation in the SEMPER model. On the other hand, the use of CORBA (and object-oriented approach in general) in the OMG model may be interpreted as the sign of a slight implementation (i.e., technology) bias. However, both models are rather generic, and the differences are minor.

As for the aspects of scope, both the OMG and SEMPER models are concerned with the functional, informational, and behavioral viewpoints (the organizational perspective is not covered). However, both models appear to favor the functional perspective over the other two.

Both OMG and SEMPER models support all transaction types. Also, both provide explicit support for all phases of business transactions.

4.3   Pragmatic Properties

In terms of extendibility, both models again perform comparatively well. Again, adding new components and/or services to the OMG model is limited by the requirement that those components and/or services are able to communicate within the CORBA-defined framework; the SEMPER model has no such limitations.

In terms of openness, both OMG and SEMPER models appear to be comparatively open. In the OMG case, interfacing to other component services should not be too difficult, provided these services use CORBA-compliant interfaces. No such limitations are present in the SEMPER case.

Overall, the OMG model does exhibit a certain technology dependence, mainly on account of its use of the Object Management Architecture, which is, in turn, implemented through CORBA. It must be noted that CORBA is essentially an interoperability framework, not tied up to any particular language or implementation technology. Systems based on different technologies, including legacy ones, can be interconnected through CORBA request brokers. Even the choice of CORBA as the underlying interoperability framework is somewhat arbitrary, since the OMG model itself does not prohibit the use of another, functionally equivalent architecture for distributed systems.

On the other hand, the SEMPER framework is essentially technology-neutral, and the actual implementation is free to use whatever technology may seem fit for the purpose.

4.4   Conclusion

In general, both OMG and SEMPER reference models present a comprehensive, yet unique, perspective of electronic commerce systems. Both models contain a lot of detail, yet they are still rather generic, and further analysis—and, possibly, factors unrelated to the models themselves—is necessary before the decision to adopt one or the other is made.

Future research in the area of reference models for electronic commerce can proceed in two directions. First, a more comprehensive analysis of both reference models and component service models and protocols could be undertaken, in order to gain deeper understanding of their respective strengths and weaknesses. Compatibility between reference models and component service models and protocols is also an important issue. In this way, application developers will be able to choose the optimal reference model according to the characteristics of the electronic commerce system they are about to design and implement.

Second, the insights from the comparative analysis of existing reference models could lead to the development of a better reference model or models, either as a modification or generalization of an existing model, or (more likely) as a combination of two or more of them. In both cases, the final goal is to use the reference models to help develop new, powerful, flexible and interoperable electronic commerce applications and systems.

5   Directions for Future Research

The quality framework outlined in this paper is far from being perfected, as there are many more details to be specified. The main problem, of course, is that quality attributes are usually too complex, yet at the same time too amorphous to be evaluated on a simple scale; furthermore, there are few, if any, quantitative measures for quality attributes. In order to cater for these deficiencies, research in this area could relate to research in architecture description languages and, by extension, in formal specification languages.

In any case, the quality framework certainly has to be made more comprehensive, by including more quality attributes and investigating their relationship and possible interdependence. At the same time, variants of the framework tailored to specific application areas could (and probably will) be developed.

The quality framework, finally, has to be validated, and subsequently refined, on different reference models.

References

1. L. Bass, P. Clements, and R. Kazman. *Software Architecture in Practice*. The SEI Series in Software Engineering. Addison-Wesley, Reading, MA, 1998.
2. E. V. Berard. *Essays on Object-Oriented Software Engineering*, volume I. Prentice-Hall, Englewood Cliffs, NJ, 1993.
3. G. Booch, J. Rumbaugh, and I. Jacobson. *The Unified Modeling Language User Guide*. Addison-Wesley, Reading, MA, 1999.
4. C. Bussler and S. Jablonski. An approach to integrate workflow modeling and organization modeling in an enterprise. In *Proceedings of 3rd IEEE Workshop on Enabling Technologies: Infrastructure for Collaborative Enterprises*, pages 81–95, Morgantown, WV, Apr. 1994.
5. P. Coad and E. Yourdon. *Object-Oriented Analysis*. Yourdon Press, Englewood Cliffs, NJ, second edition, 1991.
6. R. ElMasri and S. B. Navathe. *Fundamentals of Database Systems*. Benjamin/Cummings, Redwood City, CA, 2nd edition, 1995.
7. O. I. Lindland, G. Sindre, and A. Sølvberg. Understanding quality in conceptual modeling. *IEEE Software*, 11(2):42–49, Mar. 1994.
8. S. McConnell, M. Merz, L. Maesano, and M. Witthaut. An open architecture for electronic commerce. OSM response to OMG Electronic Commerce Domain Task Force RFP-2, OSM, 24th Feb. 1997.
9. The OMG/CommerceNet Joint Electronic Commerce Whitepaper. Technical report, OMG, 27th July 1997.
10. ISO/IEC CD 14662. Information Technology – Open EDI reference model, Draft International Standard ISO/IEC JTC1/SC 30, 1998.
11. ISO 7498. Basic Reference Model for Open Systems Interconnection, International Standard, 1984.
12. ISO/IEC 10746-3. Basic reference model of open distributed processing – part 3: Prescriptive model, International Standard ISO/IEC JTC1/SC21 N 7055, Sept. 1992.
13. B. F. Schmid and M. A. Lindemann. Elements of a reference model for electronic markets. In *Proceedings 31st Annual Hawaii International Conference on System Sciences*, volume 4, pages 193–201, Kohala Coast, HI, Jan. 1998.
14. M. Schunter and M. Waidner. Architecture and design of a secure electronic marketplace. In *Proceedings 8th Joint European Networking Conference (JENC8)*, Edinburgh, UK, June 97.
15. D. Selz and P. Schubert. Web assessment – a model for the evaluation and the assessment of successful electronic commerce applications. In *Proceedings 31st Annual Hawaii International Conference on System Sciences*, volume 4, pages 222–231, Kohala Coast, HI, Jan. 1998.
16. R. M. Soley, editor. *Object Management Architecture Guide, Revision 3.0*. John Wiley and Sons, New York, 3rd edition, June 1995.
17. V. Zwass. Electronic commerce: structures and issues. *International Journal of Electronic Commerce*, 1(1):3–23, Fall 1996.

Measures for Assessing Dynamic Complexity Aspects of Object-Oriented Conceptual Schemes

Geert Poels and Guido Dedene

MIS Group, Dept. Applied Economic Sciences, Katholieke Universiteit Leuven
Naamsestraat 69, B-3000 Leuven, Belgium
{geert.poels, guido.dedene}@econ.kuleuven.ac.be

Abstract. System developers are increasingly realising that the quality of a system must be ensured in the early stages of the development life cycle. It is in this context that a number of quality frameworks for conceptual schemes have been proposed. However, before the quality of a conceptual schema can be improved, it must be assessed. Accordingly, a number of measure suites have been proposed for measuring quality properties of conceptual schemes. In this paper we focus on one particular quality property, i.e. complexity. This property can be described as the mental burden of the persons that must understand, modify, extend, verify, implement, and reuse conceptual schemes. The proposed complexity measures for conceptual schemes have in common that they only capture the complexity of the static or structural aspects of a conceptual schema. We therefore present a complementary suite of measures that focuses on conceptual schema complexity as seen from a dynamic perspective.

1 Introduction

Conceptual modeling is an integral part of modern approaches towards system development, like Catalysis [1] and the Rational Unified Process [2]. The conceptual schema is not merely the basis for modeling the persistent system data. In object-oriented modeling, where data and process are closely linked, conceptual schemes provide the solid foundation for the design and implementation of information systems.

As an early available, key analysis artifact the quality of the conceptual schema is crucial to the success of system development. Generally, problems in the artifacts produced in the initial stages of system development propagate to the artifacts produced in later stages, where they are much more costly to identify and correct [3]. Therefore, the quality of conceptual schemes must be evaluated, and if needed improved.

This paper must be seen in the context of a measurement-based approach towards quality control for object-oriented conceptual schemes. Before quality properties can be evaluated, their values must be assessed, either by subjective expert ratings or by objective measurements. In this paper we present a formally defined measure suite to quantify various aspects related to one particular, but highly important quality property of conceptual schemes, i.e. their simplicity. We consider simplicity as a

A.H.F. Laender, S.W. Liddle, V.C. Storey (Eds.): ER2000 Conference, LNCS 1920, pp. 499-512, 2000.
© Springer-Verlag Berlin Heidelberg 2000

'quality' because its inverse, complexity, has been shown, both theoretically and empirically (e.g. [4]), to be detrimental to the ability to understand, modify, extend, verify, implement, and reuse system development artifacts.

Our measure suite differs from other suites of conceptual schema measures in the sense that it takes dynamic views on the conceptual schema into account. It extends previous work on measuring entity relationship schemes and object relationship schemes. Whereas the ER-related work focused on complexity aspects of logic data schemes, which are static by nature, the OR-related research produced complexity measures for static object schemes that result from object-oriented (domain) analysis activities. Some of these measures capture the functionality that is encapsulated in the objects, e.g. in terms of the (public) operations defined and/or inherited. In general however, no measures have been proposed to assess complexity aspects related to the functional and dynamic behaviour dimensions of conceptual schemes.

The measure suite presented in this paper is based on a formal model of object functionality and behaviour that uses the notion of event. The modeling of events and the participation of objects in these events introduces a dynamic perspective on conceptual modeling. It allows expressing complexity measures in terms of object interaction (e.g. when objects participate in the same event) and in terms of object life cycle specifications (e.g. sequence constraints on the participation in events). Our measure suite thus complements the previously proposed measures for 'static' complexity aspects of object-oriented conceptual schemes.

This paper is organised as follows. Section 2 reviews quality models for conceptual schemes and discusses the role of complexity as a quality property. Section 3 reviews previous work on conceptual schema measures. In section 4 we introduce the cornerstone of our formal model of object functionality and behaviour: the object type – event type association matrix. Next, in section 5 the measure suite is presented. Section 6 briefly discusses the complex issue of measure validation and touches upon the theoretical and empirical validity of the proposed measures. Finally, section 7 presents conclusions.

2   Quality and Complexity in Conceptual Modeling

From a systems theory point of view, a system is called complex if it is composed of many (different types of) elements, with many (different types of) (dynamically changing) relationships between them. In software engineering, it is well accepted by now that no single definition or measure can capture all possible aspects of complexity [5]. Nevertheless, the relationship between individual complexity aspects (e.g. information flow complexity [6]), also called 'internal' quality properties of a system development artifact, and 'external' quality properties like reliability, reusability and maintainability has been investigated with the purpose of building software quality prediction, evaluation and control models [7].

Compared to software engineering, the concept of quality in conceptual modeling is poorly understood [8]. Only a few comprehensive and structured quality evaluation frameworks have been proposed that provide more than a pure listing of desirable quality properties. These proposals include the frameworks of Lindland et al. [9] and Moody et al. [10]. The framework of Lindland et al. uses linguistic concepts to distinguish between three types of conceptual schema quality: syntactic quality (i.e.

the degree to which the rules of the modeling technique are adhered to), semantic quality (i.e. the degree to which the schema corresponds to the domain it models), and pragmatic quality (i.e. the degree to which users understand the schema). Within the bounds set by the complexity of the rules of the modeling technique and the complexity of the domain that must be modeled, complexity, or better, its inverse simplicity, must be seen as a pragmatic quality aspect. The framework of Moody et al. considers in its revised form eight quality factors, one of them being simplicity.

According to Moody [11] a data schema is characterised by simplicity if it contains the minimum possible constructs in terms of entities, relationships and attributes. Schema size, in terms of object types and attributes, has also been considered as an aspect of pragmatic quality by Assenova and Johannesson [8]. However, they acknowledge that the lowest possible size is not necessarily a 'quality' of a conceptual schema. Also, Lindland et al. [9] argue that structuredness might be more important for pragmatical quality than 'expressive economy'. These findings are consistent with the notions of complexity that have been used in software engineering research, where more emphasis is laid on structural aspects such as coupling, cohesion, depth and width of the inheritance lattice, etc [7].

Our model of conceptual schema complexity, its relationship with size and structure on the one hand, and 'external' quality properties on the other hand, is based upon similar models used in software engineering research [12], [4]. Fig. 1 shows for instance the complexity model of Briand et al. [12]. The structural properties of a software engineering artifact affect the 'cognitive' complexity of the artifact, i.e. the mental burden of the persons who have to deal with the artifact (e.g. developers, testers, maintainers). According to Briand et al. it is the, sometimes necessary, high complexity of an artifact which causes it to display undesirable external qualities.

The complexity model of Briand et al. is the basis for much empirical research in the area of software artifact complexity. In this paper we assume a similar model to hold for conceptual schemes, given that they are artifacts used in the initial stages of system development.

Fig. 1. The complexity model for system development artifacts of Briand et al. [12]

3   Previous Work on Conceptual Schema Measures

Lindland et al. [9] mention inspection, visualisation, animation, explanation, simulation and filtering as techniques for improving pragmatical quality. These techniques help to understand a conceptual schema without modifying it. Another technique is to transform a schema in order to improve its pragmatic quality properties (e.g. complexity) [8, 13]. Such schema transformations require pragmatic quality properties to be assessed, both before and after the schema is transformed, to evaluate the effectiveness of a transformation. It is in this context, and to assess conceptual schema quality in general, that measurement instruments in the form of rating scales and measures have been proposed.

A number of researchers have proposed measurement instruments for entity relationship schemes. Moody [11] presents twenty-five 'metrics' for assessing the quality factors of entity relationship schemes identified in [10]. These 'metrics' are a mix of rating scales, cost figures or estimates, and counts. The latter include counts based on subjective expert knowledge (e.g. number of items in the data model that do not correspond to user requirements). However, the measures proposed for simplicity (i.e. number of entities, number of entities and relationships, number of entities, relationships and attributes) are objective and 'automatable' counts. A similar suite of twelve complexity measures for ER schemes has been presented by Genero et al. [14].

Moser and Misic [15] propose size, coupling and cohesion measures for object-oriented conceptual schemes (also called business models), which are based on a formal and generic object model described in [16]. Badri et al. [17] and Genero et al. [18] present complexity measures for the object-oriented analysis schemes of more specific development methods, respectively OOA [19] and OMT [20]. Most of the object-oriented software measures proposed in the literature (e.g. the MOOSE measures [21], the MOOD measures [22]) are however design or code measures that capture aspects not relevant for conceptual modeling. Briand et al. [23, 24] have shown that at least a few of these measures could also be used as specification or analysis measures.

The complexity measures for object relationship diagrams take size and structure aspects related to object operations into account. Compared to the measure suites for ER schemes they measure more than data schema quality. However, they are based on a static object model and do not capture the complexity of the dynamic perspective. The measure suite presented in this paper is meant to remedy this situation and complements the existing measures for (object-oriented) conceptual modeling.

4   Event-Driven Object-Oriented Conceptual Modeling

Objects in a domain are affected by the occurrence of events. As an example, consider an ORDER object that can be placed, changed, delivered, invoiced, paid, etc. In event-driven conceptual modeling a dynamic perspective on the domain is taken. Objects, relationships, rules and constraints are modelled starting from the things that happen, i.e. the events. In this section we present a formal model of objects and events based on an 'archetype' method, described in [25, 26] that is sufficiently abstract and generic to capture the main aspects of event-driven conceptual modeling.

Fig. 2 shows an extract of our meta-model (in UML notation [27]) for event-driven object-oriented conceptual modeling. Modeling starts by identifying the different types of event that are relevant to the Universe of Discourse (i.e. the domain). Hereafter, a capital A is used to denote the universe of event types relevant for the UoD. It must be noted that some methods (e.g. Catalysis [1]) model events only indirectly, via the action concept. Actions are different from events in the sense that they have a duration. However, actions can easily be transformed into events: both the beginning and ending of an action qualify as events.

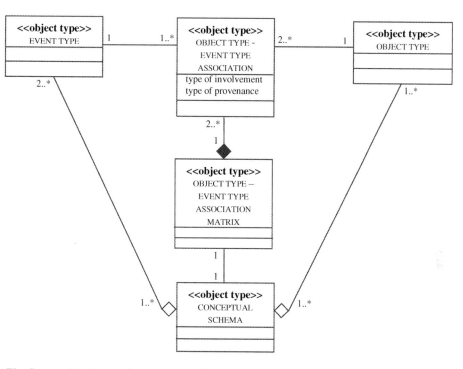

Fig. 2. A partial view on the meta-model for event-driven object-oriented conceptual modeling

The persons, things, etc. in the UoD that are involved in the occurrences of the event types in *A* are modelled as objects. Objects are described by a number of properties, which are specified in an object type. We assume that all objects identified during conceptual modeling are persistent, i.e. they have a state, represented at any moment by the values of their attributes, and they exist for a certain period of time. Objects therefore always participate in at least two events: a creating event and an ending event. The participation in the ending event does not imply that the object is physically destroyed. It means that the object can no longer participate in real-world events. The set of object types relevant to the universe of event types *A* is denoted by a capital *T*.

In event-driven conceptual modeling we specify an operation in the object type for each type of events that the instances of the object type can participate in. When an object participates in an event then the corresponding operation is triggered, which (possibly) changes the state of the object. It must be noted that in an object-oriented implementation of the conceptual schema, not all operations must effectively be implemented as methods in the class definition of the object type because of mechanisms like inheritance and delegation.

This dynamic perspective on conceptual modeling is captured in the so-called object type – event type association matrix [25]. Each conceptual schema has one object type – event type association matrix composed of object type - event type associations. Such an association relates one event type in *A* with one object type in *T* and means that instances of the object type participate in occurrences of the event type. It has two attributes. The type of involvement specifies whether an event participation creates an object (value: 'C'), ends the life of an object (value: 'E'), or just modifies the life cycle state of the object (value: 'M'). A 'modifying' event type can, but does not necessarily change the visible object state (i.e. the attribute values). The type of provenance specifies whether the object type - event type association has been inherited (value: 'I') from an ancestor, has been acquired through propagation (value: 'A'), or is a newly defined (or 'own') association (value: 'O'). The formal definition of the object type - event type association matrix is taken from [25]:

> For some UoD, let A be the universe of event types and T be the set of object types. The object type - event type association matrix is a map
> $\tau\colon A \times T \to \{O, A, I\} \times \{C, M, E\} \cup \{(\text{' '}, \text{' '})\}$.
> When $\tau(e,P) = (R,J)$ with $R \in \{O, A, I, \text{' '}\}$ and $J \in \{C, M, E, \text{' '}\}$, we write that $\tau(e,P) = R/J$.
> We define the partial maps τ_p and τ_i that return the type of provenance and the type of involvement as $\tau_p\colon A \times T \to \{O, A, I, \text{' '}\}$ and $\tau_i\colon A \times T \to \{C, M, E, \text{' '}\}$.

Table 1 shows an example object type - event type association matrix for a (simplified) library. Consistent with the rules proposed in [25], an object type - event type association is propagated from an object type P to an object type Q if objects of P are existence dependent on objects of Q.[1]

The object type - event type association matrix is the cornerstone of our formal model of object functionality and behaviour. It allows expressing complexity aspects of conceptual schemes in terms of (common) event participation. The basic underlying conjecture is that, all other things being equal, the more types of event that an object participates in, and the more objects of different types that participate in a same event, the more complexity is added to the conceptual schema. One of the motivations for this conjecture is that when objects participate in an event, interesting things happen in both the domain and the information system: business rules and constraints are checked (i.e. is the event participation allowed?), operations are triggered, object states are changed, etc. This is especially relevant when two or more objects jointly participate in an event (e.g. when a member of the library makes a reservation for a copy, an instance of MEMBER and an instance of COPY participate in a reserve event). Such a joint participation synchronises the lives of the participating objects, might lead to the creation or ending of instances of other object types (e.g. reserve creates an instance of RESERVATION), might necessitate checking additional rules (e.g. a reservation is refused if the copy is on shelf), etc.

The type of involvement values of the object type - event type associations help to derive the dynamic behaviour of objects. They specify a default life cycle. First, a choice is made between the creating event types to create an object instance. Next, the state of the object may be modified zero, one or more times, using any type of modifying event. Finally, a choice is made between the ending event types to end the life of the object. For instance, if sequence, selection and iteration are denoted using the ".", "+", and "*" symbols respectively, then a default life cycle for a RENEWABLE_LOAN object is specified by (borrow + fetch) . renew* . (return + lose).

---

[1] For a formal and elaborate definition of 'existence dependency' we refer to [26].

Table 1. Object type - event type association matrix for a simplified library

	ITEM	VOLUME	COPY	RESERVATION	MEMBER	LOAN	NOT_RENEWABLE_LOAN	RENEWABLE_LOAN
acquire	O/C	I/C	I/C					
catalogue	O/M	I/M	I/M					
sell	O/E	I/E	I/E					
reserve				A/M	O/C	A/M		
cancel				A/M	O/E	A/M		
fetch				A/M	O/E	A/M		O/C
start_membership					O/C			
end_membership					O/E			
borrow		A/M	A/M		A/M	O/C	I/C	I/C
return		A/M	A/M		A/M	O/E	I/E	I/E
lose		A/E	A/E		A/M	O/E	I/E	I/E
renew			A/M		A/M			O/M

The domain might impose additional constraints on the life cycle of objects. We might for instance require that a copy can only be borrowed if it has been catalogued first. The diagrams (e.g. Finite State Machines) or mathematical expressions used to specify life cycle schemes, other than the default ones, are not shown in the meta-model of Fig. 2. We use them to complement the object type - event type association matrix when measuring complexity aspects related to the dynamic behaviour of objects.

Complexity Measures for Event-Driven Object-Oriented Conceptual Modeling

The suite of measures presented here assesses various complexity aspects related to the size and structure of a conceptual schema from the dynamic perspective described in the previous section. Most of these measures simply require querying the object type - event type association matrix. A single one requires additional information that is contained in the object life cycle specifications. As the measure definitions are formulated in terms of object type - event type associations (i.e. event participations), instead of attributes, relationships, or operations, they complement, but do not necessarily substitute, the previously published measure suites for conceptual modeling that were reviewed in section 3.

We do not claim that the measure suite is complete. In fact, due to space limitations, we had to limit the number of measure definitions. For some extra measures, related to polymorphic behaviour aspects, we refer to [28].

For the measure definitions, assume a universally qualified conceptual schema S with universe of event types A, set of object types T, and an object type - event type association matrix τ. We use the symbol # for the cardinality of a set.

5.1   A Size Measure

The size of a schema has been defined as the number of object types and attributes [8] or the number of constructs (i.e. entities, relationships and attributes) [11]. Analogously, we define it here as the number of object type - event type associations specified in the object type - event type association matrix.

A size measure for conceptual schemes is the Level of Event Participation (LEP). It returns the count of non-empty cells in the object type - event type association matrix. The LEP measure is given by the following equation:

$$\text{LEP(S)} = \sum_{P \in T} \#\{e \in A \mid \tau(e,P) \neq \text{'}/\text{'}\} . \tag{1}$$

The value returned for the library example is 42. Note that the size of a conceptual schema is related to the size of the information system (though not necessarily in terms of data volume). The more types of event an object is involved in, the more operations must possibly (but not necessarily) be implemented in the class definition of the object type. Hence, LEP can be used to derive an early, albeit rough, estimate of the size of the information system (e.g. in terms of lines of code). Early size estimates are useful and essential for project budgeting purposes. They are the basis for effort and cost estimates, and for pricing, outsourcing and scheduling decisions.

5.2   Structure Measures

An aspect of structure that has received a lot of attention in software engineering is coupling. Coupling has been described as the degree of interdependence between system development artifacts (e.g. object classes) [7]. The main arguments in favour of low coupling are that the stronger the coupling between artifacts, (i) the more difficult it is to understand individual artifacts, and hence to maintain them; (ii) the larger the extent of (unexpected) change and defect propagation effects across artifacts, and consequently the more testing required to achieve satisfactory reliability levels; (iii) the lower the reusability of individual artifacts.

In object-oriented software, coupling has mostly been measured in terms of message passing [24]. In conceptual modeling we do not wish to decide yet whether object communication will be based on message passing. In our opinion, it might thus be useful to express coupling in terms of common event participations. Object types are then coupled if their instances participate in the same types of event.

A coupling measure for conceptual schemes is the Level of Object Type Coupling (LOTC). It counts for each object type P the number of other types of object that participate in a same type of event as the instances of P, and then adds these counts. The equation for the LOTC measure is:

$$\text{LOTC(S)} = \sum_{P \in T} \#\{Q \in T \text{-} \{P\} \mid \exists e \in A: \tau(e,P) \neq \text{'}/\text{'} \wedge \tau(e,Q) \neq \text{'}/\text{'}\} . \quad (2)$$

The value returned for the library example is 40. A normalised version of this measure is the Degree of Object Type Coupling (DOTC). It relates the actual LOTC value to the theoretical maximum LOTC value given the number of object types in the schema. The DOTC measure is given by the following equation:

$$\text{DOTC(S)} = \text{LOTC(S)} / (\#T.(\#T - 1)) . \quad (3)$$

The DOTC value for the library is $40 / 56 = 0.71$. In object-oriented analysis and design, coupling between object types has also been defined in terms of the number of associations and generalisation/specialisation relationships with other object types in the schema. This type of coupling is called association-based or static coupling [29]. In the context of conceptual modeling, association-based coupling has not been measured in terms of its effect on dynamic aspects, like the inheritance and propagation of event participations. To assess the extent of inheritance and propagation in a conceptual schema we propose the following measures.

The Level of Inheritance of Event Participation (LIEP) returns the count of inherited object type - event type associations. The Level of Propagation of Event Participation (LPEP) returns the count of object type - event type associations that have been acquired through propagation. The equations for LIEP and LPEP are:

$$\text{LIEP(S)} = \sum_{P \in T} \#\{e \in A \mid \tau_p(e,P) = I\} \quad (4)$$

$$\text{LPEP(S)} = \sum_{P \in T} \#\{e \in A \mid \tau_p(e,P) = A\} . \quad (5)$$

The respective values for library are 12 and 17. Normalised versions of these measures are the Degree of Inheritance of Event Participation (DIEP) and the Degree of Propagation of Event Participation (DPEP). The equations are:

$$\text{DIEP(S)} = \text{LIEP(S)} / \text{LEP(S)} \quad (6)$$

$$\text{DPEP(S)} = \text{LPEP(S)} / \text{LEP(S)} . \quad (7)$$

The respective values for library are $12 / 42 = 0.29$ and $17 / 42 = 0.40$. As opposed to DOTC, the values of DIEP and DPEP can never be equal to one. There must always be some object type - event type associations that are neither inherited, nor acquired through propagation.

5.3   Measures for Dynamic Behaviour Complexity

Objects synchronise their lives when they jointly participate in an event. Some of these synchronising events are special in the sense that they both create and end objects. The object types involved in such event types are in a way coupled, but this coupling is not captured by measures for static coupling, nor by the inheritance and propagation measures defined in the previous subsection. We therefore define here a measure of synchronisation-based coupling. We say that an object type P is synchronisation-based coupled with an object type Q if there is an event that ends the life of an instance of P and creates an instance of Q, or vice versa.

The Level of Synchronisation-based Coupling (LSC) measures the extent of synchronisation-based coupling in a conceptual schema. It counts for each object type P the number of other object types it is synchronisation-based coupled with, and then adds these counts. The degree of Synchronisation-based Coupling (DSC) normalises the value of LSC by relating it to the Level of Object Type Coupling. The LSC and DSC measures are given by the following equations:

$$LSC(S) = \sum_{P \in T} \#\{Q \in T - \{P\} \mid \exists e \in A : (\tau_l(e,P) = C \wedge \tau_l(e,Q) = E) \vee \quad (8)$$

$$(\tau_l(e,P) = E \wedge \tau_l(e,Q) = C)\}$$

$$DSC(S) = LSC(S) / LOTC(S) . \quad (9)$$

The values for LSC and DSC in the library example are 2 and 0.05 . Only RESERVATION and RENEWABLE_LOAN are synchronisation-based coupled through *fetch* events.

The final measure we present here is somewhat different from the rest. It compares the object life cycle specifications (e.g. Finite State Machines) with the default life cycles as specified in the object type - event type association matrix. The greater the difference between the two, the more sequence constraints apply to the participation of objects in events, and thus the higher the complexity of the dynamic behaviour of objects.

This particular aspect of complexity is called object life cycle complexity. The Object Life Cycle Complexity (OLCC) measure takes the form of a distance measure. The greater the distance between the actual life cycle specifications and the default life cycle specifications, the higher the value of OLCC. The elaboration of the definition of the OLCC measure is outside the scope of this paper and has been published previously [30]. We therefore only present an informal definition here, based on the library example.

To keep things simple, assume that in the library conceptual schema only ITEM and its specialisations have a non-default life cycle specification. The default life cycle specification for ITEM, based on the type of involvement indications in Table 1, is *acquire . catalogue* . sell*, implying that between acquiring and selling, the item can be catalogued zero, one or more times. However, the actual life cycle specification is more restricted: there must be exactly one participation in a *catalogue* event, i.e. the iteration on *catalogue* events must be dropped. For the specialisations of ITEM, i.e. VOLUME and COPY, it is required that they are catalogued before being borrowed for the first time. Hence, the respective life cycle specifications are *acquire . catalogue .*

(*borrow* + *return*)* . (*lose* + *sell*) and *acquire* . *catalogue* . (*borrow* + *renew* + *return* + *reserve* + *cancel* + *fetch*)* . (*lose* + *sell*).

In [30] we have proposed a set of elementary life cycle specification transformations for which it is proven that they can be used to express the distance between two life cycle specifications. Basically, these elementary life cycle specification transformations involve adding an event type in sequence or selection, removing an event type from a sequence or selection, and adding or removing iteration operators. Object life cycle complexity is then defined as the minimum number of such transformations needed to transform the actual life cycle specification in the default life cycle specification (or vice versa). The OLCC measure returns the sum of this minimum number of transformations, over all object types. In the example, one transformation is needed for ITEM (i.e. adding an iteration operator on *catalogue*) and two transformations are needed for VOLUME and COPY (i.e. removing *catalogue* from the sequence and adding it to the selection of 'modifying' event types). Hence, the value of OLCC for library is 5.

6   Some Observations on Measure Validity

In order to be credible and useful, proposed measures for system development artifacts must be validated, both theoretically and empirically [31]. A measure is theoretically valid if it measures what it is purported to measure. Zuse [32] advocates the use of measurement theory [33] as a reference framework for the theoretical validation of software measures. All measures presented in this paper have been validated using a specific measurement theoretic structure, i.e. the segmentally additive proximity structure [34]. Basically, all measures have initially been developed as distance measures that measure the difference with respect to the property of interest (e.g. size) between the artifact (e.g. a conceptual schema) and an hypothetical 'reference' artifact showing the theoretical lowest value for the property (e.g. an empty schema). The difference (or distance, dissimilarity) between these two artifacts is then measured by counting the minimum number of elementary transformations that are needed to transform one artifact into the other (cf. our discussion of the OLCC measure in the previous section). The details of this validation process are beyond the scope of this paper, but can be found in [35].

Equally important is the empirical validation of the measures. Basically this means that we must gather empirical evidence on the relationship between the various complexity aspects (i.e. size, structure, dynamic behaviour) that are measured and 'external' quality properties (cf. Fig. 1). As far as we know, no comprehensive empirical validation study in the area of conceptual schema quality has been published yet. Moody [11] proposes action research to refine the measures he has proposed. Genero et al. [14] use a case study to claim that some of their ER schema complexity measures correlate well with the maintainance time of the application programs that manage the data conceptually represented in the ER schemes. Another validation strategy is to use schema transformations, like the ones proposed in [13], to validate measures. The basic hypothesis underlying this type of study is that schema transformations improve the quality of the conceptual schema, and thus the complexity values returned by the measures should be lower after the transformations than they were before. It must be noted however that quality criteria, including objective measures, have also been used to show that schema transformations improve the quality of the schemes [8].

Currently, we have only gathered limited evidence of the empirical validity of our measures. Some of the complexity and distance measures have been applied in the context of a reference framework for conceptual schemes of an organisation's front-office to investigate their potential as indicators of perceived complexity and reengineering impact [36]. However, we were not able yet to draw definite conclusions regarding their empirical validity. We must note however that for many software engineering artifacts, the impact of size and structural properties on external quality properties has been demonstrated [37]. Examples include the relationship between object class size and defects found [38], the negative effect of coupling on fault-proneness and reusability in object-oriented software [12, 39, 40] and object-based software [41], the impact of the morphology of the inheritance structure on the quality of software [42], and the relationship between the extent of polymorphism in a system and its probability of containing faults [43]. Although the external validity of these empirical studies in the context of conceptual schema quality must still be properly investigated, they do provide an indication of the potential importance and relevancy of many of the complexity aspects for which measures were proposed in this paper.

7   Conclusions

This paper presents a suite of measures to assess size, structure, and dynamic behaviour aspects of the complexity of object-oriented conceptual schemes as seen from a dynamic perspective. This measure suite is based on a formal model of object functionality and behaviour that is obtained using an event-driven approach to conceptual modeling. We related the new complexity measures to existing quality frameworks for conceptual modeling and to existing measure suites for entity relationship schemes and object-oriented (domain) analysis schemes and we showed the complementary nature of our measures.

We also noted the lack of a comprehensive empirical validation study in the area of conceptual schema quality. The work presented in this paper is part of a project investigating the effect of complexity aspects of early system development artifacts on the 'external' quality properties of information systems (e.g., maintainability, reusability). The ultimate goal of this project is to build early quality prediction, evaluation and control models. Our further research therefore includes a number of empirical investigations that will use the measures presented in this paper as part of its measurement instrumentation.

Acknowledgements

Geert Poels is a Postdoctoral Fellow of the Fund for Scientific Research - Flanders (Belgium)(F.W.O.) and wishes to acknowledge the financial support of the F.W.O.

References

1. D'Souza, D.F., Wills, A.C.: Objects, Components, and Frameworks with UML: the Catalysis Approach. Addison-Wesley (1999)
2. Rational Software: Object Oriented Analysis and Design, Student Manual (1998) http://www.rational.com/
3. Boehm, B.W.: Software Engineering Economics. Prentice-Hall (1981)
4. Tegarden, D.P., Sheetz, S.D., Monarchi, D.E.: A Software Complexity Model of Object-Oriented Systems. Decision Support Systems: An Int'l J. 13 (1995) 241-262
5. Fenton, N.: Software Measurement: A Necessary Scientific Base. IEEE Trans. Software Eng. 20 (1994) 199-206
6. Shepperd, M., Ince, D.: Algebraic Validation of Software Metrics. In: Proc. 3rd European Software Eng. Conf. (ESEC'91). Milan (1991) 343-363
7. Fenton, N.E., Pfleeger, S.L.: Software Metrics: A Rigorous & Practical Approach. PWS Publishing Company, London (1997)
8. Assenova, P., Johannesson, P.: Improving Quality in Conceptual Modelling by the Use of Schema Transformations. In: Proc. 15th Int'l Conf. Conceptual Modeling (ER'96). Cottbus, Germany (1996) 277-291
9. Lindland, O.I., Sindre, G., Solvberg, A.: Understanding Quality in Conceptual Modeling. IEEE Software 11 (1994) 42-49
10. Moody, D.L., Shanks, G.G., Darke, P.: Improving the Quality of Entity Relationship Models - Experience in Research and Practice. In: Proc. 17th Int'l Conf. Conceptual Modeling (ER'98). Singapore (1998) 255-276
11. Moody, D.L.: Metrics for Evaluating the Quality of Entity Relationship Models. In: Proc. 17th Int'l Conf. Conceptual Modeling (ER'98). Singapore (1998) 211-225
12. Briand, L.C., Wüst, J., Ikonomovski, S., Lounis, H.: A Comprehensive Investigation of Quality Factors in Object-Oriented Designs: an Industrial Case Study. In: Proc. 21st Int'l Conf. Software Eng. (ICSE'99). Los Angeles (1999) 345-354
13. McBrien, P., Poulovassilis, A.: A Formal Framework for ER Schema Transformation. In: Proc. 16th Int'l Conf. Conceptual Modeling (ER'97). Los Angeles (1997) 408-421
14. Genero, M., Piattini, M., Calero, C.: An Approach to Evaluate the Complexity of Conceptual Database Models. In: Proc. 3rd European Software Measurement Conf. Madrid (2000)
15. Moser, S., Misic, V.B.: Measuring Class Coupling and Cohesion: A Formal Metamodel Approach. In: Proc. Asia Pacific Software Eng. Conf. (APSEC'97). Hong Kong (1997) 31-40
16. Misic, V.B., Moser, S.: Formal Approach to Metamodeling: A Generic Object-Oriented Perspective. In: Proc. 16th Int'l Conf. Conceptual Modeling (ER'97). Los Angeles (1997) 243-256
17. Badri, L., Badri, M., Ferdenache, S.: Towards Quality Control Metrics for Object-Oriented Systems Analysis. In: Proc. 16th Int'l Conf. Technology of Object-Oriented Languages (TOOLS-16). Versailles, France (1995) 193-206
18. Genero, M., Manso, M.E., Piattini, M., Garcia, F.J.: Assessing the Quality and the Complexity of OMT Models. In: Proc. 2nd European Software Measurement Conf. Amsterdam (1999) 99-109
19. Coad, P., Yourdon, E.: Object-Oriented Analysis. Prentice-Hall (1990)
20. Rumbaugh, J., Blaha, M., Premerlani, W., Eddy, F., Lorensen, W.: Object Oriented Modeling and Design. Prentice-Hall (1991)
21. Chidamber, S.R., Kemerer, C.F.: A Metrics Suite for Object Oriented Design. IEEE Trans. Software Eng. 20 (1994) 476-493
22. Brito e Abreu, F., Carapuça, R.: Object-Oriented Software Engineering: Measuring and Controlling the Development Process. In: Proc. 4th Int'l Conf. Software Quality (ICSQ'94). McLean, VA (1994)

23. Briand, L.C., Daly, J.W., Wüst, J.K.: A Unified Framework for Cohesion Measurement in Object-Oriented Systems. Empirical Software Eng., An Int'l J. 3 (1998) 65-117

24. Briand, L.C., Daly, J.W., Wüst, J.K.: A Unified Framework for Coupling Measurement in Object-Oriented Systems. IEEE Trans. Software Eng. 25 (1999) 91-121

25. Snoeck, M.: On a process algebra approach for the construction and analysis of M.E.R.O.DE.-based conceptual models. Ph.D. dissertation. Katholieke Universiteit Leuven (1995)

26. Snoeck, M., Dedene, G.: Existence Dependency: The Key to Semantic Integrity Between Structural and Behavioural Aspects of Object Types. IEEE Trans. Software Eng. 24 (1998) 233-251

27. Booch, G., Rumbaugh, J., Jacobson, I.: The Unified Modeling Language User Guide. Addison-Wesley (1999)

28. Poels, G., Dedene, G.: Measures for Object-Event Interactions. In: Proc. 33rd Int'l Conf. Technology of Object-Oriented Languages and Systems (TOOLS-33). Mont St. Michel, France (2000) 70-81

29. Brito e Abreu, F., Esteves, R., Goulao, M.: The Design of Eiffel Programs: Quantitative Evaluation Using the MOOD Metrics. In: Proc. 20th Int'l Conf. Technology of Object-Oriented Languages (TOOLS-20). Santa Barbara, Calif. (1996)

30. Poels, G.: On the use of a Segmentally Additive Proximity Structure to Measure Object Class Life Cycle Complexity. In: Dumke, R., Abran, A.: Software Measurement: Current Trends in Research and Practice. Deutscher Universitäts Verlag, Wiesbaden, Germany (1999) 61-79

31. Kitchenham, B., Pfleeger, S.L., Fenton, N.: Towards a Framework for Software Measurement Validation. IEEE Trans. Software Eng. 21 (1995) 929-944

32. Zuse, H.: A Framework for Software Measurement. Walter de Gruyter, Berlin (1998)

33. Roberts, F.S.: Measurement Theory with Applications to Decisionmaking, Utility and the Social Sciences. Addison-Wesley (1979)

34. Suppes, P., Krantz, D.M., Luce, R.D., Tversky, A.: Foundations of Measurement: Geometrical, Threshold, and Probabilistic Representations. Academic Press, San Diego, Calif. (1989)

35. Poels, G., Dedene, G.: Distance-based software measurement: necessary and sufficient properties for software measures. Information and Software Technology 42 (2000) 35-46

36. Poels, G., Viaene, S., Dedene, G.: Distance Measures for Information System Reengineering. In: Proc. 12th Int'l Conf. Advanced Information Systems Eng. (CAiSE*00), Stockholm (2000) 387-400

37. Briand, L., Arisholm, E., Counsell, S., Houdek, F., Thévenod-Fosse, P.: Empirical Studies of Object-Oriented Artifacts, Methods, and Processes: State of The Art and Future Directions. Tech. Rep. IESE 037.99/E, Fraunhofer IESE (1999)

38. Benlarbi, S., El Emam, K., Goel, N.: Issues in Validating Object-Oriented Metrics for Early Risk Prediction. In: Proc. 10th Int'l Symposium Software Reliability Eng. (ISSRE'99). Boca Raton, Florida (1999)

39. Basili, V.R., Briand, L., Melo, W.L.: A Validation of Object-Oriented Design Metrics as Quality Indicators. IEEE Trans. Software Eng. 22 (1996) 751-761

40. Briand, L., Daly, J.W., Porter, V., Wüst, J.: A Comprehensive Empirical Validation of Product Measures for Object-Oriented Systems. Tech. Rep. ISERN-98-07, Fraunhofer IESE (1998)

41. Briand, L.C., Morasca, S., Basili, V.R.: Defining and Validating Measures for Object-Based High-Level Design. IEEE Trans. Software Eng. 25 (1999) 722-743

42. Brito e Abreu, F., Melo, W.: Evaluating the Impact of Object-Oriented Design on Quality. In: Proc. 3rd Int'l Software Metrics Symposium (METRICS'96). Berlin (1996)

43. Benlarbi, S., Melo, W.L.: Polymorphism Measures for Early Risk Prediction. In: Proc. 21st Int'l Conf. Software Eng. (ICSE'99). Los Angeles (1999) 334-344

Measuring the Quality of Entity Relationship Diagrams

Marcela Genero [1], Luis Jiménez [2], and Mario Piattini[1]

[1]Grupo ALARCOS
[2]Grupo ORETO
University of Castilla-La Mancha
Ronda de Calatrava, 5
13071, Ciudad Real (Spain)
E-mail: {mgenero, ljimenez, mpiattin}@inf-cr.uclm.es

Abstract. Database quality depends greatly on the accuracy of the requirement specification and the greatest effort should focus on improving the early stages of database life cycle. Conceptual data models form the basis of all later design work and determine what information can be represented by a database. So, its quality has a significant impact on the quality of the database which is ultimately implemented. In this work, we propose a set of metrics for measuring entity relationship diagram complexity, because in today's database design world it is still the dominant method of conceptual modelling. The early availability of metrics allows designers to measure the complexity of entity-relationship diagrams in order to improve database quality from the early stages of their life cycle. Also we carried out a controlled experiment in order to analyse the existent relationships between each of the proposed metrics and each of the maintainability sub-characteristics. In order to analyse the obtained empirical data we propose a novel data analysis technique based on fuzzy regression trees.

1 Introduction

Database quality depends greatly on the accuracy of the requirement specification and the greatest effort should be focus on improving the early stages of database life cycle. In a typical database design a conceptual data model which specifies the requirements about the database is first built. The conceptual data model determines what information can be represented by a database [1]. So, its quality have a significant impact on the quality of the database which is ultimately implemented [2], and an even greater impact if we take into account the size and complexity of current databases. Improving the quality of conceptual data models will therefore be a major step towards the quality improvement of the database development. We will focus on entity-relationship (ER) diagrams because in today's database design world it is the dominant method of conceptual modelling [3].

In practice, evaluation of the quality of conceptual data models takes place in an *ad hoc* manner, if at all. There are no generally accepted guidelines for evaluating the quality of data models, and little agreement even among experts as to what makes a "good" data model [4].

A.H.F. Laender, S.W. Liddle, V.C. Storey (Eds.): ER2000 Conference, LNCS 1920, pp. 513-526, 2000.
© Springer-Verlag Berlin Heidelberg 2000

In general we agree with Krogstie et al. [5] in the sense that "Most literature provides only bread and butter lists of useful properties without giving a systematic structure for evaluating them". Moreover these lists are mostly unstructured, use imprecise definitions, often overlap, and properties of models are often confused with language method properties [6]. In addition to this, these lists are not generally sufficient to ensure quality in practice, because different people will have different interpretations of the same concept. It is necessary to have quantitative and objective measures to reduce subjectivity and bias in the evaluation process.

Recently, some frameworks have been proposed which attempt to address quality in conceptual modelling in a much more systematic way [4-7]. The only papers that propose metrics for conceptual data models are those proposed by Eick [8], Gray et al. [9], Moody [10] and Kesh [11.

Although all of this metric proposal is a good starting point to think about quality in conceptual modelling in a numeric scale, most of them are open-ended, subjective and lack empirical and theoretical validation. Thus, there is a need for metrics and quality models that can be applied in the early stages of database design, and we are particularly concerned with that applied to ER diagrams, to ensure that that designs have favorable internal properties that will lead to the development of quality databases.

Before thinking about how to measure the quality of ER diagrams it is essential to define a quality model. This quality model must describe each of the characteristics that compose the concept of "quality". The ISO 9126 [12] proposed a quality model that can be applied to any artifact produced at any stages of the software development life cycle. To our knowledge, not all of the characteristics proposed in that standard are suitable for ER diagrams. We focus our work on one of the most important quality characteristics, maintainability. Maintainability is influenced by the following sub-characteristics:

- Understandability: the ease with which the conceptual data model can be understood.

- Legibility: is the ease with which the conceptual data model can be read, with respect to certain aesthetic criteria [13].

- Simplicity: means that the conceptual data model contains the minimum number of constructions possible.

- Analysability: the capability of the conceptual data model to be diagnosed for deficiencies or for parts to be modified.

- Modifiability: the capability of the conceptual data model to enable a specified modification to be implemented.

- Stability: the capability of the conceptual data model to avoid unexpected effects from modifications.

- Testability: the capability of the conceptual data model to enable modifications to be validated.

As most of these maintainability sub-characteristics are in turn influenced by complexity [14], our objective in this work is to provide a set of metrics for measuring ER diagram complexity (section 2).

As in other aspects of Software Engineering, proposing techniques and metrics is not enough, it is also necessary to put them under theoretical and empirical validation in order to demonstrate their utility [15-18]

We have made a theoretical validation of some of the proposed metrics following Briand framework [19] in [20] (Briand et al., 1996) and Zuse´s framework [21] in [22]. But, we can conclude that in software measurement there is not an agreement of how to make theoretical validation. Most authors have opposing ideas of which is the best way of making a theoretical validation.

In order to empirically validate the proposed metrics we will show in section 3 how we have performed a controlled experiment. The main objective of this experiment is to ascertain the existent relationships between each of the proposed metrics and each of the maintainability sub-characteristics. In order to analyse the obtained empirical data we propose a novel data analysis technique based on fuzzy regression trees. The results demonstrate what metrics are more relevant taking into account each of the sub-characteristic mentioned above. Finally the last section draws on our conclusions, and presents our future work.

2 Metrics for ER Diagram Complexity

As complexity is a multidimensional attribute it is not advisable to try to combine different aspects of this attribute into a single measurement unless you have a model or theory to support you [23]. Henderson-Sellers [24] distinguishes three types of complexity, among which he quoted "product or structural complexity", which is our focus when we refer to the concept of complexity.

In this section we will define a set of closed-ended metrics [25] to measure ER diagram complexity, taking into account its constituent parts, such as entities, attributes and relationships. These closed-ended metrics are more useful, because they are bounded (in this case in the interval [0,1]), they can be easily visualised by graphics, and their values are easily interpreted.

2.1 RvsE Metric

This metric measures the relation that exists between the number of relationships and the number of entities in an ER diagram.

We define this metric as follows:

$RvsE = \left(\dfrac{N^R}{N^R + N^E} \right)^2$	N^R is the number of relationships in the ER diagram. N^E is the number of entities in the ER diagram. Being $N^R + N^E > 0$.

When we calculate the number of relationships (N^R), we consider the IS_A relationships and the aggregation relationship. In the case, of IS_A relationships we consider one relationship for each child-parent pair. In aggregation relationships we consider one relationship for each part-whole pair. The number of relationships per entity in the ER diagram influences this metric. Intuitively, the greater the number of relationships the greater the complexity. RvsE metric is zero when there are no relationships, it takes the value 1 when there are a lot of relationships and very few entities.

2.2 EAvsE Metric

This metric measures the relations that exist between the number of entity attributes and the number of entities in an ER diagram.

We define this metric as follows:

$$EAvsE = \left(\frac{N^{EA}}{N^{EA} + N^{E}} \right)^{2}$$	N^{EA} is the number of entity attributes within the ER diagram. N^{E} is the number of entities within the ER diagram. Being $N^{EA} + N^{E} > 0$.

When we calculate the number of entity attributes in the ER diagram (N^{EA}) we also consider composite and multivalued attributes (each of them is considered to take the value 1). The number of attributes per entity in the ER diagram influences this metric. Intuitively, the greater the number o attributes the greater their complexity.

EAvsE metric is zero when there are no entity attributes, it takes value 1 when there are a lot of entity attributes and very few entities.

2.3 RAvsR Metric

This metric measures the relations that exist between the number of relationship attributes and the number of relationships in an ER diagram.

We define this metric as follows:

$$RAvsR = \left(\frac{N^{RA}}{N^{RA} + N^{R}} \right)^{2}$$	N^{RA} is the number of relationship attributes within the ER diagram. N^{R} is the number of relationships within the ER diagram. Being $N^{AR} + NR > 0$.

When we calculate the number of relationship attributes in the ER diagram (N^{EA}) we also consider composite and multivalued attributes (each of them is considered to take the value 1).

In this metric when we calculate the number of relationships (N^{R}), we disregard IS_A relationships and aggregation relationships.

The number of attributes per relationship in the ER diagram influences this metric. Intuitively, the greater the number of attributes the greater their complexity.

RAvsR metric is zero when there are no relationship attributes, it takes value 1 when there are a lot of relationship attributes and very few relationships.

2.4 M:NRel Metric

The M:N Relationships metric measures the number of M:N relationships compared with the number of relationships in an ER diagram.

We define this metric as follows:

$M:N\,Rel = \dfrac{N^{M:NR}}{N^{R}}$	$N^{M:NR}$ is the number of M:N relationships within the ER diagram. N^{R} is the number of relationships within the ER diagram. Being $N^{R} > 0$.

In this metric when we calculate the number of relationships (N^R), we disregard IS_A relationships and aggregation relationships.

M:NRel metric scores zero when there are no M:N relationships, it scores 1 when there is a high percentage of M:N relationships.

2.5 1:NRel Metric

The 1:N Relationships metric measures the number of 1:N relationships (also include 1:1 relationships) compared with the total number of relationships in an ER diagram.
We define this metric as follows:

$1:N\,Rel = \dfrac{N^{1:NR}}{N^{R}}$	$N^{1:NR}$ is the number of 1:N relationships in the ER diagram. N^{R} is the number of relationships in the ER diagram. Being $N^{R} > 0$.

In this metric when we calculate the number of relationships (N^R), we disregard IS_A relationships and aggregation relationships.

1:NRel metric scores zero when there are no 1:N relationships, it scores 1 when there is a high percentage of 1:N relationships.

2.6 N-aryRel Metric

The N-ary Relationships metric measures the number of N-ary relationships (not binary) compared with the number of relationships in the ER diagram.

It is convenient for the number of N-ary relationships in an ER diagram to be minimal, because they contribute to increasing its complexity.
We define this metric thus:

$N-ary\,Rel = \dfrac{N^{N-aryR}}{N^{R}}$	N^{N-aryR} is the number of N-ary relationships in the ER diagram. N^{R} is the number of relationships in the ER diagram. Being $N^{R} > 0$.

In this metric when we calculate the number of relationships (N^R), we disregard IS_A relationships and aggregation relationships.

This metric is zero when the ER diagram has no N-ary relationships, it takes the value 1 when all of the relationships are N-ary.

2.7 BinaryRel Metric

The Binary Relationships metric measures the number of Binary relationships compared with the number of relationships in the ER diagram.
We define this metric thus:

$Binary\ Re\ l = \dfrac{N^{BinaryR}}{N^{R}}$	$N^{N\text{-}aryR}$ is the number of Binary relationships in the ER diagram. N^{R} is the number of relationships in the ER diagram. Being $N^{R} > 0$.

In this metric when we calculate the number of relationships (N^{R}), we disregard IS_A relationships and aggregation relationships.
This metric is zero when the ER diagram has no Binary relationships, it takes the value when all of the relationships are binary.

3 Empirical Validation of the Proposed Metrics

We are interested in ascertaining if any relationship exists between each of the proposed metrics and each of the maintainability sub-characteristics: understandability, legibility, simplicity, analysability, modifiability, stability, and testability. With that objective, we have carried out and controlled experiment.

In the remaining of this section, we will present: the experimental design, how we collect the experimental data, the technique used to analyse the empirical data, and the results of the experiment.

3.1 Experimental Design and Data Collection

Sixteen subjects (database designers) participated in the experiment. The subjects were given twenty four ER diagrams taken from different books of database design [26-27]. Each diagram have enclosed a form which includes the description of maintainability sub-characteristics, such as: understandability, legibility, simplicity, analysability, modifiability, stability, and testability. Each subject has to rate each sub-characteristic using a scale consisting of seven linguistic labels. For example for the sub-characteristic understandability we proposed the following linguistic labels:

Extremely difficult to understand	Very difficult to understand	A bit difficult to understand	Neither difficult nor easy to understand	A bit easy to understand	Very easy to understand	Extremely easy to understand

After collecting the data, we assign to each label a number in the following way:
The greater the expert´s rate the greater the difficulty to understand.

Extremely difficult to understand	Very difficult to understand	A bit difficult to understand	Neither difficult nor easy to understand	A bit easy to understand	Very easy to understand	Extremely easy to understand
7	6	5	4	3	2	1

All of the proposed metrics were automatically calculated using our metric tool called MANTICA [28].

At this point we have all the experimental data. The next step is to analyse them, which we will explain in the next subsection.

5.2 Data Analysis Technique and Results

Due to the nature of the software development process and products, one cannot expect to use in Software Engineering the same measurement data analysis techniques that are used in "exact" sciences, e.g., Physics, Chemistry, nor obtain the same degree of precision and accuracy [29]. So that, we will analyse the empirical data with a data analysis technique based on fuzzy regression trees with linguistic variables [30]. This approach provide models that allow us to discover the most relevant conceptual relationships between the data we are analysing, where the accuracy of this models is sacrificed in favour of its simplicity and easiness to understand.

In the next subsection we will describe the theoretical grounds and the methodology for building the regression fuzzy trees.

Induction Method for Building Fuzzy Regression Trees. This induction method is a generalisation of the classical regression approach.

Let $S=\{s_1, \ldots, s^n\}$ be a set of data, which are defined by the value given by a set of variables $X=\{X^1, \ldots, X^d, Y\}$, $s_i=(x^1_i, \ldots, x^d_i, y_i)$. Let's suppose there is a function F which is only known at the points of S so that $F(s_i)=y_i$. The objective of regression, which is defined classically as a parametric function F', is to minimise the distance between the sample output values y_i and the predicted value $F(s_i)$.

The regression problem differs from the classification problem in that the output variable can be continuous, where as in classification this is strictly categorical. From this perspective, classification can be thought to be a subcategory of regression. Recursive partition methods for classification problem such as ID3 decision tree have been applied, as a regression method by restrictions, in the CART program [31], developed in statistical research community.

The program CART (classification and regression tree) is based on the building of tree structure where the regions are defined by possible answers to a set of questions raised about the variables that define the problem. This approach creates a set of disjoint regions $SR=\{r_1, \ldots, r_p\}$ of the global domain of problem. These partitions are obtained according to the kind of questions and answers that we have formulated. Our method generalize the obtained kind of partitions, using the method proposed by Linares et al. [32] based on the fuzzy set theory [33]. We use Linares et al.´s idea [32] in order to built a fuzzy regression tree (FCART).

Hereafter, we will explain how to obtain the fuzzy regression tree. Let A_T be a fuzzy set defined over the set S of the node T, such as $A_T : S \rightarrow [0,1]$. In the root node this fuzzy is $A_{root} : S \rightarrow 1$. We define as output value at node T by the membership value of each point s_i, $AT (s_i)$ and the output value in this point yi.

$$F''(T) = \frac{\sum_{i=1}^{n} A_T (s_i)^m * y_i}{\sum_{i=1}^{n} A_T (s_i)^m}$$

The error estimated is defined by

$$Err(T) = \frac{\sum_{i=1}^{n} (F''(T) - y_i)^2 * A_T (s_i)^m}{\sum_{i=1}^{n} A_T (s_i)^m}$$

Now, our problem is how to create the set of questions for the node T. The questions were formulated for each variable, obtaining a binary fuzzy partition for every one. We supposed a binary fuzzy partition for the fuzzy set of node T by the fuzzy set A^j_T of variable j, this is $Q^i_T = \{B(x), C(x)\}$. This partition originates two new nodes and two new fuzzy sets for them

$$A_{T_1} (s_i) = min(A_T (s_i), B(x_i^j))$$
$$A_{T_2} (s_i) = min(A_T (s_i), C(x_i^j))$$

IF variable^j IS A_{T_1} THEN .....

ELSE IF variable^j IS A_{T_2} THEN ..

Following the CART program, we defined the proportion for B(x) and C(x) in relation to fuzzy set A^j_T as

$$P(T_1) = \frac{\sum_{i=1}^{n} B(x_i^j)}{\sum_{i=1}^{n} A_T^j (x_i^j)}$$

The quality of the partition can be estimated through the equation

$$C(T, p^j) = Err(T_1) * P(T_1) + Err(T_2) * P(T_2)$$

where p^j is the fuzzy partition of variable j.

We will select the p^* which minimises the value of C(T, p^j). This approach to create the questions originates hierarchical fuzzy partition for each variable. The partition process stops when a stop criterion is raised. In this case we look for larger estimated error. So that stop criterion can be

$$ERROR = \max_{T \in \bar{T}} Err(T) \leq \varepsilon$$

Where $\bar{T}$ is the set of leaf nodes of the tree, and each one represent a region for our solution. The output value y_i' for a input value s_i is

$$F'(s) = \frac{\sum_{T \in \bar{T}} A_T(s)^m * F''(T)}{\sum_{T \in \bar{T}} A_T(s)^m}$$

3.3 Experiment Results

Due to the sake of brevity in this paper we used the induction method shown in the previous subsection, to show only the existent relationships between our metrics and the maintainability sub-characteristic, understandability. For the other sub-characteristics the process is similar.

We establish a fuzzy regression tree for the maintainability sub-characteristic: understandability. We use a learning set with X={set of our metrics} and Y={the values of understandability obtained in our experiment}. We have obtained 384 values (24 ER diagrams and 16 subjects) of this unknown function:

F(RvsE; EAvsE; RavsR; M:Nrel; 1:Nrel; N-aryRel; BinaryRel) = Understandability.

The obtained fuzzy sets, which define the fuzzy regression tree (see fig. 1), are labelled by a set of linguistic labels. This labelling process allows us to abstract the numeric values for building a conceptual linguistic model which is highly qualitative and more closed to human minds.

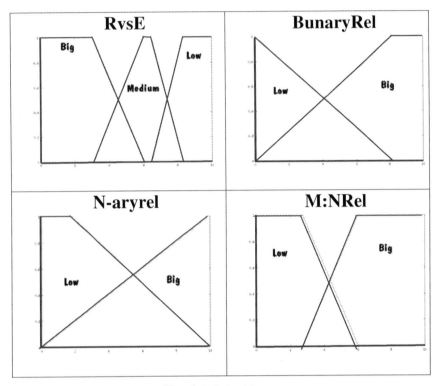

Fig. 1. Labels of fuzzy sets

We only show in fig. 1 the fuzzy representation of the values of RvsE, BinaryRel, N-aryRel and M:NRel metrics because after applying the induction method we have found that the others are not relevant for the understandability.

The obtained fuzzy regression tree is shown in fig. 2 by a nested structure if-then-else.

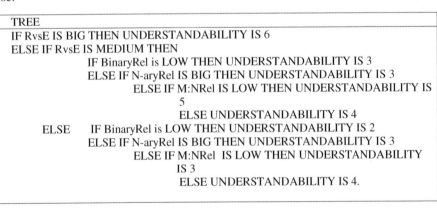

Fig. 2. A Fuzzy Regression Tree

As we can see, when RvsE is "big" then understandability is 6 (the model is "very difficult to understand"), when RvsE is "medium" then understandability value ranges between 3 and 5 (the model is "a bit easy to understand" – the model is a "bit difficult to understand"), and when RvsE is "low" then understandability value ranges between 2 and 4 (the model is "a very easy to understand" – the model is a "neither difficult nor easy to understand"). This result show that understandability is very close at RvsE value and that Rbin, Rter and MNRel show more local value for this characteristic. Good values of understandability are assigned to low value of RvsE and in opposition bad values are assigned to big values of RvsE.

4 Conclusions

Due to the growing complexity of actual databases, continuous attention to and assessment of the conceptual data models is necessary to produce quality databases. Following this idea, we have presented a set of objective and automatically computed metrics for evaluating ER diagram complexity.

We want to highlight that our proposal cannot be considered as a final proposal. Instead, it is only a starting point and we require feedback to improve it. However, due to the lack of objective metrics for measuring conceptual model quality, it serves the purpose of getting database designers to think about the quality of their conceptual data models in numeric terms.

We have presented an induction method for building fuzzy regression trees that allow us to build models for ascertaining the existent relationships between the proposed metrics and the ER diagram maintainability sub-characteristics. So, for example, the metric RvsE could be used as an understandability indicator. This induction method is highly qualitative and very closed to human mind, and above all very ease to understand. Moreover, the models built through this induction technique can also be used as a prediction model by using approximate reasoning [34].

Although the results obtained in this experiment are encouraging, we are aware that it is necessary to undertake more experiments, and also more controlled ones when different values of all the metrics are considered in different conceptual models. Also "real" case studies taken from enterprises must be carried out, with the objective of assessing these metrics as predictors of maintenance efforts, and therefore, determining whether they can be used as early quality indicators.

We cannot disregard the increasing diffusion of the object-oriented paradigm in conceptual modelling. We think that object oriented models are more appropriate than ER diagrams to describe the kind of information systems built nowadays. We have also been working on metrics for measuring OMT [35] class diagrams [36] and also UML [37] class diagrams [38].

In our knowledge, few works have been done towards measuring models that capture the dynamic aspects of an OO software systems [39-40]. As is quoted in [41] this is an area which need further investigation. So that, as a future work we will define metrics for UML dynamic diagrams, such as state diagrams or activity diagrams.

Furthermore, we will not only address the sub-characteristics of maintainability, we also have to focus our research towards measuring other quality factors as proposed in the ISO 9126[12]. Also we will focus on dynamic aspects of object modelling.

Acknowledgements

This research is part of the MANTICA project, partially supported by CICYT and the European Union (CICYT-1FD97-0168).

References

1. Feng, J. The "Information Content" problem of a conceptual data schema and a possible solution. Proceedings of the 4th UKAIS Conference: Information Systems-The Next Generation, University of York, (1999) 257-266
2. Shanks, G. and Darke, P. Quality in Conceptual Modelling: Linking Theory and Practice. Proc. of the Pacific Asia Conference on Information Systems (PACIS'97), Brisbane, (1997) 805-814
3. Muller, R. Database Design For Smarties: Using UML for Data Modeling. Morgan Kaufman, (1999)
4. Moody, D., Shanks, G. and Darke, P. Improving the Quality of Entity Relationship Models – Experience in Research and Practice. Proceedings of the Seventeenth International Conference on Conceptual Modelling (ER '98), Singapore, (1998) 255-276
5. Krogstie, J., Lindland, O. and Sindre, G. Towards a Deeper Understanding of Quality in Requirements Engineering, Proceedings of the 7th International Conference on Advanced Information Systems Engineering (CAISE), Jyvaskyla, Finland, June, (1995) 82-95
6. Lindland, O., Sindre, G. and Solvberg, A. Understanding Quality in Conceptual Modelling. IEEE Software, March, Vol. 11 N° 2, (1994) 42-49
7. Schuette, R. and Rotthowe, T. The Guidelines of Modeling – An Approach to Enhance the Quality in Information Models. Proceedings of the Seventeenth International Conference on Conceptual Modelling (ER '98), Singapore, November 16-19, (1998) 40-254
8. Eick, C. A Methodology for the Design and Transformation of Conceptual Schemas. Proc. of the 17th International Conference on Very Large Data Bases. Barcelona (1991)
9. Gray, R., Carey, B., McGlynn, N. and Pengelly A. Design metrics for database systems. BT Technology, Vol. 9 N° 4, (1991) 69-79
10. Moody, D. Metrics for Evaluating the Quality of Entity Relationship Models. Proceedings of the Seventeenth International Conference on Conceptual Modelling (ER '98), Singapore, November 16-19, (1998) 213-225
11. Kesh, S. Evaluating the Quality of Entity Relationship Models. Information and Software Technology, Vol. 37 N° 12, (1995) 681-689.
12. ISO/IEC 9126-1. Information technology- Software product quality – Part 1: Quality model. (1999)
13. Batini, C., Ceri, S. and Navathe, S. *Conceptual database design. An entity relationship approach.* Benjamin Cummings Publishing Company. (1992)
14. Li, H. and Cheng, W. An empirical study of software metrics. IEEE Transactions on Software Engineering, Vol. 13 N° 6, (1987) 679-708

15. Fenton, N. and Pfleeger, S. Software Metrics: A Rigorous Approach. 2nd. edition. Chapman & Hall, London (1997)
16. Kitchenham, B., Pflegger, S. and Fenton, N. Towards a Framework for Software Measurement Validation. IEEE Transactions of Software Engineering, Vol. 21 N° 12, (1995) 929-943
17. Schneidewind, N. Methodology For Validating Software Metrics. IEEE Transactions of Software Engineering, Vol. 18 N° 5, (1992) 410-422
18. Basili, V., Shull, F. and Lanubile, F. Building knowledge through families of experiments. IEEE Transactions on Software Engineering, Vol. 25 N° 4, (1999) 435-437
19. Briand, L., Morasca, S. and Basili, V.. Property-Based Software Engineering Measurement. IEEE Transactions on Software Engineering, Vol. 22 N° 6, (1996) 68-86
20. Genero, M., Piattini, M. and Calero, C. (2000). Formalization of Metrics for Conceptual Data Models. UKAIS 2000. Cardiff, 26-28 April, (2000) 99-119
21. Zuse, H. A Framework of Software Measurement. Walter de Gruyter, Berlin (1998)
22. Genero, M., Piattini, M., Calero, C. Serrano, M. (2000). Measures to get better quality databases. ICEIS 2000. Stafford, 4-7 July, (2000) 49-55
23. Fenton, N. Software Measurement: A Necessary Scientific Basis. IEEE Transactions on Software Engineering, Vol. 20 N° 3, (1994) 199-206
24. Henderson-Sellers, B. Object-oriented Metrics - Measures of complexity. Prentice-Hall, Upper Saddle River, New Jersey. (1996)
25. Lethbridge, T. Metrics For Concept-Oriented Knowledge bases. International Journal of Software Engineering and Knowledge Engineering Vol. 8 N° 2, (1998) 161-188
26. Ruiz, I. and Gómez-Nieto, M. Diseño y uso de Bases de Datos Relacionales. Ra-Ma, (1997) (in spanish)
27. De Miguel, A. and Piattini, M. Fundamentos y Modelos de Bases de Datos. Ra-Ma. (1997) (in spanish)
28. Calero, C., Pascual, C., Serrano, M. and Piattini, M. Measuring Oracle Database Schema. Computers and Computational Engineering in Control, (Cap. 42), World Scientific Engineering Society, (1999) 237-243
29. Morasca, S. and Ruhe, G. Guest Editors´Introduction: Knowledge Discovery From Empirical Software Engineering Data. International Journal of Software Engineering and Knowledge Engineering, Vol. 9 N° 5, (1999) 495-498
30. Zadeh, L. The Concept of Linguistic Variable and its Applications to Approximate Reasoning Part I. Information Sciences, Vol. 8, (1973) 199-249.
31. Breiman, L., Friedman, J., Olshen, R. and Stone, C.. Classification and Regression Trees. Wadsworth, Belmont, CA, (1984)
32. Linares, L., Delgado, M. and Skarmeta, A. Regression by fuzzy knowledge bases. Proceedings of the 4th European Congress on Intelligent Techniques and soft computing. Aachen, Germany, September, (1996) 1170-1176
33. Zadeh, L. Fuzzy sets. Information and control, (1965) 338-353.
34. Sugeno, M. An Introductory Survey of Fuzzy Control. Information Sciences, Vol. 36, (1985) 59-83.
35. Rumbaugh, J., Blaha M., Premerlani, W., Eddy, F., and Lorensen, W. Object-Oriented Modeling and Design. Prentice Hall, USA, (1991)

36. Genero, M., Manso, Mª E., Piattini, M. and García, F. Assessing the Quality and the Complexity of OMT Models. 2nd European Software Measurement Conference - FESMA 99, Amsterdam, The Netherlands, (1999) 99-109
37. Booch, G., Rumbaugh, J. and Jacobson, I. The Unified Modeling Language User Guide. Addison-Wesley, (1998)
38. Genero, M., Piattini, M. and Calero, C. Métricas para Jerarquías de Agregación en diagramas de clases UML. Memorias del Jornadas Iberoamericanas de Ingeniería de Requisitos y ambientes de Software, IDEAS´2000, Cancún, México, 5-7 Abril, (2000) 373-384 (in spanish)
39. Poels, G. On the use of a Segmentally Additive Proximity Structure to Measure Object Class Life Cycle Complexity. Software Measurement: Current Trends in Research and Practice. Deutscher Universitäts Verlag, (1999) 61-79
40. Poels, G. On the Measurement of Event-Based Object-Oriented Conceptual Models. 4th International ECOOP Workshop on Quantitative Approaches in Object-Oriented Software Engineering, June 13, Cannes, France. (2000)
41. Brito e Abreu, F., Zuse, H., Sahraoui, H. and Melo, W. Quantitative Approaches in Object-Oriented Software Engineering. Object-Oriented technology: ECOOP´99 Workshop Reader, Lecture Notes in Computer Science 1743, Springer-Verlag, (1999) 326-337.

Behavior Consistent Inheritance in UML

Markus Stumptner[1] and Michael Schrefl[2]*

[1] Institut für Informationssysteme, Technische Universität Wien,
mst@dbai.tuwien.ac.at
[2] School of Computer and Information Science, University of South Australia,
cismis@cs.unisa.edu.au

Abstract

Object-oriented design methods express the behavior an object exhibits over time, i.e., the object life cycle, by notations based on Petri nets or state charts. The paper considers the specialization of life cycles via inheritance relationships as a combination of extension and refinement, viewed in the context of UML state machines. Extension corresponds to the addition of states and actions, refinement refers to the decomposition of states into substates. We use the notions of observation consistency and invocation consistency to compare the behavior of object life cycles and present a set of rules to check for behavior consistency of UML state machines, based on a one-to-one mapping of a meaningful subset of state machines to Object/Behavior Diagrams.

1 Introduction

Object-oriented design methodologies such as OMT [15], OOSA [4], OOAD [2], OBD [6], and UML [16] differ from programming languages in that they represent the behavior of an object type not merely by a set of operations, which may be performed on instances of the object type, but instead provide a higher-level, overall picture on how instances of the object type may evolve over their lifetime, e.g., a hotel or car reservation, which must be requested before it can be issued and can be used only after being issued. Such a description is often referred to as "object life cycle". In this paper we consider a special variant of state charts, the state chart diagrams of the Unified Modeling language (UML), which has recently emerged as a prominent commercial object-oriented design notation. The behavior of an UML object class is represented by a state chart, which consists of states and state transitions. When mapped to OODB designs, transitions are mapped to operations and states to pre- and postconditions of these operations.

Object-oriented systems organize object types in hierarchies in which subtypes inherit and specialize the structure and the behavior of supertypes. These inheritance hierarchies provide a major aid to the designer in structuring the description of an object-oriented system, and they guide the reader who tries to understand the system by pointing out similarities between object types that are so connected. Informally, specialization means for object life cycles that the

---

* This work was partially supported by the Austrian Science Fund project N Z29-INF.

A.H.F. Laender, S.W. Liddle, V.C. Storey (Eds.): ER2000 Conference, LNCS 1920, pp. 527–542, 2000.

object life cycle of a subtype should be a "special case" of the object life cycle of the supertype. There are two ways in which an object life cycle may be made more special. One way is to add new features, which we call *extension*. For example, a "reservation with payment" extends a "reservation" in that it provides for additional features relevant for payment such as billing, paying, and refunding. The other way is to consider inherited features in more detail, which we call *refinement*. For example, a "reservation with alternative payment" refines a "reservation with payment" in that it provides for special means to pay, such as by cash, by cheque, or by credit card.

Again, extension and refinement should not be employed arbitrarily but according to certain consistency criteria in order to increase understandability and usability. Ebert and Engels [3] pointed out that object life cycles can be compared based upon what a user observes (*observation consistency*) and based upon which actions associated with transitions a user may invoke on an object (*invocation consistency*).

Informally, observation consistent specialization guarantees that if features added at a subtype are ignored and features refined at a subtype are considered unrefined, any processing of an object of the subtype can be *observed* as correct processing from the point of view of the supertype. In our example of "reservation with payment", observation consistency is satisfied if the processing of reservations with payment appears (can be observed) as a correct processing of reservations when all features relevant to payment are ignored.

Weak invocation consistency captures the idea that instances of a subtype can be used the same way as instances of the supertype. For example, if one extends a television set by a video text component, one would usually expect that the existing controls of the television set should continue to operate in the same way. An extended property, strong invocation consistency, guarantees that one can continue to use instances of a subtype the same way as instances of a supertype, even after operations (or activities) that have been added at the subtype have been executed. In our television set example, to obey strong invocation consistency means that invoking any video text function should still leave the volume control operative.

In this paper we discuss inheritance of object life cycles in the realm of UML, based on earlier results [19, 20] on specialization of Object/Behavior Diagrams (OBD's) to UML. In our earlier work, this was based on a rigorous formal description of the semantics of behavior diagrams. However, the absence of a formal definition of UML statecharts makes such a direct approach to modeling of UML behavior consistency impractical, however, we can circumvent this problem by mapping UML statecharts to OBD diagrams for which we have in the past provided just an analysis. In the past we have already treated the inheritance of object life cycles in UML in [21] in an overview manner. In this paper we provide a presentation of the mapping from UML statecharts to OBD's as a more detailed justification of the rules presented in [21].

Note beforehand that "refinement" is used in the UML documentation as a synonym for "inheritance relationship". We are of the opinion that the term "refinement" should be reserved for the case where existing features are described in more detail. It should not also be used to cover the case of newly added features, for which we use the term "extension".

2   UML Behavior Modeling from an OBD Perspective

Object/Behavior diagrams (OBD's) have been originally developed for the design of object-oriented databases [6, 18, 1]. The structure and the behavior of instances of an object type are represented by an object diagram and a corresponding behavior diagram, respectively, of which only behavior diagrams are relevant to us here. A behavior diagram of an object type represents the possible life cycles of its instances by activities, states, and arcs corresponding to transitions, places, and arcs of Petri nets. This approach combines the simple formal semantics and well-understood properties of Petri nets with the clarity and intuitive understanding of an object-centered representation. When mapped to program code, activities are mapped to operations and states to the pre- and postconditions of these operations. For reference purposes, we show an example OBD for object type RESERVATION in Figure 1. The required formal definitions of behavior diagrams as well as results gained in the OBD context can be found in [19, 20].

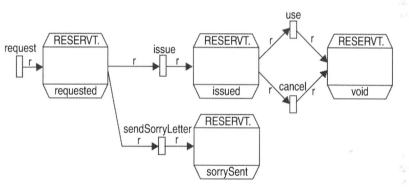

Fig. 1. Behavior diagram of object type RESERVATION

In UML, the dynamic behavior of a system is modeled by Sequence diagrams, Collaboration diagrams, and *Statechart diagrams*, which are graphs representing UML state machines. The latter show "the sequences of states that an object or an interaction goes through during its life in response to received stimuli" ([13], p.103). This diagram type corresponds most closely to behavior diagrams. It is derived from Harel's statechart formalism [5], with a number of alternations and restrictions. The basic concept is that an object's potential life cycle is represented by a set of different states in which the object may find itself. Actions may be associated with transitions, entry to or exit from a state. Each UML statechart has an initial state and one or several final states.

In object-oriented design, a primary use of UML statecharts is to specify in which order operations, represented by transitions, may be invoked on instances of the object class. Transitions are triggered by call events and their actions invoke the operation at that point [14]. Differences from standard statecharts include the property that events are typically considered to correspond directly to operations defined on classes, which means they can carry parameters and model the behavior of specific objects. No processes (activities with temporal extent) are considered to exist. States can also have an activity associated and a transition without an associated event exiting such a state will fire if the activity is finished. For the sake of brevity we exclude completion transitions and assume that an event is associated with every transition.

For space reasons, we have made a number of simplifying assumptions in the UML notation. [1]

Transitions in UML are assumed not to occur in zero time, but the purpose is primarily to represent the execution time of the code and the timespan is considered irrelevant for modeling purposes. Parts of the execution that take time must be represented in UML by a state, a transition entering this state, and a transition exiting from this state.

Since transitions are considered instantaneous, we can say that every instance of an object class is at any point in time in one or several states of its state machine. The set of states an object is situated in at a given moment is jointly referred to as *life cycle state*. As given by the UML semantics definition, a transition will be performed if its event occurs and its guard condition is satisfied. We make a number of simplifying assumptions, as we are interested in statechart diagrams as a high level design notation for database-oriented information systems. The intent is to provide guidance to a designer early in the development cycle. As a result, we are not, at the moment, interested in specifications of object behavior at the level of implemented methods. We generally consider events to be call events, i.e., method invocations. However, we do not examine the effects of such calls, i.e., their implementation, in detail and consider it only insofar as the effects result in state changes. We do not analyze the semantics of individual actions, activities, and guard conditions. (However, relative strictness of guard conditions is important and will be considered later.) For ease of presentation we also assume that at most one transition exists between each pair of states.

Based on the above considerations, we provide a simplified definition of UML state machines.

Definition 1 (UML State Machine).

A UML State Machine (USM) $U_O = (S_O, T_O, L_O, \alpha_O, \Omega_O)$ *of a class O (the subscripts are omitted if O is understood) consists of a set of states $S \neq \emptyset$ and a set of transitions T of the form $l : (S_1, S_2)$ where $S_1, S_2 \subseteq S$, and $l \in L_O$ with L_O being the set of event labels. The labels are of the form e (where e is an event name) if e is associated with one transaction, or of the form e_i (i.e., with a unique index is added), if the event name e occurs multiple times in U_O. There is a distinguished state in S, the initial state α, where for no $S' \in S : l : (S', \{\alpha\}) \in T$; and there is a set of distinguished states of S, the set of final states Ω, where for no $S \subseteq \Omega$ and no $S' \subseteq S$ it holds that $l : (S, S') \in T$ for any l. In addition, every transition t can have associated with it a guard condition (g(t)).*

For a transition $l : (S_1, S_2)$, we call S_1 the set of *source states* of the transition, and S_2 the set of *sink states*.

---

[1] The reader familiar with UML will know that UML provides special forms of states and transitions, such as history states, stubbed transitions, and branch transitions. We do not consider these advanced concepts in this paper.

Since we do not consider action, guard, and activity semantics, the situation of an object at a given point in its lifetime is described by the set of states it occupies in the USM. (The USM must contain substates if that set is not a singleton.) We refer to this situation as the life cycle state of the object.

Definition 2 (Life cycle state). *A life cycle state (LCS) σ of an object class O is a subset of $S \cup T$. We denote the* initial *LCS $\{\alpha\}$ by A.*

Definition 3 (Firing transitions). *A transition $t = l : (S, S') \in T$ can be performed on a life cycle state σ of an USM U_O, if the source states of t are contained in σ, the event e occurs (where $l = e$ if e is unique in U_O and $l = e_i$ if there are multiple occurrences of e), and the guard condition $g(t)$ evaluates to True. Performing t on LCS σ yields the life cycle state $\sigma' = \sigma \setminus S \cup S'$ via an interim (transition) state $\sigma' = (\sigma \setminus S) \cup \{l\}$.*

Note that the guard conditions have to be specified separately as they express conditions on object states (relationships, attribute values) that are outside the scope of the state definitions of the statechart. We also impose the restriction that no two operations with the same event can be used as labels on two transitions that could be possibly fired on some LCS reachable from the initial LCS.

Example 1. Figure 2 shows the statechart diagram of object class RESERVATION. Note that the event archive occurs in multiple event labels and is therefore assigned a unique index in each case. This represents the polymorphic nature of events (which are assumed to correspond to operation calls as mentioned above): the same event, happening in different life cycle states, can imply the execution of different actual actions.   □

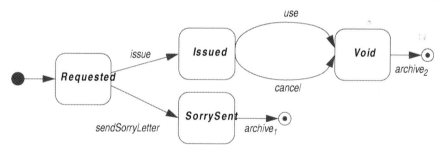

Fig. 2. UML statechart diagram of object class RESERVATION

A statechart diagram of an object class specifies all legal sequences of life cycle states. A particular sequence of life cycle states of an object class is referred to as *life cycle occurrence* of that object class.

Definition 4 (Life cycle occurrence (LCO)). *A condensed life cycle occurrence (cLCO) γ of object class O is a sequence of life cycle states $\sigma_1, \ldots, \sigma_n$, such that $\sigma_1 = A$, and for $i = 1 \ldots n - 1$ either $\sigma_i = \sigma_{i+1}$, or there exists a transition $t \in T$ such that t can be performed on σ_i and yields σ_{i+1}. A cLCO can be expanded to an* uncondensed LCO (uLCO) *by adding the interim transition state between any two consecutive states of the cLCO. Any subsequence of a LCO γ is called* partial LCO. *A LCO γ is called* complete, *if $\sigma_n \subseteq \Omega$.*

532   M. Stumptner and M. Schrefl

Example 2. A possible cLCO of object class RESERVATION is [{α}, {requested}, {issued}, {void}] (cf. Figure 2). □

Definition 5 (Reachable life cycle states). *The set of non-interim life cycle states reachable from LCS σ, written R(σ), contains every non-interim LCS σ' of O that can be reached from σ by any sequence of performed transitions.*

Example 3. Given the statechart diagram of object class RESERVATION, we g ∉ R({requested}) = {{issued}, {sorrySent}, {void}} (cf. Figure 2). □
An alternate approach to describing a life cycle occurrence γ is to denote the sequence of transitions that cause γ rather than to denote the life cycle states of a life cycle occurrence.

Definition 6 (Activation sequence). *An activation sequence μ of object class O is a sequence of event labels $\tau_1, \ldots \tau_n$ ($n \geq 0$), where $\tau_i \in L$; μ is valid on some LCS $\sigma \in R(A)$ if there is some partial cLCO $\gamma = \sigma_1 \ldots \sigma_{n+1}$ of O where for $i \in \{1 \ldots n\}$ σ_{i+1} results from performing the transition[2] $\tau_i : (S, S') \in T$ on σ_i and we say μ yields the trace γ. If $n = 0$, the activation sequence is called empty. An empty activation sequence can be applied on every LCS σ and yields trace σ.*

Example 4. A possible activation sequence for object class RESERVATION is [request, issue, use, archive₂]. □
    Note again that these activation sequences are actually operation sequences, not event sequences (by being uniquely indexed, the operations express that the same event can result in different actions being taken depending on the state of the object before the event). This very nice effect (when comparing LCO's we are comparing the actual invocation sequence of actions, not events) is the reason for the labeling restriction imposed after Definition 3.

2.1   Life Cycles and Substates

A statechart diagram can be ordered hierarchically by substates: states nested within other, so-called composite states. A composite state can be either *sequential* or *concurrent*. The former contains a separate diagram (e.g., the composite state Canceled in Figure 5). This diagram can either be entered by a transition having the parent state as its target, in which case the subdiagram needs to possess an explicit initial state, or by having individual substates explicitly given as target states for transitions entering the composite state. In the latter case (concurrent composite state), the composite state is divided into separate concurrent regions, which are entered in parallel through so-called *fork* transitions, and left in parallel through *join* transitions (see Figure 4).

    All concurrent subregions of a composite state must always be left and entered jointly. This condition is ensured in UML in that any transition that exits from a state in a concurrent subregion of a composite state, by default exits all other concurrent subregions, too.

---

[2] Remember that a transition is uniquely identified by its label.

Transitions may not only have simple states as sinks and sources, but also composite states. A transition to the boundary of a composite state is a shorthand notation for having the same transition pointing to the initial pseudostate(s) of the composite state. Transitions emerging from a composite state may be named or unnamed. An unnamed transition from a composite state is triggered when the final pseudo-state within the composite state is reached. Triggering a named transition emerging from a composite state causes the exit from all substates of the composite state. For simplicity, we do not consider short-hand notations herein and assume that all statecharts have nested state diagrams without initial and final pseudo-states.

2.2 Simplifying Statecharts

For the purpose of checking behavior consistency, since composite states have no intrinsic semantics, we transform UML statecharts into a *canonical form* such that they contain only transitions between simple states and such that concurrent regions are always explicitly exited. Note that this is an internal transformation making the comparison easier, it does not restrict the designer from drawing diagrams in any desired fashion.

We apply the following steps iteratively until no transition that can be transformed is left:

1. For any transition that exits from a state in one concurrent subregion of a composite state, but not from another subregion of that composite state: Add the concurrent substate representing this subregion as source state to that transition (if necessary convert the transition from simple to complex).[3]
2. For any transition that emerges from the boundary of a composite state: If the composite state has n disjoint substates, replace the transition by n transitions, each consuming from a different substate. If the composite state has concurrent substates, replace the transition by a complex transition consuming from each substate.

Definition 7 (Substates). *For a USM $U_O = (S_O, T_O, L_O, \alpha_O, \Omega_O)$ we define a relation sub such that for $s_1, s_2 \in S$, $sub(s_1, s_2)$ holds iff s_1 is a substate of the composite state s_2. We write $cp(s_2)$ to designate s_2 as a composite state, i.e., $cp(s_2) \Leftrightarrow \exists s_1 \in S : sub(s_1, s_2)$.*

Example 5. In the statechart diagram for object class RESERVATION_WITH_-PAYMENT shown in Figure 5, $cp(\mathsf{Issued})$ and $sub(\mathsf{Billed}_1, \mathsf{Issued})$ hold.

Since an object resides in a composite state if and only if it resides in one of the composite state's substates, composite states can be ignored for comparing behavior and are therefore not included in life cycle states.

Finally, we define a homomorphism from USM's to OBD's.

Definition 8 (U2O Mapping).

A UML to OBD Mapping (U2O Mapping) u is a one-to-one function that maps an USM $U = (S, T, L, \alpha, \Omega)$ to an OBD $B = (S', T', F, \alpha, \Omega)$, where

---

[3] Usually in UML, no arcs are drawn from the boundary of a concurrent substate of a composite state. But notice that such an arc is only temporary and will be replaced in further iteration steps.

1. $S' = \{s \in S \mid \neg cp(s)\}$ *(i.e., composite states are ignored).*
2. $T' = \{l \mid \exists l : (S, S') \in T\}$.
3. $(s,t) \in F$ *and* $(t, s') \in F$ *iff* $t : (S, S') \in T$ *s.t.* $s \in S$ *and* $s' \in S'$.

The mapping is extended to LCS's in the obvious manner, e.g., $u(\sigma) = \{u(s) \mid s \in \sigma\}$, and likewise for uLCO's. For an activation sequence $\mu = \tau_1, \ldots \tau_n$, $i = 1, \ldots, n$, $u(\mu)$ is defined as the sequence $[s(\tau_1), c(\tau_1) \ldots s(\tau_n), c(\tau_n)]$ (cf. [19]).

In other words, each USM activation sequence is mapped to the corresponding OBD activation sequence under consideration of the fact that (1) if state s of an OBD B is refined in B', then s itself is not retained in B', (2) multiple USM transitions are not active in parallel, and (3) guards and events are ignored. Due to the restrictions on substates of USM's the result is always a valid OBD.

Notice that, whereas the mapping between the static structure of USM's and OBD's is bijective, the mapping of possible activation sequences of a USM to a corresponding OBD is not surjective. Every USM can be simulated by a corresponding OBD, and an OBD in which execution is restricted in that a started activity must be completed before another one is started can be simulated by a corresponding USM.

3   Consistent Extension

For UML statechart diagrams, "extension" means adding transitions and states. Simple transitions may become complex transitions and complex transitions may receive additional sink and source states.

Example 6. Figure 3 shows a statechart diagram for object class FRIENDLY_RES-ERVATION. This statechart diagram extends the statechart diagram of object class RESERVATION shown in Figure 2.

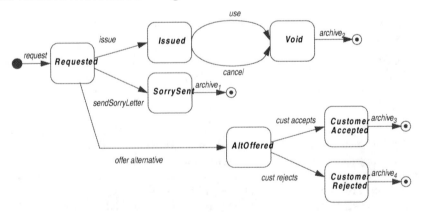

Fig. 3. UML state diagram of object class FRIENDLY_RESERVATION

Adding a "parallel path" is easily achieved in UML if the additional "parallel path" starts in all its alternatives at transitions having the same sink state and, likewise, ends in all its alternatives at transitions having a common source state. Since UML concurrency requirements trivially imply that an object always leaves and enters all concurrent subregions of a composite state simultaneously, other extensions may not be expressible by adding states and transitions to an existing statechart diagram unless behavior consistency is sacrificed. The semantics of the intended extension must then be alternatively expressed by refinement, i.e., the decomposition of states into so-called substates that provide a more detailed description of the inner workings of that state. This solution may also include the need to "duplicate" transitions and states, as shown in the following example.

Example 7. Assume we wish to extend the statechart diagram of object class RESERVATION (cf. Figure 2) by adding an additional payment process that is either ended by using the reservation or by refunding the payment if the reservation has been canceled before. Unfortunately we find that the diagram cannot be extended to cover this functionality by simply adding transitions and states. Instead, the statechart diagram for object class RESERVATION_WITH_PAYMENT (cf. Figure 5) is obtained from the statechart diagram for object class RESERVATION by using refinement rather than extension, whereby different transitions from state issued to void for canceling a reservation apply, depending on whether the reservation has still to be billed, has already been billed, or has already been paid. Notice that the statechart diagram for object class RESERVATION_WITH_PAYMENT shown in Figure 4 is no observation consistent extension of the statechart diagram of object class RESERVATION (see next subsection). The same effect could be achieved with extension (and without duplication of states or transitions) in OBD's (cf. [20]). The statechart diagram shown in Figure 5, which uses refinement, is an observation consistent specialization (see below for the definition of specialization) of the statechart diagram shown in Figure 2.

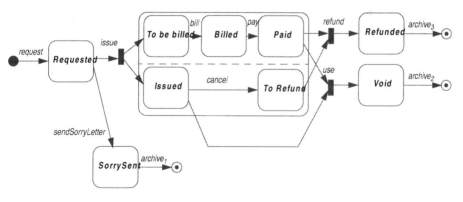

Fig. 4. UML state diagram of object class RESERVATION_WITH_PAYMENT

3.1   Kinds of Behavior Consistency

Intuitively, one expects the statechart of a subclass to be "consistent" with the statechart of a superclass. In this section, we define precisely what "consistent" means with respect to extension of UML statecharts. Similar to inheritance of operations, for which several classes of conformance have been defined, several possibilities exist to relate the statechart diagram of a subclass to the statechart diagram of a superclass.

Following the example from OBD's [19, 20], the idea naturally emerges to use the addition of states and transitions in order to define the notion of "consistent extension" of state machines.

As an example, consider the statechart diagram FRIENDLY_RESERVATION which is based on the diagram for RESERVATION (cf. Figures 3 and 2).

Based on our earlier research on Petri net-based behavior diagrams, we found two approaches are common for comparing the behavior of two Petri nets (cf. [12]): (1) Abstracting from actions, one can compare the possible sequences of sets of states in which tokens reside and (2) abstracting from states, one can compare the possible sequences in which transitions can be performed. These approaches are usually followed alternatively. In UML, as noted, we compare action sequences, not event sequences.

Comparing life cycle occurrences of statechart diagrams, both approaches coincide since we can denote a life cycle occurrence either by the sequence of its life cycle states or by the activation sequence generating it, whichever is more convenient (see above).

We are now ready for describing the three kinds of behavior consistency from [19, 20] in an UML context.

As described above, the perspective of **observation consistency** in semantic data models and object-oriented systems is that each instance of a subclass must be observable according to the structure and behavior definition given at the superclass, if features added at the subclass are ignored. A life cycle occurrence of a subclass can be observed at the level of the superclass, if activities, states, and labels added at the subclass are ignored. This is expressed by the following definition of the restriction of a life cycle occurrence.

Definition 9 (Restriction of a life cycle state). *The* restriction of a life cycle state σ' *of an object class* O' *to object class* O, *written* $\sigma'/_O^r$, *is defined as* $\sigma'/_O^r = \sigma' \cap (S \cup T)$.

Definition 10 (Restriction of a life cycle occurrence). *The* restriction of a life cycle occurrence $\gamma' = \sigma'_1, \ldots, \sigma'_n$ *of object class* O' *to object class* O, *written* $\gamma'/_O^r$, *is defined as* $\sigma_1, \ldots, \sigma_n$, *where for* $i = 1 \ldots n$: $\sigma_i = \sigma'_i/_O^r$.

Example 8. A possible life cycle occurrence of object class RESERVATION_WITH_-PAYMENT (cf. Figure 4) is [{requested}, {issued,toBeBilled}, {issued,billed}, {issued,paid}, {void}]. The restriction of this life cycle occurrence to object class RESERVATION yields [{requested}, {issued}, {issued},{issued}, {void}]. Observation consistent extension of behavior requires that each possible life cycle occurrence of a subclass is, if we disregard activities and states added at the subclass, also a life cycle occurrence of the superclass.

Definition 11 (Observation consistent extension). *We refer to a USM* $U_{O'} = U' = (S', T', L', \alpha', \Omega')$ *as an* observation consistent extension *of a USM* $U_O = U = (S, T, L, \alpha, \Omega)$, *if for every uLCO (and, hence, also every cLCO)* γ' *of object class* O', $\gamma'/_O^r$ *is an uLCO of* O.

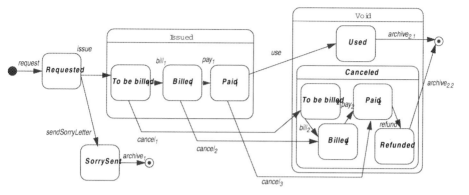

Fig. 5. Alternate UML state diagram for object class RESERVATION_WITH_PAYMENT

Observation consistency ensures that all instances of an object class (including those of its subclasses) evolve only according to its statechart diagram. This property is especially important for modeling workflows, where, for example, the current processing state of an order should always be visible at the manager's abstraction level defined by some higher-level object class. The example below, which violates observation consistency, illustrates this.

Example 9. The statechart diagram of class FRIENDLY_RESERVATION depicted in Figure 3 is no observation consistent extension of the statechart diagram of class RESERVATION depicted in Figure 2. The restriction of cLCO [{requested}, {altOffered}] of class FRIENDLY_RESERVATION to class RESERVATION yields [{requested}, {}], which is no cLCO of class RESERVATION.

As mentioned in the introduction, we distinguish two forms of invocation consistency. **Weak invocation consistency** ensures that if an object is extended with new features, e.g., a television set with video text, the object is usable the same way as without the extension.

Definition 12 (Weak invocation consistent extension).

A USM $U_{O'} = U' = (S', T', \alpha', \Omega')$ is a weak invocation consistent extension *of a USM $U_O = U = (S, T, L, \alpha, \Omega)$ if every activation sequence μ valid on A in U is also valid on A' in U' and for their respective traces γ and γ' it holds that $\gamma = \gamma'/_O^r$.*

Example 10. The statechart diagram of class FRIENDLY_RESERVATION depicted in Figure 3 is a weak invocation consistent extension of the statechart diagram of class RESERVATION depicted in Figure 2.

Example 11. The statechart diagram of object class RESERVATION_WITH_PAY-MENT depicted in Figure 4 is no weak invocation consistent extension of the statechart diagram of object class RESERVATION depicted in Figure 2. The activation sequence [request, issue, use, archive$_2$] is valid for class RESERVATION, but not for class RESERVATION_WITH_PAYMENT, as use cannot be applied on the LCS {issued,toBeBilled} reached by executing [request, issue].

Strong invocation consistency is satisfied if one can use instances of a subclass in the same way as instances of the superclass, despite using or having used new operations of the subclass.

Definition 13 (Strong invocation consistent extension).
A USM $U_{O'} = U' = (S', T', \alpha', \Omega')$ is a strong invocation consistent extension of a USM $U_O = U = (S, T, L, \alpha, \Omega)$ if
(1) every activation sequence μ valid on A in U is valid on A' in U', and (2) every activation sequence μ valid in U on $\sigma'/_O^r$ ($\sigma' \in R(A')$) is valid in U' on every life cycle state $\hat{\sigma}'$ that results from performing any activation sequence κ of activities in $T' \setminus T$ on σ', and for the traces γ and γ' generated by performing μ in U on $\sigma'/_O$ and by performing μ in U' on $\hat{\sigma}'$, respectively, it holds that $\gamma = \gamma'/_O^r$. (Note: μ and κ may be empty).

Example 12. The statechart diagram of object class FRIENDLY_RESERVATION depicted in Figure 3 is no strong invocation consistent extension of the statechart diagram of object class RESERVATION depicted in Figure 2: The call event send-SorryLetter will perform the transition sendSorryLetter:({Requested},{SorrySent}) for every instance of RESERVATION in LCS {requested}, but the event will be ignored according to UML semantics on an instance of FRIENDLY_RESERVATION if the call event offerAlternative has occurred before and the guard was true.

3.2   Checking Behavior Consistent Extension in UML

Example 7 has shown that extending the behavior of an USM cannot always be achieved by merely adding transitions and states. For those cases where it can be, we rephrase the rules for checking behavior consistency introduced for OBD's [19] in the context of USM's. This rephrasing is valid since the U2O mapping always produces an equivalent OBD, with the exception that the mapping ignores guard conditions which are part of the preconditions for UML transitions. The guard conditions therefore need to be explicitly introduced and the corresponding passages are given in italics below.

Consider two USM's U' and U of an object class O' and its superclass O, where U' extends the statechart diagram of U by additional states and transitions. Then, the rules to check for behavior consistency of statechart diagram extensions (based on the rules given for OBD's in [19]) are:

1. The *rule of partial inheritance* specifies that
   (a) the initial states of the statechart diagrams U' and U must be identical
   (b) every transition of U' which is already in U has at least the same source states and sink states than it has in U
   (c) *Since the rule of partial inheritance is in the line of covariance, for each transition t in U' that is already present in U, the guard condition $g'(t)$ in U' must be at least as strong as the guard condition $g(t)$ for t in U:*
       $g'(t) \Rightarrow g(t)$
2. The rule of *immediate definition of pre states, post states, and labels* requires that a transition of U may in U' not receive an additional source state or sink state that is already present in U.

3. The rule of *parallel extension* requires that a transition added in U' does not receive a source state or a sink state that was already present in U.
4. The rule of *full inheritance* requires that the set of transitions of U' is a superset of the set of transitions of U.
5. The rule of *alternative extension* requires that
   (a) a transition in U' which is already present in U has in U' at most the source states than the transition has in U.
   (b) *Since the rule of alternative extension is considered contra-variant, for each transition t in U' that is already present in U, the guard condition $g'(t)$ must be in U' at least as weak as the guard condition $g(t)$ in U: $g(t) \Rightarrow g'(t)$*

Rules 1, 2, and 3 are sufficient to check whether a USM U' is an observation consistent extension of another statechart diagram U. They are also necessary, provided that analogous assumptions to the safety, activity-reduced, and deadlock-free conditions introduced in [19] for behavior diagrams are taken here.

Rules 1, 2, 4 and 5 are sufficient to check whether a USM U' is a weak invocation consistent extension of another USM U. They are also necessary if conditions corresponding to those introduced for OBD's above are obeyed.

Rules 1 to 5 are sufficient to check whether a USM U' is a strong invocation consistent extension of another USM U.

4   Behavior Consistent Refinement

Since transitions in UML correspond to events that are not further decomposable in time, the only meaningful refinement in UML is to replace a transition either by another transition or by several alternative transitions, but it is not meaningful to replace a transition by a statechart diagram. A simple state may be replaced by a composite state, which may be concurrent or not.

Just as for extension, we assume that the statecharts have been transformed such that they contain only transitions between simple states and that concurrent regions are always explicitly exited (cf. Section 2).

The refinement of UML statechart diagrams may be captured by a total refinement function h analogous to the refinement function defined for behavior diagrams. For transitions, similar to the comparison of statechart diagrams for extension, function h is defined upon simple and complex transitions (and not on their lower-level components). For states, function h is defined only on simple states but not on complex states. Then, behavior consistency is defined in a manner analogous to behavior diagrams based on h.

Consider two USM U' and U of an object class O' and its superclass O, where U' refines U through refinement function h that maps transitions onto transitions and states of U' and that maps simple states of U' onto simple states of U. Intuitively, the concept of the refinement function means that if a simple state s of U has been refined to a composite state in U', then h maps the substates of s in U' and transitions between these states to s, and h maps the transitions in U' that are incident to the substates of s into transitions of U that are incident to s.

To compare life cycle states of a state machine S and a state machine U' that extends U, we used the notion of restriction of a LCS. For refinement, this notion is replaced by *abstraction of a LCS* using refinement function h.

Definition 14 (Abstraction of a life cycle state). *The* abstraction of a life cycle state σ' *of an USM of of an object class O' to an USM of object class O, written $\sigma'/_O^a$, is defined as $\sigma'/_O^a = \{h(s) \mid s \in \sigma' \}$.*

The notions of LCO and observation consistency are defined for refinement analogous to extension using $/^a$ instead of $/^r$.

Example 13. The statechart diagram of object class RESERVATION_WITH_PAY-MENT depicted in Figure 5 is an observation consistent refinement of the statechart diagram of object class RESERVATION shown in Figure 2. The refinement function h between RESERVATION_WITH_PAYMENT and RESERVATION maps each substate of issued to issued, each substate of canceled and used to void, and the remaining simple states of RESERVATION_WITH_PAYMENT to the same-named states of RESERVATION. Further, the refinement function h maps transition shown outside of the graphical region of a composite state to the same-named transition, whereby a possible index is omitted or (in the case of a nested index) truncated one level, and h maps transitions shown within composite state issued to issued, and transitions shown within composite state void to void.

4.1   Checking Behavior Consistent Refinement in UML

We verbally rephrase the rules for checking behavior consistency introduced for refinements of behavior diagrams [20] in the context of UML statechart diagrams.

Consider two UML statechart diagrams S' and S of an object class O' and its superclass O, where S' refines the statechart diagram of S through refinement function h as described above. Then, the rules to check for behavior consistent refinement are:

1. The rule of *pre- and poststate satisfaction* requires that for every transition t' in S': for every source state s of $h(t')$, there exists a state s' in S' such that $h(s') = s$, and for every sink state s of $h(t')$ there exists a sink state s' of t' in S' such that $h(s') = s$.
2. The rule of *pre- and poststate refinement* requires that for every source state s' of a transition t' in S', where s' and t' do not belong to the same refined state (i.e., $h(s') \neq h(t')$), $h(s')$ is a source state of $h(t')$, and for every sink state s' of a transition t' in S', where s' and t' do not belong to the same refined state, $h(s')$ is a sink state of $h(t')$.

Note that the definition of these rules for OBD contained additional conditions enforcing consistent behavior of multiple source and sink states for a given transition (Rules R1(a1) and R2(b1) in [20]). Since UML concurrency requirements trivially imply that an object always leaves and enters all concurrent subregions of a composite state simultaneously, no counterparts for these rules are needed. Note that it is because these rule conditions can be omitted that no labels are required in USM's (whereas labeled behavior diagrams are required to model consistent refinement correctly in OBD's).

4.2   Specialization in UML

We have shown in [20] that for behavior diagrams, behavior consistent specialization can be defined in terms of behavior consistent refinement and observation consistent extension. Specialization by concatenating refinement and extension can in principle be applied to UML statechart diagrams. Each specialization of an UML statechart diagram can be technically split into a refinement and an extension. But compared to behavior diagrams, a specialization can – due to the concurrency restrictions of UML – not always be naturally split into an extension part and a refinement part. Example 7 shows that it is not always possible to produce an intended parallel extension of an UML statechart diagram by merely adding states and transitions. The intended extension is possible, but technically it must be realized by a refinement. Thus from an application point of view, the clear distinction between extension and refinement is obscured.

5   Related Work

A number of inheritance policies are discussed in the context of the UML state diagram semantics definition [14]. They are referred to as Subtyping, Strict Inheritance, and General Refinement. (Remember that in UML, refinement is synonymous for inheritance.) Of these, the subtyping policy corresponds to weak invocation consistency. The other two policies provide neither observation nor invocation consistency and are instead geared towards implementation-level and code inheritance issues.

Other work on inheritance rules for object life cycles includes work based on state diagrams that are related via graph (homo-)morphisms [3, 17], finite automata [10], and state machines [8]. A discussion of life cycles on Petri net basis, but without completeness or sufficiency results, is contained in [22]. Restrictions on inheritance relationships in concurrent object-oriented languages were examined, e.g., in [9], and an approach that expresses subtype relations in terms of implications between pre- and postconditions of individual operations plus additional constraints was given in [7], providing explicit criteria for individual operations, but not for descriptions of complete object life cycles. A more detailed comparison can be found in our previous work [19, 20].

6   Conclusion

In this paper we have treated specialization of object life cycles by examining extension and refinement in the context of UML statecharts.

The ubiquity of UML means that despite its shortcomings any step towards capturing the relevant semantic properties of the language is of immense practical relevance. This paper has presented the first presentation of a formal scheme for describing consistency of UML lifecycle inheritance in terms of a set of explicit rules and provides a basis for incorporating further aspects of UML.

We presented a one-to-one mapping of a meaningful subset of UML statechart diagrams to OBD's which allows direct reuse of the OBD behavior consistency results [19, 20] in UML. To achieve this goal, object life cycles are compared not on the level of event calls but actually in terms of the operations performed, taking the current state of the object at the time of the event call into account, providing a comparison of actual system behavior.

References

1. P. Bichler and M. Schrefl. Active Object-Oriented Database Design Using Active Object/Behavior Diagrams. In J. Widom and S. Chakravarthy, editors, *Proc. IEEE RIDE'94*, pages 163–171, Houston, 1994.
2. Grady Booch. *Object-Oriented Analysis and Design with Applications (2nd edition)*. Benjamin Cummings, 1994.
3. J. Ebert and G. Engels. Observable or Invocable Behaviour — You Have to Choose. Technical report, Universität Koblenz, 1994.
4. D.W. Embley, B.D. Kurtz, and S.N. Woodfield. *Object-Oriented Systems Analysis: A Model-Driven Approach*. Prentice Hall, 1992.
5. David Harel. Statecharts: A visual formalism for complex systems. *Science of Computer Programming*, 8, 1987.
6. G. Kappel and M. Schrefl. Object/behavior diagrams. In *Proceedings ICDE'91*, pages 530–539, Kobe, Japan, April 1991.
7. B. Liskov and J. M. Wing. A behavioral notion of subtyping. *ACM Transactions on Programming Languages and Systems*, 16(6):1811–1841, November 1994.
8. J.D. McGregor and D.M. Dyer. A note on inheritance and state machines. *ACM SIGSOFT Software Engineering Notes*, 18(4), 1993.
9. O. Nierstrasz. Regular types for active objects. In *Proc. OOPSLA*, 1993.
10. B. Paech and P. Rumpe. A new concept of refinement used for behaviour modelling with automata. In *Proc. FME'94*, Springer LNCS 873, 1994.
11. Mike Papazoglou, Stefano Spaccapietra, and Zahir Tari, editors. *Advances in Object-Oriented Data Modelling*. MIT Press, 2000.
12. L. Pomello, G. Rozenberg, and C. Simone. *A Survey of Equivalence Notions for Net Based Systems*. LNCS 609. Springer-Verlag, 1992.
13. Rational Software Corp. *UML Notation Guide, Version 1.1*, September 1997.
14. Rational Software Corp. *UML Semantics, Version 1.1*, September 1997.
15. J. Rumbaugh, M. Blaha, W. Premerlani, and F. Eddy. *Object-Oriented Modeling and Design*. Prentice Hall, 1991.
16. J. Rumbaugh, I. Jacobson, and G. Booch. *The Unified Modeling Language Reference Manual*. Addison-Wesley Publishing Company, 1999.
17. G. Saake, P. Hartel, R. Jungclaus, R. Wieringa, and R. Feenstra. Inheritance conditions for object life cycle diagrams. In *Proc. EMISA Workshop*, 1994.
18. M. Schrefl. Behavior modeling by stepwise refining behavior diagrams. In H. Kangassalo, editor, *Proc. ER'90*, amsterdam, 1991. Elsevier North Holland.
19. M. Schrefl and M. Stumptner. Behavior Consistent Extension of Object Life Cycles. In *Proc. OOER'95*, volume 1021 of *LNCS*. Springer-Verlag, 1995.
20. M. Schrefl and M. Stumptner. Behavior consistent refinement of object life cycles. In *Proc. ER'97*. Springer-Verlag, 1997.
21. M. Schrefl and M. Stumptner. *On the Design of Behavior Consistent Specializations of Object Life Cycles in OBD and UML*. In Papazoglou et al. [11], 2000.
22. W. M. P. van der Aalst and T. Basten. Life-Cycle Inheritance — A Petri-Net-Based Approach. In *Proc. 18th Intl. Conf. on Application and Theory of Petri Nets*, LNCS. Springer, 1997.

The Viewpoint Abstraction in Object-Oriented Modeling and the UML

Renate Motschnig-Pitrik

Department of Computer Science and Business Informatics
University of Vienna, Rathausstr. 19/9
A-1010 Vienna, Austria
email: motschnig@ifs.univie.ac.at

Abstract: In object-oriented (OO) development the viewpoint abstraction has attracted by far less attention than classical abstraction mechanisms, such as classification, generalization, and aggregation. In OO databases, however, recent research has produced powerful view concepts supporting customization, schema evolution, and updates of base objects through views.
This paper discusses features of the viewpoint abstraction in the context of OO modeling and specifies extensions to the UML to support the modeling of views. We suggest employing an explicit notion of a view based on research on contexts and on OO databases in order to facilitate the customization of OO models through views. Further, the role of views to support an incremental development process will be discussed.

1 Introduction

In software development, abstraction mechanisms are known to be indispensable means to deal with complexity. While classification, generalization, and part-whole relationships have conquered well-respected positions in conceptual and analysis models, so far the viewpoint abstraction has received considerably less attention in models of early phases of software development. The term *view* is overloaded with different meanings in different approaches and tends to be defined only informally.

The viewpoint abstraction has gained recognition in- and outside of computer science. In architecture, for instance, different views such as bird's eye-, front-, side-, and various perspective views are indispensable tools of every architect for decomposing complexity in the course of modeling a building. Note that the individual views provide criteria on what objects shall be visible in one view but perhaps absent from another one. Further note, that objects, e.g. a door, window, or wall, look different depending on the view. In fact, expressing information on conceptual entities relative to some viewpoint may be based on any number of criteria, including:

- Focus of attention, aspect under consideration (realized in UML models)
- Fulfilling of a particular task or function (UML context for collaboration)
- Relevance for a particular user-group or actor (database views, user views)
- Relevance for a system extension or a customized version of the system
- Authorization and access control.

A.H.F. Laender, S.W. Liddle, V.C. Storey (Eds.): ER2000 Conference, LNCS 1920, pp. 543-557, 2000.
© Springer-Verlag Berlin Heidelberg 2000

Note that all criteria share the property of being extrinsically grounded! This means that the decision, in which view a conceptual entity be represented and which of its properties be modeled depends on the way its model is going to be used, based on some arbitrary, external purpose. This is in sharp contrast to classical mechanisms involving abstraction that typically are based on intrinsic properties of entities, such as similarity or inclusion [1], [2], [3]. In brief, viewpoints reflect what some agent considers relevant/irrelevant in a particular real-world situation and thus complement classical abstraction mechanisms by allowing one to focus attention on selected concepts and/or properties while ignoring others.

Currently, OO techniques, in particular the UML, support only special manifestations of viewpoints explicitly. As an example consider UML's supply of several complementary models and diagramming techniques that allow one to focus on structural -, behavioral -, implementation -, or distribution aspects, respectively, while suppressing others. Other manifestations, such as the customization of static structure diagrams (else called class diagrams) to reflect the view of a particular user group can be expressed in the UML, but require excessive effort. This is because construction and consistency checking need to be done manually in commercial tools and furthermore, changes in the base diagram are not automatically reflected in the views.

It is the coincidence of two phenomena that provides strong motivation for extending the UML by a powerful viewpoint mechanism. Firstly, views in the area of OODB's can be designed to be by far more powerful than their relational cousins that constitute a broadly appreciated technique in relational databases. For example, the updateability of OODB views along with the existence of efficient algorithms for the maintenance of capacity-augmenting views [4] make OODB view mechanisms elegant platforms for schema evolution and the customization of systems without necessitating updates to the base schema. Secondly, class diagrams such as UML's static structure diagram can play versatile roles: They can be interpreted as conceptual models of OODB's and hence provide an immediate solution to the processing of persistent data. In this role, they enjoy the full benefits from advanced OODB view mechanisms. But even if other ways are used to implement persistent data--currently the rule rather than the exception--the fact that UML's static structure diagram can represent a schema graph allows one to exploit all those benefits that concern the intentional (schema-level) aspects of OODB view mechanisms, such as logical data independence and the customization of schemas for specific user groups. In brief, results from research on OODB view mechanisms strongly suggest studying their application in OO analysis.

Given the high potential of improvement by extending OOA techniques and, in particular, UML static structure diagrams by incorporating a viewpoint mechanism, this paper sketches the process and structures involved in the derivation of views, discusses various features of the viewpoint abstraction, and specifies the necessary extensions to the UML in order to support views. Essentially, our solution to the modeling of views within the UML consists of the provision of appropriate stereotypes, operations, and two major workflows. The first is devoted to the derivation of individual view classes while the second workflow is concerned with the construction of the class diagram to hold the view classes. Major advantages resulting from this approach are that new requirements can be built into the system without affecting the base schema and all applications running on it. Also, experimental changes to the base schema can first be tested on a view schema before being incorporated into an application. This feature significantly facilitates incremental development as advocated by the Unified Software Development Process [5].

Our work is based on the results from [6] and [7] that analysed view concepts in various areas of software development and, respectively, derived requirements on view mechanisms for OODB's. The current paper builds upon the view mechanism accompanying the MultiView System [8] and its evolution into the TSE System (Transparent Schema Evolution [9]). TSE supports capacity-augmenting views and has been implemented on top of the commercial OODB System Gemstone.

Previous research on manifestations of the viewpoint abstraction in software development, databases, programming languages and AI [6] points to the fact that several related terms, such as perspective, view, virtual schema, view schema, context, space, world, role, proxy-object, and topic are employed to capture notions we chose to subsume under the term viewpoint abstraction. Due to the absence of a broadly agreed-upon terminology, below we introduce one in the course of illustrating our approach via an example. This shall prepare the reader for a more precise discussion of individual features of the viewpoint abstraction in the context of OO modeling in Section 3. Section 4 proceeds by embedding the proposed viewpoint mechanism into the UML. The final Section summarizes related work and concludes the paper.

2 Preliminaries: Terminology and Introductory Example

To illustrate our approach, consider a simplified fragment of an analysis model of a university administration system as depicted in Fig. 1. The figure shows a UML static structure diagram (also referred to as a class diagram) containing five interconnected classes. The class "Person" is the target of three generalization links that, respectively, interconnect the subclasses "Student", "SupportStaff", and "TeachingStaff" with their superclass "Person". Further, the classes "TeachingStaff" and "Course" are interconnected by an association called "mayTeach".

In the context of discussing viewpoint mechanisms, we refer to a class whose instances are stored (e.g. the class "Student" in the example) as a base class. A class diagram--like that in Fig. 1--whose classes are all base classes is called a base context. In contrast, a class that is derived from other classes by an OO query is called a derived class. A view context is a class diagram that, in general, is composed of base and derived classes. All classes making up the view context, i.e. base- and derived classes, are called view classes. The term *view* will be used in a generic sense--such as in view-technology--to subsume the terms *view class* and *view context*.

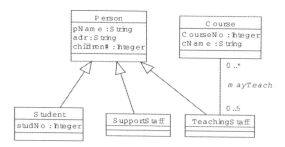

Fig. 1: Fragment of a university database as static structure diagram in the UML

Returning to the example, imagine that the university intends to establish a child bureau running a day-care center and organizing other child-care activities. Thus, they would like to keep all staff with kids in various age groups informed about vacancies and ongoing activities. Suppose, the child bureau plans to develop its own application programs on the university administration system sketched in Fig. 1. They find out, that each instance from "SupportStaff" as well as from "TeachingStaff" should carry information about the ages of children in the case that the value of the attribute "children#" is greater than 0. Also, an operation (say, "childStatistics()") to determine the percentage of staff having children shall be available. Further, staff with at least one child shall be associated with an operation that supports the distribution of relevant information to parents.

In traditional systems, the developers need to assess the impact of the changes on other use-case realizations or programs based on the same class diagram. With view technology, the child bureau can be provided a customized view context that reflects all the desired changes to the base context, without affecting the latter. Thus, our goal is to construct a view context that satisfies the requirements from the child bureau.

We start by deriving the required classes to be included in the view context later-on. Since the child bureau does not need to distinguish between support- and teaching staff and wants to perform child statistics on all staff, an appropriate "Staff" class is derived. This is achieved by concatenating two object-algebra operations as follows:

Staff' : = **union** (SupportStaff, TeachingStaff)
Staff : = **refine** childStatistics() **for** Staff'

The next task is to derive a class that holds all parent staff, such that the latter can be sent appropriate information. For this reason we select all staff that have a non-zero value in their "children#" attribute and associate them with the operation "parentInfo()":

ParentStaff' : = **select from** Staff **where** (children# > 0)
ParentStaff" : = **refine** parentInfo() **for** ParentStaff'

So far, we derived classes from base classes and refined them with operations, without adding new, capacity-augmenting features (i.e. attributes and operations). However, the capability to handle capacity-augmenting (i.e. stored) features significantly increases the expressive power of view mechanisms, as illustrated here and discussed below. Therefore, the next object-algebra expression shows the addition of the capacity-augmenting attribute "kidsBornOn" to the class "ParentStaff" in order to provide the child bureau with information on the age classes of children for all staff who are parents:

ParentStaff : = **refine** kidsBornOn: List **for** ParentStaff"

Individual view classes being derived, the target view context needs to be constructed. In the literature, several solutions to this task have been proposed [7]. The simplest approach mimics the relational model, where a view schema is simply a flat collection of tables resulting from queries. In this approach, the view context is constructed by making all classes constituting the view direct subclasses of one root class thus resulting in a flat structure [10]. The primary disadvantage of this approach is the fact that features of classes need to be duplicated and all advantages of OO subclassing get lost. Other solutions to constructing OO view contexts suggest to associate derived classes in proper places of the base context or to construct a separate subclass lattice consisting of all classes making up the view context. Proponents of the first alternative argue that view classes should use as much information as possible from base classes and hence favor a tight coupling between the base- and the derived classes resulting in a combined context. Although this approach prevents introducing inconsistencies between base- and view context, the combined context quickly becomes overloaded. A strong argument in support of the separation of the view- and the base context is the fact that the view context serves as the interface between the base context and its customized variants, typically requested by external clients. In this respect, a separate view context tends to be by far less complex and provides an explicit layer of indirection resulting in complete logical data independence [11].

Understandability being a key concern of OO modeling, we wish the view mechanism to provide means to construct and to maintain separate view contexts and ideally take care of consistency between base- and view contexts. The OODB view mechanism designed for the MultiView System [8], [9] meets these requirements and hence will be adopted here. MultiView combines the advantages of inserting derived classes into the base-context and providing separate view contexts by introducing an intermediate, global context (called global schema in MultiView) into which all derived classes are automatically integrated [12].

In the following, we illustrate the construction of the global context using the university administration example and postpone further discussion on MultiView's viewpoint mechanism till Section 3. Above, the classes "Staff" and "ParentStaff" have been derived to become part of the child bureau's view. Thereby we created intermediate classes such as "Staff' ", "ParentStaff' ", and "ParentStaff'' ". This happened for didactic reasons only, to illustrate individual object algebra operations in separation. In practice, nesting of respective operations would have been applied.

Next, the derived classes "Staff" and "ParentStaff" need to be integrated into the global context that initially (i.e. before the integration of view classes) is a copy of the base context. The global context, at each point in time, exhibits the property of being *maximally classified*. This means that inheritance is exploited whenever possible such that features (attributes and operations) of classes are never duplicated. The result from inserting the two classes into the global context is depicted in the left-hand side of Fig. 2. As can be seen from the figure, the algorithm for classifying the "Staff" class needs to introduce an intermediate class, call it "Staff' ", in order to keep the global context maximally classified.

Once the view classes are integrated into the global context, the customized view context can be constructed. For this reason, the user simply selects those classes that shall join the view context. In MultiView, a view specification language is provided for this purpose [8]. Although the same textual language could be used for the specification of view contexts in OO modeling, we find it more practical to allow the user to create a view context and to select the appropriate classes via direct manipulation techniques. The result of selecting the "Staff" and "ParentStaff" class from the global schema (see left-hand side of Fig. 2) results in the view context depicted on the right-hand side of Fig. 2.

Summarizing, powerful OO view mechanisms need to take care of two issues. Besides allowing one to *derive view classes*, they have to support the *construction of view contexts* that provide customized interfaces to the base system and further are consistent with the latter.

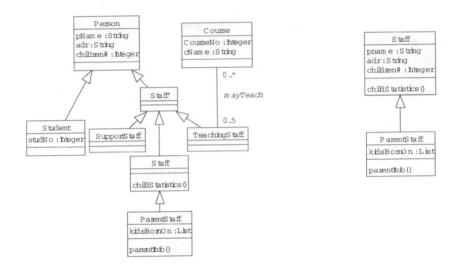

Fig. 2: The global context (left) and the child bureau's view context (right)

Using an intermediate global schema to hold all base and derived classes makes the above requirements feasible. Support of automated classification makes the construction of view contexts simple and avoids inconsistencies caused by human error. Furthermore, since all classes are integrated within one context, the merging of versions that are expressed as views is facilitated. This is particularly helpful for incremental development in so far, as additional requirements or use-cases could be specified as views on the base core system first and integrated into it later, e.g. after approval in the respective iteration workflow [5].

3   OO View Mechanisms: Features and Trade-Offs

Relational Versus OO View Mechanisms. Like in the relational data model, the query language plays a vital role in the derivation of views in OO models. Compared with the relational model, however, where every expression in relational algebra-- such as a query--returns a relation that may act as a view, the derivation of view classes in OO models is more sophisticated. This is primarily due to the fact that objects can be constructed by arbitrary combinations of type constructors such as sets, tuples, lists. Thus, class customization should have the capability of producing different compositions of type constructors. A further major difference between relational and OO views is the fact that in the relational model a view context is simply a flat collection of relations, whereas OO view contexts are class lattices. They tend to be restructured as a result from class customization such that finding the proper position of a view class in a class lattice needs to be considered.

Class Customization as Derivation of View Classes. Class customization has to provide means for the restructuring of features contained in base classes yielding view classes with derived information content. In the case that the restructuring is based on classes and objects already existing in the base context, it is possible to associate these objects with customized view classes by making them instances of the latter. This kind of restructuring is called *object-preserving* since it preserves base objects rather that generating new objects from existing ones. As has been argued in [9], object-preserving operators are sufficient in practice to support the majority of customization demands. The category of object-preserving queries further can be characterized by an essential property: View contexts whose classes are derived by object-preserving operators have been shown to be generally *updateable* [13]. More precisely, this means that updates to instances initiated in some view context can be propagated to respective base objects in a unique way. This is an essential advantage over relational views that can be updated only, if they contain the key of their (single) underlying base relation [14].

Capacity-Augmenting Views. A further significant increase in expressive power can be achieved by allowing views to be capacity augmenting, i.e. by providing means to support view classes that contain original, non-derived features (attributes and methods). As an example consider the "kidsBornOn" attribute in the view class "ParentStaff" shown in Fig. 2. Clearly, capacity augmenting views call for some degree of materialization since objects are extended by features that need to be stored. Fortunately, applying implementation techniques like object-slicing [15], [9], efficient algorithms for the maintenance of materialized, object-preserving views have been designed and successfully applied [4].

Query Language Style. The query language is the central means by which view classes are derived. OO view mechanisms differ in the style of the query language as well as in the degree to which amendments to pure queries are proposed. In particular, object-generating queries tend to be extended by mechanisms that assign oid's to newly created objects in order to make them accessible. Regarding language style of recent proposals, SQL-related languages based on object algebra outnumber predicate-based and rule-based query languages.

Object Algebra Operators. Although the embedding of views into OO modeling is independent of the particular object algebra used for class customization, the approach presented here relies on the object algebra from MultiView. It assumes an object model based on features generally agreed upon in the literature, such as encapsulation, subclassing, subtyping, full inheritance and multiple inheritance [16], [17].

In the sequel, we distinguish types from classes in the following sense (see also [16], [13]). Whereas *types* are specifications of (structural and behavioral) *features*, *classes* are containers of objects--type extents--sharing the same type. For each class C, there is a set of objects, called its *extent* (extent(C)). The elements of that set are *instances* of C's type (type(C)). *Subtyping* is defined such that a type T1 *is a subtype of* another type T2 iff all features defined for T2 are also defined for T1. In particular, further features may be added to the subtype. Subtyping defines a partial order on types and is a precondition for subclassing as defined in the following. A class C1 *is a subclass of* a class C2, iff it holds that type(C1) is a subtype of type(C2) and extent(C1) ⊆ extent(C2).

In a nutshell, MultiView's object algebra encompasses the usual set-based operators for *union, difference,* and *intersection,* yielding, respectively, a view class whose type is the lowest common supertype of the input types, a subset of its first argument with the same type, and the greatest common subtype of the input types. Further, the algebra provides the *select, hide* and *refine* operators. The *select* operator returns a subset of the input set of objects satisfying a given predicate with the type of the resulting set being unchanged. The *hide* operator is used to blend out certain properties of a type and hence results in a type that is a supertype of the input type. Finally, the *refine* operator yields a result that is a subtype of the input type since it allows features (attributes or operations) to be added to the input-type definition. Thereby attributes can either be derived or original, in other words, capacity-augmenting. All operators are object-preserving thus resulting in updateable views.

The main reason for selecting MultiView's class derivation mechanisms for our proposal is the follow-up support for the second major issue in OO view mechanisms, namely the construction of view contexts that provide logical data independence. Whereas the overall approach has been illustrated and motivated in the previous section, some accompanying issues are discussed in more detail below.

The Construction of View Contexts. In the literature, OO view mechanisms differ most widely with respect to the way of assembling customized classes into some view context. As indicated above, solutions range from proposing a simple, flat (or two-level) structure at the price of loss of inheritance, to inserting view classes into the base schema in order to retain all information (via inheritance) from the latter but thereby giving up logical data independence. In the author's view, logical data independence as well as inheritance can be retained by introducing an intermediate global context to hold all classes and deriving independent view contexts from it. Below, we adopt the workflow proposed in [8] for constructing a view context from base- and view classes in three steps:

- Automated view classification, meaning the insertion of view classes into the global context;
- Selection of classes from the global context that shall join the view context;
- Automated construction of the view context as a subgraph of the global context.

Insertion of View Classes into a Uniquely Identified Global Context. The latter is a maximally classified context initialized with the base context and serving as an intermediate structure that encompasses all view classes [8]. Every view context is a subgraph of the global context in so far, as the set of classes (nodes in the graph) in each view context is a subset of the set of classes in the global context. Further, each view context maintains only generalization relationships among its view classes that are directly derivable from the global context. Given a global context GC = (V, E), than for each view context VC = (VV, VE) it holds that VV ⊆ V and VE ⊆ transitive-closure (E).

Automated View Classification. Algorithmic support for placing view classes into the global context is particularly desirable, since it avoids inconsistencies and relieves the view designer from acquiring detailed knowledge about the whole static structure model that can become quite complex. Unfortunately, since the classification problem is not decidable in general because it involves the comparison of arbitrary predicates and operations [18], the proposed algorithm is sound but incomplete [12]. It proposes a correct, although not optimal solution in the worst case, placing a class too high up in the lattice. The algorithm assumes that predicates in the where clauses of select operations are restricted to be conjunctions of atomic predicates. Further, the algorithm resolves conflicts caused by composing the subtype- and the subset hierarchies into one generalization hierarchy by introducing intermediate classes resulting in a maximally classified structure of the class lattice, thus exploiting inheritance and reuse to a maximum degree. Since the intermediate classes exist only in the global context, the users of views are provided with view contexst that are tailored to suit their needs. The classification algorithm has been optimized for the particular object algebra operators. Its complexity is quadratic for the hide- and intersect operators and linear for all others.

Class Selection. While MultiView proposes a view specification language to support the next step--class selection--, our approach is to extend OO modeling tools such as Rose98i [19] by interactive means of selecting classes from the global context to join the view context.

Construction of the View Context. The third step, namely the construction of the view context as a subgraph of the global context, again can be done automatically. The respective algorithms need to take care of the consistency of the generalization hierarchy and of the closure property with respect to association relationships [8], [9]. The first subtask is accomplished by an algorithm that generates a generalization lattice from a set of view classes. The *closure property* of a view schema with respect to association relationships requires that every association emanating from a class in the view schema be defined in the view schema. In the case that a view schema is not closed, the minimal set of classes to be added to it shall automatically be suggested. Further details on the algorithm such as its linear complexity are discussed in [8].

As Much Analogous Functionality Between Base- and View Context as Possible. Ideally, the user should not see any difference as to whether (s)he is dealing with the base- or a view context. In particular, all operations on classes, including the creation of base class instances through view classes should be supported. An interesting point is the provision of multiple classification in the view context. Given that more that one view class is derived from the same base class, a situation--known as *multiple classification*--arises, where objects may happen to be direct instances of two classes. Since prohibiting such situations could result in an explosion of "artificial" view contexts, we rather suggest to allow for multiple classification and to apply implementation techniques such as *object-slicing* that have been applied for the implementation of *role objects* [15], [20]. Aiming for as few differences as possible among base- and view contexts, the provision of multiple classification in base contexts would be a natural consequence. Though the object model underlying the UML allows for multiple classification, the lack of support for the latter by commercial OO languages has the effect that UML's standard semantics is tuned to single classification.

Another aspect of deviation concerns updates of structural (type) information. Whereas updates of types in the base context potentially effect all view contexts, updates of types in derived contexts are not allowed in order to meet the invariant condition that each view context is a subgraph of the global context. Hence, instead of updating structural information in view contexts new views need to be derived.

View Independence. A feature derivable directly from the view context construction process and the fact that updates to structural information in view contexts are not allowed is view independence [8]. It says that the specification of existing view contexts is not effected by the definition of new view contexts.

Authorization. Unlike in the relational approach, to the author's knowledge, authorization models based on OO views have not (yet) been discussed in the literature on OO views. A flexible, predicate- and content-based authorization model for the more general notion of a *context* has been proposed in [21].

4 Embedding Views into the UML

4.1   Features of the UML that Support Aspects of the Viewpoint Abstraction

The UML already provides a number of concepts for the modeling of specific aspects of the viewpoint abstraction. Full-scale support of a powerful viewpoint mechanism, however, calls for a number of extensions. In this respect, UML's notion of stereotypes appears well suited to introduce the required extensions in a straightforward way. Below, we discuss current features of the UML in the light of view modeling and proceed by specifying stereotypes to complement them.

ClassifierRole. A "ClassifierRole" defines a role to be played by an object within a collaboration. The role describes the type of object that may play the role by listing required attributes, operations and relationships to other roles. In fact, the role is a customized type, but it is restricted to a subset of features of its source class(es) and, furthermore, it may appear in collaborations only.

Derived Element. In the UML, modeling elements that can be computed from other elements can be distinguished by being prefixed with a slash. While, in the general case, the derivation rule is arbitrary, constraints on the syntax and semantics of the derivation can be associated with it. As illustrated on the left-hand side of Fig. 3, the combination of the notational elements *derived package*, *dependency relationship*, and *constraint* are well suited to specify the notion and stereotype of a view context.

Views. The notion of a view has been present in the UML from the very beginning, although it is implicit and, with the exception of architectural views, specialized [22]: A view is a subset of UML modeling constructs that represent one aspect of a system. One to two kinds of diagrams provide a notation for the concepts in each view. UML distinguishes Static-, Use Case-, State Machine, Activity-, Interaction-, Physical-, and Model Management views. Architectural views typically are subsets of the above views modeling architecturally relevant aspects of a system. Note that the architectural view of a static structure diagram could well be modeled as a view context as defined above. Further note, that the newly introduced notion of a view context is consistent with and aims to extend UML's native notion of a view. In this paper, the extension is targeted at static structure diagrams leaving other UML views for further research.

Contexts. Similar to the notion of a view, the notion of a context has been present in the UML, although in an implicit and restricted form only. It is best known under the term *collaboration context* appearing in *collaboration diagrams*. More generally, in the UML a context is defined as a view of a set of modeling elements that are related for a purpose, such as to execute an operation or form a pattern [22]. Again, our notion of a view context precisely adheres to the above definition, but raises view contexts to an existence as explicit, stereotyped packages.

Interfaces. Interfaces are appreciated mechanisms for identifying and specifying collections of operations that are offered by a class, a package, or a component and together fulfill a particular service. Like view classes, interfaces implement the viewpoint abstraction in so far as they provide means for customization by abstracting from features that are irrelevant for a particular task or viewpoint. Unlike view classes, interfaces are restricted to collecting operation signatures such as to support encapsulation. View classes, on the other hand, are associated with view contexts in order to provide complete, customized user views.

4.2   Stereotypes for Modeling View Classes and View Contexts

In the UML, stereotypes serve as an extensibility mechanism by allowing one to define new kinds of model elements based on an existing model element. A stereotype represents a variation of an existing model element in so far, as it may be associated with additional constraints and/or may extend the semantics of the model element it originates from. The left-hand side of Fig. 3 specifies the way a class with the stereotype <<view class>> relates to its original model element, a class: Every view class is derived from a number of base classes via an object algebra operation.

Since packages provide separate name spaces for the elements they contain, view contexts are modeled as stereotypes of packages. In this way view independence is achieved. The package stereotypes <<global context>> and <<view context>> are defined with the following constraints and semantics:

- Each class contained in a global- or a view context is either a class from the base static structure model or it is a view class;
- No structural changes to classes contained in a global- or a view context are allowed to be performed directly;
- Changes affected in the base static structure diagram are reflected in the global- and the view contexts;
- The global- and the view contexts are closed (see Section 3) and fully classified.

In addition, packages of the <<global context>> stereotype are constrained to hold all base classes (the default for a class) and view classes. Further, every <<view context>> package is constrained to be derived from the <<global context>> package (as modeled on the right-hand side of Fig. 3) such that the derivation produces a subgraph of the global context.

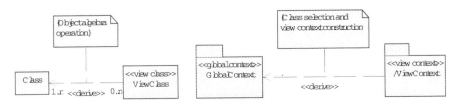

Fig. 3: Introducing stereotypes for view classes and view contexts

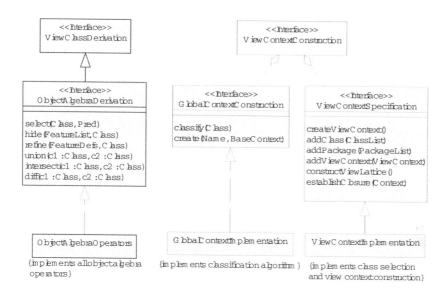

Fig. 4: Interfaces and Operations for class derivation (left) and view context construction (right)

4.3   Operations and Implementation

Besides the definition of stereotypes, operations need to be provided for the two major workflows: view class derivation and view context construction. Fig. 4 illustrates a candidate architecture for integrating the operations described in Section 3 into UML applications employing views. The proposed architecture is designed for generality and modifiability with respect to the embedding of other viewpoint mechanisms.

Currently, the UML extensions are being implemented as a Master's Thesis [23] on top of the tool Rose98 [19] using Rose's extensibility link. The prototype implementation deals with a restricted set of UML modeling constructs, consisting of classes and relationships, leaving remaining issues for further iterations.

5   Conclusion and Related Work

Due to the broad applicability of the viewpoints, related work is spread across several areas of computer science. Results from the analysis of various viewpoint mechanisms can be found in [6]. In the closer context of object orientation, views have been proposed in Embley's OOSA (Object Oriented Structured Analysis) [24] to distinguish between high-level and lower-level representations, in order to provide the proper level-of-detail for users and developers, respectively. The majority of previous work addresses view mechanisms for OODB's. focusing on the derivation of view classes via a query language. Unlike most other approaches, the view mechanism incorporated in the MultiView System [8] encompasses the construction of view contexts as adopted above. A detailed comparison of view mechanisms for OODB's appeared as [7]. The use of capacity-augmenting views for implementing schema evolution as an extension to the MultiView System resulting in the TSE System (Transparent Schema Evolution) is discussed in [9], while [4] presents efficient algorithms for maintaining the resulting materialized views. Another area in which viewpoints play a significant role is requirements engineering [25]. There, requirements stemming from agents that employ different representations and occupy individual viewpoints are elicited and need to be integrated such that the research question appears to be complementary to the derivation of views as discussed in this paper. Finally, the notion of a context--essentially a generalization of the concept of a view context-- has been proposed e.g. in [26], [27], [21] to model information base decompositions.

This paper has proposed to extend OO modeling by a simple yet powerful viewpoint mechanism, based on MultiView [8]. The basic motivation has been to allow for customized variants of (static structure) models reflecting the needs of particular applications and of system increments. The goal thereby is to keep the base model simple rather than constantly increase its complexity by incorporating additional or changed requirements. More generally, we propose to complement traditional OO reuse techniques by a viewpoint mechanism targeted at the reuse of collections of customized classes. Finally, we specified the necessary extensions to the UML to embed the proposed viewpoint mechanism.

Further research will address the extension of our prototype to deal with further UML constructs as well as the provision of high-level operations, such as merging or difference, for constructing new view contexts from existing ones.

Acknowledgements: Special thanks are due to Martin Hitz for insightful discussions on the UML and to Markus Schett for implementing the prototype.

References

1. Storey V.,C.: Understanding Semantic Relationships. Very Large Database Journal, 2(4), (1993) 455-488
2. Motschnig-Pitrik R., Mylopoulos, J.: Classes and Instances. Int. Journal on Intelligent and Cooperative Information Systems, 1(1), (1992) 61-92
3. Motschnig-Pitrik R., Kaasboll J.: Part-Whole Relationship Categories and Their Application in Object-Oriented Analysis. IEEE TSE 11(5), (1999) 779-797
4. Kuno H., Rundensteiner E.: Incremental Maintenance of Materialized Object-Oriented Views in MultiView: Strategies and Performance Evaluation. IEEE TKDE, 10(5), (1998)
5. Jacobson I., Booch G., Rumbaugh J.: The Unified Software Development Process. Addison-Wesley, Object Technology Series, (1999)
6. Motschnig-Pitrik R.: "An Integrating View on the Viewing Abstraction: Contexts and Perspectives in Software Development, AI, and Databases"; Journal of Systems Integration, Kluwer, 5 (1), (1995) 23-60
7. Motschnig-Pitrik R.: Requirements and Comparison of View Mechanisms for Object-Oriented Databases. Information Systems, 21(3), (1996) 229-252

8. Rundensteiner E.: MultiView: A Methodology for Supporting Multiple Views in Object-Oriented Databases. Proc. of 18th Int. Conf. on Very Large Databases , Vancouver, (1992)
9. Ra Y-G., Rundensteiner E., A.: A Transparent Schema-Evolution System Based on Object-Oriented View Technology. IEEE TKDE, 9(4), (1997), 600-624
10. Kim W.: A Model of Queries in Object-Oriented Databases. In: Proc. of the Internat. Conf. on Very Large Databases, (1989) 423-432
11. Date C., J.: An Introduction to Database Systems. Vol.1, 5th ed., Addisson-Wesley, (1990)
12. Rundensteiner E., A.: A Classification Algorithm for Supporting Object-Oriented Views. Proc. of the Int. Conf. on Information and Knowledge Management, (1994) 18-25
13. Scholl M., H., Laasch C., Tresch M.: Updateable Views in Object Oriented Databases. In: Proc. of the 2nd Conf on DOOD, Munich, (1991)
14. Gottlob G., Paolini P., Zicari R.: Properties and Update Semantics of Consistent Views. ACM TODS, 13(4), (1988) 486 - 521
15. Martin J., Odell J.:Object-Oriented Analysis and Design. Prentice Hall, (1992)
16. Beeri, C.: New Data Models and Languages - the Challenge. In: Proc. of PODS 92, (1992)
17. Scholl M., Schek H., Tresch: Object-algebra and views for multi-objectbases.In: Oezsu et al. (ed.), Distributed Object Management, Morgan Kaufmann, (1993) 352-373
18. Nebel B.: Terminological Reasoning is Inherently Intractable. Artificial Intelligence 43, (1990) 235-249
19. Rational Software Corporation: Rational Rose. http://www.rational.com, (1999)
20. Gottlob G., Schrefl M., Röck B.: Extending Object-Oriented Systems with Roles. ACM TOIS, 14(3), (1996) 268-296
21. Motschnig-Pitrik R.: A Generic Framework for the Modeling of Contexts and its Applications. Data & Knowledge Engineering 32, (2000) 145-180
22. Rumbaugh J., Jacobson I., Booch G.: The Unified Modeling Language Reference Manual. Addison-Wesley, 1999.
23. Schett M.: Development of a Prototype for Embedding Views in the UML. Master's Thesis; University of Vienna, Dept. of Computer Science and Business Informatics; (2000)
24. Embley D., W., Kurtz B., D., Woodfield S., N.: Object-Oriented Systems Analysis--A Model-Driven Approach. Prentice Hall, Englewood Cliffs, (1992)
25. Nuseibeh B., Kramer J., Finkelstein A.: A Framework for Expressing the Relationships Between Multiple Views in Requirements Specifications. IEEE TSE 20(10), (1994) 760-773
26. Mylopoulos J., Motschnig-Pitrik R.: Partitioning Information Bases with Contexts. In: Proc. of the 3rd Internat. Conference on Cooperative Information Systems, Vienna, (1995) 44-54
27. Theodorakis M., Constantopoulos P.: Context-Based Naming in Information Bases. International Journal of Cooperative Information Systems, 6(3&4), (1997) 269-292

XML Conceptual Modeling Using UML

Rainer Conrad[1], Dieter Scheffner, and J. Christoph Freytag

Department of Computer Science
Humboldt-Universität zu Berlin, Germany,
{rainer.conrad|dieter.scheffner|freytag}@informatik.hu-berlin.de

Abstract. *The eXtensible Markup Language (XML) is increasingly finding acceptance as a standard for storing and exchanging structured and semi-structured information. With its expressive power, XML enables a great variety of applications relying on such structures - notably product catalogs, digital libraries, and electronic data interchange (EDI). As the data schema, an XML Document Type Definition (DTD) is a means by which documents and objects can be structured. Currently, there is no suitable way to model DTDs conceptually. Our approach is to model DTDs and thus classes of documents on the basis of UML (Unified Modeling Language). We consider UML to be the connecting link between software engineering and document design, i.e., it is possible to design object-oriented software together with the necessary XML structures. For this reason, we describe how to transform the static part of UML, i.e. class diagrams, into XML DTDs. The major challenge for the transformation is to define a suitable mapping reflecting the semantics of a UML specification in a DTD correctly. Because of XML's specific properties, we slightly extend the UML language in a UML-compliant way. Our approach provides the stepping stone to bridge the gap between object-oriented software design and the development of XML data schemata.*

1 Introduction

Because XML is widely accepted for storage and exchange of information, it enables a great variety of applications that depend on semistructured data within different fields of usage. This variety of applications demands various document structures. Document Type Definitions (DTDs) provide a grammar for creating document structures. They allow the designer to specify tailor-made structures, e.g. product catalogs, digital libraries, and electronic data interchange (EDI) in business-to-business e-commerce.

However, even the design of simple DTDs may cause difficulties, partly due to the textual form of the grammar itself. As we experienced in some cases, understanding DTDs is not intuitive. A DTD in its current textual form commonly lacks clarity and readability, therefore erroneous design and usage are inevitable.

As the data schema for documents, or objects respectively, XML DTDs suggest themselves to be modeled conceptually. The main goal of conceptual modeling is to separate the designer's intention from implementation details. In addition, the model's inherent visualization contributes to a better understanding

A.H.F. Laender, S.W. Liddle, V.C. Storey (Eds.): ER2000 Conference, LNCS 1920, pp. 558–571, 2000.

of the design. We propose to use a subset of UML for DTD design in particular and XML schemata in general. Using UML for document design, we are able to combine object-oriented software design with the XML document structures. Hence, conceptual modeling with UML helps to improve a redesign and to reveal possible structural weaknesses in the document design. Being extensible and adaptable, UML retains the expressiveness of target languages properly. One of XML's specific properties for which we extend UML in a compliant way, is the concept of order.

The structure of our paper is as follows: First, we give a brief overview of XML followed by a summary of related work. We then describe relevant modeling concepts of UML and its transformation into DTDs. With XML documents only having a static structure, we currently focus on the static model. Thereafter, we discuss our extensions necessary for the implementation dependent concepts of the modeling. Due to space limitation, our presentation focuses on the major aspects of our transformation model. An extended version of this paper can be found in [2].

2   XML

We briefly describe the XML concepts used for our transformation. First, we show how DTDs are arranged in the context of XML documents in order to give a better understanding of the DTD constructs afterwards.

2.1   XML Documents

An XML document has a logical and a physical structure [10]. The physical structure of a document is made up of *entities* that are ordered hierarchically. The logical structure is explicitly characterized and described by "markups" which comprise declarations, elements, comments, character references, and processing instructions. Essentially, the document structure consists of an optional document type declaration containing the Document Type Definition (DTD), and a document instance. The purpose of a DTD is to provide a grammar for a class of documents.

2.2   XML DTD Constructs

According to the XML specification, DTDs consist of markup declarations namely element declarations, attribute-list declarations, entity declarations, notation declarations, processing instructions, and comments [10]. As for these declarations, they are the elementary building blocks on which a DTD can be designed.

Element Type and Attribute-List Declarations make up the kernel of DTDs. Together, they declare the *valid* structures of a document instance, namely the nested element tags with their additional attributes. An element type declaration associates the element name with the element content. XML provides a variety of facilities for the construction of the element content, namely *sequence* of elements, *choice* of elements, cardinality constructors (?, *, +), the types EMPTY, ANY, #PCDATA, and *mixed* content. *sequence* requires elements to have a fixed order, whereas *choice* expresses element alternatives. An EMPTY element has no content, whereas ANY indicates that the element can contain data of type #PCDATA or *any* other element defined in the DTD (comp. [5]). Mixed is useful when elements are supposed to contain character data (#PCDATA), optionally interspersed with child elements.

The name of the attribute list must match the name of the corresponding element. The list of attribute declarations consists of the attribute names, their types and default declarations.

Entity declarations serve the reuse of DTD fragments and text as well as the integration of unparsed data. An entity declaration binds an entity to an identifier. Being external entities, unparsed entities always have notation references.

Notation declarations provide a name for the format of an unparsed entity. They might be used as reference in entity declarations, and in attribute-list declarations as well as in attribute specifications.

Processing instructions (PIs) play an important role while checking integrity constraints of valid document instances. PIs have to be checked while parsing a document instance. The XML parser validates the document instance first and consumes the processing instructions known to XML. Then an application can handle more specific PIs.

3   Related Work

In this paper, we examine the mapping of UML to XML. However, we are aware of other data models for semistructured data. Since semistructured data are often regarded to be "schemaless", most of the following relevant models apply to instances and do not give an instance independent description of the data. We discuss some of these models, namely the Object Exchange Model, the Document Object Model, and the XML Data Model as described by the W3C.

The Object Exchange Model (OEM) was developed for the exchange of data between heterogeneous sources at Stanford University before it became a model for semistructured data [4]. OEM represents the hierarchical structure of document data using graphs. The graph based approach makes it possible to query document instances. Such queries handle various types of corresponding data from different documents. However, documents often are clustered in document classes like product catalogs, books, etc. OEM does not take this into account.

With the idea of providing an API for manipulating document instances, the W3C has defined the Document Object Model (DOM) [9]. A programmer equipped with DOM is in a position to create, navigate the structure, add, modify, or delete elements and content of any documents. DOM proposes an implementation neutral and language independent API that assigns elements and attributes to interface definitions formulated in OMG's IDL. XML's logical structure parts are handled as objects in the same way. These objects are arranged in a graph to reflect the document structure. As a side effect, the DOM comes up with an instance based model that defines the logical structure of documents.

To visualize the structures of documents, W3C designed the XML Data Model [3]. The lexical structure is represented by a tree. The tree is extended by additional links representing references within the tree, thus generating a graph. The node types correspond to XML's logical structure. Being very simple and providing no more than a baseline, the XML Data Model supports conceptual modeling of document instances.

The models reviewed above only deal with modeling on instance level. Apart from this, modeling of more general structures on schema level, namely document classes, is desirable. This would facilitate a more abstract view on document structures according to database modeling, or software engineering respectively. By conceptually modeling DTDs with UML, the focal point of our approach is the schema level.

Our approach should not be confused with the XML Metadata Interchange (XMI) format [7]. The XMI describes the exchange format for UML class diagrams in an XML style. This way, XMI enables the exchange of schema information for cooperative work. However, our interest focuses on the development of DTDs by means of modeling of UML class diagrams and their mapping into XML DTDs.

4   Relevant Modeling Concepts and Their Transformation

According to Rumbaugh, Jacobson, and Booch, *the Unified Modeling Language (UML) is a general-purpose visual modeling language that is used to specify, visualize, construct, and document the artifacts of a software system* [8, p. 3]. For this reason, UML is widely accepted. Furthermore, UML incorporates an easy-to-understand graphical notation and visualization of interrelations, and correlations supporting the modeling of static structure and dynamic behavior of systems.

For our transformation approach, relevant constructs are those of UML's Static View and Model Management View. The Static View consists of classes and their relationship such as association, generalization, and various kinds of dependencies. The Model Management View describes the organization of the model itself, i.e., a model consists of a set of packages which in return are made up of model elements, e.g., classes as well as packages, recursively.

In the following we describe the UML constructs while explaining the transformation of these constructs into DTD fragments. Considering the pure conceptual view, this chapter provides an intuitive transformation into DTDs. For requirement reasons, it is the designer's responsibility to tune the result. Additionally, the next chapter addresses the implementation dependent extensions which integrate tuning aspects into the diagrams.

4.1   Classes

Classes, characterizing objects with the same properties, consist of a class name, attributes, and methods. All objects must satisfy the constraints given by their class. Analogously, the content of an XML element is constrained by its type declaration. Thus, we transform UML classes into XML element type declarations.

The class names become the names of the element types. The attributes are transformed into element content description. This is motivated by considering class attributes to be the composite parts of a class. Therefore, dealing with attributes is analogous to aggregations as described below. However, UML aggregations do not support order, thus we determine the order to be from top to bottom.

The name of an attribute provides the name for the element type in content specification. We notice, the semantics of the UML attribute construct and XML content model for elements diverge. In UML attribute names are mandatory whereas the attribute types are optional. In contrast, an element content only consists of type names. There are no access names bound to these type names as in programming languages. For this reason, attribute names imply their attribute type names in our approach. If there is no class representing a suitable declaration for an attribute type, the attribute type is assumed to be an element whose content type is #PCDATA. Multiplicity specifications of attributes are mapped into cardinality specifications (with specifiers ?, *, +) used for element content construction, e.g., [0..1] maps into ?. Currently, class attributes and initial values are not supported. Figure 1 depicts a simple example of an author class with its transformation result.

```
<!ELEMENT author (firstname, middlename*,
                  lastname, suffix?) >

<!ELEMENT firstname  (#PCDATA) >
<!ELEMENT middlename (#PCDATA) >
   . . .
   . . .
```

author

firstname
middlename [0..*]
lastname
suffix [0..1]

Fig. 1. Sample transformation

4.2   Aggregation and Composition

An aggregation specifies a whole-part relationship between an aggregate — a class that represents the whole — and a constituent part. Aggregation is a more general form of composition where constituent parts directly depend on the whole part; they cannot exist independently. Composition mainly applies to attribute composition described above.

The only DTD construct that expresses a part-of relationship is the element content. XML element types can appear in several element contents. In general, it is impossible to restrict this appearance to exactly one content. Therefore, the concept of aggregation might be the predominant form that applies to XML element's content.

An aggregation might be refined with multiplicity. This multiplicity specification is semantically as rich as the cardinality specification of elements in element type content defined in XML. Thus, all forms of cardinality can be handled. Figure 2 shows an alternative modeling of Figure 1 using aggregation. The order in the sequence is implicit given by the notation itself.

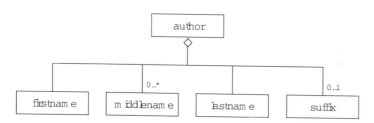

Fig. 2. Simple Aggregation

4.3 Generalization

The generalization concept defines a relationship between classes to build a taxonomy of classes: One class is a more general description of a set of other classes.

Figure 3 shows the class person specialized as employee or student, respectively. In addition, UML allows to constrain generalizations as follows: If subclasses have no common instances a comment with the keyword {disjoint} expressing the constraint has to be placed near the generalization symbol. Otherwise, the generalization is assumed to be overlapping which can be made explicit by providing the comment {overlapping}. If each instance shall be assigned to at least one subclass, the generalization must be constrained to {complete}. The counterpart of {complete} is {incomplete} being the default. Being orthogonal to one another, each pair reflects a different aspect of constraint.

Our approach adopts the idea of the transformation of the generalization construct into the Relational Model. To reflect the constraints' characteristics in overlapping generalizations, we propose to transform the general class into a superelement, i.e., the element type originating from the superclass (Figure 3). Such a superelement receives an additional "artificial" ID attribute declared #REQUIRED. Similarly, each subelement is augmented by a #REQUIRED IDREF attribute. Thus, a subelement instance references its general component. Note, the proposed transformation naturally applies to any generalization specified

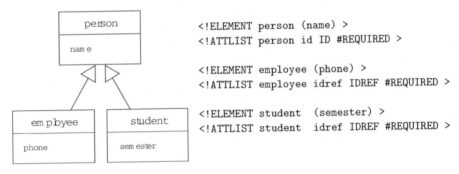

```
<!ELEMENT person (name) >
<!ATTLIST person id ID #REQUIRED >

<!ELEMENT employee (phone) >
<!ATTLIST employee idref IDREF #REQUIRED >

<!ELEMENT student  (semester) >
<!ATTLIST student  idref IDREF #REQUIRED >
```

Fig. 3. Generalization transformation

as {overlapping} and {incomplete}. To ensure the {complete} constraint in this context, we must prevent unreferenced superelement instances, therefore an additional integrity check of a document is necessary.

For generalizations specified as disjoint, we transform each class into its referring element type. Hereby, the instances are disjoint by nature. If such a generalization is further constrained by {complete}, the element type declaration of the superelement must be omitted in order to deny an element instance of that kind.

4.4 Association

Associations are relationships that describe connections among class instances. An association is a more general relationship than aggregation or generalization. A role may be assigned to each class taking part in an association, making the association a directed link. Associations are transformed into links between instances of two or more document classes in the sense of XML Linking Language (XLink) [11]. Because of the draft status of XLink, at this point, we only introduce some first ideas.

A simple link expresses a binary association. The following part of an attribute-list declaration for the referencing element is useful for representing an association.

```
xlink:type    (simple)     #FIXED "simple"
xlink:href    CDATA        #IMPLIED
xlink:title   CDATA        #IMPLIED
xlink:show    (new|replace|embed|undefined)  #FIXED "replace"
xlink:actuate (onLoad|onRequest|undefined)   #FIXED "onRequest"
```

For n-ary associations and for binary associations with multiplicity n:m, we suggest to use extended links. The element of xlink:type extended could contain element types of xlink:types locator, arc, resource, and title. The usage of them and their attributes depends on the kind of association. A locator-type or a `resource`-type element type is necessary to specify the resource instances related to the association. `arc`-typed elements direct the link. A `title`-typed element type provides a name for the link.

4.5   Packages

A package is a concept for combining several concepts such as classes and packages into groups. Nesting, generalization, and import is applicable to packages such that packages may contain other packages, one or more packages can be the specialization of other packages, and packages can import other ones. Actually, packages also have names.

The only concept for modularization that XML offers is the entity concept. We distinguish complex and simple entities. Complex entities are always a set of complete markup declarations that again may contain entity declarations themselves, thus nesting of entities is supported. This nesting corresponds to the nesting of packages.

The packages appearing in a generalization relationship are transformed into single files. The specific subpackage declares an external entity (`<!ENTITY % su-`*perpkg* `SYSTEM "` *superpkg*`.dtd">`) and resolves the entity reference (`%`*superpkg-name*`;`) in its DTD. Parts of a specific subpackage overwrite parts of the superpackage by declaring these corresponding parts as entities of the same name. The entity reference of the overwriting part must be processed first, because an entity reference cannot be overwritten once it is resolved.

The usage of the import dependency between two packages leads to semantic losses at transformation into DTDs. The non-transitive extension of the namespace is not manageable in XML. Name clashes must be avoided. The transformation of the import dependency is similar to the transformation of the generalization concept.

5   Implementation Dependent Concepts

We have to extend UML to take advantage of all facets that DTD concepts offer. Additionally, these extensions enable the tuning of the resulting models for guiding the transformation process. It is important for the extensions to be UML compliant and as minimal as possible.

5.1   Classes

We extend UML with diverse class and attribute stereotypes. They are useful for implementation dependent modeling of element attributes, element content, entity declarations, and notation declarations.

Meta Data Attributes. Up to now in our transformation a class attribute refers to element content. XML, however, allows additionally to constrain an element type by means of (element) attributes, i.e. attributes providing additional (meta) information for the element itself. In order to model such *meta data attributes*, we mark them «meta». Thus, we are able to distinguish between "normal" data attributes and meta data attributes.

Meta data attributes are transformed into attributes of the element's attribute-list. Specifically, the name of a class attribute and the type are transformed into their equivalent counterpart. UML does not constrain types. Thus, we can use specific XML attribute types like CDATA, ID, and even enumeration types.

Multiplicity specifications and initial values yield default declarations. For attributes, we restrict multiplicity to [0..1]. This specification translates into either an attribute with the default declaration #IMPLIED or an attribute with a given default value. Both forms can be distinguished, because an #IMPLIED attribute must not have a default value, but attributes with default must have one. If attribute multiplicity is not explicit, the attribute is assumed to be #REQUIRED. Being valid for each instance of the corresponding class, class attributes refer to #FIXED attributes.

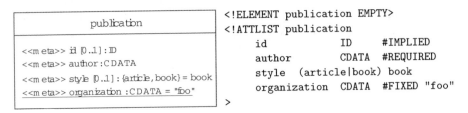

Fig. 4. An example transformation

Figure 4 depicts a simple example of a publication class with its transformation result. Note, in contrast with XML, a UML class diagram emphasizes the logical interrelation of element type declarations with their attribute-list declarations.

The Content Stereotype. To model the element content types ANY and #PCDATA, we create the «content» stereotype (see Figure 5). There is no need to model the content type EMPTY. It is up to the designer to use a UML comment.

Fig. 5. Class stereotype content

The Entity Stereotype. The «entity» stereotype refers to parsed and unparsed simple entities. In order to distinguish global entities and parameter entities on modeling, we only separate them by name, i.e., parameter entities, only visible on DTD level, get the prefix % (see Figure 6 (a)).

```
<!ENTITY % list "ol | ul | dl" >    <!ENTITY article SYSTEM "./article.tex"
                                                         NDATA latex >
```

(a) parameter entity   (b) unparsed entity

Fig. 6. Simple entities

Complex entities are always a set of complete markup declarations. In contrast, simple entities are represented by classes stereotyped «entity». Being either an *entity value* or an external identifier, the *entity definition* as well as the reference to the corresponding notation declaration is a UML comment as shown in Figure 6 (b).

The Notation Stereotype. In contrast to other markup declarations, notation declarations have a special meaning. Notations refer to helper applications which are capable to process data. A notation declaration binds an external identifier for a given notation to a name that can be used as a reference in entity declarations, and in attribute specifications of attribute-list declarations.

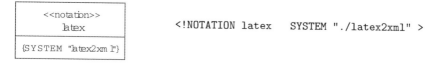

```
<!NOTATION latex    SYSTEM "./latex2xml" >
```

Fig. 7. The notation declaration

Referring to helper applications, a notation suggests itself to be a «utility» class. For this reason, we introduce the stereotype «notation» to the class latex in the example of Figure 7. The external application required is identified by its external identifier {SYSTEM "latex2xml"}.

5.2   Aggregation

In Section 4.2 we proposed a standard transformation for aggregations using element content. Nonetheless, to model XML content structures like *sequence*, *choice*, and *mixed* properly, still causes some problems.

Sequence. UML lacks the support of ordering of classes, attributes etc. explicitly, because it is not necessary to have aggregations or compositions ordered in programming objects. By default, we consider an aggregation to be transformed into XML *sequence*. This implies an implicit order given by the model. To model a sequence explicitly, the comment {sequence} may be used. We are able to specify the order of elements with the help of comments (see Figure 8). The first element is marked as {1}, the second as {2}, and so forth. Moreover, a sequence may obtain cardinality, e.g. {sequence : 1} (see Figure 8). In case of the trivial cardinality (1), both the colon and the 1 can be dropped.

We emphasize that UML propagates the comment {ordered} for usage in aggregations. In contrast to class sequences, the specifier {ordered} only applies to the order of objects of the same class. Hereby, {ordered} appears with cardinality greater than one.

Choice. Unfortunately, there is no appropriate concept available in UML that tackles the problem of handling alternative classes which are members of other classes. Often, choices originate from generalizations as shown in Section 5.3. If nevertheless a choice should be modeled, e.g., to be closer to the implementation, it is expressible with a clumsy structure of some helper classes and a generalization.

Choices describe alternative element contents that are mutual exclusive. Principally, the {xor} constraint is applicable for this kind of content. However, it conflicts with the repetitive use of the choice constructor due to the cardinality specification. Therefore, we propose the {choice} constraint of which the use is analogous to the mapping of the sequence construct. Figure 8 shows an application example taken from ETDML DTD [6]. With this semantics, we are given an elegant method to describe an element's *choice* content. The only difference between *sequence* and *choice* is found in the keywords *sequence* and *choice* which describe the type of aggregation. Analogous to *sequence*, *choice* can have its own cardinality — in Figure 8 it is expressed by {choice : 0..*} and this directly refers to the * in (p | citation | table | mm)*.

It is easy to combine the concepts of sequence and choice with one another in any way that is necessary for the model. Figure 8 illustrates that the structure of model schema follows the grammar rule provided by the element definition naturally, and is therefore easy to understand on the one hand, and on the other hand easy to handle while designing a model.

Mixed Content. The *mixed content* is subdivided into simple mixed content, and complex mixed content. We speak of complex mixed content when elements of that type of content may contain #PCDATA, optionally interspersed with child

```
<!ELEMENT chapter
   ( head?, (p | citation | table | mm )*, section*)
>
```

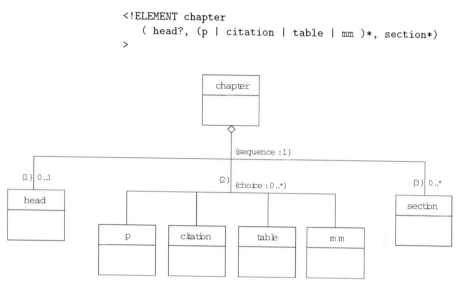

Fig. 8. A more complex example taken from ETDML DTD

elements. Using the *choice* modeling concept introduced above, the *mixed content* has to be expressed by a choice with cardinality 0..*. Appearing once within the choice, the «content» class #PCDATA is the crucial feature which makes *mixed content* different from *choice content*. Simple mixed content is only made up of #PCDATA.

5.3   Generalization

The mapping of the generalization concept reviewed in Section 4.3 results in different element types corresponding to their classes. For some DTDs, it might be desirable to collapse the generalization structure into one class definition. Therefore, we adapt the idea of universal relations for element type declarations. To express this transformation, the generalization specification obtains a {discriminator} constraint. A single element type combines all variant properties found in the generalization, i.e., all element attributes as well as the element contents have to be merged in one attribute-list declaration or one element content specification, respectively.

With application dependent processing instructions to check integrity constraints, we support attribute alternatives. Attribute alternatives require attributes to have null values which are produced by #IMPLIED default declarations. In our approach discriminator attributes indicate the class an instance belongs to. For example,

```
uml:discriminator:author (author-person|author-org) #REQUIRED
```
contained in the attribute-list declaration of author supplies the discriminator attribute for the generalization in Figure 9.

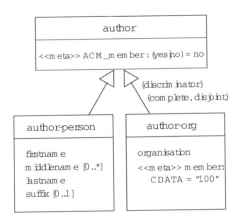

Fig. 9. Generalization using discriminator

The following processing instruction handles variant attributes of the sub-classes; the suffix of the qualified attribute (`author:author-org`) must match a value of the discriminator attribute:

```
<?UML:DISCRIMINATOR author
    <!ATTLIST author:author-org  member CDATA "100">
?>
```

The merging of element contents depends on the generalization constraints as presented in the table below; $cont_i$ denotes the element content of the transformed class i, with "1" representing the superclass:

	incomplete	complete
disjoint	$cont_1,(cont_2\|cont_3)?$	$cont_1,(cont_2\|cont_3)$
overlapping	$cont_1,cont_2?,cont_3?$	$cont_1,((cont_2,cont_3?)\|cont_3)$

6   Conclusion

To bridge the gap between object-oriented software design and the development of XML data schemata, we use essential parts of static UML to model XML data schemata. The focal point of our approach is to find a suitable mapping from UML into XML DTDs. Being close to the real world, UML concepts for DTD design improve the design process and clarify the understanding of DTD semantics.

Starting from the abstract modeling, we develop an almost complete mapping. To exploit all DTD constructs, we slightly extended UML, thus providing a more flexible modeling approach. All proposed extensions are UML compliant and as minimal as possible.

For future work, the transformation of UML concepts must be validated in case studies. Furthermore, we plan to adapt our approach to XML Schema. That is a DTD based method that allows the user to design DTDs with additional constraints. In previous work we applied our conceptual modeling approach to information integration [1]. We plan to pursue the UML-XML mapping approach in the context of information integration. For practical purposes, a CASE tool adopting the transformation is desirable. So far, we ignored the dynamic aspects when defining UML classes. We further have to investigate if and how class specific dynamic specifications are treated.

We believe that a combination of UML design with a UML to XML mapping is important enough to support document design in XML, thus influencing future developments of XML.

References

1. R. Conrad and D. Scheffner. Conceptual Modeling of XML for Integration Purposes (in German). In R.-D. Kutsche, U. Leser, and J.C. Freytag, editors, *4. Workshop "Föderierte Datenbanken"*. TU Berlin, 1999. Forschungsberichte des Fachbereichs Informatik.

2. R. Conrad and D. Scheffner. XML Conceptual Modeling using UML. Technical report, Institute of computer science, HU Berlin, 2000. Extended version; to appear.

3. World Wide Web Consortium. The XML Data Model. http://www.w3.org/XML/Datamodel.html, Jan 2000.

4. Roy Goldman and Jennifer Widom. Dataguides: Enabling query formulation and optimization in semistructured databases. In Matthias Jarke, Michael J. Carey, Klaus R. Dittrich, Frederick H. Lochovsky, Pericles Loucopoulos, and Manfred A. Jeusfeld, editors, *VLDB'97, Proceedings of 23rd International Conference on Very Large Data Bases*, pages 436–445. Morgan Kaufmann, 1997.

5. Michel Goosens and Janne Saarela. A Practical Introduction to SGML. In *TUGboat*, volume 16(3), 1995.

6. Neill Kipp, Zaodat Rahman, and Marc Bjorklund. Electronic Thesis and Dissertation Markup Language Version 2. Technical Report Version 2.0 Beta, Virginia Polytechnic Institute and State University, December 1998.

7. OMG. XML Metadata Interchange (XMI). Technical report, OMG, Oct 1998.

8. James Rumbaugh, Ivar Jacobson, and Grady Booch. *The Unified Modeling Language Reference Manual*. Addison Wesley, 1999.

9. World Wide Web Consortium. Document Object Model (DOM) Level 1 Specification, Version 1.0. Technical Report REC-DOM-Level-1-19981001, W3C, October 1998.

10. World Wide Web Consortium. Extensible Markup Language (XML), Version 1.0. Technical Report REC-xml-19980210, W3C, February 1998.

11. World Wide Web Consortium. XML Linking Language (XLink). Technical Report WD-xlink-2000021, W3C, February 2000.

Metadata Engineering for Corporate Portals Using XML

Peter Aiken [1] and Kathi Hogshead Davis [2]

[1] Department of Information Systems, Virginia Commonwealth University
1015 Floyd Avenue, Richmond, VA 23284-4000
{paiken@acm.org}

[2] Department of Computer Science, Northern Illinois University
Dekalb, IL 60115
{kdavis@cs.niu.edu}

Abstract. Careful analysis and preparation is required in order to prepare for XML-based delivery of data via Corporate Portals. This process is refereed to as Metadata Engineering. This presentation describes the use of the metadata model to guide the metadata engineering as a precursor to metadata implementation in preparation for XML-based delivery.

In metadata engineering, logical models representing the "as is" system data are developed by reverse engineering the data. Once derived this metadata is typically maintained using entity relationship diagrams. Metadata about entity relationship diagrams can be maintained with a many to many association between two metadata entities: LOGICAL DATA ENTITY and LOGICAL DATA ATTRIBUTE. The two metadata entities form the basis of a metadata model that can be used as a structure facilitating the subsequent metadata implementation. Understanding the requirements of metadata engineering is a necessary prerequisite to delivering data via Corporate Portals via XML.

Reusable metadata have proven to be valuable business and/or system engineering tools. Metadata management and reuse is key to effective strategic systems implementation such as: corporate portals/XML; data warehouses; e-commerce solutions; ERP implementations; information logistics networking; legacy system migration/integration; organizational repositories, etc.

Within this broad scope of work, metadata engineering initiatives address pervasive and significant issues impeding the ability of organizations to manage their data predictably—on time, within expected cost, and with expected functionality. If organizational data or metadata is not maintained, it cannot be useful for systems evolution. Ideally, organizational metadata should be built or rebuilt once and maintained continuously thereafter. Consequently, organizations have had difficulty gaining metadata-engineering experience.

The evolution of strategic data use from an ad hoc labor-intensive activity to an engineering-based discipline is supported by appropriate technologies. The approach

A.H.F. Laender, S.W. Liddle, V.C. Storey (Eds.): ER2000 Conference, LNCS 1920, pp. 572-573, 2000.
© Springer-Verlag Berlin Heidelberg 2000

(illustrated below left) consists of two complimentary activities: 1) extracting, understanding, and improving organizational metadata and 2) effectively incorporating the metadata into organizational business and systems engineering efforts. The approach, built on the robust, information-engineering base, extends elements of the Zachman framework to develop a core set of organizational metadata models (illustrated below right).

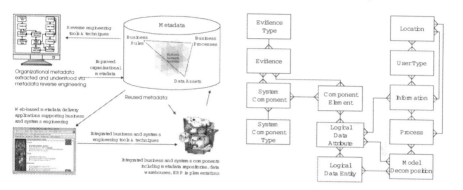

After identifying this core metadata model, our engineering approach has initially focused in two areas: 1) Improving the ability of organizations to predict and control quality, schedule, cost, cycle time and productivity when performing data engineering activities. 2) Improving data engineering capabilities. We are actively researching these areas.

The Role of Information Resource Management in Managing a Corporate Portal

Arvind D. Shah

Performance Development Corporation
ads@perfdev.com

Abstract. The Corporate Portal is a central gateway to the processes, databases, systems and workflows of an enterprise. When personalized to the job responsibilities of employees via the Intranet, the corporate portal provides a seamless, single point of access to all of the resources that employees need to do their jobs.

When further personalized securely via the Internet and Extranets to the interests of suppliers, customers and business partners, the corporate portal becomes the integrating conduit of the many disparate databases, systems and workflows each enterprise uses to carry out business with others. It also becomes a single place to manage rapid enterprise change.

Implementation of a corporate portal requires interfaces with legacy systems and data warehouses. An application of the enterprise architecture technique lends itself to a logical design for an enterprise portal. The portal grows through evolutionary stages. It is therefore, essential that all portal planning and development be based on the adaptive architecture, which is systematically maintained. The configuration of the portal continuously changes as the ebusiness changes. Metadata initiated during the architecture stage will play a key role in maintaining and managing the corporate portal on an ongoing basis. The presentation also covers issues Data Administration has to address in order to assure successful functioning of a corporate portal.

Outline:

- Corporate portal – evolutionary stages of its growth
- Corporate portal in relation to legacy databases and data warehouses
- Key components of an enterprise portal
- Corporate portal architecture development – process, data and technology
- Corporate portal change management
- The role of metadata in portal management
- Problems and pitfalls to avoid
- The role of Information Resource Management in managing a portal

What You Will Learn:

- What is a corporate portal and how does it grow?
- How are ERP systems and data warehouses interfaced with a portal?
- How do enterprise architecture techniques help in developing and managing an effective portal?
- What are the roles of DA, DBA and IRM groups in managing a portal?

A.H.F. Laender, S.W. Liddle, V.C. Storey (Eds.): ER2000 Conference, LNCS 1920, p. 574, 2000.
© Springer-Verlag Berlin Heidelberg 2000

The Five-Tier Five-Schema Concept

Michael H. Brackett

President, DAMA International

Introduction

Many types of data models are being developed using a variety of different methods and notations. Some of these data models are oriented toward physical implementation, others toward understanding business data, and a few toward business managers. Many of the data models are overloaded with detail and much of that detail is not relevant to the intended audience. The result is a confusion about what the data models represent, their intended audience, and their purpose. This abstract provides a framework for understanding the different types of data models through a five-tier five-schema concept developed by the author.

Three-Schema Concept

An initial two-schema concept was promoted when physical databases were first being developed. The concept consisted of an internal schema that represented the way data were stored in databases and an external schema that represented the way data were used by applications. The problem with this concept was that the external schema seldom matched the internal schema and it was difficult for many different applications to use one internal schema. This led to the development of many different databases with internal schema that matched the external schema.

A three-schema concept resolved the discrepancy between the internal and external schema. A conceptual schema was added that was the common denominator between the internal schema and the external schema. It became the basis for the evolution of database management systems, and much of the conceptual and logical data modeling that is done today.

Five-Schema Concept

As business clients became involved in data modeling to capture their business knowledge, they began asking what data were being normalized. That question could not be answered based on the three-schema concept. It became very clear after many discussions with business clients that there was a business schema representing the transactions used by the business. The business schema were normalized into the external schema.

The author defined a four-schema concept that included a business schema and changed the names of the schema to avoid the strong physical perspective. The external schema were renamed to data view schema, the conceptual schema were renamed to logical schema, and the internal schema were renamed to physical schema.

The emergence of client / server technology and its evolution into networks opened up the concept of partitioning data and deploying data to different data sites. The author defined a five-schema concept that consists of business schema, a data view schema, a logical schema, a deployment schema, and a physical schema. The

A.H.F. Laender, S.W. Liddle, V.C. Storey (Eds.): ER2000 Conference, LNCS 1920, pp. 575-576, 2000.
© Springer-Verlag Berlin Heidelberg 2000

business schema are normalized to data view schema, the data view schema are optimized to logical schema, the logical schema are deoptimized to deployment schema, and the deployment schema are denormalized to physical schema.

Three-Tier Five-Schema Concept

The five-schema concept worked well at a detail level, but another problem emerged when managers and executives were included in the data modeling process. The detailed logical data models were not appropriate for managers or executives and conceptual data models were used to generalize the data needed to support the business. Data modeling tools were emerging, but they could not automatically denormalize data from the logical schema to the physical schema, or support a deployment schema. The data models that were developed were largely physical data models. A true logical data model was often referred to as a conceptual data model.

The author developed a three-tier concept to resolve this conceptual data model confusion. The strategic tier is a logical schema that addresses the executive audience, and contains information about business objects and business events in the real world that are relevant to the business. The tactical tier is a logical schema that addresses middle managers, and contains more detail about the business objects and events than the strategic tier. The detail tier is a logical schema that contains all the detail necessary to develop a database that supports the business activities. The combination of these three tiers and five schema in the detail tier provide a three-tier five-schema concept for data models.

Five-Tier Concept Five-Schema Concept

Data warehousing and data mining have recently evolved resulting in many discussions about the development of data models. The author enhanced the three-tier five-schema concept to a five-tier five-schema concept to bring formality and understanding to the development of data models. The third tier of the three-tier five-schema concept became an operational tier for the management of data to support day-to-day business operations and operational decision making. The new fourth tier is an analytical tier for proving or disproving known or suspected trends and patterns and supports management decision making. The new fifth tier is a predictive tier for finding unknown or unsuspected trends and patterns, and generally includes data mining.

The schema development for the analytical and predictive tiers is down the logical schema to the desired tier, then to the right through the deployment schema to the physical schema. This approach resolves the issues about data normalization by introducing an operational data normalization, an analytical data normalization, and a predictive data normalization. The data are normalized differently for different purposes in the logical schema and are then denormalized to the physical schema in that tier.

Many details are still being worked out for the data models representing the seventeen schema in the five-tier five-schema concept, particularly for the schema in the fourth and fifth tiers. The five-tier five-schema concept provides a framework for formalizing the development of data models, their contents, their purpose, and their intended audiences. It will help bring clarity, consistency, and understanding to the development of all data models.

Documenting Meta Data Transformations

Alex Friedgan, Ph.D.

Data Cartography, Inc. (www.datacarta.com)
alex.friedgan@usa.net

Abstract. The adequate cross-referencing of a physical database and its logical model is crucial for software maintenance support. The paper describes approach, which eliminates the need for the logical model to closely follow the physical design. It allows modernizing of the logical model with new ideas and concepts without losing information needed for the support of an installed base of applications. Examples of meta-data transformations and corresponding documentation requirements are included.

1 Introduction

With the high turnover rate of the industry, adequate documentation is a must during software maintenance. This need gives an excellent opportunity to expand the Data Administration function into software maintenance support.

The maintenance enhancements don't lead to the design of a completely new database, but are limited instead to the confines of existing data structures. With limited project scope the developers have no motivation to understand the logical data model. Abstraction of the existing tables into the logical model and consecutive interpretation of the new elements back into the physical world becomes the Data Administrator's (DA) burden. An adequate physical-logical mapping is crucial here.

For a number of reasons the physical implementation differs from the logical data model. The usual performance improvements for the operational data store lead to denormalization. The attributes, inherited from the supertype, are repeated in multiple subtype tables. The atomic data warehouse tables add the overhead of keeping value changes over time. The data mart design collapses a chain of parent-child relationships into a dimension table.

2 Base Attribute Mapping

The simplest way to map physical to logical is to identify an attribute, which was the source for each column. It is called the base attribute.

However, the base attribute mapping is imprecise and generic. It's difficult to help developers who deal with details if all you have is a big picture vaguely correlated to the real world. Any change still requires additional interpretation of the physical schema. This time-consuming effort is repeated over and over again.

A.H.F. Laender, S.W. Liddle, V.C. Storey (Eds.): ER2000 Conference, LNCS 1920, pp. 577-578, 2000.
© Springer-Verlag Berlin Heidelberg 2000

Tailoring of the logical model after the physical data structures minimizes this effort, so the models used for maintenance have a tendency to closely follow the physical schema.

3   Transformation Mapping

The objective is to strike a balance between a need for information, and an ability to maintain the documentation up to date. This can be accomplished by building it as an extension to the base attribute mapping; and keeping it relatively simple, limiting it to the frequently used scenarios.

Some of the transformations and corresponding requirements for documenting the physical-logical mapping are described below:

- Redundant Attribute(s). To improve application performance, attribute(s) from a parent entity might be redundantly stored in a child entity. The transformation could be documented by referencing the base attribute, and all relationships within the access path from the parent to the child entity. The resultant column definition can be compiled by tracing the access path from the child entity back to the base attribute.
- Pivoting Transformation. Pivoting transforms a relational table into a 'spreadsheet'. It creates a repeating attribute (or repeating group of attributes). The table grows sideways, and shrinks in the number of rows. The transformation documentation includes the base attribute, the pivoting attribute and the value of the pivoting attribute.
- Data Warehouse Snapshots, etc.

Meta Data Transformations Tool
A tool was developed to perform Meta Data Transformations. It accepts data from the CASE tool and the database catalog, and provides a graphical interface for mapping.[1]

Now, all the change requests were immediately decoded to the logical model constructs. In that context, they were much easier to understand and steer towards a more relational resolution.

Advantages of Documenting the Meta Data Transformations
- Better documentation provides for easier knowledge transfer, streamlines the software maintenance process.
- Full life-cycle software support promotes the Data Administrator's function within the IT organization.
- Flexible cross-referencing eliminates the need for a logical model to closely follow the physical design. It allows the modernization of the logical model with new ideas and concepts without losing information needed for the support of an installed base of applications.

---

[1] Application developed by Mark A. Friedgan hubrix@hubrix.com

Advanced Data Model Patterns

David C. Hay

Essential Strategies, Inc.
davehay@essentialstrategies.com

The book *Data Model Patterns: Conventions of Thought*[1] describes a set of standard data models that can be applied to standard business situations. These patterns, it turns out, occur on several levels. At the basic level are models of the things seen in business. The patterns in the book are a bit more abstract than conventionally seen, but they do describe things that are easily recognizable to anyone: people and organizations, products, contracts, and so forth.

There is a more abstract level of modeling, however, which is necessary when the things being modeled don't fall into these tidy categories. This level, also described in the book, is the subject of this paper.

There is a problem with entities that describe products, for example. For PRODUCT to be an entity the attributes for all occurrences of each must be the same. This simply is not always true.

The attributes of a compressor are quite different from the attributes of a computer or a barrel of crude oil. We would like to have a single concept for "Product", but that concept has many different flavors.

We could define a sub-type for each PRODUCT TYPE, but when new product types are being invented all the time, the data management task would be impossible.

This paper presents an approach to addressing this problem. After summarizing the standard business level models, it proposes a model with PARAMETER itself as an entity, then adding an entity for assigning PARAMETERS to each kind of product. A PRODUCT, then,

This approach has wide applicability. It was originally defined for a bank that had trouble with its marketing department. They were adding new product types every week, and the attributes of each were different. It was then used in a laboratory, with different laboratory test types collecting different kinds of information. Its most ambitious use was in pharmaceutical clinical research, where every clinical study is defined differently.

The technique has its disadvantages, and these must be addressed. Not the least of the problems is that of mapping legacy systems to a database with this architecture. You are not really mapping old columns to new attributes anymore. You are mapping old columns to new PARAMETER ASSIGNMENTS. This gets tricky.

From this work, however, it turns out that this model is very close to the model of the relational database itself. The paper will show how this works.

If this isn't advanced enough, and there is time, the author's model of double-entry bookkeeping can also be presented.

---

[1] David Hay, *Data Model Patterns: Conventions of Thought,* Dorset House Publishers, Inc. (New York: 1996). This article is largely derived from this book.

A.H.F. Laender, S.W. Liddle, V.C. Storey (Eds.): ER2000 Conference, LNCS 1920, p. 579, 2000.
© Springer-Verlag Berlin Heidelberg 2000

Information Quality at Every Stage of the Information Chain

Elaine Stricklett

Vice President

Acton Burnell, Inc.
1500 N. Beauregard St.
Suite 210
Alexandria, VA 22311

email: estricklett@actonb.com

Larry English tells us that "quality is free." The lack of quality is what is costly. Ensuring the quality of the data in an organization involves those who are involved in every stage of an organization's entire information chain – starting with the planning and conceptual modeling of the data. Establishing information quality requires understanding – and practicing – the principles of information and data quality at every stage of the information chain.

Until data warehouses emerged in the early 90's, most people just did not see the need to worry too much about dirty data. Information systems professionals could work wonders with those physical fields, records, and files to keep those operational systems humming. However, as soon as someone pulls a piece of data out of that operational system and gives it to a decision-maker, the scenario changes. Where did this data come from? What does this field mean? Does this amount include the employee's dependent health care costs or just the employee's health care costs? And the questions continue. Often, decision-makers take the data at face value and act on the "facts" placed before them.

Unless an enterprise has made extraordinary efforts, it can expect data (field) error rates of approximately 1-5% (Error rate = number of erred fields/number of total fields.)

That which doesn't get measured, usually doesn't get managed, so the enterprise should expect that it has other serious data quality problems as well. Quite often, enterprises are bedeviled by redundant and inconsistent data on their databases and they do not have the information they really need.

Virtually everyone in an organization touches data in some way. Everyone (including senior managers, process and data owners, and information professionals) must understand that the principles of information and data quality involve very real, practical procedures that can be carried out at every point in the information chain.

A.H.F. Laender, S.W. Liddle, V.C. Storey (Eds.): ER2000 Conference, LNCS 1920, pp. 580-581, 2000.
© Springer-Verlag Berlin Heidelberg 2000

Implementing data quality improvements often requires a change in the basic culture of the organization. Data quality improvements that result from a change in the culture of the organization are frequently more sweeping and more enduring.

In this presentation, we will look at examples of dirty data from research and from my own experience (just in case there are any skeptics in the room!), the consequences of dirty data, and methods by which we can ensure quality data. We will also examine how information professionals can effect real and permanent improvements in the quality of the data and information in their own organizations.

A Fact-Oriented Approach to Business Rules

Terry Halpin

Microsoft Corporation, USA
TerryHa@microsoft.com
www.orm.net

Abstract. Effective database applications, business rules management, data warehousing, enterprise modeling and re-engineering all depend on the quality of the underlying data model. To properly exploit relational, object-relational or object database technology, a clear understanding is needed as to how to create conceptual business models, transform them to logical database models for implementation on the chosen platform, and query the populated models. Fact-orientation provides a truly conceptual way to accomplish these tasks, facilitating communication between the modeler, the domain expert and the application. This presentation provides insights into the fact-oriented approach for modeling and querying information systems, focusing on verbalization and instantiation of data use cases for capturing business rules, including recent work on negative and default rule verbalizations.

1   Introduction

The Unified Modeling Language (UML) is widely used for object-oriented code design, and can be used for database modeling, where its class diagrams provide an extended version of Entity-Relationship modeling (ER) [4]. Although facilitating the transition to object-oriented code, UML's implementation concerns and de-emphasis of value-based identification and temporal aspects render it less suitable for developing and validating a conceptual data model with domain experts. UML advocates a use-case-driven modeling process, but since use cases focus on process rather than data, the move from use cases to class diagrams is often little more than a black art. While ER techniques take a more realistic approach to object identification and persistence, like UML they use an attribute-based, graphical notation that is less than ideal for capturing and validating business rules.

One effective way to supplement UML and ER approaches for data modeling is to exploit the potential of *data use cases*, or visualized examples of the kinds of data that the information system is expected to manage (including the possibility of changes between the as-is and to-be models). Although experienced modelers have long used data examples, their full potential has been rarely tapped outside the Object-Role Modeling (ORM) community, where data use cases play a central role in the modeling process, and multiple-instantiation is uniformly supported by the notation [2, 3]. ORM is a fact-oriented modeling method with rigorous foundations that provides a simple but expressive means of capturing and validating business rules in diagrams and/or natural language [1]. ORM models map in a fairly transparent way to UML and ER models, so can be used in conjunction with them.

A.H.F. Laender, S.W. Liddle, V.C. Storey (Eds.): ER2000 Conference, LNCS 1920, pp. 582-583, 2000.
© Springer-Verlag Berlin Heidelberg 2000

2   Data Use Cases, Verbalization, and Instantiation

A data use case is an example of data that is to be input to or output from the information system, and may be represented in a variety of ways (e.g. forms, tables or graphs). During analysis, the domain expert verbalizes facts from the example, using natural language. The modeler reformulates these as elementary facts, then abstracts these to fact types, expressed in a natural but formal language (e.g. Book has BookTitle; Room at Time is booked for Activity). The populations of these fact types are constrained by business rules, which the modeler validates with the domain expert by (a) verbalizing the rules, and (b) populating the fact types with sample populations that satisfy the rules as well as with counter-examples that illustrate violations of the rules.

ORM differs from ER and UML in expressing all facts as sentences, where objects play roles (parts in relationships). ORM's graphical syntax depicts a sentence type as a logical predicate or sequence of one or more roles (shown as boxes), each connected to its object type (shown as a named ellipse). Fact types may be populated with tables of facts where each role corresponds to a column, facilitating validation by population. In contrast, attribute-based approaches such as ER and UML are problematic for verbalizing and multiply instantiating fact types other than simple binary relationships, and their use of attributes introduces null values.

ORM's data modeling graphical syntax is richer than UML's (e.g. it covers general subset, exclusion, disjunctive mandatory and ring constraints), and can be automatically verbalized. Its textual language now allows positive, default and negative verbalization of business rules. *Positive verbalizations* emphasize constraints that do apply, e.g. **each** Book has **at most one** BookTitle. *Default verbalizations* indicate what is possible because of the absence of an explicit constraint, e.g. **it is possible that more than one** Book has **the same** BookTitle. *Negative verbalizations* indicate what is impossible because of the presence of a constraint, e.g. **it is impossible that the same** Room at **the same** Time is booked for **more than one** Activity. Negative verbalizations are best used in conjunction with counter-examples. Seeing the impact of a rule's violation helps domain experts to confirm whether the rule applies. In practice, this overall approach has proven effective for developing correct, clear and complete data models.

References

1.   Halpin, T.A. 1995, *Conceptual Schema and Relational Database Design*, 2nd edn (revised), WytLytPub, Bellevue USA. A third edition will be published by Morgan Kaufmann in 2001.
2.   Halpin, T.A. 1999, 'Fact-orientation before object-orientation: the case for data use cases', *DataToKnowledge Newsletter*, vol. 27, no. 6, (Nov./Dec. 1999).
3.   Halpin, T.A. & Bloesch, A.C. 1999, 'Data modeling in UML and ORM: a comparison', *Journal of Database Management*, Idea group Publishing Company, Hershey, USA.
4.   OMG 1999, *OMG Unified Modeling Language Specification*, version 1.3, UML Revision Task Force, available online from www.omg.org.

Personalized Digests of Sports Programs Using Intuitive Retrieval and Semantic Analysis

Takako Hashimoto, [1,2] Yukari Shirota, [1,2] Atsushi Iizawa, [1,2]
and Hideko S. Kunii [2]

[1] Information Broadcasting Laboratories, Inc.
1-1-1 Nishi-Asakusa, Taito-ku, Tokyo, 111-0035 JAPAN
email: {takako, shirota, izw}@ibl.co.jp
[2] Software Research Center, Imaging System Business Group, Ricoh Company, Ltd.
1-1-17 Koishikawa, Bunkyo-ku, Tokyo, 112-0002 JAPAN
email: {takako, shirota, izw, hkunii}@src.ricoh.co.jp

Abstract. Recently, digital broadcasting has experienced rapid growth. Digital broadcasting can deliver additional data as program attachments, which viewers can use to flexibly browse and retrieve parts of the program on their TV-receiving terminals. They can also make personalized digests from the broadcast TV programs. This function is particularly useful for viewers of sports programs because the digests reflect varied viewer preferences, such as favorite teams and players. This paper presents a method for making personalized digests for sports programs using additional data attachments.

A major problem with currently available multimedia retrieval systems is that, because semantic descriptions must be added manually, such descriptions require an excessive amount of time. In addition, the descriptions themselves are dependent on the person describing the events, which may mean that other viewers will not be able to retrieve events of importance to them. To solve these problems, we have developed a method, which uses *intuitive retrieval* and *semantic analysis* to create personalized digests.

(A) *Intuitive Retrieval*: A broadcaster can easily and quickly adds specific primitive events, such as "beginning of game" or "additional runs." To make personalized digests, however, ordinary viewers might prefer to use more abstract intuitive descriptions, such as "important offense," "counterattack trigger" and "fine save," to specify what they are looking for. We have created a process called *intuitive retrieval* to correlate these intuitive descriptions with the more primitive events. For *intuitive retrieval*, we introduced *abstract events* to add intuitive descriptions. The viewer can then use *abstract events* to retrieve video scenes and make personalized digests.

(B) *Semantic Analysis*: Personalized digests must incorporate information about each viewer's preferences. To analyze the semantic importance of the video scenes, we have therefore introduced *status parameters* as elements to indicate the semantic importance of video scenes. The *status parameters* can represent various game characteristics, such as a play's importance to offense and defense. In addition, *status parameter* values may be adjusted to reflect viewers' preferences.

We describe our digest-making method. Figure 1 is an ER diagram of our method with four entities: "video scene," "frame," "event," and "status parameter." A video scene is a time series of frames $f_1 f_2 ... f_n$. The frames f_1 and f_n are the first and last frames of the video scene. Each frame f includes its frame number:

$f=(fno)$.

A video scene s is expressed in our method as follows:
$s=(sid,\ type,\ ffno,\ lfno,\ c_1...c_n)$

A.H.F. Laender, S.W. Liddle, V.C. Storey (Eds.): ER2000 Conference, LNCS 1920, pp. 584-585, 2000.

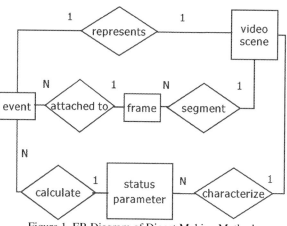

Figure 1 ER Diagram of Digest Making Method

where *sid* is a scene ID, *type* is a scene type, *ffno* and *lfno* are the first and last frame numbers, and $c_1...c_n$ are status parameters.

The flow is as follows. The broadcaster transmits a TV program -- a video scene with primitive event information. These events are added to frames. The relationship between a frame *f* and an event *e* is one-to-many (See Figure 1). Some frames have multiple events and others have no events. The set of events denoted by *E* consists of two kinds of events: *primitive events* Ep and *abstract events* Ea. Primitive events *Ep* express primitive event entities marked by a broadcaster. The *abstract events Ea* are defined by a state transition pattern that consists of multiple primitive events. Both events have the following common structure where *eid* is an event ID, *fno* is the corresponding frame number, *type* is an event type that the event expresses, and $attr_1,..., attr_n$ are event attributes:

$e=(eid, \ fno, \ type, \ attr_1,...,attr_n)$

An *abstract event* is defined using the following rules where *type* is an *abstract event* type, *parent_type* is a parent scene type, and *pattern* is the state transition pattern:

$r=(type, parent\_type, pattern).$

Within the range of the scene defined by *parent_type*, if the given pattern is found, then a new *abstract event* will be generated.

At the initial state, there is just one video scene containing only primitive events. The digest-making process then automatically generates *abstract events* that correspond with frames. Using the set of events *E*, the given frame series is logically segmented into sub-video scenes.

Each video scene has a *status parameter* list $c_1 ...c_n$ where each parameter value is calculated using events. Semantic importance of the video scene is calculated using the *status parameter* values. Important scenes with high semantic importance are extracted to make digests as a subset of the given video scenes.

The following illustrates our method as applied to an individual sports program, first using baseball as an example. A baseball game has a clear structure based on the following consist-of relationships -- "game" / "inning" / "half inning" / "at bat" / "pitch" -- from which abstract events such as "reversal" and "counter attack trigger" are generated. By comparison, the structure of a soccer game is not clear. In addition, the score information is scarce and insufficient to express the game contents. Therefore, we have introduced new *abstract events* and *status parameters* specific to soccer games.

To show that our method can be applied to various kinds of sports programs, we demonstrate our digest-making prototype systems for baseball games and soccer games.

Author Index

Lecture Notes in Computer Science

For information about Vols. 1–1844
please contact your bookseller or Springer-Verlag